T0332754

Analytic Projective Geometry

Projective geometry is the geometry of vision, and this book introduces students to this beautiful subject from an analytic perspective, emphasising its close relationship with linear algebra and the central role of symmetry. Starting with elementary and familiar geometry over real numbers, readers will soon build upon that knowledge via geometric pathways and journey on to deep and interesting corners of the subject. Through a projective approach to geometry, readers will discover connections between seemingly distant (and ancient) results in Euclidean geometry. In mixing results from the past 100 years with the history of the field, this text is one of the most comprehensive surveys of the subject and an invaluable reference for undergraduate and beginning graduate students learning classic geometry, as well as young researchers in computer graphics. Students will also appreciate the worked examples and diagrams throughout.

JOHN BAMBERG is Associate Professor of Mathematics at the University of Western Australia, where he previously obtained his PhD under the auspices of Cheryl Praeger and Tim Penttila. His research interests include finite and projective geometry, group theory, and algebraic combinatorics. He was a Marie Skłodowska-Curie fellow at Ghent University from 2006 to 2009 and a future fellow at the Australian Research Council from 2012 to 2016.

TIM PENTTILA is an Australian mathematician whose research interests include geometry, group theory, and combinatorics. He was an academic at the University of Western Australia for 20 years and a professor at Colorado State University for 10 years.

"This book provides a lively and lovely perspective on real projective spaces, combining art, history, groups, and elegant proofs."

– William M. Kantor

"This book is a celebration of the projective viewpoint of geometry. It gradually introduces the reader to the subject, and the arguments are presented in a way that highlights the power of projective thinking in geometry. The reader surprisingly discovers not only that Euclidean and related theorems can be realised as derivatives of projective results, but there are also unnoticed connections between results from ancient times. The treatise also contains a large number of exercises and is dotted with worked examples, which help the reader to appreciate and deeply understand the arguments they refer to. In my opinion this is a book that will definitely change the way we look at the Euclidean and projective analytic geometry."

– Alessandro Siciliano, Università degli Studi della Basilicata

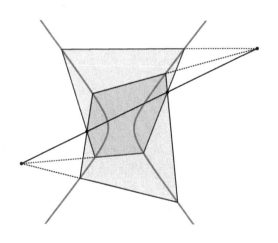

Analytic Projective Geometry

John Bamberg
University of Western Australia

Tim Penttila
University of Adelaide

Shaftesbury Road, Cambridge CB2 8EA, United Kingdom

One Liberty Plaza, 20th Floor, New York, NY 10006, USA

477 Williamstown Road, Port Melbourne, VIC 3207, Australia

314–321, 3rd Floor, Plot 3, Splendor Forum, Jasola District Centre, New Delhi – 110025, India

103 Penang Road, #05–06/07, Visioncrest Commercial, Singapore 238467

Cambridge University Press is part of Cambridge University Press & Assessment, a department of the University of Cambridge.

We share the University's mission to contribute to society through the pursuit of education, learning and research at the highest international levels of excellence.

www.cambridge.org
Information on this title: www.cambridge.org/9781009260596

DOI: 10.1017/9781009260626

First published 2023

A catalogue record for this publication is available from the British Library.

A Cataloging-in-Publication data record for this book is available from the Library of Congress

ISBN 978-1-009-26059-6 Hardback

Contents

Preface

Projective geometry is the geometry of vision. Yet Arthur Cayley saw that it is *all* geometry. The mathematical historian Morris Kline called it the 'science born of art', and the very early history of its development from that origin is documented in the book *The Geometry of an Art* by the later mathematical historian Kirsti Andersen. Some of those developments (and some later ones) appear in Chapters 11 and 12, and what Cayley meant is explained in Chapters 7 through 10. Felix Klein also advocated for the centrality of projective geometry, but is better known for bringing out the central role of symmetry in geometry in his *Erlangen Programme*. Our treatment of most of the topics in this book emphasises this central role of symmetry, through the prominent place we assign groups, and we also explain Klein's view on transformation geometry in Chapter 9. Moreover, this whole subject has a close relationship with linear algebra, and this underpins our treatment. What Jürgen Richter-Gebert calls 'the beauty that lies in the rich interplay of geometric structures and their algebraic counterparts' is a recurring theme of our book. Finally, we try to illustrate some of the advantages gained by taking a projective approach to geometry in Chapter 10, where we obtain connections between seemingly distant (and ancient) results in Euclidean geometry via the perspective hard-won in the earlier chapters.

This book is an introduction to projective geometry, and our coordinates are mostly over the real numbers. However, there is advanced and novel material for the practician. Chapter 6 examines one of the leitmotifs of this book – *involutions* and their role in projective geometry. This is taken much further in Section 10.2, where we begin with an old result of Pappus, and explore the more modern theorems of Ferrers, Jeřábek, Lehmer & Daus, Gardner & Gale, Robson & Strange, and their astounding synergy.

xi

We find that an approach that teaches the subject conceptually while also sketching its development resonates with us as teachers and authors, and also hope that it will find sympathetic vibrations in students and readers.

Part I

The Real Projective Plane

...no branch of mathematics competes with projective geometry in originality of ideas, coordination of intuition in discovery and rigor in proof, purity of thought, logical finish, elegance of proofs and comprehensiveness of concepts. The science born of art proved to be an art.

Morris Kline (1955, p. 86)

1

Fundamental Aspects of the Real Projective Plane

> Whereas Euclidean geometry describes objects as they are, projective geometry describes objects as they appear.

Kristen R. Schreck (2016, p. 159)

Three-dimensional Euclidean space, $\mathbb{R}^3$, is perhaps the most familiar and natural geometry to the lay person. In this introductory section, we will show how we can build a 'new world out of nothing' (to use János Bolyai's asseveration) from the interplay between perpendicularity and parallelism, of lines and planes together. This interplay leads to the *real projective plane* and *duality*.

1.1 Parallelism

In three-dimensional space, 'parallelism' applies to two different types of object – to lines and to planes. A line can be parallel to a plane, a plane can be parallel to another plane, and a line can be parallel to a line (see Figure 1.1). In particular, a line ℓ is parallel to a line ℓ' if they are, firstly, lying in a common plane (i.e., *coplanar*) and, secondly, non-intersecting. Two non-intersecting lines that are not coplanar are *skew*. Two planes are parallel if they are non-intersecting. A line ℓ and a plane π are parallel if they are non-intersecting or ℓ lies within π.

1.2 Perpendicularity

Similarly, 'perpendicularity' is a relation that works for lines and planes alike. We really only need to understand perpendicularity of lines to understand what happens when we introduce planes. Two lines are perpendicular if they are coplanar and perpendicular in the common plane. A line ℓ is perpendicular to

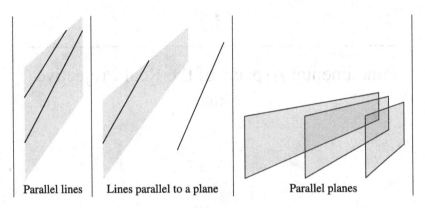

| Parallel lines | Lines parallel to a plane | Parallel planes |

Figure 1.1 Parallel lines and planes.

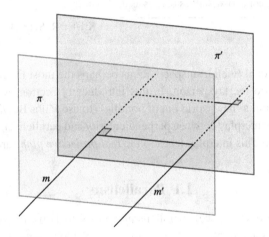

Figure 1.2 Interrelationship of perpendicularity and parallelism.

a plane π if ℓ is perpendicular to every line of π that it intersects. Two planes π and π' are perpendicular if they meet in a line ℓ, and π is perpendicular to some line of π'.

We will see soon that the most illuminating property of $\mathbb{R}^3$ is the inter-relationship of perpendicularity and parallelism arising from the following property:

PARALLEL–PERPENDICULAR PROPERTY: Let m, m' be lines and let π, π' be planes. If $m \parallel m'$, $\pi \parallel \pi'$, and $\pi \perp m$, then $\pi' \perp m'$ (see Figure 1.2).

We can now elevate to the next level of abstraction. For each line ℓ, the **parallel class** $[\ell]$ of ℓ is the set of all lines parallel to ℓ (including ℓ itself). Similarly, write $[\pi]$ for the planes parallel to π. The first observation we make is the following:

Property 1.1

Let ℓ and π be a line and a plane (respectively). Then either

- *no element of $[\ell]$ is parallel to any element of $[\pi]$, or*
- *every element of $[\ell]$ is parallel to every element of $[\pi]$, and we say that $[\ell]$ is **parallel to** $[\pi]$.*

So it makes sense to write $[\ell] \parallel [\pi]$. This relation of parallelism between parallel classes of lines and planes is symmetric, and could abstractly be an *incidence relation* between two different types of objects. With this in mind, we make a second observation:

Property 1.2

- *For any two different parallel classes of lines $[\ell]$ and $[\ell']$, there is a unique parallel class of planes that is parallel to both $[\ell]$ and $[\ell']$.*
- *For any two different parallel classes of planes $[\pi]$ and $[\pi']$, there is a unique parallel class of lines that is parallel to both $[\pi]$ and $[\pi']$.*

We can manufacture a geometry $\mathcal{G}$ out of these parallel classes of lines and planes. The geometry we create will be *planar* in the sense that we have only two types of object, which we might as well temporarily call[1] *pistettä* (singular: *piste*) and *linjat* (singular: *linja*). This new geometry will consist only of objects and an incidence relation between them — no distance, no midpoints, no angles, no parallelism.

Pistettä	Parallel classes of lines of $\mathbb{R}^3$
Linjat	Parallel classes of planes of $\mathbb{R}^3$
Incidence	A parallel class of lines is 'incident' with a parallel class of planes if and only if they are parallel.

So from what we have discussed above, the incidence relation here is symmetric: two pistettä lie on a unique linja and two linjat have a unique piste in

---

[1] For some readers, the use of the words 'points' and 'line' will interfere with their understanding, so to make it clear that we are defining new points and new lines, we temporarily adopt another language for these new points and new lines.

common. Therefore, there cannot be 'parallel' linjat in this geometry $\mathcal{G}$; two linjat are always concurrent. This geometry is an example of a non-Euclidean geometry – a **projective plane**.

1.3 Duality

Let $\mathcal{L}$ be the set of parallel classes of lines and let Π be the set of parallel classes of planes. There is a natural correspondence between $\mathcal{L}$ and Π: if $[\ell]$ is a parallel class of lines, then we take the set $\ell^{\perp}$ of all planes that are perpendicular to ℓ. By the parallel–perpendicular property, this set of planes is a parallel class of planes and did not depend on the representative we took from ℓ. Conversely, if $[\pi]$ is a parallel class of planes, we map to the set $\pi^{\perp}$ of all lines that are perpendicular to π. Thus we have the following correspondence:

$$[\ell] \longrightarrow [\ell]^{\perp} := [\ell^{\perp}]$$
$$[\pi] \longrightarrow [\pi]^{\perp} := [\pi^{\perp}]$$

Note that if we apply $\perp$ twice, then our objects are left invariant. For example, if we take all of the planes perpendicular to a line ℓ, and then take all of the lines perpendicular to those planes, it will result in the parallel class of ℓ. Moreover, this correspondence respects parallelism between elements of $\mathcal{L}$ and Π:

$$\left([\ell]^{\perp}\right)^{\perp} = [\ell], \qquad\qquad \left([\pi]^{\perp}\right)^{\perp} = [\pi],$$
$$[\ell] \parallel [\pi] \iff [\ell]^{\perp} \parallel [\pi]^{\perp}.$$

Finally, let's see what the perpendicularity correspondence $\perp$ does to $\mathcal{G}$. We saw above that it preserves incidence. So if P is a piste of $\mathcal{G}$ and m is a linja, then $P^{\perp}$ is incident with $m^{\perp}$ if and only if P is incident with m. We also saw that $\perp$ maps a piste to a linja, and then a linja to a piste, in such a way that if performed twice, it left the objects invariant. Such a map is called a **polarity**. This polarity also has the property that no piste P is incident with its image $P^{\perp}$; but we will return to this later once we have investigated projective planes more thoroughly.

1.4 Two Models of the Real Projective Plane

We have already seen that parallel classes of lines and planes of $\mathbb{R}^3$ form a projective plane – an incidence geometry of points and lines such that any pair of points determine a unique line, and two distinct lines always meet in a point. Each parallel class of lines has a representative passing through the origin O of $\mathbb{R}^3$. So we can replace each parallel class by a one-dimensional subspace

Table 1.1 *The real projective plane.*

Points	1-dimensional subspaces of $\mathbb{R}^3$
Lines	2-dimensional subspaces of $\mathbb{R}^3$
Incidence	inclusion

of $\mathbb{R}^3$. Likewise, the parallel classes of planes can be simulated by taking two-dimensional subspaces of $\mathbb{R}^3$. Formally, the **real projective plane** PG(2, $\mathbb{R}$) is the incidence structure defined in Table 1.1.

Theorem 1.3

(i) Any two points of PG(2, $\mathbb{R}$) *are incident with a unique line.*
(ii) Any two lines PG(2, $\mathbb{R}$) *are incident with a unique point.*

Proof The proof follows from elementary linear algebra. In particular, for (i), note that any two 1-dimensional subspaces of $\mathbb{R}^3$ span a unique 2-dimensional subspace. For (ii), we observe that any two 2-dimensional subspaces of $\mathbb{R}^3$ meet in a unique 1-dimensional subspace. □

We denote a point of PG(2, $\mathbb{R}$) by homogeneous coordinates (x, y, z), $x, y, z \in \mathbb{R}$, not all zero. This means that we are simply dropping the angled brackets from the subspace $\langle(x, y, z)\rangle$ of $\mathbb{R}^3$; since

$$\langle(x, y, z)\rangle = \langle(cx, cy, cz)\rangle, \quad \text{for all } c \in \mathbb{R}\backslash\{0\},$$

we have $(x, y, z) = (cx, cy, cz)$. This is what we mean by saying that the coordinates are *homogeneous*, and it will be clear from the context that we are describing a point of PG(2, $\mathbb{R}$) and not a vector of $\mathbb{R}^3$. Note that $(0, 0, 0)$ is not the homogeneous coordinates of any point of PG(2, $\mathbb{R}$).

We denote a line with equation $ax + by + cz = 0$ by homogeneous line coordinates $[a, b, c]$, $a, b, c \in \mathbb{R}$, not all zero. (Again, note that $k(ax+by+cz) = 0$ for $k \in \mathbb{R}$ with $k \neq 0$ yields the same line, so these coordinates are indeed *homogeneous*.) A point (x, y, z) is incident with the line $[a, b, c]$ if and only if $ax + by + cz = 0$. Note that $[0, 0, 0]$ is not the homogeneous line coordinates of any line of PG(2, $\mathbb{R}$).

Another way to define the real projective plane is to take the real Euclidean plane and enlarge it a little bit. Each line is equipped with an additional point – its *point at infinity* – which is simply the parallel class of that line. This ensures that two parallel lines now become two intersecting lines in the larger geometry. One extra line is introduced, and it is simply the set of all points at infinity – the *line at infinity*. Formally, the **extended Euclidean plane** is the incidence structure defined in Table 1.2.

Table 1.2 *The extended Euclidean plane.*

Points	points of $\mathbb{R}^2$ parallel classes of lines (*points at infinity*)
Lines	lines of $\mathbb{R}^2$ the line at infinity
Incidence	inherited from $\mathbb{R}^2$; a point at infinity is incident with every line of the corresponding parallel class and with the line at infinity

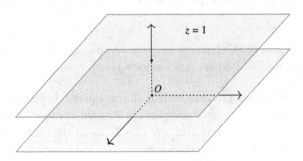

Figure 1.3 The Euclidean plane embedded in $\mathbb{R}^3$.

We have insinuated from the beginning that these models of incidence geometries are *the same*, and we have already said that they are models of *the* real projective plane. To make this mathematically correct, we define two incidence geometries (of points and lines) to be **isomorphic** if there is a bijection between their sets of points that respects their lines. In other words, there is a bijection ϕ mapping points of one incidence geometry onto the points of the other, such that if ℓ is a line of the first geometry, then the image of the points incident with ℓ (under ϕ) is precisely the set of points of a line of the second incidence geometry.

Theorem 1.4
The extended Euclidean plane and the real projective plane are isomorphic.

Proof Embed the Euclidean plane in $\mathbb{R}^3$ as the plane $z = 1$ (see Figure 1.3). Consider the projection via the origin O from a non-parallel plane π not on the origin to $z = 1$. The line ℓ that is the intersection of π and $z = 0$ will have no image and the line m that is the intersection of the plane parallel to π with $z = 1$ will have no preimage. The points of ℓ are in one-to-one correspondence with parallel classes of lines of $z = 1$: namely a point P corresponds to the parallel class of lines of $z = 1$ parallel to OP. Moreover, given a point Q of m, each

Table 1.3 *From the real projective plane to the extended Euclidean plane.*

Real projective plane	Extended Euclidean plane
$\langle(a,b,c)\rangle, c \neq 0$	$\left(\frac{a}{c}, \frac{b}{c}, 1\right)$
$\langle(a,b,0)\rangle$	parallel class $\{bx - ay = c : c \in \mathbb{R}\}$ of the plane $z = 1$.

line ℓ on Q is the image of a line of π and these lines arising from Q form a parallel class in π. Moreover, the image of a line of π (other than ℓ) is a line of the plane $z = 1$.

Now, removing π from the picture, we have a one-to-one correspondence between the 1-dimensional subspaces of $\mathbb{R}^3$ of the points and parallel classes of the plane $z = 1$, and the 2-dimensional subspaces of $\mathbb{R}^3$ and the lines and the line at infinity of the plane $z = 1$ that preserves incidence. Therefore, the extended Euclidean plane and the real projective plane are isomorphic. It is worthwhile detailing this isomorphism in Table 1.3.

Composing this with the isomorphism $(a, b, 1) \mapsto (a, b)$ of the plane $z = 1$ with $\mathbb{R}^2$, we obtain:

- $\langle(x, y, z)\rangle, z \neq 0$, in the real projective plane corresponds to (X, Y) in $\mathbb{R}^2$ if and only if $X = \frac{x}{z}, Y = \frac{y}{z}$;
- $\langle(1, -m, 0)\rangle$ in the real projective plane corresponds to the parallel class of lines of slope m in $\mathbb{R}^2$;
- $\langle(0, 1, 0)\rangle$ in the real projective plane corresponds to the parallel class of vertical lines in $\mathbb{R}^2$. $\qquad\square$

Moving between Cartesian (X, Y) coordinates and homogeneous (x, y, z) coordinates via the equations $X = x/z, Y = y/z$ is due to Hesse (1842). This was adopted by Cayley (1870) and generalised to n dimensions.

We say that a set of points is **collinear** if they are all incident with the same line. Likewise, a set of lines is **concurrent** if they are all incident with the same point. Using determinants, there is a simple test for when three points are collinear or three lines are concurrent.

Theorem 1.5

Three points $(x_1, y_1, z_1), (x_2, y_2, z_2), (x_3, y_3, z_3)$ *are collinear if and only if*

$$\begin{vmatrix} x_1 & y_1 & z_1 \\ x_2 & y_2 & z_2 \\ x_3 & y_3 & z_3 \end{vmatrix} = 0.$$

Proof $(x_1, y_1, z_1), (x_2, y_2, z_2), (x_3, y_3, z_3)$ are collinear if and only if the matrix in the above determinant has rank two.   □

Theorem 1.6
Three lines $[a_1, b_1, c_1], [a_2, b_2, c_2], [a_3, b_3, c_3]$ *are concurrent if and only if*

$$\begin{vmatrix} a_1 & b_1 & c_1 \\ a_2 & b_2 & c_2 \\ a_3 & b_3 & c_3 \end{vmatrix} = 0.$$

Proof $[a_1, b_1, c_1], [a_2, b_2, c_2], [a_3, b_3, c_3]$ are concurrent if and only if the matrix in the above determinant has nullity one.   □

The **dual** of a statement about the plane is the statement that results after interchanging *point* and *line*, *collinear* and *concurrent*, *intersection* and *join* and making the necessary linguistic adjustments.

THE PRINCIPLE OF DUALITY (Poncelet, 1822; Gergonne, 1825/6)
The dual of every theorem about $\mathrm{PG}(2, \mathbb{R})$ is also a theorem.

See also Poncelet (1995a, 1995b). For a proof of the *principle of duality*, note that the map taking a point (a, b, c) to a line $[a, b, c]$ preserves incidence. For an example of the principle of duality at play, note that Theorems 1.5 and 1.6 are dual.

1.5 Recap: The Real Projective Plane as Involving Points and Lines

When we think of the Euclidean space $\mathbb{R}^3$, we think of the incidence structure of points, lines, and planes. Now let us be more abstract and instead think of the incidence structure $\mathcal{J}$ whose 'points' are the parallel classes of lines in Euclidean space, and whose 'lines' are the parallel classes of planes in Euclidean space. The incidence relation would be derived from the natural incidence relation of class representatives.

new points	parallel classes of lines in Euclidean space
new lines	parallel classes of planes in Euclidean space

Now delete a parallel class Π of planes and all lines parallel to it to obtain a new incidence structure $\mathcal{A}$. Choose a plane π of Π, and a point P not on π.

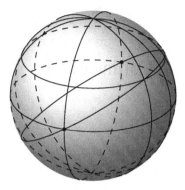

Figure 1.4 The real projective plane modelled on the sphere.

Consider the map ρ that takes a parallel class of lines to the intersection of its unique member on P with π, and a parallel class of planes to the intersection of its unique member on P with π. Since the parallel classes have parallel representatives if and only if the unique members on P are contained in one another, it follows that these parallel classes of lines and planes have parallel representatives precisely when the images under ρ are contained in one another. Therefore, $\mathcal{A}$ is isomorphic to the *affine plane* π. (The deleted objects thus naturally give the points at infinity of π as lines on P parallel to π and a line at infinity as the plane on P parallel to π.)

So we see here that the real projective plane can be realised as an extension of the incidence structure $\mathcal{A}$, and this incidence structure is just the parallel classes of lines and planes in Euclidean space, minus one class of planes.

Exercises

1.1 Consider the unit sphere S^2 in $\mathbb{R}^3$. Let $\mathcal{P}$ be the set of pairs of antipodal points of S^2, and let $\mathcal{L}$ be the set of great circles of S^2. Define incidence between elements of $\mathcal{P}$ and $\mathcal{L}$ by natural inclusion: a pair of antipodal points is incident with a great circle if both points lie on the great circle (compare with Figure 1.4). Show that this incidence geometry is isomorphic to $\mathsf{PG}(2, \mathbb{R})$.

1.2 Let (x_1, y_1, z_1) and (x_2, y_2, z_2) be two points of $\mathsf{PG}(2, \mathbb{R})$, written in homogeneous coordinates. Show that the unique line lying on these two points can be computed using the vector cross product.

1.3 A **quadrangle** is a set of four points, no three collinear, and a **quadrilateral** is the dual object of a quadrangle. What is a quadrilateral when expressed in terms of lines?

2

Collineations

The dominant role of group theory in mathematics was long unsuspected; for eighty years, the concept of a group was ignored. It was
Galois who first had a clear notion of groups; but it is only since the
work of Klein and especially Lie that we have begun to see that there
is almost no mathematical theory where this notion does not hold an
important place.

Henri Poincaré (1921, p. 137).[1]

2.1 The Projective General Linear Group

First we introduce some notation. If a point P is incident with a line ℓ, then we
write $P \mathrel{I} \ell$. If P is not incident with ℓ, then we write $P \mathrel{\not I} \ell$. A **collineation**, or
automorphism, of $\mathsf{PG}(2, \mathbb{R})$ is a bijection on the points taking lines to lines.[2]
In other words, if we consider the action of a collineation g on the set S of
points $\{P \colon P \mathrel{I} \ell\}$ incident with a line ℓ, then $S^g = \{P^g \colon P \mathrel{I} \ell\}$ is the set of points
of some line. The set of all collineations of $\mathsf{PG}(2, \mathbb{R})$ forms a group under
the composition operation, and we denote this group by $\mathrm{Aut}(\mathsf{PG}(2, \mathbb{R}))$. This
group is 'large' in the sense that it acts 2-transitively on points, but it also has
other properties that translate to important symmetry traits for the underlying
geometry.

Given a 3×3 invertible matrix A, the map $(x, y, z) \mapsto (x, y, z)A$ is a
collineation. Two such matrices A, B define the same collineation if and only
if $A = cB$, for $c \in \mathbb{R}$ with $c \neq 0$, because the coordinates are homogeneous. So
this action of the group $\mathsf{GL}(3, \mathbb{R})$ of all 3×3 invertible matrices has a non-trivial

---

[1] Reprinted by permission from Springer Nature: *Acta Mathematica*. Rapport sur les travaux de
M. Cartan. Poincaré, H. © 1921.

[2] If you are familiar with the previous chapter, then a collineation is an isomorphism of $\mathsf{PG}(2, \mathbb{R})$
to itself.

kernel Z. This normal subgroup Z consists of the non-zero scalar matrices, and the factor group $GL(3, \mathbb{R})/Z$, which we denote by $PGL(3, \mathbb{R})$, is the **projective general linear group**. We can think of $PGL(3, \mathbb{R})$ as 3×3 invertible matrices that are identified up to a scalar; that is, two elements are equal if the matrix representation of one of them is a scalar multiple of the other's matrix representation.

Given the way we introduced the real projective plane as the extended Euclidean plane, you could be forgiven for thinking the line at infinity was special. But this is not the case – all lines of the real projective plane have the same status.

Theorem 2.1
$PGL(3, \mathbb{R})$ *acts transitively on the lines of* $PG(2, \mathbb{R})$.

Proof Given two lines ℓ, m, take a basis $\{v_1, v_2\}$ for ℓ and extend it to a basis $\{v_1, v_2, v_3\}$ for $\mathbb{R}^3$, and also take a basis $\{w_1, w_2\}$ for m and extend it to a basis $\{w_1, w_2, w_3\}$ for $\mathbb{R}^3$. There is an invertible matrix A with $v_i A = w_i$, for $i = 1, 2, 3$, and this induces a collineation of $PG(2, \mathbb{R})$ taking ℓ to m. □

This proof shows that $PGL(3, \mathbb{R})$ acts 2-transitively on points of $PG(2, \mathbb{R})$.

The real affine plane $AG(2, \mathbb{R})$ is basically the geometry of $\mathbb{R}^2$ when we forget about distance and angle. It is the *incidence geometry* of $\mathbb{R}^2$, with its salient properties being the parallelism of lines and the invariance of affine ratio. We can recover the real affine plane $AG(2, \mathbb{R})$ from the real projective plane by deleting a line and all points incident with it, reversing the procedure we first used to construct the real projective plane.

Theorem 2.2
Let ℓ be a line of $PG(2, \mathbb{R})$. Then the incidence structure of points not on ℓ and lines other than ℓ, with the restricted incidence, is isomorphic to $AG(2, \mathbb{R})$.

Proof This is an immediate consequence of Theorems 1.4 and 2.1. □

The concept of *permutational isomorphism* allows us to formally describe when two group actions are *the same*. Essentially, two actions are permutationally isomorphic if the groups are isomorphic and one of the actions can be seen as simply a relabelling of the set acted upon in the other action. If G is a group acting on a set Ω, and H is a group acting on a set Γ, we say that these two actions are **permutationally isomorphic** if there exists a group isomorphism $\varphi \colon G \to H$ and a bijection $\beta \colon \Omega \to \Gamma$ such that

$$\beta(\omega^g) = \beta(\omega)^{\varphi(g)} \qquad (2.1)$$

for all $\omega \in \Omega$ and $g \in G$. The pair (β, φ) is a **permutational isomorphism** of the two actions, and permutational isomorphism is an equivalence relation on group actions. We will often use the notation (set, group) for a group action, and then use the shorthand '(Ω, G) is permutationally isomorphic to (Γ, H)'.

Worked Example 2.3 (The action on lines)
Let g be a collineation of $PG(2, \mathbb{R})$ induced by a matrix A of $GL(3, \mathbb{R})$. If $[a, b, c]$ are the homogeneous line coordinates of ℓ, then what is the effect of A on these coordinates when we take the image of ℓ under g? You might think that ℓ^g has coordinates $[a, b, c]A$, but this is incorrect. As a set of points, ℓ^g is

$$\{(x, y, z)A : (x, y, z) \cdot [a, b, c] = 0\}.$$

Let $A^{-\top}$ denote the matrix $\left(A^{-1}\right)^{\top}$. By placing the identity matrix into the middle of the predicate, we can realise this set another way:

$$(x, y, z) \cdot [a, b, c] = 0 \iff (x, y, z)AA^{-1}[a, b, c]^{\top} = 0$$
$$\iff (x, y, z)A\left([a, b, c]A^{-\top}\right)^{\top} = 0$$
$$\iff (x, y, z)A \cdot [a, b, c]A^{-\top} = 0.$$

Therefore, if we substitute (x', y', z') for $(x, y, z)A$, we see that ℓ^g is the set of points

$$\left\{(x', y', z') : (x', y', z') \cdot [a, b, c]A^{-\top} = 0\right\}.$$

So the homogeneous line coordinates for ℓ^g are those resulting from applying $A^{-\top}$ to the right of $[a, b, c]$. This action of $GL(3, \mathbb{R})$ is sometimes known as the **adjoint action** of $GL(3, \mathbb{R})$, and the map $A \mapsto A^{-\top}$ is an outer automorphism of $GL(3, \mathbb{R})$. Indeed, via this outer automorphism, we see that the action of $GL(3, \mathbb{R})$ (resp. $PGL(3, \mathbb{R})$) on the points of $PG(2, \mathbb{R})$ is permutationally isomorphic to the action of $GL(3, \mathbb{R})$ (resp. $PGL(3, \mathbb{R})$) on the lines of $PG(2, \mathbb{R})$. $\qquad\square$

One of the most instructive examples of permutational isomorphism arises from the identification of the points of $PG(2, \mathbb{R})$ not incident with a line, with the points of the affine plane $AG(2, \mathbb{R})$. The affine group $AGL(2, \mathbb{R})$ is the group of *affine transformations* of the real affine plane $AG(2, \mathbb{R})$.

Theorem 2.4
Let ℓ be a line of $PG(2, \mathbb{R})$. Then the stabiliser of ℓ in $PGL(3, \mathbb{R})$ acting on the points not on ℓ is permutationally isomorphic to $AGL(2, \mathbb{R})$ acting on the points of $AG(2, \mathbb{R})$.

Proof By Theorem 2.1, we need only prove this for the line $\ell\colon z = 0$. The stabiliser of $z = 0$ in $\mathsf{PGL}(3, \mathbb{R})$ consists of collineations induced by matrices of the form

$$\begin{bmatrix} a & b & 0 \\ c & d & 0 \\ e & f & 1 \end{bmatrix},$$

where $ad - bc \neq 0$. Using the correspondence between $\mathbb{R}^2$ and $\mathsf{PG}(2, \mathbb{R})$, so that (X, Y) corresponds to (x, y, z) for $z \neq 0$ where $X = \frac{x}{z}$ and $Y = \frac{y}{z}$, we see that a matrix of this form yields the map

$$(X, Y) \mapsto (X, Y) \begin{bmatrix} a & b \\ c & d \end{bmatrix} + (e, f),$$

which is an element of $\mathsf{AGL}(2, \mathbb{R})$, and all elements of $\mathsf{AGL}(2, \mathbb{R})$ are of this form. So, for emphasis, we have an isomorphism $\varphi\colon \mathsf{PGL}(3, \mathbb{R})_\ell \to \mathsf{AGL}(2, \mathbb{R})$ (where $\mathsf{PGL}(3, \mathbb{R})_\ell$ is the stabiliser of ℓ in $\mathsf{PGL}(3, \mathbb{R})$) given by

$$\begin{bmatrix} a & b & 0 \\ c & d & 0 \\ e & f & 1 \end{bmatrix} \xrightarrow{\varphi} \left((X, Y) \mapsto (X, Y) \begin{bmatrix} a & b \\ c & d \end{bmatrix} + (e, f) \right),$$

and a bijection β from the points (x, y, z) of $\mathsf{PG}(2, \mathbb{R})$ not on ℓ to the points $(x/z, y/z)$ of $\mathsf{AG}(2, \mathbb{R})$, such that Equation (2.1) evidently holds. $\qquad\square$

2.2 The Fundamental Quadrangle

A **quadrangle** is a set of four points, no three collinear. The **fundamental quadrangle** is

$$(1, 0, 0), (0, 1, 0), (0, 0, 1), (1, 1, 1).$$

Theorem 2.5
$\mathsf{PGL}(3, \mathbb{R})$ *acts regularly on ordered quadrangles.*

Proof First note that $\mathsf{GL}(3, \mathbb{R})$ acts transitively on ordered bases of $\mathbb{R}^3$, since if $\{u, v, w\}$ is a basis of $\mathbb{R}^3$, then the matrix having rows u, v, w (in order) maps $(1, 0, 0)$, $(0, 1, 0)$, $(0, 0, 1)$ to u, v, w in turn (when applied to the right). Hence, $\mathsf{PGL}(3, \mathbb{R})$ is transitive on ordered triangles. To show transitivity on ordered quadrangles, it suffices to show that the stabiliser S of $(1, 0, 0)$, $(0, 1, 0)$, $(0, 0, 1)$ in $\mathsf{PGL}(3, \mathbb{R})$ is transitive on the points on no side of the triangle; that is, on the points with no coordinate zero. Now S is induced by diagonal matrices, and given any point (a, b, c) with $abc \neq 0$, the map $(x, y, z) \mapsto (ax, by, cz)$ lies in S and takes $(1, 1, 1)$ to (a, b, c). The stabiliser of $(1, 1, 1)$ in S is induced by the scalar matrices; it consists just of the identity collineation.   □

Theorem 2.6 (Fano's property)
The diagonal points (i.e., intersections of opposite sides) of any quadrangle are not collinear.

Proof By Theorem 2.5, we may assume that our quadrangle is the fundamental one: A: $(1, 0, 0)$, B: $(0, 1, 0)$, C: $(0, 0, 1)$, D: $(1, 1, 1)$. The diagonal points can easily be calculated, and they are $AB \cap CD = (1, 1, 0)$, $AC \cap BD = (1, 0, 1)$, and $AD \cap BC = (0, 1, 1)$. The matrix having the diagonal points as rows clearly has full rank, and so, by Theorem 1.5, the diagonal points are not collinear.   □

Just as for matrices, there are passive and active views of collineations. The passive (or *alias*) view is as a change of coordinates. The active (or *alibi*) view is as a transformation.

Coordinates may be introduced in the real projective plane by choosing the points of a quadrangle and assigning them the coordinates of the fundamental quadrangle.

Theorem 2.7 (Mac Lane, 1936)
Consider eight (distinct) points $A, B, \ldots, H$ in $PG(2, \mathbb{R})$ such that the following eight triples are collinear: ABD, BCE, CDF, DEG, EFH, FGA, GHB, HAC. Then all eight points lie on a line.

Möbius (1828) had asked whether there exists a pair of polygons with n sides each, having the property that the vertices of one polygon lie on the lines through the edges of the other polygon, and vice versa. Mac Lane's theorem says that there is no solution for $n = 4$ in the real plane, but Kantor (1881) found a solution for $n = 4$ in the complex projective plane. This gives an 8_3 configuration called the *Möbius–Kantor 8_3 configuration*. In fact, it can be shown that it embeds in a projective plane over a division ring if and only if that division ring has an element t with $t^2 + t + 1 = 0$.

Proof By Theorem 2.5, we choose coordinates so that B: $(0, 0, 1)$, G: $(1, 0, 1)$, F: $(0, 1, 1)$, D: $(1, 1, 1)$. This implies that A: $(1, 1, 2) = BD \cap GF$. Let H: $(1, 0, -t)$. Then

$$C: (1 + t, t, t) = DF \cap AH,$$

$$E: (1 + t, t, 1 + t) = BC \cap DG.$$

Now E, F, H collinear implies $t^2 + t + 1 = 0$.

Conversely, if $t^2 + t + 1 = 0$, then the above assignment of coordinates gives a Möbius–Kantor 8_3 configuration that is not on a line. (A, B, D lie on the line $x = y$, B, C, E on the line $tx = (1 + t)y$, C, D, F on the line $y = z$, D, E, G on the line $x = z$, E, F, H on the line $-tx + y - z = 0$, F, G, A on the line $x + y - z = 0$, G, H, B on the line $y = 0$, H, A, C on the line $-tx + (2 + t)y - z = 0$.)   □

So this simultaneously proves the truth of Mac Lane's configuration theorem over the real numbers (since $\mathbb{R}$ contains no zero of $x^2 + x + 1$) and its falsity over the complex numbers (i.e., it gives Kantor's embedding).

The proof is adapted from Ostrom and Sherk (1964, proof of Theorem 2), a result that was pointed out as holding more generally than for just finite fields by Rigby (1965, Theorem 6.1) (even extending to a skew field). The connection between the two subjects is that the Möbius–Kantor 8_3 configuration is isomorphic to the affine plane of order 3 with both a pencil of lines deleted and the point carrying the pencil deleted. It turns out that the argument of Ostrom and Sherk did not need the whole affine plane of order 3, but merely the Möbius–Kantor 8_3 configuration. MacLane's theorem is valid in $\mathsf{AG}(2, \mathbb{R})$ and, as such, was rediscovered by mathematicians working in automatic deduction in geometry, namely S.-C. Chou, X.-S. Gao, and others (see Conti and Traverso, 1995).

2.3 The Fundamental Theorem of Projective Geometry

We can prove the Fundamental Theorem of Projective Geometry (FToPG, Theorem 2.9) by using the Fundamental Theorem of Affine Geometry (FToAG), that is, $\mathrm{Aut}(\mathsf{AG}(2, \mathbb{R})) = \mathsf{AGL}(2, \mathbb{R})$. We will also need the following simple result in permutation group theory.[3]

Theorem 2.8
Given a transitive subgroup H of a permutation group G on a set Ω and some $\omega \in \Omega$, $G = H$ if and only if $G_\omega = H_\omega$.

Theorem 2.9 (Fundamental theorem of projective geometry)
$\mathrm{Aut}(\mathsf{PG}(2, \mathbb{R})) = \mathsf{PGL}(3, \mathbb{R})$.

---

[3] It is a consequence of the standard result known as the Orbit–Stabiliser theorem in the theory of group actions.

Proof Since $PGL(3, \mathbb{R})$ is a subgroup of $\mathrm{Aut}(PG(2, \mathbb{R}))$ and acts transitively on lines, it suffices (by Theorem 2.8) to show that $PGL(3, \mathbb{R})_\ell$ is equal to $\mathrm{Aut}(PG(2, \mathbb{R}))_\ell$ for some line ℓ. As noted above, every element of the stabiliser $\mathrm{Aut}(PG(2, \mathbb{R}))_\ell$ induces an automorphism of $AG(2, \mathbb{R})$ and conversely: indeed, if X is the set of points of $PG(2, \mathbb{R})$ not on ℓ, then $(X, \mathrm{Aut}(PG(2, \mathbb{R}))_\ell)$ is permutationally isomorphic to $(AG(2, \mathbb{R}), \mathrm{Aut}(AG(2, \mathbb{R})))$. Since we already noted that $(X, PGL(3, \mathbb{R})_\ell)$ is permutationally isomorphic to $(AG(2, \mathbb{R}), AGL(2, \mathbb{R}))$ (Theorem 2.4), the result now follows from the FToAG.   □

We have a number of remarks concerning the FToPG:

1. First of all, every collineation of $PG(2, \mathbb{R})$ is induced by a linear map.
2. We will see later (Chapter 3) that from the FToPG we can classify *dualities* and *polarities*.
3. By the FToPG, it follows that a bijection of the points of the real projective plane that preserves collinearity of triples preserves *cross-ratio*, which we will explore in Chapter 4. Thus, cross-ratio is intrinsic to real projective geometry.
4. Observe from the proof just now that the FToAG implies the FToPG. Conversely, if we assume that the FToPG holds, then we proceed to show that an automorphism of $AG(2, \mathbb{R})$ extends to an automorphism of $PG(2, \mathbb{R})$ fixing the line at infinity, and hence $\mathrm{Aut}(AG(2, \mathbb{R})) = AGL(2, \mathbb{R})$. Therefore, the FToAG and the FToPG are equivalent theorems.

A point P is **fixed** by a collineation g if $P^g = P$. A line ℓ is **fixed** by a collineation g if $\ell^g = \ell$.

Theorem 2.10
Let g be a collineation of $PG(2, \mathbb{R})$.

(i) The join of two fixed points of g is a fixed line of g.
(ii) The intersection of two fixed lines of g is a point fixed by g.

Proof

(i) Let P and Q be two fixed points of g, and let ℓ be the line joining P and Q. By definition of a collineation, ℓ^g is incident with P^g and Q^g, and since $P^g = P$ and $Q^g = Q$, we see that ℓ^g is the line joining P and Q. By uniqueness, $\ell = \ell^g$.
(ii) Let ℓ and m be two fixed lines of g, and let $P = \ell \cap m$. By definition of a collineation, P^g is incident with ℓ^g and m^g, and since $\ell^g = \ell$ and $m^g = m$, we see that $P^g = \ell \cap m = P$.   □

Theorem 2.11

Any collineation of PG(2, ℝ) *has a fixed point and a fixed line.*

Proof A fixed point of the collineation induced by a matrix M corresponds to a row eigenvector of M (because we have homogeneous coordinates). The characteristic polynomial of M is cubic and so has a zero over ℝ. Thus, M has a row eigenvector and so the collineation has a fixed point. The second statement is dual to the first, so follows by the *principle of duality*. □

An **axis** of a collineation is a line fixed point-wise. Dually, a **centre** of a collineation is a point such that every line on it is fixed.

Theorem 2.12

(i) *A collineation has a centre if and only if it has an axis.*

(ii) *A non-identity collineation has at most one centre and at most one axis.*

Proof We first prove (ii). Let σ be a non-identity collineation, and let P and Q be two points fixed line-wise by σ. If $P \neq Q$, then a point of intersection of a line on P and a line through Q, not equal to PQ, will be fixed by σ. There are at least two such points fixed by σ, thus creating an ordered quadrangle that is fixed by σ; a contradiction by Theorem 2.5. Therefore, $P = Q$. The dual statement follows from the principle of duality.

Now to prove (i). Let σ be a collineation having an axis ℓ. Now σ can fix at most one point not on ℓ, since otherwise σ would fix an ordered quadrangle (which contradicts Theorem 2.5). Suppose σ fixes a point P not incident with ℓ, and suppose Q is a point on ℓ. Then $(PQ)^\sigma = P^\sigma Q^\sigma = PQ$ and so σ fixes every line on P. Therefore, P is a centre in this case. Otherwise, if σ fixes no point not incident with ℓ, then for any point $P \nmid \ell$ we have that PP^σ is a line. Let $Q := PP^\sigma \cap \ell$. Since P, Q, and P^σ are collinear, we have

$$QP = QP^\sigma = Q^\sigma P^\sigma = (QP)^\sigma.$$

Hence, $PQ = PP^\sigma$ is a fixed line. If X is a point of a line other than QP or ℓ, then XX^σ is fixed. Therefore, $PP^\sigma \cap XX^\sigma$ is a fixed point and so lies on the line ℓ, and so $PP^\sigma \cap XX^\sigma = Q$. Therefore, Q is a centre of σ. So a collineation having an axis will also have a centre. By the principle of duality, the result holds. □

We say that a collineation is **central** if it has a centre. A **homology** is a central collineation having its centre not lying on its axis. An **elation** is a central collineation having its centre incident with its axis.

Example 2.13 (An elation)

Let $c \in \mathbb{R}^*$. Then the collineation

$$(x, y, z) \mapsto (x + cz, y, z)$$
$$[a, b, c] \mapsto [a, b, c - ac]$$

is an elation with centre $(1, 0, 0)$ and axis $[0, 0, 1]$. To see why, every line incident with $(1, 0, 0)$ is of the form $[0, b, c]$ and is readily seen to be fixed by
the collineation. Moreover, every point incident with $[0, 0, 1]$ is of the form
$(a, b, 0)$, which we can see again is fixed by the collineation.

Example 2.14 (A homology)

Let $c \in \mathbb{R}^*$ with $c \neq 1$. Then $(x, y, z) \mapsto (cx, y, z)$ is a homology with centre
$(1, 0, 0)$ and axis $[1, 0, 0]$. We leave the details for the reader as an exercise.

The terminology for elations is due to Sophus Lie (1893). That for
homologies traces its way back to Poncelet (1822).

Central collineations play a fundamental role in projective geometry and
they come in seven flavours.

Theorem 2.15 (Classification of collineations)

*The fixed points and lines of a collineation fall into one of the following
classes:*

(Ia) a triangle
(Ib) a non-incident point-line pair
(II) a pair of points and a pair of lines
*(III) all points on a line and all lines on a point, with the line and the point not
  incident*
(IV) an incident point-line pair
*(V) all points on a line and all lines on a point, with the line and the point
  incident*
(VI) all points and lines of the plane.

Proof The collineation is induced by a matrix A and the characteristic polynomial of A is cubic. If the characteristic polynomial is the cube of a linear
polynomial and there is an eigenspace of dimension 3, then A is a scalar matrix
and the collineation is the identity, so we are in Class (VI). If the characteristic
polynomial is the cube of a linear polynomial and there is an eigenspace of
dimension 2, then we are in Class (V). If the characteristic polynomial is the

cube of a linear polynomial and there is an eigenspace of dimension 1, we are in Class (IV). If the characteristic polynomial is the product of the square of a linear polynomial and another linear polynomial, then we are in Class (III) if there is an eigenspace of dimension 2 and in Class (II) if not. If the characteristic polynomial is the product of three distinct linear polynomials, we are in Class (Ia). If the characteristic polynomial is the product of a linear polynomial and an irreducible quadratic polynomial, then we are in Class (Ib). This exhausts the possibilities.   □

Note that Class (III) is the set of non-identity homologies, Class (V) is the set of non-identity elations, and Class (VI) contains only the identity collineation. The next result shows that a central collineation is determined by its axis and its centre, plus the image of one additional point.

Theorem 2.16
Let C and ℓ be a point and a line, respectively. Let A and B be points not equal to C and not incident with ℓ. There is at most one central collineation with centre C, axis ℓ, and mapping A to B.

Proof Let X be any point not incident with ℓ. Suppose first that X is not incident with AC. Let $Y := XA \cap \ell$ and $m := XC$. Then X is equal to the intersection of m and AY and, hence, if α is a central collineation with centre C, axis ℓ, and mapping A to B, then

$$X^\alpha = (AY \cap m)^\alpha = A^\alpha Y^\alpha \cap m^\alpha = BY \cap m.$$

So X^α is completely determined.

Suppose now that X is incident with AC. Let Y be a point not incident with AC or ℓ. Then X does not lie on YC, but we do have that Y^α is completely determined. Now α maps Y to Y^α, and so, by the argument above, X^α is completely determined: namely, $X^\alpha = Y^\alpha(XY \cap \ell) \cap XC$. Therefore, α is the unique central collineation with centre C, axis ℓ, and mapping A to B.   □

Second proof of Theorem 2.16 Suppose ϕ and ϕ' are central collineations with centre C, axis ℓ, and mapping A to B. Then $\phi'\phi^{-1}$ fixes C, ℓ and

$$A^{\phi'\phi^{-1}} = B^{\phi^{-1}} = A.$$

Let m be a line incident with A. Then m meets ℓ in a point X. Since every point on ℓ is fixed by $\phi'\phi^{-1}$, it follows that m is fixed as $m = AX$. So C and A are both centres for $\phi'\phi^{-1}$; which implies by Theorem 2.12 that $\phi'\phi^{-1} = 1$. Therefore, $\phi' = \phi$.   □

Theorem 2.17
Let C and ℓ be a point and a line, respectively. Let A and B be points not equal to C and not incident with ℓ, with A, B, C collinear. Then there is a unique central collineation with centre C and axis ℓ, mapping A to B.

Proof By Theorem 2.16, there is at most one central collineation with centre C, axis ℓ, and mapping A to B. We need to show that there is at least one such collineation. Suppose first that C is incident with ℓ. Since $\mathrm{PGL}(3, \mathbb{R})$ acts transitively on *flags* (by Exercise 2.1), we may suppose without loss of generality that $\ell: z = 0$ and $C: (1,0,0)$. So we can write $A = (a, t, 1)$ and $B = (b, t, 1)$. It is straightforward to see that the collineation having representative matrix

$$\begin{bmatrix} 1 & 0 & 0 \\ 1 & 1 & 0 \\ b-a-t & 0 & 1 \end{bmatrix}$$

maps A to B, and fixes ℓ point-wise and C line-wise.

Next, suppose that C is not incident with ℓ. Since $\mathrm{PGL}(3, \mathbb{R})$ acts transitively on *anti-flags* (by Exercise 2.2), we may suppose without loss of generality that $\ell: z = 0$ and $C: (0,0,1)$. This time we can write $A = (a_1, a_2, 1)$ and $B = (b_1, b_2, 1)$, but with $(a_1, a_2), (b_1, b_2) \neq (0,0)$ and $a_1 b_2 = a_2 b_1$ (in order for A, B, C to be collinear). It is straightforward to see that the collineation having representative matrix

$$\begin{bmatrix} a_2 & 0 & 0 \\ b_1 - a_1 & b_2 & 0 \\ 0 & 0 & a_2 \end{bmatrix}$$

maps A to B, and fixes ℓ point-wise and C line-wise. $\square$

Theorem 2.18 (Pappus (c.340) (compare with Pappus of Alexandria, 1986))
Let X, Y, Z be points of a line ℓ and X', Y', Z' be points of another line ℓ' of PG(2, $\mathbb{R}$), all not equal to $\ell \cap \ell'$. Then $XY' \cap YX'$, $XZ' \cap X'Z$, $YZ' \cap Y'Z$ are collinear.

Proof Let $O = \ell \cap \ell'$ (see Figure 2.1). Let a be the line joining $V = XY' \cap YX'$ and $W = YZ' \cap Y'Z$. Consider the central collineation ϕ with centre O and axis a which takes X to Y, and the central collineation ψ with centre O and axis a which takes Y to Z. Then $(Y')^\phi$ lies on ℓ' and on $(XY')^\phi = (XV)^\phi = YV$, so $(Y')^\phi = YV \cap \ell' = X'$. Similarly, $(Z')^\psi$ lies on ℓ' and on $(YZ')^\psi = (YW)^\psi = ZW$, so $(Z')^\psi = ZW \cap \ell' = Y'$. But ϕ and ψ have the same centre and the same axis, and so commute (by Exercises 2.3 and 2.4). So with $\zeta = \psi\phi$, we have

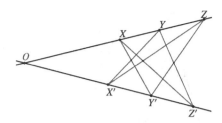

Figure 2.1 Pappus' configuration.

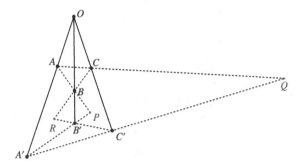

Figure 2.2 Desargues' configuration.

$$X^\zeta = X^{\psi\phi} = X^{\phi\psi} = Y^\psi = Z$$

and $(Z')^\zeta = (Z')^{\psi\phi} = (Y')^\phi = X'$. Thus $(XZ')^\zeta = X'Z$, giving $(XZ' \cap X'Z)^\zeta = XZ' \cap X'Z$. Thus $XZ' \cap X'Z$ lies on a.   □

As an exercise in using the principle of duality, we state the dual of Pappus' theorem (Theorem 2.18):

Let x, y, z be lines incident with a point L, and let x', y', z' be lines of another point L' of $\mathsf{PG}(2, \mathbb{R})$, all not equal to the line LL'. Then the lines joining $x \cap y'$ and $y \cap x'$, $x \cap z'$ and $x' \cap z$, and $y \cap z'$ and $y' \cap z$ are concurrent.

Theorem 2.19
If the sides of a hexagon lie alternatingly on two points, then the diagonals are concurrent.

Proof This follows from Theorem 2.18 and the principle of duality.   □

We say that ordered sets of points $\{P_1, \ldots, P_n\}$ and $\{Q_1, \ldots, Q_n\}$ are in *perspective from a point O* if O, P_i, Q_i are collinear for all $i \in \{1, \ldots, n\}$. The point O is often called the *centre of perspectivity*.

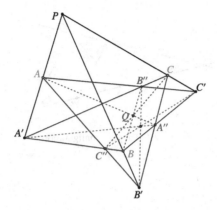

Figure 2.3 Theorem 2.21.

Theorem 2.20 (Desargues (1648) (see Desargues, 1951))
In PG(2, ℝ), *given triangles* A, B, C *and* A', B', C' *in perspective from a point*
O, *the lines* AB *and* $A'B'$ *meet in a point* P, *the lines* AC *and* $A'C'$ *meet in a*
point Q, *the lines* BC *and* $B'C'$ *meet in a point* R, *and* P, Q, *and* R *are collinear.*

Proof Let $O := AA' \cap BB'$ (see Figure 2.2). Let a be the line joining P and Q.
Consider the collineation ϕ with centre O and axis a taking A to A'. Now ϕ fixes
BB' and P, so takes AP to $A'P = A'B'$, and therefore $B^\phi = BB' \cap A'B' = B'$.
Hence, ϕ fixes CC' and Q, so takes AQ to $A'Q = A'C'$, and therefore $C^\phi =$
$CC' \cap A'C' = C'$. Therefore, $(BC)^\phi = B'C'$, and so $BC \cap a = B'C' \cap a$: R is on
the axis a of ϕ.   □

The line containing P, Q, and R in Theorem 2.20, is often called the *axis*
of perspectivity. We can dualise the definition of *perspectivity from a point* so
that we obtain perspectivity from a line, and Desargues' theorem can simply
be stated the following way: '*Two triangles are in perspective from a point if*
and only if they are in perspective from a line'.

Theorem 2.21 (van Lamoen, 1999)
Let ABC *and* $A'B'C'$ *be two triangles in perspective from a point* P. *Let* $A'' :=$
$BC' \cap B'C$, $B'' := CA' \cap C'A$ *and* $C'' := AB' \cap A'B$. *Then* ABC *and* $A''B''C''$
are in perspective from a point Q, $A'B'C'$ *and* $A''B''C''$ *are in perspective from*
a point R, *and* P, Q, *and* R *are collinear (see Figure 2.3).*

Proof The following proof is from Grinberg (n.d.). Let $X := BC \cap B'C'$, $Y :=$
$CA \cap C'A'$ and $Z := AB \cap A'B'$. The triangles $AB'C'$ and $A'BC$ are in perspective
from P, so, by Desargues' theorem (Theorem 2.20), $C'' := AB' \cap A'B$, X and
$B'' := C'A \cap CA'$ are collinear. Thus BC, $B'C'$ and $B''C''$ are concurrent in

X. With $A'' := B'C \cap BC'$, similarly, we have CA, $C'A'$ and $C''A''$ concurrent in Y, and AB, $A'B'$, $A''B''$ concurrent in Z. By Desargues' theorem (Theorem 2.20), since ABC and $A'B'C'$ are in perspective, X, Y, and Z are collinear. Thus, by the dual of Desargues' theorem (Theorem 2.20), ABC and $A''B''C''$ are in perspective from a point Q, and $A'B'C'$ and $A''B''C''$ are in perspective from a point R.

Since the lines BC, $B'C'$, and $B''C''$ have a common point X, the triangles $BB'B''$ and $CC'C''$ are in perspective. By Desargues' theorem (Theorem 2.20), $P = BB' \cap CC'$, $Q = BB'' \cap CC''$, and $R = B'B'' \cap C'C''$ are collinear.   □

Theorem 2.21 will turn out to be a simple consequence of a property of three-dimensional projective space, by projecting from a pair of *desmic tetrahedra* (see Section 14.2).

Theorem 2.22 (Eggar, 1998)
Let A, B, C, D, S be five points, no three collinear in $\mathsf{PG}(2,\mathbb{R})$. Let $X = AD \cap BS$, $Y = BD \cap AS$. Let Z be any point on DC and let $R = CX \cap AZ$, $Q = CY \cap BZ$, $P = RY \cap SZ$. Then X, P, Q are collinear.

Proof The triangles RCQ and ADB are in perspective from Z and so, by Desargues' theorem (Theorem 2.20), X, Y, K are collinear, where $K = AB \cap RQ$. The triangles RPQ and ASB are also in perspective from Z and so, by Desargues' theorem (Theorem 2.20), Y, K, $PQ \cap SB$ are collinear. Thus, $X = PQ \cap SB$ since both these points are the intersection point of the lines YK and BS.   □

In Section 4.2, we will examine in detail *perspectivities* and *projectivities*, but for the moment, it suffices to know that a *projection* from one line ℓ to another m, via a point O, is simply the map $P \mapsto PO \cap m$. The following beautiful application of Desargues' theorem (Theorem 2.20), has its earliest attribution in 1976, as far as the authors are aware.

Theorem 2.23 (Kárteszi, 1976, p. 170)
Let $A_1A_2A_3A_4A_5$ be a pentagon in $\mathsf{PG}(2,\mathbb{R})$. Let a_i be the side opposite A_i, so that $a_1 = A_3A_4$, $a_2 = A_4A_5$, $a_3 = A_5A_1$, $a_4 = A_1A_2$, $a_5 = A_2A_3$. Then for any point B_1 on a_1, the sequence of projections from a_1 to a_4 with centre A_5, from a_4 to a_2 with centre A_3, from a_2 to a_5 with centre A_1, from a_5 to a_3 with centre A_4, and from a_4 to a_1 with centre A_2, when composed, fixes B_1.

Proof By the first projection we obtain point B_5 from point B_1, by the second B_4 from B_5, by the third B_3 from B_4, by the fourth B_2 from B_3. This derivation implies that the triangle $B_2A_2B_3$ is in perspective with the triangle $A_5B_5B_4$ from the centre A_1. But then the axis of perspectivity is the line $A_3A_4 = a_1$ and this is cut by the side B_5A_5 of the triangle $A_5B_5B_4$ in a point B_1. By Desargues' theorem (Theorem 2.20) the side A_2B_2 of the triangle $B_2A_2B_3$ must cut the axis in the same point. But this implies that the projection from a_4 to a_1 with centre A_2 carries the point B_2 onto the original point B_1.   □

Theorem 2.24 (Chasles 1852, art. 389, p. 284)
Three triangles in perspective from the same point have concurrent axes of perspective.

Proof Let O be the common centre of perspectivity for the three triangles, and let a, b, c be the three axes of perspective for each pair of triangles. By the dual of Theorem 2.16, there exists a homology ϕ_a with axis a, centre O, and such that $c^{\phi_a} \neq b$. We also know there exists a homology ϕ_b with axis b, centre O, and mapping c^{ϕ_a} to c. The collineation $\phi_a\phi_b$ fixes O and all the lines incident with O, because each homology does. So $\phi_a\phi_b$ is a central collineation. Now

$$c^{\phi_a\phi_b} = \left(c^{\phi_a}\right)^{\phi_b} = c$$

and so c is the axis of $\phi_a\phi_b$, because c is not incident with the centre O of $\phi_a\phi_b$. Moreover, a and b are lines incident with O, and so $\phi_a\phi_b$ fixes $a \cap b$. This means that $a \cap b$ is incident with the axis of $\phi_a\phi_b$ (which is c), and so a, b, c are concurrent.   □

We also have the dual result.

Corollary 2.25 (Chasles 1852, art. 388, p. 283)
Three triangles in perspective from the same line have collinear centres of perspective.

2.4 Interlude: Homogenisation of Polynomials

A line in the affine plane $\mathsf{AG}(2, \mathbb{R})$ is of the form $ax + by = c$ where the indeterminates are x and y. This line can be identified with the set of points (x, y) of $\mathsf{AG}(2, \mathbb{R})$ that satisfy this equation. Alternatively, we can write this line as the span of two points: if $b \neq 0$, then we can write the line as

$$\langle (1, (c - a)/b), (-1, (c + a)/b) \rangle,$$

and if $b = 0$, then we could write $\langle (c/a, 0), (c/a, 1) \rangle$ instead. In passing to the projective plane $\mathsf{PG}(2, \mathbb{R})$, we can also describe a line as the points collinear with two given points. The two points $(1, (c - a)/b)$ and $(-1, (c + a)/b)$ of the affine plane become $(b, c - a, b)$ and $(-b, c + a, b)$ in homogeneous coordinates for $PG(2, \mathbb{R})$. Then a point (x, y, z) is collinear with these two points if the following holds (by Theorem 1.5):

$$\begin{vmatrix} x & y & z \\ b & c - a & b \\ -b & c + a & b \end{vmatrix} = 0.$$

A simple calculation shows that this equation becomes $ax + by = cz$, the *homogenisation* of the original affine equation $ax + by = c$. Indeed, we can take the homogenisation of any polynomial in x and y to obtain a polynomial in x, y, and z, where the total degree of each term is the same. For instance, the equation of a parabola $y^2 - x = 0$ has homogenisation $y^2 - xz = 0$; we simply multiply terms by sufficient powers of the indeterminate z in order for the total degree of each term to achieve the degree of the original polynomial.

Next, consider affine equations for conics:

Ellipse	Parabola	Hyperbola
$\frac{x^2}{a^2} + \frac{y^2}{b^2} = 1$	$y^2 = 4a^2 x$	$\frac{x^2}{a^2} - \frac{y^2}{b^2} = 1$

The homogeneous form for the equation of an ellipse is

$$\frac{x^2}{a^2} + \frac{y^2}{b^2} = z^2$$

and we see immediately that there are no solutions in x and y if z is zero. This means that an ellipse has no point on the line at infinity. However, the homogeneous form for the equation of a parabola is $y^2 = 4a^2 xz$, and if we let $z = 0$ in this equation, we see that $y = 0$ and x can be any non-zero real number. Since this is now an equation that describes a set of homogeneous coordinates for points on this parabola, there is just one point with $z = 0$ on this parabola, namely $(1, 0, 0)$. So a parabola has one point at infinity. Finally, the hyperbola has homogeneous form

$$\frac{x^2}{a^2} - \frac{y^2}{b^2} = z^2$$

which has two points on the line at infinity: $(a, b, 0)$ and $(a, -b, 0)$.

Fermat was the first to prove the identity between degree 2 equations in two variables and conic sections, in the *Isagoge* (Fermat, 1999, pp. 91–103).

Finally, a *projective algebraic curve* is the set of points with homogeneous coordinates (x, y, z) that are zeros of a homogeneous polynomial in the variables x, y, and z.

Theorem 2.26 (Waring, 1772, pp. 23–5; 1762, p. 82)
Collineations of $PG(2, \mathbb{R})$ *take projective algebraic curves to projective algebraic curves and preserve the degree.*

Proof A collineation $g \colon (x, y, z) \mapsto (x', y', z')$ of $PG(2, \mathbb{R})$ has the form $x' = ax+by+cz$, $y' = dx+ey+fz$, $z' = gx+hy+iz$. Given a homogeneous polynomial $p(x, y, z)$ in x, y, and z of degree n, the curve $p(x, y, z) = 0$ is mapped by g to the curve $p(x', y', z') = 0$, and $p(ax + by + cz, dx + ey + fz, gx + hy + iz)$ is homogeneous of degree n. $\qquad\qquad\square$

Exercises

2.1   A **flag** is an incident point and line pair. Show that $PGL(3, \mathbb{R})$ acts transitively on the flags of $PG(2, \mathbb{R})$.

2.2   An **anti-flag** is a non-incident point and line pair. Show that $PGL(3, \mathbb{R})$ acts transitively on the anti-flags of $PG(2, \mathbb{R})$.

2.3   Let ϕ and ψ be homologies sharing a common centre and axis. Prove $\phi\psi = \psi\phi$.

2.4   Let ϕ and ψ be elations sharing a common centre C.
    (a) Show that $\phi\psi$ is an elation with centre C.
    (b) Let σ and τ be elations with a common centre C, and let a and b be their axes, respectively. Let m be a line not incident with C. Show $m^{\sigma\tau} = m^{\tau\sigma}$. (*Hint:* Consider the line on $a \cap m^{\tau}$ and $b \cap m^{\sigma}$.)
    (c) Prove that two elations with a common centre commute.

2.5   Show that a collineation that fixes three collinear points has an axis.

2.6   Let ℓ be a line. Show that if a collineation g fixes three distinct points of ℓ, then it fixes every point of ℓ.

2.7   Show that $PGL(3, \mathbb{R})$ is generated by the set of homologies of $PG(2, \mathbb{R})$. (You may use the fact that $GL(3, \mathbb{R})$ is generated by any union of conjugacy classes that contains elements of all non-zero determinants and does not consist only of scalar matrices.)

2.8   (Eggar, 1998, Proposition 2.1) Let X, Y, Z, A_1 be any four points in $PG(2, \mathbb{R})$. Choose points A_2 on XA_1, W on XZ, R_1 on A_1Z, U on XY, and S_1 on A_1Y. Let $R_2 = WR_1 \cap A_2Z$, $S_2 = US_1 \cap A_2Y$, $P_1 = R_1Y \cap S_1Z$, and $P_2 = R_2Y \cap S_2Z$. Show that P_1, P_2, $UZ \cap WY$ are collinear.

2.9 Let A, B, C, D be a set of four distinct points of $\mathsf{PG}(2, \mathbb{R})$, no three collinear. Consider the *diagonal triangle* formed by $E := AB \cap CD$, $F := AC \cap BD$, and $G := AD \cap BC$. Show that EFG is in perspective with any of the triangles whose vertices are three of the vertices of the quadrangle $ABCD$.

2.10 Let A_1, A_2, A_3, A_4 be a quadrangle and let $\{\ell_1, \ell_2, \ell_3, \ell_4\}$ be a set of four lines, no three concurrent. If five of the six intersections $\ell_i \cap \ell_j$ lie on the side opposite to $A_i A_j$, prove that the sixth intersection lies on the remaining side. If $\{m_1, m_2, m_3, m_4\}$ is another set of four lines, no three concurrent with the same property, prove that the points $\ell_i \cap m_i$ ($i = 1, 2, 3, 4$) are collinear.

2.11 Let m_1, m_2, m_3 be lines on A, A' be a point not on m_1, m_2 or m_3, and n_1, n_2, n_3 be lines on A'. Show that $(m_1 \cap n_2)(m_2 \cap n_1)$, $(m_3 \cap n_1)(m_1 \cap n_3)$ and $(m_3 \cap n_2)(m_2 \cap n_3)$ are concurrent.

2.12 Let A, B, C, D, E be five points of $\mathsf{PG}(2, \mathbb{R})$, no three collinear, ℓ be the line joining $AB \cap CD$ and E, m be lines on $AD \cap BC$ and E, $F := \ell \cap BC$, $G := \ell \cap AD$, $H := m \cap AB$, $I := m \cap CD$. Show that FI, GH, and BD are concurrent.

2.13 (Nehring, 1942) Let ABC be a triangle, A_1 be on BC, B_1 on CA, C_1 on AB with AA_1, BB_1, CC_1 concurrent. Let X_1 be a point on BC, $X_2 := X_1 B_1 \cap AB$, $X_3 := X_2 A_1 \cap CA$, $X_4 := X_3 C_1 \cap BC$, $X_5 := X_4 B_1 \cap AB$, $X_6 := X_5 A_1 \cap CA$. Then C_1, X_6, and X_1 are collinear.

2.14 Lines AB and CD intersect in U and AC, BD intersect in V. Define $F := UV \cap AD$, $G := UV \cap BC$, $H := BF \cap AC$. Show that GH, CF, and AU are concurrent.

2.15 A, B, C, D are vertices of a quadrangle and $BC \cap AD = X$, $CA \cap BD = Y$, $AB \cap CD = Z$, $BC \cap YZ = L$, $BD \cap ZX = M$, $CD \cap XY = N$. Show that L, M, and N are collinear.

2.16 The sides AB and CD of quadrilateral $ABCD$ meet at point P, and the sides BC and AD meet at point Q. Through point P a line is drawn that intersects sides BC and AD at points E and F. Prove that the intersection points of the diagonals of quadrilaterals $ABCD$, $ABEF$, and $CDFE$ lie on a line that passes through point Q.

2.17 (Dao Thanh Oai) In ABC, the collinear points D, E, F lie on BC, CA, AB, respectively. $AFA'E$, $BFB'D$, $CEC'D$ are three parallelograms. Show that A', B', C' are collinear.

2.18 Let $ABCD$ be a parallelogram, F a point on no side of $ABCD$, ℓ the line on F parallel to AD, m the line on F parallel to AB, $E := \ell \cap AB$, $G := m \cap AD$, $H := m \cap BC$, $K := \ell \cap CD$. Show that AH, EC, and DF are concurrent.

2.19 The extensions of the sides AB and CD of the quadrilateral $ABCD$ meet
     at the point P, and the extensions of the sides BC and AD meet at the
     point Q. Through point P a line is drawn that intersects sides BC and AD
     at points E and F. Prove that the intersection points of the diagonals of
     quadrilaterals $ABCD$, $ABEF$, and $CDFE$ lie on a line that passes through
     point Q.

3

Polarities and Conics

If the Greeks had not cultivated Conic Sections, Kepler could not have superseded Ptolemy; if the Greeks had cultivated Dynamics, Kepler might have anticipated Newton.

Whewell (2015, p. 311)

Kepler's discovery would not have been possible without the doctrine of conics. Now contemporaries of Kepler – such penetrating minds as Descartes and Pascal – were abandoning the study of geometry ... because they said it was so UTTERLY USELESS. There was the future of the human race almost trembling in the balance; for had not the geometry of conic sections already been worked out in large measure, and had their opinion that only sciences apparently useful ought to be pursued, the nineteenth century would have had none of those characters which distinguish it from the *ancien régime*.

Charles Sanders Peirce, 'Lessons from the History of Science: The Scientific Attitude' (1896), in Peirce et al. (1931, Section 11 'The study of the useless', p. 75)[1]

In the Euclidean plane, we can define a map that swaps points and lines, with respect to a circle of radius r. As we will see, this map extends to a duality of the extended Euclidean plane, and this duality has a reciprocal nature, in that applying it twice results in the same thing you started with.

Given an external point, we can draw two tangents to it, and then we draw a line across the circle that joins the two points of tangency (see Figure 3.1). This line is the image of our external point; conversely, the external point is the image of the secant line by reversing this construction. If instead we take

---

[1] Collected Papers of Charles Sanders Peirce, Volumes I – II, edited by Charles Hartshorne and Paul Weiss, Cambridge, MA: The Belknap Press of Harvard University Press, Copyright ©1931, 1932, 1959, 1960 by the President and Fellows of Harvard College. Used by permission. All rights reserved.

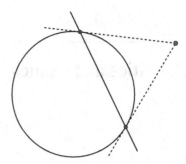

Figure 3.1 The pole–polar relationship for a circle.

an internal point P, then we take the diameter d through P, and then the unique line perpendicular to d that is at a distance $r^2/d(O,P)$ from the centre O of the circle. If P lies on the circle, then we take the tangent at P. This is the *pole–polar* relationship of a circle, where the point in question is the *pole* of its *polar* line. The point O does not have a polar line in the Euclidean plane, but in the projective plane, this mapping is complete. Moreover, every conic has such a duality arising from it, and these are known as *polarities*.

3.1 Dualities and Polarities

An incidence-preserving map from the points to the lines of $PG(2, \mathbb{R})$ is called a **duality**. For example, the map interchanging point and line coordinates is a duality, called the **standard duality**. Indeed, if g is a collineation of $PG(2, \mathbb{R})$ and δ is the standard duality, then their composition $g\delta$ is a duality. A duality of order two, a so-called **polarity**, has special significance – as we shall see. The dualities together with the collineations of $PG(2, \mathbb{R})$ are **correlations** of $PG(2, \mathbb{R})$: a permutation of the entire set of points and lines of $PG(2, \mathbb{R})$ that preserves incidence. The correlations form a group which we naturally call the **correlation group** of $PGL(3, \mathbb{R})$. Note that the product of two dualities is a collineation, and hence the product of two correlations that are both not collineations is a collineation.

Lemma 3.1
Every duality of $PG(2, \mathbb{R})$ *is the product of the standard duality δ with an element of* $PGL(3, \mathbb{R})$. *Moreover, the correlation group of* $PG(2, \mathbb{R})$ *is the semidirect product* $PGL(3, \mathbb{R}) \rtimes \langle \delta \rangle$.

Proof Let Δ be a duality of $PG(2, \mathbb{R})$ and let δ be the standard duality. Then the product $g = \delta\Delta$ of Δ with the standard duality δ is a collineation, so by the FToPG Theorem 2.9, $g \in \mathrm{PGL}(3, \mathbb{R})$. Therefore, $\Delta = \delta g$ and the result follows. □

Given a polarity, a point is **absolute** if it is incident with its image under the polarity.

Theorem 3.2

There are two conjugacy classes of polarities of $\mathrm{PG}(2, \mathbb{R})$ *with representatives the standard duality* $(a, b, c) \mapsto [a, b, c]$ *(with no absolute points), and the orthogonal polarity* $(a, b, c) \mapsto [a, b, -c]$ *(with infinitely many absolute points).*

Proof Suppose that Δ is a polarity of $\mathrm{PG}(2, \mathbb{R})$ and let $g = \delta \Delta$ (where δ is the standard duality). By Lemma 3.1, there exists $A \in \mathrm{GL}(3, \mathbb{R})$ with g mapping the point (x, y, z) to the point $(x, y, z)A$. Now g has the action $[a, b, c] \mapsto [a, b, c]A^{-\top}$ on lines, and so $\Delta = \delta g$ maps the point (d, e, f) to the line $[d, e, f]A$ and the line $[a, b, c]$ to the point $(a, b, c)A^{-\top}$. Since Δ has order two, it follows that $AA^{-\top}$ is a scalar multiple of the identity: $AA^{-\top} = \lambda I$. As A is invertible, we have

$$\lambda^3 = \det(\lambda I) = \det(AA^{-\top}) = \det(A)\det(A^{-\top}) = \det(A)/\det(A) = 1.$$

Therefore, $\lambda = 1$ and $A = A^\top$; that is, A is symmetric. Now a symmetric real matrix can be orthogonally diagonalised: there exists an invertible matrix B such that $B^\top A B$ is a diagonal matrix D. (You will recall that we can take B to be a matrix whose columns are normalised eigenvectors of A, and so $B^{-1} = B^\top$.) Suppose d_1, d_2, d_3 are the diagonal entries of D, and note that the d_i are non-zero as A is invertible. Next, we let

$$C := \begin{bmatrix} 1/\sqrt{|d_1|} & 0 & 0 \\ 0 & 1/\sqrt{|d_2|} & 0 \\ 0 & 0 & 1/\sqrt{|d_3|} \end{bmatrix}$$

and hence $A' := C^\top B^\top A B C$ has diagonal entries ± 1. Now by permuting the columns (by conjugating by a permutation matrix) and replacing A' by $-A'$ if necessary, we may assume that either $A' = I$ or

$$A' = \begin{bmatrix} 1 & 0 & 0 \\ 0 & 1 & 0 \\ 0 & 0 & -1 \end{bmatrix}.$$

Then the conjugate of Δ (in the correlation group) we have obtained is either the standard polarity or the polarity $(a, b, c) \mapsto [a, b, -c]$. The standard polarity has no absolute points as $x^2 + y^2 + z^2 = 0$ implies $x = y = z = 0$. In contrast, the second duality has the absolute points (x, y, z) with $z^2 = x^2 + y^2$, which has infinitely many solutions, even counting scalar multiples as the same. (There are no points (x, y, z) of $\mathsf{PG}(2, \mathbb{R})$ with $z^2 = x^2 + y^2$ and $z = 0$, so we may put $X = \frac{x}{z}$ and $Y = \frac{y}{z}$ so that $(x, y, z) = (X, Y, 1)$ and now we want $X^2 + Y^2 = 1$. That is, the absolute points correspond to the points of the unit circle in $\mathbb{R}^2$.)   □

3.2 Conics

A **conic** is the set of zeroes in $\mathsf{PG}(2, \mathbb{R})$ of a real homogeneous quadratic equation $Q(x, y, z) = 0$ in three variables. We will often refer to the function Q as a **quadratic form**. The conic is non-degenerate if $Q(x, y, z)$ doesn't factor over $\mathbb{C}$. It may be empty. We usually mean a non-empty, non-degenerate conic when we say conic. In particular, a nondegenerate conic does not contain a line.

Example 3.3
Consider the quadratic form $Q(x, y, z) := xy + z^2$. If we suppose that the line at infinity is $z = 0$, then the affine points $(x, y, 1)$ satisfying this homogeneous quadratic equation form the curve $xy = -1$; a hyperbola. This conic is non-degenerate as $Q(x, y, z)$ does not factor over $\mathbb{C}$.

Apollonius of Perga (in French, de Perge) wrote a book *Conics* devoted to conics (Apollonius de Perge, 2009), which consists of geometry more advanced than that of Euclid's *Elements*. The historian of mathematics Carl Boyer called Apollonius the 'greatest geometer of antiquity'.

APOLLONIUS OF PERGA (262 BC–c.190 BC) was called the great geometer by his contemporaries. His most famous work is the *Conics* in eight books, all but the last of which are extant. The first book is on cones, diameters, and ordinates of conics; the second on asymptotes of hyperbolas, conjugate opposite hyperbolas, chords, and diameters; and third on areas, material underlying the power of a point with respect to a conic, pole, and polar, and material underlying the Chasles–Steiner definition of a conic and foci. The fourth is on pole and polar, intersections and tangencies of conics. The fifth is on normals of conics, including curvature of evolutes. The sixth is on congruence and similarities of conics, and conics on a cone. The seventh is on conjugate diameters, rectus latum, and rectus transversum. Most other work of Apollonius is lost: Cutting of a ratio, Cutting of an area, Determinate section, Tangencies, Inclinations, Plane loci, On the burning glass, On the cylindrical helix, et cetera, with a few surviving fragments in Arabic translation (with Cutting of a ratio the best-preserved).

Define f by
$$f((x, y, z), (x', y', z')) := Q(x + x', y + y', z + z') - Q(x, y, z) - Q(x', y', z').$$

Then f is a bilinear form,[2] and can be written as

$$f((x, y, z), (x', y', z')) = (x, y, z)A(x', y', z')^\top,$$

for a 3-by-3 symmetric matrix A. We say that A is the **Gram matrix** of f. Whilst the matrix A for f is uniquely defined, there are many matrices that we can choose for writing Q in the following way:

$$Q(x, y, z) = (x, y, z)M(x, y, z)^\top.$$

Example 3.4

Consider the quadratic form $Q(x, y, z) := xy + z^2$. In a matrix representation, we could choose

$$M := \begin{bmatrix} 0 & 1 & 0 \\ 0 & 0 & 0 \\ 0 & 0 & 1 \end{bmatrix} \quad \text{or} \quad M := \begin{bmatrix} 0 & \frac{1}{2} & 0 \\ \frac{1}{2} & 0 & 0 \\ 0 & 0 & 1 \end{bmatrix}.$$

However, the bilinear form f arising from Q has Gram matrix

$$A := \begin{bmatrix} 0 & 1 & 0 \\ 1 & 0 & 0 \\ 0 & 0 & 2 \end{bmatrix},$$

that is, $f((x, y, z), (x', y', z')) = xy' + x'y + 2zz'$.

By the definition of f, we see that $A = M + M^\top$, no matter the choice M for the matrix for Q. We obtain Q from f by evaluating f on the diagonal: $Q(x, y, z) = \frac{1}{2}f((x, y, z), (x, y, z))$. There is a simple way to test for non-degeneracy of a conic $\mathcal{K}$ by taking determinants.

Theorem 3.5

Let $\mathcal{K}$ be a conic with defining quadratic form Q and Gram matrix A for the associated bilinear form arising from A. Then $\mathcal{K}$ is non-degenerate if and only if $\det(A) \neq 0$.

Proof Suppose $Q(x, y, z)$ factors over $\mathbb{C}$ into two linear factors:

$$Q(x, y, z) = (k_1 x + k_2 y + k_3 z)(\ell_1 x + \ell_2 y + \ell_3 z)$$
$$= \left((x, y, z)(k_1, k_2, k_3)^\top\right)\left((\ell_1, \ell_2, \ell_3)(x, y, z)^\top\right).$$

Let $K = (k_1, k_2, k_3)$ and $L = (\ell_1, \ell_2, \ell_3)$. So we can write a matrix representative M for Q by the Kronecker product $M = K^\top L$. It is then clear that M has rank 1. Now consider the Gram matrix A for the bilinear form arising from Q. Now

---

[2] A map $f \colon \mathbb{R}^2 \times \mathbb{R}^2 \to \mathbb{R}$ is a bilinear form if it is linear in each coordinate.

$A = M + M^\top$, and we will see now that A is singular. If K and L are linearly independent, then the *cross product* $K \times L$ is orthogonal to both K and L, and therefore

$$(K \times L)(M + M^\top) = (K \times L)(K^\top L + L^\top K)$$
$$= (K \times L)K^\top L + (K \times L)L^\top K$$
$$= ((K \times L) \cdot K)L + ((K \times L) \cdot L)K = 0.$$

So A is singular when K and L are linearly independent. If $K = \lambda L$, for some $\lambda \in \mathbb{R}^*$, then $A = 2\lambda K^\top K$, which has rank 1 and is hence singular. Therefore, $\det(A) = 0$.

Conversely, suppose $\det(A) = 0$. Then 0 is an eigenvalue of A and there exists a point $P: (p_0, p_1, p_2)$ such that $PA = 0$. Then $Q(p_0, p_1, p_2) = \frac{1}{2} PAP^\top = 0$, and hence P lies on the conic $\mathcal{K}$. So suppose $\mathcal{K}$ is more than a single point. Let R be a point of $\mathcal{K}$ apart from P, and let $\lambda \in \mathbb{R}^*$. Then

$$(P + \lambda R)A(P + \lambda R)^\top = PAP^\top + 2\lambda PAR^\top + \lambda^2 RAR^\top = 0.$$

Therefore, $\mathcal{K}$ contains the line PR, and is thus degenerate. For the remaining case, where $\mathcal{K}$ consists of just one point, we refer to Exercise 6.1. □

Suppose $\mathcal{K}$ is a non-degenerate conic. If a line ℓ meets $\mathcal{K}$, then we obtain a quadratic equation describing the points of intersection (see Theorem 3.14). Thus lines of $\mathsf{PG}(2, \mathbb{R})$ meet $\mathcal{K}$ in 0, 1, or 2 points, and are correspondingly called **external**, **tangent**, or **secant** lines to $\mathcal{K}$. We will see in Theorem 3.7 that every point of $\mathcal{K}$ lies on a unique tangent line to $\mathcal{K}$.

Example 3.6

Let $\mathcal{K}$ be a non-degenerate conic with associated 3×3 symmetric Gram matrix A. Let $P: (p_0, p_1, p_2)$ be a point of $\mathcal{K}$. Then the line $p: [a, b, c]$, whose coordinates satisfy $(p_0, p_1, p_2)A = (a, b, c)$, is a tangent line to $\mathcal{K}$ incident with P. To see this, suppose that there is another point $P': (p'_0, p'_1, p'_2)$ of $\mathcal{K}$ incident with p. Then

$$0 = [a, b, c] \cdot (p'_0, p'_1, p'_2) \cdot = (p_0, p_1, p_2)A(p'_0, p'_1, p'_2)^\top = PA(P')^\top.$$

Now consider a point Q on the line joining PP', not equal to P. Then, $Q = \lambda P + P'$ for some $\lambda \in \mathbb{R}$. So, if we think of Q in its homogeneous coordinate representation, we find that

$$QAQ^\top = (\lambda P + P')A(\lambda P + P')^\top$$
$$= \lambda^2 PAP^\top + \lambda\left(PA(P')^\top + P'AP^\top\right) + P'A(P')^\top$$
$$= \lambda\left(PA(P')^\top + P'AP^\top\right)$$
$$= 0$$

because $P'AP^\top = PA^\top(P')^\top = PA(P')^\top = 0$. Therefore, the whole line PP' is contained in $\mathcal{K}$; a contradiction since $\mathcal{K}$ is non-degenerate. Therefore, p only intersects $\mathcal{K}$ in the point P.

Theorem 3.7
Let $\mathcal{K}$ be a non-degenerate (non-empty) conic. Then for every point P of $\mathcal{K}$, there is a unique tangent line passing through P.

Proof Let A be a Gram matrix for τ. Since $\mathcal{K}$ is non-degenerate, A is invertible. Let P be a point of $\mathcal{K}$. From Example 3.6, there exists at least one tangent line p through P, which is the result of multiplying the homogeneous coordinates of P by A (i.e., PA) and changing round brackets to square ones to obtain homogeneous coordinates for p.

Now suppose ℓ is a line on P that is a tangent line to $\mathcal{K}$. We use a similar argument to that of Example 3.6. Since ℓ is not contained in $\mathcal{K}$, there exists a point Q, incident with ℓ, satisfying $QAQ^\top \neq 0$. Now every point of ℓ, apart from P, is of the form $\mu P + Q$ for some $\mu \in \mathbb{R}$. Since ℓ is a tangent line, none of these points lies on $\mathcal{K}$, and hence for all $\mu \in \mathbb{R}$, we have

$$0 \neq (\mu P + Q)A(\mu P + Q)^\top = \mu^2 PAP^\top + 2\mu PAQ^\top + QAQ^\top$$
$$= 2\mu PAQ^\top + QAQ^\top.$$

Since this holds for all μ, it follows that $PAQ^\top = 0$ and hence Q lies on p. Therefore, $\ell = PQ = p$. $\qquad\square$

The above result establishes a bijection between the points of a non-degenerate conic $\mathcal{K}$ and its tangent lines, which thus leads to the notion of the **dual conic** of $\mathcal{K}$. Moreover, in Exercise 3.8 at the end of this chapter, we see that the dual conic indeed has the property that no three lines of it are concurrent.

A point not on $\mathcal{K}$ is **internal** if it does not lie on any tangent line to $\mathcal{K}$ and **external** if it lies on two tangent lines to $\mathcal{K}$. We may choose to replace Q by $-Q$, if necessary, so that $\det(A) > 0$. Then a point (x, y, z) is internal if $Q(x, y, z) > 0$, on $\mathcal{K}$ if $Q(x, y, z) = 0$ and external if $Q(x, y, z) < 0$.

Example 3.8
The unit circle of the Euclidean plane is the locus of the equation $X^2 + Y^2 = 1$. Projectively, we write the circle in homogeneous coordinates as $Q(x, y, z) = 0$ where

$$Q(x, y, z) := z^2 - x^2 - y^2.$$

In matrix form, we have $Q(x, y, z) = (x, y, z)A(x, y, z)^\top$ *where*

$$A = \begin{bmatrix} -1 & 0 & 0 \\ 0 & -1 & 0 \\ 0 & 0 & 1 \end{bmatrix}.$$

Note that $\det(A) = 1$. *In the Euclidean plane, we think of the points* (X, Y) *satisfying* $X^2 + Y^2 < 1$ *as the points enclosed by the circle; the internal points. In* $\mathsf{PG}(2, \mathbb{R})$, *these points appear as* $(X, Y, 1)$ *and* $Q(X, Y, 1) = 1 - X^2 - Y^2 > 0$. *So an internal point point has* $Q(X, Y, 1) > 0$ *and an external point has* $Q(X, Y, 1) < 0$.

The map $\tau \colon (x, y, z) \mapsto [u, v, w]$, where $(u, v, w) = (x, y, z)A$, is a map from the points of $\mathsf{PG}(2, \mathbb{R})$ to the lines of $\mathsf{PG}(2, \mathbb{R})$. When $\mathcal{K}$ is non-degenerate, a point–line pair interchanged by τ are said to be **pole** and **polar** (respectively). The polar of a point P on $\mathcal{K}$ is the tangent line to $\mathcal{K}$ at P (see Example 3.6); the pole of a tangent line ℓ to $\mathcal{K}$ is $\ell \cap C$ (its **point of contact** to $\mathcal{K}$).

The pole–polar relationship with respect to a conic goes back to Apollonius of Perga (c.262 BC–c.190 BC) and his treatise *Conics*.

Theorem 3.9 (de la Hire, 1685)
Let $\mathcal{K}$ *be a conic of* $\mathsf{PG}(2, \mathbb{R})$. *If* P *lies on the polar of* P', *then* P' *lies on the polar of* P.

Proof Let A be the symmetric Gram matrix of the bilinear form underlying $\mathcal{K}$, and let $P(x, y, z)$, $P'(x', y', z')$ be points of $\mathsf{PG}(2, \mathbb{R})$. Then P is on the polar of P' if and only if $(x, y, z)A(x', y', z')^\top = 0$, which is equivalent to $(x', y', z')A(x, y, z)^\top = 0$. Hence, P is on the polar of P' if and only if P' is on the polar of P. □

When this happens, we say that P and P' are **conjugate** (and that the polar of P and the polar of P' are **conjugate**). Recall (from the paragraph preceding Theorem 3.2) that a polarity is a duality of order two.

Theorem 3.10 (Poncelet, 1817–18)
Let $\mathcal{K}$ *be a non-degenerate conic of* $\mathsf{PG}(2, \mathbb{R})$. *The pole–polar relationship with respect to* $\mathcal{K}$ *defines a polarity* τ *of* $\mathsf{PG}(2, \mathbb{R})$.

Proof Let A be the symmetric matrix of the bilinear form underlying $\mathcal{K}$. Consider the map ρ defined by

$$(x, y, z) \mapsto [a, b, c], \quad \text{where } (x, y, z)A = (a, b, c),$$
$$[a, b, c] \mapsto [x, y, z], \quad \text{where } (a, b, c)A^{-\top} = (x, y, z).$$

Then clearly ρ is a duality and for all points (x, y, z) we have

$$(x, y, z)^{\rho^2} = ((x, y, z)A)^{\rho} = (x, y, z)AA^{-\top} = (x, y, z)AA^{-1} = (x, y, z).$$

Therefore, ρ has order 2 and is thus a polarity. □

We will see later (Theorem 3.14) that the polarity τ interchanges points P of $\mathcal{K}$ with the tangent line to $\mathcal{K}$ at P, external points E with the secant line joining the points of contact of the two tangents at E, and internal points with external lines. A point P of $\mathsf{PG}(2, \mathbb{R})$ is **absolute** with respect to a polarity τ if and only if P is incident with P^τ. The set of absolute points of the polarity arising from a conic is precisely the set of points of that conic.

Theorem 3.11 (von Staudt, 1847)
Let τ be a polarity of $\mathsf{PG}(2, \mathbb{R})$ with absolute points. Then the set of absolute points of τ is a (possibly empty) conic of $\mathsf{PG}(2, \mathbb{R})$, and if it is non-empty, then τ is the polarity arising from that conic.

Proof Simply apply Theorem 3.2. □

Worked Example 3.12 (The group actions on conics and polarities)
Let $\mathcal{K}$ be a non-degenerate conic of $\mathsf{PG}(2, \mathbb{R})$, and let τ be the polarity arising from $\mathcal{K}$. Let $g \in \mathsf{PGL}(3, \mathbb{R})$. Then $g^{-1}\tau g$ is a duality of order 2, and so is a polarity. What is the conic $\mathcal{K}'$ of absolute points of this polarity? We simply write down the absolute points set-theoretically:

$$\begin{aligned}
\mathcal{K}' &= \{P \colon P^{g^{-1}\tau g} \, \mathrm{I} \, P\} = \{P \colon \left(P^{g^{-1}}\right)^\tau \mathrm{I} \, P^{g^{-1}}\} \\
&= \{P^{g^{-1}} \colon \left(P^{g^{-1}}\right)^\tau \mathrm{I} \, P^{g^{-1}}\}^g \\
&= \{R \colon R^\tau \, \mathrm{I} \, R\}^g \qquad\qquad \text{by substituting } R \text{ for } P^{g^{-1}} \\
&= \mathcal{K}^g.
\end{aligned}$$

Therefore, the conjugation action of $\mathsf{PGL}(3, \mathbb{R})$ on polarities is permutationally isomorphic to the natural action of $\mathsf{PGL}(3, \mathbb{R})$ on conics (as sets of points). □

Theorem 3.13
$\mathsf{PGL}(3, \mathbb{R})$ *acts transitively on non-empty, non-degenerate conics.*

Proof By Theorem 3.2, there are two conjugacy classes of polarities of $\mathsf{PG}(2, \mathbb{R})$, with respect to the conjugation action of $\mathsf{PGL}(3, \mathbb{R})$. Only one class consists of polarities with absolute points, and hence the polarities arising from non-empty, non-degenerate conics forms a conjugacy class. Next observe (as in Worked Example 3.12) that the action of $\mathsf{PGL}(3, \mathbb{R})$ on conics, as sets of points, is permutationally isomorphic to the action of $\mathsf{PGL}(3, \mathbb{R})$ by conjugation on the polarities arising from conics.   □

Theorem 3.14
Let $\mathcal{K}$ be a nondegenerate conic of $\mathsf{PG}(2, \mathbb{R})$, and let τ be the polarity arising from $\mathcal{K}$. Then τ interchanges a point P of $\mathcal{K}$ with the tangent line to $\mathcal{K}$ at P, external points E with the secant line joining the points of contact of the two tangents at E and internal points with external lines.

Proof By Theorem 3.13, we may assume that $\mathcal{K}$ is the conic with equation $x^2 + y^2 = z^2$, and hence a point $P: (p_0, p_1, p_2)$ is mapped to the line $[p_0, p_1, -p_2]$, by τ. Suppose $P: (p_0, p_1, p_2)$ is a point of $\mathsf{PG}(2, \mathbb{R})$ not in $\mathcal{K}$. A point (x, y, z) in the intersection of P^τ and $\mathcal{K}$ also satisfies $p_0 x + p_1 y - p_2 z = 0$. Therefore, $p_2^2 z^2 = (p_0 x + p_1 y)^2$, and so

$$(p_0 x + p_1 y)^2 = p_2^2(x^2 + y^2).$$

Hence, we have

$$x^2 \left(p_2^2 - p_0^2\right) - 2xy p_0 p_1 + y^2 \left(p_2^2 - p_1^2\right) = 0. \tag{3.1}$$

The discriminant for this quadratic (in x) is

$$\Delta := 4y^2 p_2^2(p_0^2 + p_1^2 - p_2^2) = (2y p_2)^2 \cdot Q(p_0, p_1, p_2)$$

Suppose $p_2 \neq 0$. Equation 3.1 has no solutions when $\Delta < 0$, or equivalently $Q(p_0, p_1, p_2) < 0$. That is, P^τ is an external line if and only if P is an internal point. Equation 3.1 has a unique solution when $\Delta = 0$, which is equivalent to $Q(p_0, p_1, p_2) = 0$; that is, $P \in C$. Finally, Equation 3.1 has two solutions when $Q(p_0, p_1, p_2) > 0$; that is, P^τ is a secant line if and only if P is an external point. We leave the case $p_2 = 0$ to the reader.   □

Exercises

3.1 Two conics meet at points A, B, C, D. Let $P := AB \cap CD$, $Q := BC \cap AD$, and $R := AC \cap BD$. Show that the intersection points of the common tangent lines to the conics meet on the lines PQ, QR, PR.

3.2 Suppose four points A, B, X, Y lie on a conic. Denote the pole of AB by C, and the pole of XY by Z. Let $P = AX \cap BY$ and $Q = AY \cap BX$.

    (a) Prove that C, Z, P, Q are collinear, and that P, Q are conjugate points with respect to the conic.

(b) Suppose another conic passes through A, B, P, Q. Show that a common secant of our two conics passes through Z.

3.3 Suppose XA, XB, YC, YD are four tangents to a conic, with their respective points of contact being A, B, C, D. Prove that XY is a side of the diagonal triangle of $ABCD$, and that the six points A, B, C, D, X, Y lie on a conic.

3.4 Assume we have four lines through a point O which are all secants to a conic at the points A, A'; B, B'; C, C'; D, D'. Suppose AB and CD meet at L, and $A'B'$ and $C'D'$ meet at M. Prove that O, L, M are collinear.

3.5 Suppose ABC is a triangle inscribed in a conic, and T is the pole of AB. Assume there is a line on T meeting BC in M and AC in N. Show that M and N are conjugate with respect to the conic.

3.6 If a pentagon $ABCDE$ circumscribes a conic $\mathcal{K}$ of $\mathsf{PG}(2, \mathbb{R})$ and if CE is tangent to $\mathcal{K}$ at D, show that AD, BE, CF are concurrent.

3.7 If a quadrangle is inscribed in a conic of $\mathsf{PG}(2, \mathbb{R})$, prove that the tangents at its vertices meet in pairs on the sides to the diagonal triangle of the quadrangle.

3.8 Prove that three tangent lines to a conic of $\mathsf{PG}(2, \mathbb{R})$ cannot be concurrent.

3.9 Show that if A, B, C, D are points on a conic and a, b, c, d are the respective tangents to the conic then all four diagonals of $ABCD$ and $abcd$ are concurrent.

3.10 Suppose that POP', QOQ', ROR' are three concurrent chords of a conic S of $\mathsf{PG}(2, \mathbb{R})$, containing P, P', Q, Q', R, R', and X is any other point of S. Let $L = QR \cap XP'$, $M = RP \cap XQ'$, $N = PQ \cap XR'$. Show that L, M, N lie on a line through O.

3.11 Let $\mathcal{K}$ be a conic of $\mathsf{PG}(2, \mathbb{R})$, A, B, C, D, E be distinct points of $\mathcal{K}$, ℓ be the tangent line to $\mathcal{K}$ at A. Let $L := AB \cap DE$, $M := BC \cap EA$, $N := \ell \cap CD$. Show that L, M, N are collinear.

3.12 Let ABC be a triangle of $\mathsf{PG}(2, \mathbb{R})$. Let A_1, A_2 be points on the line BC, let B_1, B_2 be points on the line CA, and let C_1 and C_2 be points on the line AB. Show that there is a conic D such that the lines AA_1, AA_2, BB_1, BB_2, CC_1, and CC_2 are tangent to D if and only if the points A_1, A_2, B_1, B_2, C_1, and C_2 lie on the same conic.

3.13 Let $\mathcal{K}$ be a conic of $\mathsf{PG}(2, \mathbb{R})$ inscribed in the triangle $A_1 A_2 A_3$, and let $P_i \in A_{i+1} A_{i+2}$ (subscripts modulo 3) be such that the lines $P_i A_i$ are concurrent. Let t_i be the second tangent from P_i to C, the first being $A_{i+1} A_{i+2}$. Prove that the points Q_1, Q_2, Q_3 defined by $Q_i := t_i \cap P_{i+1} P_{i+2}$ are collinear.

3.14 Let $\mathcal{K}$ be a conic of $PG(2, \mathbb{R})$, and let A, B, P be distinct points of $\mathcal{K}$. Let M be the point where AP meets the tangent to $\mathcal{K}$ at B, and N be the point where BP meets the tangent to $\mathcal{K}$ at A. Show that the tangent to $\mathcal{K}$ at P is the line joining P to the point $Q = MN \cap AB$.

3.15 Let $\mathcal{K}$ be a nondegenerate conic of $PG(2, \mathbb{R})$. Show that no two internal points of $\mathcal{K}$ can be conjugate with respect to $\mathcal{K}$.

4

Cross-Ratio

The world is neither Euclidean nor projective – it simply is. We may look at it only through certain filters of perception. . . . [A] kind of hybrid thinking, where one takes a projective or Euclidean viewpoint, whichever seems to be more appropriate, can lead to interesting proofs and generalizations of theorems.

Jürgen Richter-Gebert (2011, p. 368)[1]

4.1 Introduction to Cross-Ratio

We can distinguish Euclidean congruence geometry, Euclidean similarity geometry, affine geometry, and projective geometry by the invariants that must be preserved by the allowable transformations. Euclidean congruence geometry has the strongest laws and restrictions: distances are preserved, angles are preserved, parallelism is preserved. In Euclidean congruence geometry, we can measure the usual Euclidean distance between any pair of distinct points, but when we transition to similarity geometry, we lose the ability to make this measurement. Instead, we can determine when two distances are equal, and when one is larger than another. Capturing the notion of distance in affine geometry requires a higher level of abstraction. In this setting, we can determine ratios of lengths and midpoints of line segments. Moreover, *affine ratio* is preserved for three collinear points. That is, if A, B, C are three collinear points and g is an affine transformation, then we have

$$\frac{d(A, B)}{d(A, C)} = \frac{d(A^g, B^g)}{d(A^g, C^g)}.$$

[1] Reprinted by permission from Springer Nature: *Perspectives on Projective Geometry: A Guided Tour Through Real and Complex Geometry*. Richter-Gebert, Jürgen. © 2011.

Figure 4.1 Ambrogio Lorenzetti (c. 1290–1348), *The Annunciation*,[2] Pinacoteca Nazionale (Siena), 1344. The affine ratio of three successive horizontal lines in the tiled floor is 0.53.

We will see now that in projective geometry, we lose the ability to determine a ratio of lengths, and this is a subtle point that will be a theme in the forthcoming sections. For instance, it took some time for Renaissance painters and mathematicians to realise and accept that the affine ratio of the distances between three successive parallels of a pavement is not preserved when taking a perspective image. The painting in Figure 4.1 by Lorenzetti in 1344, long before Br and Alberti's pioneering work on perspective, shows that the horizontal lines of a tiled floor recede in a way that assumes a preservation of affine ratio. However, affine ratio is certainly not preserved by *central projection*, which we will revisit in Section 4.4.

Thus, in projective geometry, the science of central projection, we must abandon the possibility that a quantity such as the affine ratio is preserved. As we shall see, the best we can do is measure the *cross-ratio* of four points on a line.

Let ℓ be a line of $\mathsf{PG}(2, \mathbb{R})$, and let $O(x, y, z)$ and $I(x', y', z')$ be distinct points of ℓ. These points will serve as reference points for ℓ where we will write the other points of ℓ as a linear combination of O and I.

The **cross-ratio** $\mathsf{R}(P, Q; R, S)$ of four points P, Q, R, S of ℓ, where

$$P(x + px', y + py', z + pz'), \qquad Q(x + qx', y + qy', z + qz'),$$
$$R(x + rx', y + ry', z + rz'), \qquad S(x + sx', y + sy', z + sz')$$

---

[2] This is in the public domain.

is defined to be

$$\frac{(r-p)(s-q)}{(r-q)(s-p)}.$$

Moreover, by definition,

$$R(P, Q; R, I) = R(Q, P; I, R) = (r-p)/(r-q),$$
$$R(P, I; R, Q) = R(I, Q; R, P) = (p-r)/(p-q).$$

Cross-ratio is the closest thing to length we have in projective geometry.

A line of $PG(2, \mathbb{R})$ can be naturally associated with $PG(1, \mathbb{R})$; the projective line. The projective line is the set of one-dimensional subspaces of $\mathbb{R}^2$, and we can assign homogeneous coordinates to points, just as we did for $PG(2, \mathbb{R})$. So a point of $PG(1, \mathbb{R})$ has homogeneous coordinates $(1, n)$ (where $n \in \mathbb{R}$) or $(0, 1)$. The point O is mapped to $(1, 0)$, and the point I is mapped to $(0, 1)$. Then the point $T(x + tx', y + ty', z + tz') = O + tI$ is mapped to $(1, t)$. So we will associate P, Q, R, S with the images $(1, p), (1, q), (1, r)$ and $(1, s)$, respectively. We define the **bracket** $[\cdot]$ of two vectors $v_1 = (x_1, y_1)$ and $v_2 = (x_2, y_2)$ by

$$[v_1, v_2] = \det\begin{pmatrix} x_1 & x_2 \\ y_1 & y_2 \end{pmatrix}.$$

Notice that $[P, R] = \det\begin{pmatrix} 1 & 1 \\ p & r \end{pmatrix} = r - p$. Therefore,

$$R(P, Q; R, S) = \frac{(r-p)(s-q)}{(r-q)(s-p)} = \frac{[P, R] \cdot [Q, S]}{[Q, R] \cdot [P, S]}.$$

Now suppose we use another pair of points O' and I' to define a cross-ratio R' with O' and I' as the reference points. Since $PGL(2, \mathbb{R})$ acts 2-transitively[3] on $PG(1, \mathbb{R})$, there exists a matrix $M \in GL(2, \mathbb{R})$ such that $(O', I') = (O \cdot M, I \cdot M)$. Therefore,

$$
\begin{aligned}
R(P \cdot M, Q \cdot M; R \cdot M, S \cdot M) &= \frac{[P \cdot M, R \cdot M] \cdot [Q \cdot M, S \cdot M]}{[Q \cdot M, R \cdot M] \cdot [P \cdot M, S \cdot M]} \\
&= \frac{\det(M)^2 [P, R] \cdot [Q, S]}{\det(M)^2 [Q, R] \cdot [P, S]} \\
&= \frac{[P, R] \cdot [Q, S]}{[Q, R] \cdot [P, S]} \\
&= R(P, Q; R, S).
\end{aligned}
$$

---

[3] We leave this as a simple exercise for the reader: all that is needed is to adapt the proof of Theorem 2.1.

Property 4.1

If the coordinates of ℓ are changed, then the cross-ratio is unchanged; its value doesn't depend on which basis is used for ℓ.

Worked Example 4.2

Recall that we can associate a line of $PG(2, \mathbb{R})$ with $PG(1, \mathbb{R})$ so that O and I are mapped to $(1, 0)$ and $(0, 1)$, respectively. Let us now abuse notation and write t for the point with homogeneous coordinates $(1, t)$ and write ∞ for the point $(0, 1)$. Thus we can write $R(p, q; r, s)$ for $R(O + pI, O + qI; O + rI, O + sI)$, with the exceptional case that $O + \infty I$ is understood to be I. Arithmetic in $\mathbb{R}$ can be extended to include ∞. To see what extra rules we need, notice that $1 \cdot O + \infty \cdot I$ is the point $I = 0 \cdot O + 1 \cdot I$. So 'dividing' by ∞ ought to be the same as multiplying by 0. Hence $1/\infty$ is defined to be 0. Likewise, $1/0 := \infty$. The following identities can be deduced from these two rules:

$$\infty + t = \infty, \quad \text{for all } t \neq \infty;$$

$$\infty \cdot t = \infty, \quad \text{for all } t \neq 0.$$

So recall we have the formula

$$R(p, q; r, s) = \frac{(r - p)(s - q)}{(r - q)(s - p)},$$

which, with our extended arithmetic, does not require us to stipulate special cases. Moreover, for all $a \in \mathbb{R}$, we have

$$R(0, \infty; a, 1) = \frac{(a - 0)(1 - \infty)}{(a - \infty)(1 - 0)} = \frac{a \cdot (1/\infty - 1)}{(a/\infty - 1) \cdot 1} = \frac{a \cdot (-1)}{-1} = a.$$

We can also recover affine ratio by fixing one of the points to be '∞':

$$R(a, \infty; b, c) = \frac{(b - a)(c - \infty)}{(b - \infty)(c - a)} = \frac{(b - a)(c/\infty - 1)}{(b/\infty - 1)(c - a)} = \frac{a - b}{a - c}. \qquad \square$$

The next result is immediate from the definition of the cross-ratio; however, we will be using it often enough that it is useful to have it as a lemma.

Lemma 4.3

Given three collinear points A, B, C and a real number a distinct from 0 and 1, there is a unique point D collinear with $A, B,$ and C such that $R(A, B; C, D) = a$.

Permuting the coordinates of cross-ratio results in some interesting fractional transformations of the value of the original cross-ratio. Let V_4 be the Klein 4-group as a normal subgroup of the permutation group S_4 of degree 4.

Recall that $V_4 = \{(), (1\,2)(3\,4), (1\,3)(2\,4), (1\,4)(2\,3)\}$. The four permutations of V_4 keep the cross-ratio fixed, and so we only need to evaluate what happens to the cross-ratio when we take the six coset representatives under V_4:

Lemma 4.4
Let A, B, C, D be four collinear distinct points, and suppose $R(A, B; C, D) = a$. Let σ be a permutation of $\{A, B, C, D\}$. Then the values of $R(A^\sigma, B^\sigma; C^\sigma, D^\sigma)$, in terms of a, are accordingly:

Coset rep. of S_4/V_4	$R(A^\sigma, B^\sigma; C^\sigma, D^\sigma)$
$()$	a
$(2\,4)$	$a/(a-1)$
$(2\,3)$	$1-a$
$(3\,4)$	$1/a$
$(2\,3\,4)$	$1/(1-a)$
$(2\,4\,3)$	$(a-1)/a$

Theorem 4.5
Given eight collinear points A, B, C, D, P, Q, R, S, the identity

$$R(A, B; R, S)R(B, C; P, S)R(C, A; Q, S)$$
$$R(P, Q; C, D)R(Q, R; A, D)R(R, P; B, D) = 1$$

holds.

Proof The proof is a straightforward application of the defining formula. □

4.2 Perspectivities

Given distinct lines ℓ, ℓ' of π, and a point O incident with neither of them (the **vertex**), the map $X \mapsto XO \cap \ell'$ is a bijection from the points of ℓ to the points of ℓ', called a **perspectivity**. Given another perspectivity from the points of ℓ' to the points of ℓ'', with vertex O', we may compose the perspectivities, setting up a one-to-one correspondence between the points of ℓ and those of ℓ''. We remark that every perspectivity between lines ℓ, ℓ' via a point O is induced by a central collineation with centre O (see Figure 4.2). This will be of importance later on.

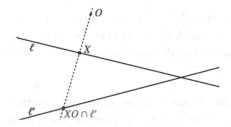

Figure 4.2 The perspectivity defined by ℓ, ℓ' and O.

A **projectivity** from the points of a line ℓ to the points of a line m is a composition of finitely many perspectivities. If $P, Q, R, \ldots$ are points of ℓ which map to $P', Q', R', \ldots$, respectively, under the projectivity from ℓ to ℓ', we write $(P, Q, R, \ldots) \overline{\wedge} (P', Q', R', \ldots)$, and say that ℓ and ℓ' are **projectively related**. If they are perspectively related via O, we write

$$(P, Q, R, \ldots) \overset{O}{\overline{\wedge}} (P', Q', R', \ldots),$$

and if they are perspectively related for some O, we write

$$(P, Q, R, \ldots) \overline{\wedge} (P', Q', R', \ldots).$$

The '$\overline{\wedge}$' notation is due to von Staudt.

Lemma 4.6
Let P, Q, R be collinear points in $\mathsf{PG}(2, \mathbb{R})$ and let P', Q', R' also be collinear points. Then there is a projectivity taking P to P', Q to Q', and R to R'.

Proof Let ℓ be the line containing P, Q, R, and let ℓ' be the line containing P', Q', R'.

Case 1: $\ell \neq \ell'$. We may assume without loss of generality that $\{P, P', Q, Q'\}$ form a quadrangle; that is, they are distinct points. Let O_1 be a point on PP' different from P and P'. Let $C = O_1 R \cap P'Q$. Then

$$(P, Q, R) \overset{O_1}{\overline{\wedge}} (P', Q, C).$$

Note that $R' \neq C$, since otherwise P' would be incident with ℓ. Next, let $O_2 = R'C \cap QQ'$. Then

$$(P', Q, C) \overset{O_2}{\overline{\wedge}} (P', Q', R').$$

Therefore, we have shown by composing these two perspectivities (see Figure 4.3) that there is a projectivity from (P, Q, R) to (P', Q', R').

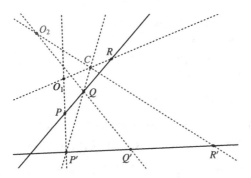

Figure 4.3 Two perspectivities defining a projectivity.

Case 2: $\ell = \ell'$ Let n be a line meeting ℓ in a point not equal to any of our six given points. By Case 1, there is a projectivity from (P, Q, R) to three points (P_0, Q_0, R_0) on n. Also by Case 1, there is a projectivity from (P_0, Q_0, R_0) to three points (P', Q', R'), and hence composing the two projectivities gives us the desired result. $\quad\square$

The following result says that cross-ratio is invariant under projection. This was essentially proved by Pappus for the absolute value of the cross-ratio in c. 340 for the case $P = P'$ (in the notation of Theorem 4.7) with Desargues removing the restriction $P = P'$ in 1648 and Möbius giving the result for signed cross-ratio in 1827.[4] Neither Pappus nor Desargues composed perspectivities, but Möbius was aware of invariance under projectivities. Carnot had introduced signed cross-ratio in 1806, in particular, in Carnot (1806, Théorème VI, in Essai sur la théorie des transversales, pp. 76–77). (This is the formula involving sines of angles.) Other early references for signed cross-ratio are Poncelet (1822, 1995a, p. 12), Möbius (1827, p. 244, para. 182), Steiner (1832, p. 11), and Chasles (1837). But it looks as though cross-ratio didn't really take off until its treatment in Michel Chasles' *Traité de géométrie supérieure* (Chasles, 1852). Poncelet and Steiner preferred to use harmonic separation and Möbius had limited influence.

---

[4] The cross-ratio was unsigned until the 19th century. One direction of the Chasles–Steiner theorem (see Theorem 6.61) requires the cross-ratio to be signed. So until negatives were fully incorporated into geometry (which Carnot did in 1803, when he also introduced signed cross-ratio), it was impossible to see conics in this way.

Theorem 4.7 (Pappus (c. 340) (see Pappus of Alexandria, 1986); Desargues (1648) (see Desargues, 1951; Möbius, 1827))
Let P, Q, R, S be collinear points in $\mathsf{PG}(2, \mathbb{R})$ and let P', Q', R', S' also be collinear points in $\mathsf{PG}(2, \mathbb{R})$. Then

$$(P, Q, R, S)\overline{\wedge}(P', Q', R', S') \iff R(P, Q; R, S) = R(P', Q'; R', S').$$

Proof For the forward direction, it is sufficient to show that

$$(P, Q, R, S)\overline{\overline{\wedge}}(P', Q', R', S') \implies R(P, Q; R, S) = R(P', Q'; R', S').$$

Suppose P, Q, R, S span a line ℓ and P', Q', R', S' span a line ℓ', and let O be a point not on ℓ or ℓ'. Let $X := \ell \cap \ell'$. Let L be a point of ℓ, not equal to X, and let L' be the perspective image of L on ℓ', via O. We will consider cross-ratio R on ℓ with reference points L and X, and similarly for ℓ', but with reference points L' and X. So $P = L + pX$, $P' = L' + p'X$, et cetera. Now any point T on ℓ can be written as $L + tX$ for some $t \in \mathbb{R} \cup \{\infty\}$. The perspective image T' of T (via O) can be written as $L' + t'X$ for some $t' \in \mathbb{R} \cup \{\infty\}$. Now O is incident with TT' and a straightforward calculation shows that $t' = \theta t$ where

$$\theta = \frac{(L' \times X) \cdot X}{(L \times X) \cdot X}.$$

(Recall that '$\times$' denotes the vector cross-product.) In particular, θ is independent of T. Therefore,

$$\begin{aligned} R(P', Q'; R', S') &= \frac{(r' - p')(s' - q')}{(r' - q')(s' - p')} = \frac{(\theta r - \theta p)(\theta s - \theta q)}{(\theta r - \theta q)(\theta s - \theta p)} \\ &= \frac{(r - p)(s - q)}{(r - q)(s - p)} \\ &= R(P, Q; R, S). \end{aligned}$$

Conversely, if $R(P, Q; R, S) = R(P', Q'; R', S')$, then, by Lemma 4.6, there is a projectivity g taking P to P', Q to Q', and R to R'. Let S'' be the image of S under g. Then $R(P, Q; R, S) = R(P', Q'; R', S'')$, so $S'' = S'$, by Lemma 4.3 above.   □

The above proof shows that a perspectivity is induced by a linear transformation, and so the first part also follows automatically from Property 4.1.

Corollary 4.8 (von Staudt, 1847)
There is a unique projectivity taking any given ordered triple of collinear points of $\mathsf{PG}(2, \mathbb{R})$ to any other given ordered triple of collinear points.

Proof Let (A_1, A_2, A_3) be a triple of points lying on a line ℓ, and let (B_1, B_2, B_3) be a triple of points on a line ℓ'. By Lemma 4.6, there exists a projectivity g mapping A_i to B_i for all $i \in \{1, 2, 3\}$. Given $(A_1, A_2, A_3) \overline{\wedge} (B_1, B_2, B_3)$, the requirement by Theorem 4.7, that if X corresponds to X' then $R(A_1, A_2; A_3, X) = R(B_1, B_2; B_3, X')$, uniquely determines the image X' of any point X, by Lemma 4.3. □

Corollary 4.9
A projectivity $\ell \to \ell'$, $\ell \neq \ell'$ fixing $\ell \cap \ell'$ is a perspectivity.

Proof Suppose the projectivity ϕ maps A to A', B to B', and fixes $\ell \cap \ell'$, with $A, B, \ell \cap \ell'$ distinct. Let $P = AA' \cap BB'$. Then the composition of ϕ with the perspectivity via P from ℓ' to ℓ fixes $A, B, \ell \cap \ell'$, so, by Theorem 4.8, it is the identity. Hence ϕ is the perspectivity via P from ℓ to ℓ'. □

Corollary 4.10 (Pappus (c. 340) (see Pappus of Alexandria, 1986))
Let A, B, C be distinct points on a line ℓ, A', B', C' be distinct points on another line ℓ', $P = \ell \cap \ell'$. Then $R(P, A; B, C) = R(P, A'; B', C')$ if and only if AA', BB', CC' are concurrent.

Proof If AA', BB', CC' are concurrent in V then the perspectivity via V from ℓ to ℓ' fixes P and maps A to A', B to B', and C to C', so $R(P, A, B, C) = R(P, A', B', C')$. Conversely, if $R(P, A, B, C)$ is equal to $R(P, A', B', C')$ then, by Theorem 4.7, there is a projectivity fixing P and mapping A to A', B to B', and C to C', so by Corollary 4.9, this is a projectivity from a point V. Hence AA', BB', CC' are concurrent in V. □

Pappus used Corollary 4.10 to prove his hexagon Theorem (2.18); see Exercise 4.11. Pappus' original result on the invariance of cross-ratio has only the case where $P = P'$ (i.e., $R(P, Q; R, S) = R(P, Q'; R', S')$). Al-Mu'taman ibn Hud (*Book of Perfection*) has the general case.

Yusuf al-Mu'taman ibn Hud (d. 1085), who was King of Zaragoza (Spain) from 1081 to 1085, was the original discoverer of Ceva's theorem (more than 600 years before Ceva) and wrote his great work *Kitab al Istikmal* (*The Book of Perfection*) in c. 1069–81, before ascending to the throne. The book, which is unfinished, includes an attempt to provide the parallel postulate.

However, as always, the history of Corollary 4.10 is more complicated than it seems. In Proposition 5 of Book III of *Sphaerica* (see Hermiz, 2015 for an

English translation of the work), Menelaus assumes the following: If A, B, C, D are points of a great circle K, P is a point not on this circle, and we draw arcs of great circles PA, PB, PC, PD meeting another great circle K; at A', B', C', D' ,then

$$\frac{2d(A, B) \cdot 2d(C, D)}{2d(A, C) \cdot 2d(B, D)} = \frac{2d(A', B') \cdot 2d(C', D')}{2d(A', C') \cdot 2d(B', D')}.$$

This has the same relationship to the invariance of cross-ratio in the plane under a perspectivity as the plane theorem of Menelaus (see Theorem 4.34) bears to the spherical theorem of Menelaus (*Sphaerica*, Book III, Proposition 1).[5] Indeed, he derived the spherical theorem III.1 by similarly assuming the plane theorem and replacing every line segment by a chord of twice the arc.

Worked Example 4.11

Let ℓ be a line. Then the set of projectivities of ℓ to itself forms a group Y_ℓ. To see this, note that the identity map on ℓ is a perspectivity, and the composition of two projectivities is a finite composition of perspectivities, and so is a projectivity. Reversing a product of perspectivities results in the inverse projectivity, and so we have a group. Moreover, by Corollary 4.8, Y_ℓ acts sharply 3-transitively on the points of ℓ. □

We now look at collineations and show that they preserve cross-ratio.

Theorem 4.12

Every perspectivity $\ell \to \ell'$ extends to a central collineation of $\mathrm{PG}(2, \mathbb{R})$.

Proof Let φ be a perspectivity from ℓ to ℓ', with centre O, and let $X := \ell \cap \ell'$. Let A be a point of ℓ, not equal to X. By Theorem 2.17, there is a unique elation g with centre O, axis OX and mapping A to $\varphi(A)$. We claim that the restriction of g to ℓ is φ. Since $\ell = AX$, we have

$$\ell^g = (AX)^g = A^g X^g = \varphi(A)X = \ell'.$$

So now take another point B of ℓ. Note that if B is O or A, then B^g is trivially equal to $\varphi(B)$. Now B^g is incident with ℓ^g but we also know that B^g lies on OB, as O is the centre of the elation g. Hence, $B^g = OB \cap \ell^g = \varphi(B)$ and we have shown that g and φ agree on points of ℓ. □

---

[5] Four points A, B, C, D on a great circle of a sphere with centre O have *spherical cross-ratio* the cross-ratio of $(OA, OB; OC, OD)$ in Euclidean 3-space. Menelaus in *Sphaerica*, in the proof of Proposition 71, implicitly used invariance of spherical cross-ratio under projection along great circles. This is two centuries before Pappus! Abu Nasri Mansur ibn Ali ibn Iraq (960–1036) *The Rectification of a Proposition of the Book of Menelaus on the Spherics* proves this result explicitly.

Corollary 4.13
Every projectivity $\ell \to \ell'$ extends to a collineation of PG(2, ℝ).

Theorem 4.14
If g is a collineation, and ℓ a line, then the restriction of g to ℓ is a projectivity from ℓ to ℓ^g.

Proof Let X be the intersection of ℓ and ℓ^g. Let P, Q, and R be three points of ℓ. By Theorem 4.8, there is a unique projectivity τ taking (P, Q, R) to (P^g, Q^g, R^g). By Corollary 4.13, there exists a collineation h that restricts to τ on ℓ. Now gh^{-1} fixes P, Q, R, and so gh^{-1} fixes every point of ℓ (by Exercise 2.4(2.6)). Therefore, g and h agree on all points of ℓ, and hence g and τ agree on the points of ℓ. ☐

There are many collineations that induce the same projectivity. Indeed, if Y_ℓ is the group of projectivities on ℓ, and g is a collineation, then the right coset $Y_\ell g$ yields infinitely many collineations inducing the same projectivity as g.

If G is a permutation group on a set Ω, and if Δ is a subset of Ω, then G_Δ is the setwise stabiliser of Δ in G, and G_Δ^Δ is the permutation group induced on Δ, which is isomorphic to $G_\Delta/G_{(\Delta)}$ where $G_{(\Delta)}$ is the kernel of the action of G_Δ on Δ.

Corollary 4.15
The collineations of PG(2, ℝ) *preserve cross-ratio.*

Theorem 4.16
The group PGL(3, ℝ)$_\ell^\ell$ *induced on a line ℓ of* PG(2, ℝ) *is permutationally isomorphic both to the group induced by projectivities mapping ℓ to ℓ, and to* (PG(1, ℝ), PGL(2, ℝ)).

Proof See Exercise 4.4. ☐

Theorem 4.17
The stabiliser of cross-ratio in the symmetric group on PG(1, ℝ) *is* PGL(2, ℝ). *That is, the subgroup of* Sym(PG(1, ℝ)) *consisting of those permutations g such that*

$$R(P^g, Q^g; R^g, S^g) = R(P, Q; R, S)$$

for all distinct P, Q, R, $S \in$ PG(1, ℝ) *is equal to* PGL(2, ℝ).

Proof Let G be the subgroup of Sym(PG(1, ℝ)) consisting of those permutations g such that $R(P^g, Q^g; R^g, S^g) = R(P, Q; R, S)$ for all distinct P, Q, R,

$S \in \mathrm{PG}(1, \mathbb{R})$. Then by Theorem 4.7, $\mathrm{PGL}(2, \mathbb{R}) \leqslant G$. Hence both groups here are 3-transitive. By Corollary 2.8, it remains to show that the stabiliser of three distinct points for both groups is trivial. But by Lemma 4.3, every cross-ratio value occurs uniquely for a given triple of points, and so if g fixes three distinct points and stabilises cross-ratio, then g fixes all points; that is, $g = 1$.   □

We will often refer to an **involution** as an involutory map on points that preserves cross-ratio (i.e., a projectivity). Given a point P, the **pencil** of lines on P is simply the set of lines incident with P, denoted pencil(P).

Lemma 4.18
(i) *If g is a polarity, and ℓ a line that is not absolute, then the map $\ell \to \ell$ defined by $P \mapsto P^g \cap \ell$ is an involution of ℓ induced by g.*

(ii) *If g is a polarity, and P a point that is not absolute, then the map from the pencil of lines on P to itself defined by $\ell \mapsto P\ell^g$ (the line joining P and ℓ^g), is an involution of the pencil of lines on P induced by g.*

Proof See Exercise 4.5.   □

A special case of (i) is notable: the restriction of conjugation with respect to a conic to a line ℓ that is not tangent to the conic is an involution of ℓ. Moreover, there is a converse: given a line ℓ and an involution t on ℓ, there exists a conic $\mathcal{K}$ such that the restriction of conjugation with respect to $\mathcal{K}$ to ℓ is t.

Theorem 4.19 (Ibrahim ibn Sinan, *al-Masā'il al-mukhtāra*, c. 930)
Given three lines ℓ, m, n and three collinear points G, D, K, on none of the lines, show that there is a point L on ℓ such that $GL \cap m = T$, $DL \cap n = I$ and such that T, I, and K are collinear.

Proof Consider the map f defined by $a \mapsto (a \cap n)D \cap (a \cap m)G$. Then f is a duality, and so the image of pencil(K) is the set of points of a line b. Let $L := \ell \cap b$. Then if c is the preimage of L, $c \cap m = T$, $c \cap n = I$ and so T, I, and K lie on c.   □

Consider five collinear points P, Q, R, S, T, and apply the formula for cross-ratio where we have parameters p, q, r, s, t, respectively, for these points:

$$
\begin{aligned}
\mathrm{R}(P, T; R, S)\mathrm{R}(T, Q; R, S) &= \frac{(r-p)(s-t)}{(r-t)(s-p)} \cdot \frac{(r-t)(s-q)}{(r-q)(s-t)} \\
&= \frac{(r-p)}{(s-p)} \cdot \frac{(s-q)}{(r-q)} \\
&= \mathrm{R}(P, Q; R, S).
\end{aligned}
$$

This interesting *cocycle* identity on five collinear points will also appear in Section 4.5 when we consider the projective analogues of the theorems of Ceva and Menelaus (4.33, 4.34).

> There is another way we can view this *cocycle* identity. For each $t \in \mathbb{R} \cup \{\infty\}$, we define a transformation ϕ_t on triples of elements of $\mathbb{R} \cup \{\infty\}$ as follows: $\phi_t(x, y, z) := (x, y, w)$ where w is such that $R(x, y; z, w) = t$. Then the identity here translates to $\phi_{t+u} = \phi_t \circ \phi_u$ and so is equivalent to the ϕ_t's forming a one-parameter group of transformations.

Now Lemma 4.4 implies that applying the permutation (1 4) to the coordinates of a cross-ratio has the following effect: $R(S, Q; R, P) = 1 - a$ where $R(P, Q; R, S) = a$. Therefore,

$$R(P, Q; R, S) = 1 - R(S, Q; R, P).$$

According to Labourie (2008), the converse is true.

Theorem 4.20
A set endowed with a function b of quadruples of points satisfying the identities

(i) $b(x, y, z, t) = b(x, w, z, t)b(w, y, z, t),$
(ii) $b(x, y, z, t) = 1 - b(t, y, z, x).$

can be realised as a subset of the projective line so that the function b is the restriction of the projective cross-ratio.

> Recall that we can use the bracket notation $[\cdot, \cdot]$ to define cross-ratio. Then (ii) in Theorem 4.20 becomes
>
> $$[x, y][z, t] - [x, z][y, t] + [x, t][y, z] = 0$$
>
> which also happens to be the Plücker relation (of the Klein quadric). See Corollary 16.2 in Chapter 16.

Worked Example 4.21
Suppose we have five collinear points P, Q, R, S, T. Then we will show using the cocycle identity that S and T are **harmonic conjugates**[6] with respect to P and Q if and only if $R(P, Q; R, S) = -R(P, Q; R, T)$. By the cocycle identity, we have

---

[6] We will be covering harmonic conjugacy in depth in Section 4.5, but for this worked example, we need to know only that the cross-ratio of the four points considered is -1.

$$R(P, Q; R, S)R(P, Q; S, T) = R(P, Q; R, T)$$

and hence $R(P, Q; R, S) = -R(P, Q; R, T)$ if and only if $R(P, Q; S, T) = -1$; that is, S and T are harmonic conjugates with respect to P and Q. ☐

4.3 Cross-Ratio of Lines

Corollary 4.10 allows us to define *cross-ratio of four concurrent lines*. Let a, b, c, d be four lines meeting in a point P. Let ℓ be any line not incident with P, and let $A := a \cap \ell$, $B := b \cap \ell$, $C := c \cap \ell$, $D := d \cap \ell$. By Corollary 4.10, the value of $R(A, B; C, D)$ does not depend on the choice of ℓ. Thus, we define the **cross-ratio of four concurrent lines** $R(a, b; c, d)$ to be the value $R(A, B; C, D)$ (for some line ℓ not incident with P; see Figure 4.4). some sense, cross-ratio of lines is the closest thing to angle measure that we have in projective geometry.

Worked Example 4.22
Using the law of sines, we have an alternative way to see that cross-ratio is invariant under projectivities. It is sufficient to show that it is invariant under perspectivities. Suppose we have four collinear points A, B, C, D and a centre V of perspectivity not on their common line, and another line ℓ such that

$$VA \cap \ell = A', \quad VB \cap \ell = B', \quad VC \cap \ell = C', \quad VD \cap \ell = D'.$$

With respect to reference points O and I on AB, let $A := O + aI$, $B := O + bI$, $C := O + cI$, $D := O + dI$. Therefore, by the law of sines:

$$R(A, B; C, D) = \frac{(c - a)(d - b)}{(c - b)(d - a)} = \frac{\sin(\angle CVA)\sin(\angle DVB)}{\sin(\angle CVB)\sin(\angle DVA)}.$$

Similarly,

$$R(A', B'; C', D') = \frac{\sin(\angle C'V'A')\sin(\angle D'V'B')}{\sin(\angle C'V'B')\sin(\angle D'V'A')}$$

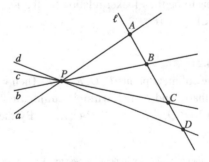

Figure 4.4 Cross-ratio of lines.

Therefore, $R(A, B; C, D) = R(A', B'; C', D')$ as the corresponding angles are the same. Indeed, the cross-ratio of four concurrent lines a, b, c, d is

$$\frac{\sin(\angle ab)\sin(\angle ad)}{\sin(\angle cd)\sin(\angle cb)},$$

where the angles are directed.

Later, we will be examining cross-ratio on a conic (Section 5.4) and from the above argument, the constancy of cross-ratio on a circle is apparent. Angles subtended on a circle by a chord are congruent, and cross-ratios of four points on a circle measured from a fifth are well-defined. Then, we can make use of a projective transformation to allow the transference of this property to conics in the real projective plane. □

Lemma 4.23
If $R(PX, PY; PZ, PW) = R(QX, QY; QZ, QW)$ and X, Y, Z are collinear, then W is on the line XYZ.

Proof Let $XYZ \cap PW = A$, $XYZ \cap QW = B$. Then

$$R(X, Y; Z, A) = R(PX, PY; PZ, PW) = R(QX, QY; QZ, QW) = R(X, Y; Z, B),$$

so $A = B$. Thus W lies on XYZ. □

Lemma 4.24
If $R(PA, PB; PC, PD) = R(QA, QB; QC, QD)$ and P, Q, A are collinear, then B, C, D are collinear.

Proof This follows from Lemma 4.23. We can replace A by $A' = BC \cap PA$, and then A', B, C are collinear, so by Lemma 4.23, D is on $A'BC$. □

The above two observations offer a simple proof of Desargues' Theorem 2.20. This proof can also be found in Godfrey and Siddons (1908, pp. 132–3 and 146–7).

Theorem 4.25 (Desargues' theorem)
If triangles ABC, $A'B'C'$ are in perspective from O, let $P = BC \cap B'C'$, $Q = CA \cap C'A'$, $R = AB \cap A'B'$. Then P, Q, R are collinear.

Proof Let $S := OA \cap BC$, $S' := OA \cap B'C'$. Then

$$R(P, B; S, C) = R(OP, OB; OS, OC) = R(P, B'; S', C').$$

Therefore, $R(AP, AB; AS, AC) = R(A'P, A'B'; A'S', A'C')$, that is,

$$R(AP, AR; AO, AQ) = R(A'P, A'R; A'O, A'Q).$$

By Lemma 4.24, P, Q, R are collinear.   □

Worked Example 4.26
We will demonstrate that the standard duality $\tau\colon [X, Y, Z] \mapsto [x, y, z]$ preserves cross-ratio, and so we have another way to view the cross-ratio of four concurrent lines. Let A, B, C, D be collinear points. By Theorem 4.15 and Theorem 4.8, we may suppose that $A = (1, 0, 0)$, $B = (0, 1, 0)$, $C = (1, 1, 0)$, and $D = (1, d, 0)$ for some $d \neq 0$. If we suppose that A and B are the reference points for the cross-ratio, then we have $R(A, B; C, D) = 1/d$. Let $X = (0, 0, 1)$. By definition, $R(XA, XB; XC, XD) = R(A, B; C, D)$. Let us calculate the four lines in question, take their image under τ, and calculate the cross-ratio of these four points:

$$XA = [0, 1, 0] \longrightarrow (0, 1, 0) = B, \quad XC = [1, -1, 0] \longrightarrow (1, -1, 0) =: C',$$

$$XB = [1, 0, 0] \longrightarrow (1, 0, 0) = A, \quad XD = [-d, 1, 0] \longrightarrow (-d, 1, 0) =: D'.$$

Now we calculate the cross-ratio of $R(B, A; C', D')$ directly from the defining formula:

$$R(B, A; C', D') = 1/R(A, B; C', D') = \frac{1}{d}.$$

Therefore, $R(A, B; C, D) = R((XA)^\tau, (XB)^\tau; (XC)^\tau, (XD)^\tau)$ as expected.   □

Theorem 4.27
If Δ is a duality of $\mathrm{PG}(2, \mathbb{R})$, and P_1, P_2, P_3, P_4 are distinct collinear points, then P_1^Δ, P_2^Δ, P_3^Δ, P_4^Δ are distinct concurrent lines and

$$R(P_1, P_2; P_3, P_4) = R(P_1^\Delta, P_2^\Delta; P_3^\Delta, P_4^\Delta).$$

Proof Let ℓ be the line spanned by the P_i. Since Δ is incidence-preserving (by definition), it follows that P_i^Δ is incident with ℓ^Δ for each $i \in \{1, 2, 3, 4\}$. Therefore, $P_1^\Delta, P_2^\Delta, P_3^\Delta, P_4^\Delta$ are concurrent in the point ℓ^Δ. Worked Example 4.26 shows that the standard duality $\tau\colon [X, Y, Z] \mapsto [x, y, z]$ preserves cross-ratio and hence by Theorem 4.15 and Lemma 3.1, the result follows.   □

The above theorem can also be derived from Exercise 4.10.

4.4 Interlude: From Affine Ratio to Cross-Ratio

Central projection, between planes that are not necessarily parallel, does not preserve midpoints as can be seen in Figure 4.5. So central projection can't preserve affine ratio.

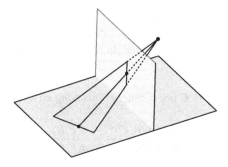

Figure 4.5 Central projection does not preserve midpoints of line segments.

Recall from Theorem 4.8 that we can compose a finite sequence of perspectivities to map any triple of collinear points on one line to any other triple of collinear points on another line. So we must use four collinear points in order to define an invariant of central projection between planes that are not necessarily parallel.

Consider four equally spaced collinear points A, B, C, D. So

$$\frac{\text{ratio}(D, B, A)}{\text{ratio}(C, B, A)} = \frac{d(D, B)/d(D, A)}{d(C, B)/d(C, A)} = \frac{2/3}{1/2} = \frac{4}{3}.$$

Now take perspective images A', B', C', D' of A, B, C, D, and suppose that D' is at infinity (see Figure 4.6). (So PD is parallel to the line $A'B'$.) You will notice in the figure that $\text{ratio}(C', A', B') = 4/3$. This will become apparent in the following discussion.

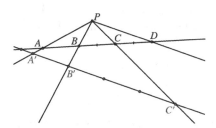

Figure 4.6 Perspective projection of four equally spaced points.

The law of sines states that if we have a triangle ABC, then

$$\frac{\sin(\angle BCA)}{d(A, B)} = \frac{\sin(\angle CAB)}{d(B, C)} = \frac{\sin(\angle ABC)}{d(C, A)}.$$

We can use affine ratio to synthesise this statement in terms of affine ratio. Define the **sine ratio** of three concurrent distinct lines ℓ_1, ℓ_2, ℓ_3 by

$$\sin(\ell, m, n) = \frac{\sin \angle \ell m}{\sin \angle \ell n}$$

where the angles are oriented, as usual, in the anticlockwise sense.

Theorem 4.28

Suppose that points A, B, C are distinct points lying on a line ℓ, and that point P is not on ℓ. Then

$$\frac{\text{ratio}(A, B, C)}{\sin(PA, PB, PC)} = \frac{d(P, B)}{d(P, C)}.$$

Proof By the law of sines we have

$$\frac{\sin(\angle BPA)}{d(A, B)} = \frac{\sin(\angle PAB)}{d(B, P)} \quad \text{and} \quad \frac{\sin(\angle CPA)}{d(A, C)} = \frac{\sin(\angle PAC)}{d(C, P)}.$$

Now

$$\begin{aligned}
\frac{\text{ratio}(A, B, C)}{\sin(PA, PB, PC)} &= \frac{d(A, B)}{d(A, C)} \cdot \frac{\sin \angle CPA}{\sin \angle BPA} \\
&= \frac{d(B, P)}{\sin(\angle PAB)} \frac{\sin \angle CPA}{d(A, C)} \\
&= \frac{\sin(\angle PAC)}{d(C, P)} \cdot \frac{d(B, P)}{\sin(\angle PAB)} \\
&= \frac{d(B, P)}{d(C, P)}
\end{aligned}$$

since $\angle PAC = \angle PAB$. $\qquad\qquad\square$

Suppose we have four collinear points A, B, C, D, not necessarily equally spaced, but A', B', C' are the perspective images of A, B, C from P on a line ℓ parallel to PD (so that D is mapped to infinity). By a number of applications of the law of sines, we have

$$\frac{\sin \angle DPB}{\sin \angle DPA} = \frac{d(P, A')}{d(P, B')}.$$

Also note that $\angle CPA = \angle C'PA'$ and $\angle CPB = \angle C'PB'$. Therefore,

$$\begin{aligned}
\frac{\text{ratio}(D, B, A)}{\text{ratio}(C, B, A)} &= \frac{\sin(PD, PB, PA) \cdot d(P, B)/d(P, A)}{\sin(PC, PB, PA) \cdot d(P, B)/d(P, A)} \\
&= \frac{\sin(PD, PB, PA)}{\sin(PC, PB, PA)} \\
&= \sin(PC', PA', PB') \frac{\sin \angle DPB}{\sin \angle DPA} \\
&= \sin(PC', PA', PB') \frac{d(P, A')}{d(P, B')} \\
&= \text{ratio}(C', A', B').
\end{aligned}$$

By the proof of Lemma 4.6, any projectivity is the composition of two perspectivities. So suppose four collinear points A, B, C, D are projected between ℓ_1 and ℓ_2 via ℓ from P_1 and P_2 to obtain A'', B'', C'', D'' on ℓ_2 via A', B', C', D' on ℓ in such a way that $P_1 D$ is parallel to ℓ (so D' is at infinity). Then

$$\frac{\text{ratio}(D, B, A)}{\text{ratio}(C, B, A)} = \text{ratio}(C', A', B')$$

and

$$\frac{\text{ratio}(D'', B'', A'')}{\text{ratio}(C'', B'', A'')} = \text{ratio}(C', A', B').$$

Therefore,

$$\frac{\text{ratio}(D, B, A)}{\text{ratio}(C, B, A)} = \frac{\text{ratio}(D'', B'', A'')}{\text{ratio}(C'', B'', A'')}$$

and so this 'double ratio' is invariant once we take two perspectivities, where the intervening line has the fourth point mapped to infinity.

Here is an alternative synthetic treatment that uses Thales' theorem (Theorem 7.5). The first to assert that cross-ratio becomes affine ratio when one point is projected to infinity was Brook Taylor (1719; see also the corollary in Andersen, 1992, p. 183). Given O and collinear points A, B, C, D and a line ℓ with OD parallel to ℓ, let A', B', C' be the respective images of A, B, C by projection from O onto ℓ. Let m be the line parallel to ℓ on B. Let E be the projection of A from O onto m, and F be the projection of C from O onto m.

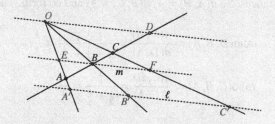

Then AEB, AOD are similar triangles (by the alternate interior angles theorem and angle-angle-angle), as are CDO and CBF. By Thales' theorem (Theorem 7.5),

$$d(A, B)/d(A, D) = d(E, B)/d(D, O), \quad d(C, D)/d(C, B) = d(D, O)/d(B, F).$$

Since ℓ and m are parallel, by Thales' theorem (Theorem 7.5),

$$d(E, B)/d(B, F) = d(A', B')/d(B', C').$$

Combining these gives

$$|R(A, C; B, D)| = \frac{d(B, A) \cdot d(C, D)}{d(B, C) \cdot d(A, D)} = \frac{d(E, B) \cdot d(D, O)}{d(D, O) \cdot d(B, F)}$$

$$= \frac{d(E, B)}{d(B, F)} = \frac{d(A', B')}{d(B', C')}$$

$$= |\text{ratio}(B', A', C')|.$$

(As for the sign, recall that $R(A, C; B, D) < 0$ if and only if $\{A, C\}$ separates $\{B, D\}$. This can happen only if B' is between A' and C', which means $\text{ratio}(B', A', C') < 0$.)

Worked Example 4.29

Consider the sequence $1, 2, 3, 4, 5, \ldots$ and the sequence of reciprocals

$$1, \frac{1}{2}, \frac{1}{3}, \frac{1}{4}, \frac{1}{5}.$$

We can take the four successive distances between these reciprocals:

$$\frac{1}{2}, \frac{1}{6}, \frac{1}{12}, \frac{1}{20}.$$

Now suppose we have collinear points A, B, C, D, E of $\mathbb{R}^2$ having these successive distances. So $d(A, B) = \frac{1}{2}$, $d(B, C) = \frac{1}{6}$, and so forth. Then

$$\text{ratio}(D, B, A) = \frac{d(D, B)}{d(D, A)} = \frac{\frac{1}{6} + \frac{1}{12}}{\frac{1}{2} + \frac{1}{6} + \frac{1}{12}} = \frac{1}{3},$$

$$\text{ratio}(C, B, A) = \frac{d(C, B)}{d(C, A)} = \frac{\frac{1}{6}}{\frac{1}{2} + \frac{1}{6}} = \frac{1}{4}.$$

So

$$\frac{\text{ratio}(D, B, A)}{\text{ratio}(C, B, A)} = \frac{4}{3}.$$

Likewise,

$$\text{ratio}(E, C, B) = \frac{\frac{1}{12} + \frac{1}{20}}{\frac{1}{6} + \frac{1}{12} + \frac{1}{20}} = \frac{4}{9}, \qquad \text{ratio}(D, C, B) = \frac{\frac{1}{12}}{\frac{1}{6} + \frac{1}{12}} = \frac{1}{3}.$$

Therefore,

$$\frac{\text{ratio}(E, C, B)}{\text{ratio}(D, C, B)} = \frac{4}{3}.$$

Thus, the sequence of reciprocals of integers is related to the perspective image of uniformly spaced points. Now recall the following facts about numbers in arithmetic and harmonic progression:

- Numbers are in arithmetic progression if and only if each is the arithmetic mean of its predecessor and its successor.
- Numbers are in harmonic progression if and only if each is the harmonic mean of its predecessor and its successor.
- Non-zero numbers are in arithmetic progression if and only if their reciprocals are in harmonic progression.

So we leave the proof of the following as an exercise for the reader:

Let $A, X_0, X_1, \ldots, X_n$ be points of a line ℓ in $\mathbb{R}^2$. Then $d(A, X_0)$ is the harmonic mean of $d(A, X_1), \ldots, d(A, X_n)$ if and only if

$$R(X_1, X_0; P, A) + R(X_2, X_0; P, A) + \cdots + R(X_n, X_0; P, A) = n,$$

for any point P on ℓ. $\qquad\qquad\qquad\qquad\qquad\qquad\qquad\qquad\qquad\qquad\qquad$ $\square$

We may extend the definition of affine ratio to points that need not be distinct, by defining $\text{ratio}(A, B, B) = 1$, $\text{ratio}(A, B, A) = 0$, and $\text{ratio}(A, A, C) = -1$ ($\text{ratio}(A, A, A)$ is left undefined). This gives an extension of cross-ratio to points that need not be distinct, by the relationship

$$R(A, B; C, D) = \frac{\text{ratio}(C, A, B)}{\text{ratio}(D, A, B)}.$$

Thus, $R(A, A; C, D) = 1$, $R(A, B; B, D) = 0$, $R(A, B; C, A) = 0$, $R(A, B; A, D) = \infty$, $R(A, B; C, B) = \infty$, $R(A, B; C, C) = 1$.

Now consider a quadrangle $ABCD$ and the pencil of conics on that quadrangle. There are three degenerate conics in that pencil: $(AD)(BC)$, $(AB)(CD)$, and $(AC)(BD)$. For any point E not on one of these three degenerate conics, there is a unique non-degenerate conic $\mathcal{K}$ containing E in the pencil, and, by the Chasles–Steiner theorem (Theorem 6.61), for every point X of $\mathcal{K} \setminus \{A, B, C, D\}$, $R(XA, XB; XC, XD)$ is a constant. With the extension of the values of cross-ratio above, we have

$$X \in (AD)(BC) \setminus \{A, B, C, D\}, \implies R(XA, XB; XC, XD) = 0;$$
$$X \in (AB)(CD) \setminus \{A, B, C, D\}, \implies R(XA, XB; XC, XD) = 1;$$

$$X \in (AC)(BD) \setminus \{A, B, C, D\}, \implies R(XA, XB; XC, XD) = \infty.$$

Thus, if AB has equation $F = 0$, CD has equation $G = 0$, AC has equation $H = 0$, and BD has equation $K = 0$, then the pencil consists of $HK = 0$ and $FG + aHK = 0$, $a \in \mathbb{R}$, and the cross-ratio of the conic $FG + aHK = 0$ is the constant a.

4.5 Harmonic Quadruples

On the real line, we have a simple way to determine when a point x is between two other points a and b; we simply check that $a < x < b$. This is equivalent to the expression

$$\frac{x - a}{x - b} < 0.$$

Moreover, we can use the quantity on the left-hand side of this inequality to determine when x is the midpoint of the segment $[a, b]$: note that

$$\frac{x - a}{x - b} = -1$$

if and only if $x = (a + b)/2$.

For the projective line, we can recapture the notions of separation and 'midpoint' for four points rather than three. We say that A, B **separates** C, D if $R(A, B; C, D) < 0$; and that $(A, B; C, D)$ is **harmonic** if $R(A, B; C, D) = -1$. In the latter case, we also say that A, B **harmonically separates** C, D.

Before we present Pappus' famous harmonic configuration theorem, we will first look at one of the fundamental properties of the affine plane: the diagonals of a parallelogram bisect one another. Let P, Q, R, S be four points in $\mathrm{PG}(2, \mathbb{R})$, no three collinear. Let $A := PQ \cap RS$ and $C := PS \cap QR$ and consider the line ℓ joining A and C. If we remove ℓ so that it becomes the line at infinity of $\mathrm{AG}(2, \mathbb{R})$, then $PQ \parallel RS$ and $PS \parallel QR$; that is, $PQRS$ is a parallelogram. Now the diagonals of this parallelogram, PR and QS, intersect in a point B that bisects each diagonal. In other words, B is the midpoint of $\overline{PR}$ and the midpoint of $\overline{QS}$ (see Figure 4.7).

Consider the parallel projection π with parallel class of PS and mapping

$$PR \longrightarrow PQ.$$

Now π maps P to P, B to X, and R to Q. Recall that parallel projection is an affine map and so preserves midpoints. Since B is the midpoint of P and R, it follows that X is the midpoint of P and Q. Therefore, we have:

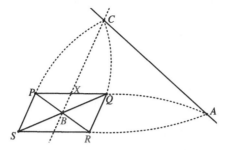

Figure 4.7 A parallelogram embedded in the projective plane – an artistic impression.

Theorem 4.30

Let PQRS be a parallelogram of AG(2, ℝ), *B be the intersection of the diagonals, ℓ be the line parallel to PS on B, and X := ℓ ∩ PQ. Then X is the midpoint of P and Q.*

The projective version of Theorem 4.30 is Pappus' harmonic quadruples theorem.

Theorem 4.31 (Pappus (c.340) (see Pappus of Alexandria, 1986))
Let P, Q, R, S be four points in PG(2, ℝ), *no three collinear. Let A = PQ∩RS, B = PR∩QS, C = PS ∩ QR, and X = BC∩PQ. Then (P, Q, A, X) is harmonic.*

Proof Let ℓ be the line joining $A = PQ \cap RS$ and $C = QR \cap PS$. Then $(P, Q; A, X)$ is harmonic if, with ℓ at infinity, X is the midpoint of P and Q in the affine plane. The rest follows from Theorem 4.30.   □

We give a second proof using homogeneous coordinates.

A second proof of Theorem 4.31 By Theorem 2.5, we may assume that *PQRS* is the fundamental quadrangle: $P(1, 0, 0)$, $Q(0, 1, 0)$, $R(0, 0, 1)$, $S(1, 1, 1)$. Then $[0, 0, 1] \cap [1, -1, 0]$ is $A(1, 1, 0)$; likewise, we have $B(1, 0, 1)$ and $C(0, 1, 1)$. So *BC* is $[1, 1, -1]$ and *PQ* is $[0, 0, 1]$, giving $X(1, -1, 0)$. Now *Q* is $(1, 0, 0) - (1, -1, 0)$ and *A* is $(1, 0, 0) - \frac{1}{2}(1, -1, 0)$, so $R(P, Q; A, X) = \frac{1}{2}/(\frac{1}{2} - (-1)) = -1$.   □

On a line ℓ of PG(2, ℝ), we say that two points *Y* and *Y′* of ℓ are harmonic conjugates with respect to two distinct points *Z* and *Z′* on ℓ, if (Z, Z', Y, Y') are harmonic. The above result can be used to find the **harmonic conjugate** of a point on a line ℓ, given three points on ℓ. The line ℓ is *PQ*, and we consider a third point *A* on ℓ. For the construction of the harmonic conjugate *X* of *A* (with respect to *P* and *Q*), we follow the following steps (see Figure 4.8):

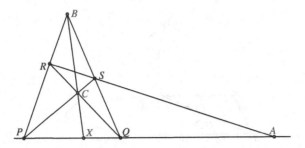

Figure 4.8 Constructing a harmonic conjugate X of A, with respect to P and Q.

- Let B be a point not on ℓ.
- Let R be a point on BP not equal to B or P.
- Construct the points $S := AR \cap BQ$ and $C := QR \cap PS$.
- Let X be the intersection of the lines BC and ℓ.

$$* * *$$

In terms of cross-ratios, we can give equations that express Ceva's concurrency condition and Menelaus' collinearity condition. To begin, let ABC be a triangle and let ℓ be a line not incident with any vertex of the triangle. Suppose D on BC (not equal to B or C), E on CA (not equal to A or C), and F on AB (not equal to A or B). Then the cross-ratio expressions that we will refer to are:

Ceva's concurrency condition AD, BE, and CF are concurrent:

$$R(B, C; D, \ell \cap BC) \cdot R(C, A; E, \ell \cap CA) \cdot R(A, B; F, \ell \cap AB) = -1.$$

Menelaus' collineation condition D, E, and F are collinear:

$$R(B, C; D, \ell \cap BC) \cdot R(C, A; E, \ell \cap CA) \cdot R(A, B; F, \ell \cap AB) = 1.$$

Theorem 4.32
If three points on the sides of a triangle satisfy Ceva's concurrency condition, their harmonic conjugates with respect to the vertices of the triangle satisfy Menelaus' collinearity condition, and vice versa.

Proof Let ABC be the given triangle, and let ℓ be a line not incident with any vertex of the triangle. Suppose D on BC (not equal to B or C), E on CA (not equal to A or C), and F on AB (not equal to A or B). Let $\bar{D}$, $\bar{E}$, $\bar{F}$ be the harmonic conjugates of D, E, F, respectively, with regard to the vertices of the triangle that they are associated with (e.g., $R(B, C; D, \bar{D}) = -1$). By the cocycle identity, we have

$R(B, C; D, BC \cap \ell) = R(B, C; D, \bar{D}) \cdot R(B, C; \bar{D}, BC \cap \ell) = -R(B, C; \bar{D}, BC \cap \ell),$

and likewise, we have $R(C, A; \bar{E}, CA \cap \ell) = -R(C, A; E, CA \cap \ell)$, and also $R(A, B; \bar{F}, AB \cap \ell) = -R(A, B; F, AB \cap \ell)$. Therefore,

$$R(B, C; \bar{D}, BC \cap \ell) \cdot R(C, A; \bar{E}, CA \cap \ell) \cdot R(A, B; \bar{F}, AB \cap \ell)$$
$$= -R(B, C; D, BC \cap \ell) \cdot R(C, A; E, CA \cap \ell) \cdot R(A, B; F, AB \cap \ell).$$

Hence, $R(B, C; D, BC \cap \ell) \cdot R(C, A; E, CA \cap \ell) \cdot R(A, B; F, AB \cap \ell) = \pm 1$ if and only if $R(B, C; \bar{D}, BC \cap \ell) \cdot R(C, A; \bar{E}, CA \cap \ell) \cdot R(A, B; \bar{F}, AB \cap \ell) = \mp 1$. □

We now prove the projective version of Ceva's theorem, which, as a by-product, yields a proof of the projective version of Menelaus' theorem.

Theorem 4.33 (Projective Ceva theorem)
Let A, B, C, O be four points in PG$(2, \mathbb{R})$*, no three collinear, and ℓ be a line, through none of A, B, C, O. Let $L = BC \cap \ell$, $M = CA \cap \ell$, $N = AB \cap \ell$. Then, if $D = AO \cap BC$, $E = BO \cap CA$, $F = CO \cap AB$,*

$$R(B, C; D, L) \cdot R(C, A; E, M) \cdot R(A, B; F, N) = -1.$$

Conversely, suppose that ABC is a triangle and that points D on BC (not equal to B or C), E on CA (not equal to A or C), and F on AB (not equal to A or B) satisfy

$$R(B, C; D, L) \cdot R(C, A; E, M) \cdot R(A, B; F, N) = -1.$$

Then AD, BE, and CF are concurrent.

Proof Choose coordinates so that $A(1, 0, 0)$, $B(0, 1, 0)$, $C(0, 0, 1)$, $O(1, 1, 1)$. Now BC is $x = 0$ and L is on BC, and not equal to B or C; let $L(0, 1, s)$ with $s \neq 0$. Also, CA is $y = 0$ and M is on CA, and not equal to A or C; let $M(t, 0, 1)$ with $t \neq 0$. Then ℓ is $x + sty - tz = 0$, and AB is the line $z = 0$, so N is the point with coordinates $(-st, 1, 0)$.

Suppose $D = AO \cap BC$, $E = BO \cap CA$, and $F = CO \cap AB$. Then $D(0, 1, 1)$, $E(1, 0, 1)$, and $F(1, 1, 0)$, which gives: $R(B, C; D, L) = 1/s$, $R(C, A; E, M) = 1/t$, $R(A, B; F, N) = -st$. So, $R(B, C; D, L) \cdot R(C, A; E, M) \cdot R(A, B; F, N) = -1$.

Conversely, let $O = AD \cap BE$ and suppose $R(B, C; D, L) \cdot R(C, A; E, M) \cdot R(A, B; F, N) = -1$. Now letting $D = AO \cap BC$ and $E = BO \cap CA$ gives $D(0, 1, 1)$ and $E(1, 0, 1)$ (as before) and $R(B, C; D, L) = 1/s$, $R(C, A; E, M) = 1/t$. Thus $R(A, B; F, N) = -st$, which implies that F has coordinates $(1, 1, 0)$. So F is on CO; thus AD, BE, and CF are concurrent in O. □

Theorem 4.34 (Projective Menelaus' theorem)

Let a, b, c, o be four lines in $\mathrm{PG}(2,\mathbb{R})$, no three concurrent, let $A := b \cap c$, $B := c \cap a$, $C := a \cap b$, $D := o \cap a$, $E := o \cap b$, $F := o \cap c$, and ℓ be a line, on none of A, B, C, D, E, F. Let $L := a \cap \ell$, $M := b \cap \ell$, $N := c \cap \ell$. Then

$$\mathrm{R}(B, C; D, L) \cdot \mathrm{R}(C, A; E, M) \cdot \mathrm{R}(A, B; F, N) = 1.$$

Conversely, suppose that ABC is a triangle and that points D on BC (not equal to B or C), E on CA (not equal to A or C), and F on AB (not equal to A or B) satisfy $\mathrm{R}(B, C; D, L) \cdot \mathrm{R}(C, A; E, M) \cdot \mathrm{R}(A, B; F, N) = 1$. Then D, E, and F are collinear.

Dually, we can define harmonic quadruples of concurrent lines: $\{a, b, c, d\}$ is a harmonic quadruple of lines if they are concurrent and $\mathrm{R}(a, b; c, d) = -1$. Harmonic conjugacy of lines is considered in Exercise 5.4.

Recall from Theorem 2.16 that a homology h is determined by its centre O, its axis a and a pair of corresponding points P and $P' = P^h$. What happens if h is an involution? First, note that P, P', and O are collinear since h must fix the line OP, and hence $OP^h \cap a = OP \cap a$. We know that h fixes O and the line a, but it also preserves cross-ratio:

$$\begin{aligned}
\mathrm{R}(O, OP \cap a; P, P^h) &= \mathrm{R}(O^h, O^h P^h \cap a^h; P^h, P^{h^2}) \\
&= \mathrm{R}(O, OP \cap a; P^h, P) \\
&= \mathrm{R}(O, OP \cap a; P, P^h)^{-1}
\end{aligned}$$

by Lemma 4.4. Therefore, $\mathrm{R}(O, OP \cap a; P, P^h)$ satisfies the quadratic $X^2 = 1$, and hence, $\mathrm{R}(O, OP \cap a; P, P^h) = 1$ or $\mathrm{R}(O, OP \cap a; P, P^h) = -1$; the first of which is impossible. Therefore, involutory homologies h have the interesting property that $(O, OP \cap a; P, P^h)$ is harmonic, and so are determined by centre and axis only, as P' is the harmonic conjugate of P with respect to O and $OP \cap a$. So, to summarise, a **harmonic homology** of $\mathrm{PG}(2,\mathbb{R})$ is a homology h such that for every point P not on the axis a and not equal to the centre O of h, we have $(O, OP \cap a, P, P^h)$ harmonic.

Theorem 4.35

Every involutory collineation is a harmonic homology.

Proof We have taken this beautiful proof from Coxeter (1949, §5.31).

Let g be an involution such that $g^2 = 1$. Suppose g interchanges the points A and A', and another pair B and B' not lying on the line AA'. The lines AA' and BB' are fixed by g and intersect in a point O which is also fixed by g. Now note that g also interchanges the lines AB and $A'B'$, and it interchanges the lines AB' and $A'B$. So the points $P := AB \cap A'B'$ and $Q := AB' \cap A'B$ are fixed by

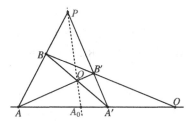

Figure 4.9 Proof of Theorem 4.35.

g. Moreover, the lines AA' and PQ meet in a fixed point A_0 on PQ (see Figure 4.9). Hence, g is either an elation or a homology, with centre O and axis PQ. Since O does not lie on PQ, g is a homology, and, by the argument above, it is a harmonic homology.   □

Ibrahim ibn Sinan ibn Thabit ibn Qurra in *On Drawing the Three Conic Sections* (c. 930) gave a construction of a hyperbola from a circle. Take the circle K with diameter AB (see Figure 4.10). For each point E on K, let D be the intersection of the tangent to K at E with AB. On the same side of AB as E, choose O on the perpendicular to AB at D with $d(D, O) = d(D, E)$. Then we let E vary and the points O mark out a hyperbola $\mathcal{H}$, and the map $K \to \mathcal{H}$ with $E \mapsto O$ is the restriction of a harmonic homology with centre A and axis the tangent to K at B.

The following result shows that often we can associate a well-defined quantity to a projectivity on a line, which will be called the **characteristic cross-ratio** of the projectivity.

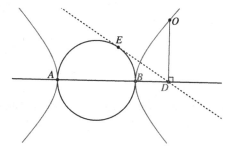

Figure 4.10 Ibrahim ibn Sinan ibn Thabit ibn Qurra's construction of a hyperbola.

Theorem 4.36

If g is a projectivity of ℓ, with distinct fixed points M and N, then
$$R(M, N; A, A^g) = R(M, N; B, B^g)$$
for all A, B on ℓ.

Proof Let $A' := A^g$, $B' := B^g$. Let $m \neq \ell$ be a line in M, S and S' be distinct points on m, with $S \neq M \neq S'$. Let $A'' := SA \cap S'A'$ and $B'' := SB \cap S'B'$. Then the composition of the perspectivity with centre A'' from ℓ to m and the perspectivity with centre B'' from m to ℓ takes M to M, A to B, and B to B', and so equals g. Therefore it fixes N and thus $R(M, N; A, A') = R(M, N; B, B')$. □

Since every projectivity g of a line has fixed points over $\mathbb{C}$, we can associate the cross-ratio in Theorem 4.36 with g, provided the fixed points of g are distinct, and we call it the **characteristic cross-ratio** of g. Under this proviso, the characteristic cross-ratio of g and its fixed points M and N determine g. The characteristic cross-ratio of a projectivity is -1 if and only if the projectivity is an involution.

Theorem 4.37 (von Staudt, 1847)

In $PG(2, \mathbb{R})$, a map from a line ℓ to itself is induced by a projectivity if and only if it sends harmonic quadruples onto harmonic quadruples.

Proof By Theorem 4.16, $PGL(3, \mathbb{R})_\ell^\ell$ is permutationally isomorphic both to the group induced by projectivities mapping ℓ to ℓ, and to $PGL(2, \mathbb{R})$ acting on the projective line $PG(1, \mathbb{R})$. Moreover, a bijection from $PG(1, \mathbb{R})$ to $PG(1, \mathbb{R})$ is in $PGL(2, \mathbb{R})$ if and only if it maps harmonic quadruples to harmonic quadruples. Certainly, $PGL(2, \mathbb{R})$ preserves cross-ratio, so maps harmonic quadruples to harmonic quadruples. By Corollary 2.8 we need only consider the stabiliser of 0, 1, and ∞, using non-homogeneous coordinates on $PG(1, \mathbb{R})$ (i.e., (x, y) corresponds to $\frac{x}{y}$). So let $f : \mathbb{R} \to \mathbb{R}$ be induced by a map $PG(1, \mathbb{R}) \to PG(1, \mathbb{R})$ that takes harmonic quadruples to harmonic quadruples. We wish to show that f is the identity. Now $(\frac{x+y}{2}, \infty; x, y)$ is harmonic, and so $f(\frac{x+y}{2}) = \frac{f(x)+f(y)}{2}$. Hence, by putting $x = 0$, $y = 2z$, we have $f(2z) = 2f(z)$. Therefore, $f(x + y) = f(x) + f(y)$. Now $(-x, 0; x, \infty)$ is harmonic, so $f(-x) = -f(x)$, and $(-x, x; 1, x^2)$ is harmonic, so $f(x^2) = f(x)^2$. Expanding $f((x + y)^2)$ in two ways, we see that $f(x)^2 + 2f(x)f(y) + f(y)^2 = f(x)^2 + 2f(xy) + f(y)^2$), so that $f(xy) = f(x)f(y)$. Thus, f is a field automorphism of $\mathbb{R}$. Hence, f is the identity, and so a map on the line taking harmonic quadruples to harmonic quadruples is induced by a projectivity. □

Apollonius (c. 200 BC; see Apollonius de Perge, 2008: [hereafter *Conics*]) studied pole and polar with respect to conics. That his definition and ours match is the content of the next theorem.

Theorem 4.38 (Apollonius' theorem (Apollonius, *Conics*, Book I, Prop. 36))
Let $\mathcal{K}$ be a (non-empty, non-degenerate) conic of $\mathsf{PG}(2, \mathbb{R})$ *and P be a point
of* $\mathsf{PG}(2, \mathbb{R})$ *not on $\mathcal{K}$. Let A be a point on $\mathcal{K}$, and let B be such that $PA \cap$
$\mathcal{K} = \{A, B\}$. Then (P, Q, A, B) is harmonic if and only if Q is on the polar line
to P.*

Proof We may take $\mathcal{K}$ to be $y^2 = xz$ and P to be $(1, 0, \pm 1)$. The line joining
P and the point A: $(1, t, t^2)$ on $\mathcal{K}$ is $[\mp t, -t^2 \pm 1, t]$ which meets $\mathcal{K}$ at A and
at B: $(t^2, \mp t, 1)$. For Q: $(1 + a, at, \pm 1 + at^2)$, we have $\mathrm{R}(P, Q; R, S) = -1$ if
and only if $1 + a = \pm 1 + at^2$, which is equivalent to the incidence of Q with
$[1, 0, -1]$; the polar line of P with respect to $\mathcal{K}$.   □

Second proof of Theorem 4.38 Write $P = A + pB$ and $Q = A + qB$ for some
non-zero real numbers p, q. Then

$$\mathrm{R}(P, Q; A, B) = \frac{-p(1 - q/\infty)}{-q(1 - p/\infty)} = p/q.$$

So $(P, Q; A, B)$ is harmonic if and only if $p/q = -1$. Now suppose G is the
Gram matrix for the bilinear form associated to $\mathcal{K}$. Then, if we write P and Q
in homogeneous coordinates, we have

$$PGQ^{\mathsf{T}} = (A + pB)G(A + qB)^{\mathsf{T}} = AGA^{\mathsf{T}} + pBGA^{\mathsf{T}} + qAGB^{\mathsf{T}} + pqBGB^{\mathsf{T}}$$
$$= pAG^{\mathsf{T}}B^{\mathsf{T}} + qAGB^{\mathsf{T}} = pAGB^{\mathsf{T}} + qAGB^{\mathsf{T}} = (p + q)AGB^{\mathsf{T}}.$$

So P and Q are conjugate if and only if $p + q = 0$. The result follows by noting
that $p + q = 0$ if and only if $p/q = -1$.   □

A beautiful proof that uses Pappus' involution theorem (Theorem 6.12) will
appear later (Section 6.3). The reverse direction component of Theorem 4.38
(i.e., that P and Q are conjugate with respect to $\mathcal{K}$ implies they are harmonically conjugate) first appeared in Archimedes' *'On Tangent Circles'*, which
has only survived in Arabic.[7] It is not known whether or not it is truly based
on a work by Archimedes. It was translated from the unique surviving Arabic
text into Russian (Veselovskii and Rosenfeld, 1962) and then from Russian
into German (Dold-Samplonius et al., 1975; the converse to Theorem 4.38 is
in §14, p.51).

Let $\mathcal{K}$ be a conic. Then the **polar triangle** of a triangle ABC, with respect
to $\mathcal{K}$, is the triangle whose sides are the poles of A, B, C. A triangle ABC is
self-polar if it is equal to its polar triangle.

---

[7] Fi'l-dawā'ir al-mutamāssa li-Arshimidis, Hyderabad, Osmania Oriental Publications Bureau,
   1948 (source: the manuscript Patna, Bankipore 2468).

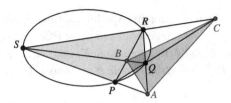

Figure 4.11 The diagonal triangle is self-polar.

Theorem 4.39 (Brianchon, 1817)
If a quadrangle is inscribed in a conic, its diagonal triangle is self-polar.

Proof Let the diagonal points of the inscribed quadrangle *PQRS* be $A = PS \cap QR$, $B = QS \cap RP$, $C = RS \cap PQ$ (see Figure 4.11). The line *AB* meets the sides *PQ* and *RS* in points C_1 and C_2 such that $(P, Q; C, C_1)$ and $(R, S; C, C_2)$ are harmonic. Both C_1 and C_2 are conjugate to *C*. Thus, the line *AB*, on which they lie is the polar of *C*. Similarly for the other diagonal points: each is the pole of the join of the other two.   □

Theorem 4.40 (Seydewitz, 1847)
If a triangle is inscribed in a conic, any line conjugate to one side meets the other two sides in conjugate points.

Proof Consider an inscribed triangle *PQR*. Any line conjugate to *PQ* is the polar of some point *C* on *PQ*. Let *RC* meet the conic again in *S*. By Theorem 4.39, the diagonal points of the quadrangle *PQRS* form a self-polar triangle *ABC* whose side contains the conjugate points *A* and *B*: one on *QR* and the other on *RP*.   □

Now suppose we want to find a self-polar triangle *PQR* for a conic of the form $\{x : xAx^\top = 0\}$, where *A* is a symmetric 3×3 matrix. So we require that $PAQ^\top = 0$, $QAR^\top = 0$, and $RAP^\top = 0$, where we associate each point *P*, *Q*, *R* with its homogeneous coordinates. We can place *P*, *Q*, *R* as rows into a matrix *M*, and since *PQR* is a triangle, *M* is invertible. We find that

$$MAM^\top = \begin{bmatrix} P \\ Q \\ R \end{bmatrix} A \begin{bmatrix} P^\top & Q^\top & R^\top \end{bmatrix} = \begin{bmatrix} p & 0 & 0 \\ 0 & q & 0 \\ 0 & 0 & r \end{bmatrix}$$

where $p = PAP^\top$, $q = QAQ^\top$, and $r = RAR^\top$. So, finding a self-polar triangle for a conic amounts to orthogonally diagonalising a symmetric matrix *A*.

Theorem 4.41 (von Staudt, 1831, p. 25)
Two non-degenerate conics with four common tangents have their eight points of contact on a (possibly degenerate) conic.

Proof Let the conics be $\mathcal{K}_1$, $\mathcal{K}_2$, the tangents be ℓ_i, $i = 1, \ldots, 4$. Let the point of contact of ℓ_i with $\mathcal{K}_1$ be P_i and with $\mathcal{K}_2$ be Q_i. Then the diagonal triangle of $P_1 P_2 P_3 P_4$ is self-polar with respect to $\mathcal{K}_1$ (Theorem 4.39). Hence, it is the diagonal triangle of $\ell_1 \ell_2 \ell_3 \ell_4$. This is self-polar with respect to $\mathcal{K}_2$ (Theorem 4.39); so is the diagonal triangle of $Q_1 Q_2 Q_3 Q_4$. Thus, $P_1 P_2 P_3 P_4$ and $Q_1 Q_2 Q_3 Q_4$ have the same diagonal triangle. Hence, $P_1 P_2 P_3 P_4$ and $Q_1 Q_2 Q_3 Q_4$ lie on a conic, by Theorem 4.39. □

4.6 Interlude: Some Invariant Theory

The isometry group Isom($\mathbb{R}^2$) of the Euclidean plane acts transitively on points, but not on pairs of distinct points. Each orbit on pairs yields an **invariant** binary relation on points. In fact, the orbits of Isom($\mathbb{R}^2$) on pairs of distinct points are completely characterised by the distance relation, so, in effect, there is just *one* binary invariant. To be more concrete, suppose we have a function f from the ordered pairs of $\mathbb{R}^2$ to some set X, which is invariant under the isometry group. So,

$$f(P^g, Q^g) = f(P, Q)$$

for all $P, Q \in \mathbb{R}^2$ and all $g \in$ Isom($\mathbb{R}^2$). Let p be the single-variable function in $\mathbb{R} \to X$ defined by $p(t) := f(O, (t, 0))$ for all $t \in \mathbb{R}$, where O is the origin $(0, 0)$. For each $(P, Q) \in \mathbb{R}^2 \times \mathbb{R}^2$, there exists an isometry h mapping (P, Q) to $(O, (d(P, Q), 0))$. Since f is an invariant, we have

$$f(P, Q) = f(P^h, Q^h) = f(O, (d(P, Q), 0)) = p(d(P, Q))$$

and hence $f = p \circ d$. Therefore, every (binary) Isom($\mathbb{R}^2$)-invariant is a *function of distance*.

We have a similar result for the similarity group, with an analogous proof. This time, Sim($\mathbb{R}^2$) is 2-transitive on points, so it only has trivial binary invariants. So we must consider triples of points, which can be collinear or noncollinear. These two cases yield the affine ratio and angle as invariants. Similarly, affine ratio is the fundamental non-trivial invariant for the affine group AGL$(2, \mathbb{R})$.

Theorem 4.42
Every ternary invariant for Sim($\mathbb{R}^2$) is a function of affine ratio and angle.

Theorem 4.43
Every ternary invariant for $\mathsf{AGL}(2, \mathbb{R})$ *is a function of affine ratio.*

For the projective plane, it is a little more complicated. Let us just contend with the projective line $\mathsf{PG}(1, \mathbb{R})$. The group $\mathsf{PGL}(2, \mathbb{R})$ acts 3-transitively on the points of $\mathsf{PG}(1, \mathbb{R})$, but intransitively on quadruples of distinct points. Fix three of our favourite points: $O := (1, 0)$, $I := (1, 1)$, $\infty := (0, 1)$.

Suppose we have an invariant f for $\mathsf{PGL}(2, \mathbb{R})$ with distinct quadruples of points as input, and a real number as output. Consider four distinct points P, Q, R, S and stipulate that $\lambda := f(\infty, O, I, (1, \lambda))$. (Note that any point distinct from ∞ can be written in the form $(1, \lambda)$.) Then there is an element of g of $\mathsf{PGL}(2, \mathbb{R})$ such that $\infty^g = P$, $O^g = Q$, $I^g = R$. Without loss of generality, we may suppose that $(1, \lambda)$ maps to S under g. Now g can be represented as a 2×2 matrix:

$$\begin{bmatrix} a & b \\ c & d \end{bmatrix}.$$

So $(P, Q, R, S) = ((c, d), (a, b), (a + c, b + d), (a + \lambda c, b + \lambda d))$, and therefore, $f(P, Q, R, S) = f(\infty, O, I, (1, \lambda)) = \lambda$. Let's assume for the moment that none of our four points is equal to ∞. We can write our points in terms of O and ∞:

$$P = O + p\infty, \qquad\qquad Q = O + q\infty,$$
$$R = O + r\infty, \qquad\qquad S = O + s\infty.$$

Then we have a system of linear equations, which we can solve for λ:

$$p = d/c, \qquad\qquad q = b/a,$$
$$r = \frac{b + d}{a + c}, \qquad\qquad s = \frac{b + \lambda d}{a + \lambda c}$$

and so

$$\lambda = \frac{sa - b}{d - sc} = \frac{a(s - q)}{d - sc} = \frac{a(s - q)}{c(p - s)}.$$

Some more rearranging shows that $a/c = (r - p)/(q - r)$. Therefore,

$$f(P, Q, R, S) = \lambda = \frac{(r - p)(s - q)}{(r - q)(s - p)}.$$

So we have shown that the values of f are determined once we fix O, I, ∞, and that the cross-ratio is essentially deduced as this invariant.

Theorem 4.44 (Clebsch, 1861)
Every projective invariant is a function of cross-ratio.

Moreover, let us now consider more than four variables $P_1, \ldots, P_n$. By 3-transitivity, we can normalise so that $P_1 = I$, $P_2 = \infty$, and $P_3 = O$. It turns out that P_l then becomes $R(P_1, P_2; P_3, P_k)$, showing that our original function in n-variables is a function of the $n - 3$ cross-ratios $R(P_1, P_2; P_3, P_k)$ for $k = 4, \ldots, n$.

ALFRED CLEBSCH (1833–72) was a German mathematician who created and inspired a German school of algebraic geometry and invariant theory, even though his doctoral dissertation and early papers were on problems in hydrodynamics. He co-founded the now prestigious journal *Mathematische Annalen*. Sadly, his life and work were cut dramatically short; he died of diptheria at the age of 39.

4.6.1 *j*-invariant

Recall that if we permute (collinear) four points P, Q, R, S and examine the values of their cross-ratio, we will have the following possibilities: λ, $1 - \lambda$, $1/\lambda$, $1/(1 - \lambda)$, $\lambda/(\lambda - 1)$, $(\lambda - 1)/\lambda$. where $\lambda = R(P, Q; R, S)$. The so-called *j*-**invariant** is the function[8]

$$j(\lambda) := \frac{(\lambda^2 - \lambda + 1)^3}{\lambda^2(1 - \lambda)^2}$$

and it satisfies $j(\lambda) = j(1 - \lambda) = j(1/\lambda) = j(1/(1 - \lambda)) = j(\lambda/(\lambda - 1))$.

4.6.2 *h*-invariant

A product of ratios of directed distances, where all the indicated points lie in $\mathrm{PG}(2, \mathbb{R})$, is called an *h*-**expression** if it has the following properties:

1. In each ratio the points that occur are collinear.
2. Each point appears in the numerator of the product exactly as many times as it does in the denominator.

So, for example, cross-ratio is an *h*-expression. If A, B, C, D, E, F are collinear points, then we define

$$S(A, B, C, D) := \frac{d(A, B) \cdot d(C, D) \cdot d(E, F)}{d(B, C) \cdot d(D, E) \cdot d(F, A)}.$$

Then S is an *h*-expression.

---

[8] The usual expression for the *j*-invariant has a coefficient of 2^8 so that it can work over an arbitrary field, particularly in characteristic 2 or 3, but this is beyond the scope of this book.

Theorem 4.45 (Eves, 1972, Theorem 6.2.2, p. 248)
Every h-expression is a projective invariant.

Proof It suffices to show that it is invariant under any homology. Given a homology g with centre V, if AB is one of the directed distances in the h-expression, then $d(A, B)d(V, AB) = d(V, A)d(V, B)\sin(\angle AVB)$. Replace every directed distance AB in the h-expression by

$$\pm d(V, A)d(V, B)\sin(\angle AVB)/d(V, AB).$$

By Condition (2), all expressions $d(V, X)$ cancel. By Condition (1), all expressions $d(V, XY)$ cancel. So the h-expression is rewritten as $\pm$ a product of terms of the form $\sin(\angle AVB)$. Now angles $\angle AVB$ and $\angle A^g VB^g$ are congruent, so $\sin(\angle AVB) = \sin(\angle A^g VB^g)$. Hence, the h-expression is unchanged by g. □

Shephard (1999, p. 1280) says: 'We feel that Eves' theorem has never been given the recognition it deserves and should be regarded as one of the fundamental results of projective geometry.' See Frantz (2011) for applications of Theorem 4.45.

4.7 Some Applications of Cross-Ratio

The following example uses the fact that four collinear points are equally spaced if and only if they have cross-ratio equal to $\frac{4}{3}$. But before we look at this nice application to the skyline of Battersea (Figure 4.12), we remark that this result on equally spaced points has its origins in *De Prospectiva Pingendi* (c. 1474–c. 1482, Book I, between Proposition 11 and Proposition 12) by Piero della Francesca ([c. 1474] 2021). He says that for collinear points on a line to look equally spaced in a perspective drawing, they should be placed as points of distances

$$2/1, 3/2, 4/3, 5/4, 6/5, 7/6, 8/7, 9/8, 10/9, \ldots$$

and (Piero didn't say this, of course!) any four successive points with these coordinates along a line have cross-ratio 4/3. So once you have cross-ratio, you can immediately prove that Piero is right! And, by varying the unit of measurement, Piero has covered all cases. See Migliari and Salvatore (2015).

Worked Example 4.46 (The skyline of Battersea)
In Volume 258 of M500[9] (June 2014), as problem 258.1, David Singmaster posed the question of where to view Battersea Power Station so that the

---

[9] The M500 Society is a mathematical society for students, staff, and friends of the Open University. By publishing M500 and organising residential weekends, the Society aims to promote a better understanding of mathematics, its applications, and its teaching.

Figure 4.12 Battersea Power Station. Available at `https://commons.wiki-media.org/wiki/File:Battersea_Power_Station_copy.jpg` under a Creative Commons Attribution-ShareAlike 3.0.

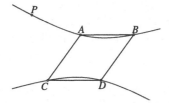

Figure 4.13 A hyperbola on the parallelogram *ABCD*.

four chimneys appear regularly spaced along the skyline. This leads to the question:

Given points *A*, *B*, *C*, *D*, no three collinear in $\mathbb{R}^2$, find the locus of all such viewpoints *P*, such that, from *P*, *A*, *B*, *C*, *D* (in that order) appear regularly spaced.

Since four collinear points appear regularly spaced from a viewpoint if and only if the join of the viewpoint to the points has cross-ratio $\frac{4}{3}$, the question is asking for the locus of *P* with R(*PA*, *PB*; *PC*, *PD*) = 4/3, which by the Chasles–Steiner theorem (Theorem 6.61) is a conic. If we know that (as for the Battersea chimneys), *ABCD* is a parallelogram arranged so that *AC* ∥ *BD* and *AB* ∥ *CD*, then two of the points are at infinity: the point at infinity on the line joining the midpoint of $\overline{AC}$ and *B* and the point at infinity on the line joining the midpoint of $\overline{BD}$ and *A*. Thus, we have a hyperbola (see Figure 4.13).

Dropping the condition of the order amounts to permuting the labels of the vertices, which gives values for the cross-ratio 4/3, 4, 3/4, −3, 1/4, −1/3.

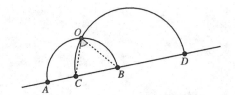

Figure 4.14 The Casey angle of four collinear points.

We cannot use the negative values because we want $\{PA, PB\}$ to not separate $\{PC, PD\}$. Thus, we could also use

$$R(PA, PB; PC, PD) = 4, \quad R(PA, PB; PC, PD) = 3/4,$$
$$R(PA, PB; PC, PD) = 1/4.$$

For a parallelogram $ABCD$, the first value (i.e., 4) corresponds to a hyperbola, and the last two are ellipses. So there are three conics through the four points where standing on a point of the conic works.[10] We refer to Exercise 4.8 for more on this problem.   □

The following concept is named after the Irish geometer John Casey. Consider any four distinct points A, C, B, D on a line, in the order specified (see Figure 4.14). So $\{A, B\}$ separates $\{C, D\}$. Let O be an intersection point of the circles with diameters $d(A, B)$ and $d(C, D)$. The angle $\theta = \angle COB$ is the **Casey angle** of the four points.

Theorem 4.47 (Casey's Theorem (Casey, 1871, pp. 703, 704; Cremona, 1885, p. 61, para. 72, n. V; Casey, 1889, p. 158, misc. ex. 20; Casey, 1892, p. 288, ex. 58, 59))
The set Λ of the cross-ratio values of four distinct collinear points can be written $\Lambda = \{\sin^2 \theta, \cos^2 \theta, \csc^\theta, \sec^2 \theta, -\tan^2 \theta, -\cot^2 \theta\}$ where θ is the Casey angle of the four points.

Proof Consider four distinct points A, C, B, D on a line, in the order specified, and let O be an intersection point of the circles with diameters $d(A, B)$ and $d(C, D)$. Recall that the Casey angle is $\theta = \angle COB$. By Worked Example 4.22,

$$R(A, B; C, D) = \frac{\sin(\angle COA) \sin(\angle DOB)}{\sin(\angle COB) \sin(\angle DOA)}.$$

---

[10] On the hyperbola according to cross-ratio $\frac{4}{3}$, you also want the plane of the film in the camera to be parallel to the asymptote in shot in order for the chimneys in the photograph to be evenly spaced, rather than being a perspective image of evenly spaced chimneys.

However, AB and CD subtend right angles at O (by Thales' theorem (Theorem 8.31), which we visit in Section 8.2). Therefore,

$$R(A, B; C, D) = \frac{\sin(\pi/2 - \theta)\sin(\pi/2 - \theta)}{\sin(\theta)\sin(\theta - \pi)} = -\cot^2(\theta).$$

The remaining five of six values follow by permuting. □

Worked Example 4.48

The following problem can be found in Frantz (2012).

Problem Given a photograph of two rectangular books lying on a table, determine the angle at which they are splayed.

Solution The books give rise two four vanishing points. Since the table has a planar surface, these vanishing points are collinear.

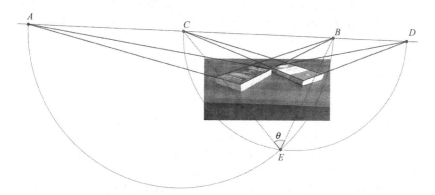

The six cross-ratios of these points in all orders are of the form $\sin^2\theta$, $\cos^2\theta$, $\csc^2\theta$, $\sec^2\theta$, $-\tan^2\theta$, $-\cot^2\theta$, by Theorem 4.47; θ is the angle in question. Moreover, θ can be calculated as follows: if A, B, C, D are the vanishing points and $\{A, C\}$ separates $\{B, D\}$, then intersect two semicircles on AC and BD on the same side of the vanishing line to obtain E. Then θ is angle $\angle BEC$. □

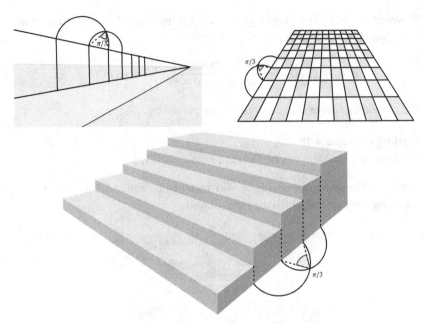

Figure 4.15 The Casey angle used for verification of correct perspective drawings of regularly spaced fence posts, rows of tiles, and stairs.

Worked Example 4.49

We have already seen (e.g., Worked Example 4.46) that four equally spaced points A, C, B, D (in that order) have cross-ratio $R(A, C; B, D) = 4/3$. The fraction $4/3$ elicits a trigonometric quantity that we commit to memory in secondary school, and is a simple consequence of Pythagoras' theorem: the fact that $\sin(\pi/3) = \sqrt{3}/2$.

$$\frac{4}{3} = \left(\frac{\sqrt{3}}{2}\right)^{-2} = \csc^2(\pi/3).$$

Therefore, if A, C, B, D are equally spaced, their Casey angle is $\pi/3$.

Since the Casey angle is a projective invariant (by Theorem 4.47), we can use it in verifying whether four points of an image are equally spaced. Figure 4.15 shows show how we can draw equally spaced fence posts, rows of tiles, and staircases. For more, we refer the reader to Marc Frantz's (2012) excellent article.   □

Exercises

4.1   Let P, Q be two points of $\mathrm{PG}(2, \mathbb{R})$ and let ℓ be a line not incident with P or Q. Consider the following *elementary correspondences*:

$$\alpha: \text{pencil}(P) \to \text{points of } \ell: x \mapsto x \cap \ell$$
$$\beta: \text{pencil}(Q) \to \text{points of } \ell: y \mapsto y \cap \ell.$$

Let g be a projectivity on the points of ℓ. Show that the composition $\alpha g \beta$ is a projectivity from pencil(P) to pencil(Q).

4.2 Give a method, such as that described in Figure 4.8, to find the harmonic conjugate line to three concurrent lines a, b, c.

4.3 Brannan et al. (2012, p. 200) show that:
- $R(A, B; C, D) = \text{ratio}(D, B, C)$, if A is the point at infinity,
- $R(A, B; C, D) = \text{ratio}(C, A, D)$, if B is the point at infinity,
- $R(A, B; C, D) = \text{ratio}(B, D, A)$, if C is the point at infinity, and
- $R(A, B; C, D) = \text{ratio}(A, C, B)$, if D is the point at infinity.

4.4 Prove Theorem 4.16 by using Theorem 4.8.

4.5 Prove Lemma 4.18.

4.6 (Hatton, 1913, p. 97) Let A, B, C, D be collinear points, and let O be a point not on that line. Let ℓ be another line on D, not on O. Let $A' := OA \cap \ell$, $B' := OB \cap \ell$, $C' := OC \cap \ell$, $O' := BC' \cap OA$, $E := O'B' \cap AB$. Show that if $R(A, B; C, D) = -1$, then $R(A, B; E, C) = -2$.

4.7 Show that the map $[0, s, t] \mapsto [t, 0, s]$ preserves cross-ratio.

4.8 Let $ABCD$ be a parallelogram, $E := AC \cap BD$, ℓ be the parallel to AB on E, $F := \ell \cap BC$, m be the parallel to BC on E, $G := m \cap CD$, $H := FG \cap AC$. For any point P not on AC, show that, with ℓ' the line on P parallel to AC, there is a line n such that $\{n \cap \ell', n \cap PA, n \cap PH, n \cap PC\}$ is a set of four equally spaced points on n.

4.9 Ever wondered where the term **harmonic mean** comes from? The harmonic mean of n real numbers $x_1, \ldots, x_n$ is

$$\frac{n}{\frac{1}{x_1} + \cdots + \frac{1}{x_n}}.$$

(i) Suppose A, B, C, D are collinear points of $\mathbb{R}^2$, lying on a line ℓ. Show that $d(A, B)$ is the harmonic mean of $d(A, C)$ and $d(A, D)$ if and only if $R(C, B; P, A) + R(D, B; P, A) = 2$, for any point P on ℓ.

(ii) Suppose A, B, C, D are collinear points on a line endowed with a direction. Show that $R(A, B; C, D) = -1$ if and only if the directed length $\overrightarrow{AB}$ is the harmonic mean of the directed lengths $\overrightarrow{AC}$ and $\overrightarrow{AD}$.

4.10 We can have projectivities and perspectivities mapping to the *dual* projective plane. So a **perspectivity** from a line ℓ to a pencil of lines on a point P is a one-to-one mapping such that each point X on ℓ is incident with its image line x on P. A **projectivity** from a line to a pencil is a finite product of perspectivities. Prove the following:

(i) If g is a duality, and ℓ a line, then the restriction of g to ℓ is a projectivity from ℓ to the pencil of lines on ℓ^g.

(ii) If g is a duality, and P a point, then the restriction of g to the pencil of lines on P is a projectivity from the pencil of lines on P to P^g.

4.11 Use Theorem 4.10 to prove Pappus' theorem (Theorem 2.18).

4.12 An artist draws an image of a yardstick on a mural. They reduce the first foot of the yardstick to 21 centimetres and the second foot to 14 centimetres. How long should they make the image of the third foot?

4.13 Show that if P, Q, R_1, R_2, R_3, R_4 are points on a conic, then the four points $S := PR_1 \cap QR_2$; $T := PR_2 \cap QR_1$; $U := PR_3 \cap QR_4$; $V := PR_4 \cap QR_3$ lie on a conic passing through the points P and Q. (This was a question Cayley (1864, May) proposed.)

4.14 A line intersects the sides of a triangle ABC in the points A_1 on BC, B_1 on AC, C_1 on BC, and (A_1, B, A_2, C), (B_1, A, B_2, C), (C_1, A, C_2, B) are harmonic. Show that the triples A_1, B_2, C_2; B_1, C_2, A_2; C_1, A_2, B_2 are each collinear and that the triples AA_2, BB_2, CC_2; AA_1, AB_2, AC_2; BB_1, BC_2, BA_2; CC_1, CA_2, CB_2 are each concurrent.

4.15 (Maclaurin, 1735) Let ℓ, m be lines of $\mathrm{PG}(2, \mathbb{R})$, and Z be a point not on ℓ or m. Let τ be the perspectivity with centre Z from ℓ to m. Let X be a fixed point on ℓ and Y be a fixed point on m, neither equal to $\ell \cap m$. Show that $\{XP^\tau \cap YP : P \in \ell\}$ is a conic.

4.16 Suppose A, B, C are three points on a line ℓ, and A', B', C' are three points on another line ℓ', which meets ℓ in O. Let $L = MC' \cap B'C$, $M = CA' \cap C'A$, $N = AB' \cap A'B$. Suppose I is a point on ℓ with $\mathrm{R}(A, B; C, I) = \mathrm{R}(A', B'; C', O)$. Show that I, L, M, and N are collinear.

5

The Group of the Conic

Today anyone can come forward, take some known truth, and submit it to various general principles of transformation; he will derive from it other truths, different or more general; and these, in turn, will be susceptible to similar operations; so that one can multiply, almost to infinity, the number of new truths deduced from the first: not all of them, it is true, will deserve to see the light of day, but a certain number of them will be able to offer interest and even lead to something very general. Therefore, in the present state of science, whosoever wishes to may generalise and create in Geometry; genius is no longer indispensable to add a stone to the edifice.

Michel Chasles (1837, pp. 268–9)

5.1 The Projective Line and the Conic

One of the leading themes of this book is about the various equivalent actions of $\mathrm{PGL}(2, \mathbb{R})$ in different geometric contexts:

- $\mathrm{PGL}(2, \mathbb{R})$ acting on the points of $\mathrm{PG}(1, \mathbb{R})$;
- $\mathrm{PGL}(2, \mathbb{R})$ acting on the points of a conic of $\mathrm{PG}(2, \mathbb{R})$;
- $\mathrm{PGL}(2, \mathbb{R})$ acting on the points of a twisted cubic of $\mathrm{PG}(3, \mathbb{R})$ (which we will see in Chapter 16).

The first one is the natural action. It is the permutation group induced by the action of invertible 2×2 matrices (i.e., $\mathrm{GL}(2, \mathbb{R})$) on the projective line. We can also think of this group as **fractional linear transformations** of the form

$$x \mapsto \frac{ax + b}{cx + d}, \quad ad - bc \neq 0$$

on $\mathbb{R} \cup \{\infty\}$.

83

It is the second of these actions, $PGL(2, \mathbb{R})$ acting on the points of a conic of $PG(2, \mathbb{R})$, that we now elucidate. Consider the conic

$$\mathcal{K}: y^2 = xz.$$

Let G be the stabiliser of $\mathcal{K}$ in $PGL(3, \mathbb{R})$. Then we can give a three-dimensional representation of $PGL(2, \mathbb{R})$ via the homomorphism $\phi: PGL(2, \mathbb{R}) \to G$ defined by

$$\begin{bmatrix} a & b \\ c & d \end{bmatrix}^{\phi} = \begin{bmatrix} a^2 & ab & b^2 \\ 2ac & ad + bc & 2bd \\ c^2 & cd & d^2 \end{bmatrix}.$$

Each point of $\mathcal{K}$ is of the form $(1, t, t^2)$, for some $t \in \mathbb{R}$, or $(0, 0, 1)$.

Note that

$$(1, t, t^2) \begin{bmatrix} a & b \\ c & d \end{bmatrix}^{\phi} = \left(1, \frac{b + dt}{a + ct}, \left(\frac{b + dt}{a + ct} \right)^2 \right)$$

$$(0, 0, 1) \begin{bmatrix} a & b \\ c & d \end{bmatrix}^{\phi} = \left(c^2, cd, d^2 \right)$$

and so $\operatorname{Im} \phi \leqslant G$. If we let $f: PG(1, \mathbb{R}) \to \mathcal{K}$ be the map defined by

$$f(1, t) = (1, t, t^2), \quad f(0, 1) = (0, 0, 1)$$

then (f, ϕ) gives rise to a permutational isomorphism of $(PG(1, \mathbb{R}), PGL(2, \mathbb{R}))$ with $(\mathcal{K}, \operatorname{Im} \phi)$. We will see in Corollary 5.11 that $\operatorname{Im} \phi = G$.

Theorem 5.1

(i) *The stabiliser of a non-empty, non-degenerate conic in $PGL(3, \mathbb{R})$ is 3-transitive on the points of the conic.*

(ii) *The stabiliser of a non-empty, non-degenerate conic in $PGL(3, \mathbb{R})$ has three orbits on lines: secant, tangent, and external lines.*

Proof We may assume (by Theorem 3.13) that the conic is $\mathcal{K}: y^2 = xz$. Let G be the stabiliser of $\mathcal{K}$ in $PGL(3, \mathbb{R})$. Consider the homomorphism $\phi: PGL(2, \mathbb{R}) \to G$ defined above. By the remark above, (f, ϕ) is a permutational isomorphism of $(PG(1, \mathbb{R}), PGL(2, \mathbb{R}))$ with $(\mathcal{K}, \operatorname{Im} \phi)$. Hence, G acts 3-transitively on $\mathcal{K}$. It follows immediately that secant lines and tangent lines each form an orbit of G. It only remains to prove that internal points form an orbit (for the polar lines are external lines). The orbit of $(1, 0, 1)$ under $\operatorname{Im} \phi$ is

$$\{(a^2 + c^2, ab + cd, b^2 + d^2): a, b, c, d \in \mathbb{R}, ad - bc \neq 0\}.$$

An internal point (x, y, z) has $Q(x, y, z) = xz - y^2 > 0$, and so an internal point can be written in the form $(1, t, t^2 + u^2)$ for some $t, u \in \mathbb{R}$ with $u \neq 0$. For a point of the form $(a^2 + c^2, ab + cd, b^2 + d^2)$ in the orbit above, setting $a = 1$, $c = 0$, $b = t$, and $d = u$ gives a point precisely of the form $(1, t, t^2 + u^2)$ with $u \neq 0$. Therefore, the orbit of $(1, 0, 1)$ under $\text{Im} \, \phi$ is the entire set of internal points. $\qquad\qquad\square$

Here is a *synthetic* way to obtain the above permutational isomorphism called **skew projection**. The setting is projective 3-space, which we will consider more rigorously in Chapter 15, but for now, we will extrapolate what we know from the projective plane and develop it casually here to the higher dimensional setting. Let π be a plane in projective 3-space $\mathsf{PG}(3, \mathbb{R})$ and ℓ, m be skew lines. Then every point P not on ℓ or m lies on a unique transversal t_P to ℓ and m: the plane $P\ell$ meets the plane Pm in the line t_P. We can then define a map ρ from points P (not on ℓ or m) to points of π by $P \mapsto t_P \cap \pi$. This is an example of a **quadratic transformation**.

Theorem 5.2
The skew projection map ρ defined above takes lines skew to the given two lines, to conics of π.

We will leave the proof of this result to you, the reader, once you have encountered Chapter 15.

Worked Example 5.3
Instead of the conic $y^2 = xz$, we will use the conic $x^2 + y^2 = z^2$ to display the action of $\mathsf{PGL}(2, \mathbb{R})$. In order to change conics, we consider the projectivity defined by the following matrix:

$$\rho = \begin{bmatrix} -1 & 0 & 1 \\ 0 & 1 & 0 \\ 1 & 0 & 1 \end{bmatrix}.$$

In order to see that this transformation does what we want it to, consider the following product:

$$\begin{bmatrix} -1 & 0 & 1 \\ 0 & 1 & 0 \\ 1 & 0 & 1 \end{bmatrix} \begin{bmatrix} 0 & 0 & -\frac{1}{2} \\ 0 & 1 & 0 \\ -\frac{1}{2} & 0 & 1 \end{bmatrix} \begin{bmatrix} -1 & 0 & 1 \\ 0 & 1 & 0 \\ 1 & 0 & 1 \end{bmatrix}^{\mathsf{T}} = \begin{bmatrix} 1 & 0 & 0 \\ 0 & 1 & 0 \\ 0 & 0 & -1 \end{bmatrix}.$$

Notice that the matrix in the middle of the left-hand side is the Gram matrix for the quadratic form $(x, y, z) \mapsto y^2 - xz$, whereas the right-hand side is the Gram matrix for $(x, y, z) \mapsto x^2 + y^2 - z^2$.

So, if $g \in \mathsf{PGL}(2,\mathbb{R})$, then $\rho\phi(g)\rho^{-1}$ preserves the quadratic form $(x,y,z) \mapsto$ $x^2 + y^2 - z^2$ that defines the 'unit circle'. So we obtain an equivalent embedding ϕ' of $\mathsf{PGL}(2,\mathbb{R})$ in $\mathsf{PGL}(3,\mathbb{R})$:

$$\begin{bmatrix} a & b \\ c & d \end{bmatrix} \mapsto \begin{bmatrix} a^2 - b^2 - c^2 + d^2 & 2(-ab - cd) & -a^2 - b^2 + c^2 + d^2 \\ 2(-ac + bd) & 2(bc + ad) & 2(ac + bd) \\ -a^2 + b^2 - c^2 + d^2 & 2(ab + cd) & a^2 + b^2 + c^2 + d^2 \end{bmatrix}.$$

Let's take the stabiliser of the line at infinity, $z = 0$, in the image ϕ'. Now

$$\begin{bmatrix} a^2 - b^2 - c^2 + d^2 & 2(-ab - cd) & -a^2 - b^2 + c^2 + d^2 \\ 2(-ac + bd) & 2(bc + ad) & 2(ac + bd) \\ -a^2 + b^2 - c^2 + d^2 & 2(ab + cd) & a^2 + b^2 + c^2 + d^2 \end{bmatrix} \begin{bmatrix} 0 \\ 0 \\ 1 \end{bmatrix}$$

$$= \begin{bmatrix} -a^2 - b^2 + c^2 + d^2 \\ 2(ac + bd) \\ a^2 + b^2 + c^2 + d^2 \end{bmatrix}$$

and, hence, $a^2 + b^2 = c^2 + d^2$ and $ac + bd = 0$; that is, $(c,d) = (\pm b, \mp a)$. So to summarise, we obtain transformations of the form:

$$\begin{bmatrix} a^2 - b^2 & -2ab & 0 \\ \pm 2ab & \pm(a^2 - b^2) & 0 \\ 0 & 0 & a^2 + b^2. \end{bmatrix}$$

Therefore, by stabilising a conic (the 'unit circle') and the line at infinity, we see that the 'origin' $(0,0,1)$ is also forced to be fixed. Note that if we let $C := (a^2-b^2)/(a^2+b^2)$ and $S := -2ab/(a^2+b^2)$, then $C^2+S^2 = 1$ and the projectivity above can be represented by:

$$\begin{bmatrix} C & S & 0 \\ \mp S & \pm C & 0 \\ 0 & 0 & 1. \end{bmatrix}$$

So we have shown that the stabiliser of the unit circle in $\mathsf{AGL}(2,\mathbb{R})$ is equal to the stabiliser $\mathsf{O}(2,\mathbb{R})$ in the isometry group $\mathsf{Isom}(\mathbb{R}^2)$.   □

5.2 Frégier Involutions

Let K be a non-degenerate, non-empty conic of $\mathsf{PG}(2,\mathbb{R})$ and let P be a point not lying on K. Define the following map f_P on the points of K. We let $f_P(X) = X$ if PX is tangent to K otherwise, $f_P(X)$ is the other point of PX on K (see Figure 5.1);

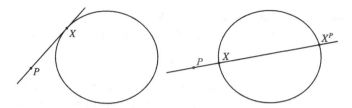

Figure 5.1 Frégier involutions.

Note that f_P is an involution and so it must be a permutation of the points of K; we call it the **Frégier involution** of K with **Frégier point** P. Let us now employ the simpler notation X^P for the point $f_P(X)$; we will be considering the group generated by these involutions, and so X^{PQ} will mean the composition of two Frégier involutions (with Frégier points P and Q, respectively).

Theorem 5.4 (Frégier's Theorem (Frégier, 1816/17))
Let K be a non-degenerate conic and ℓ be a line, not tangent to K. Let P be the pole of ℓ with respect to K. Then the harmonic homology h with centre P and axis ℓ stabilises K, and is a Frégier involution on K. Moreover, P is the Frégier point of h.

Proof Since h is a harmonic homology, if Q is on K, and $PQ \cap K = \{Q, Q'\}$ with $P' := PQ \cap \ell$, then (P, P', Q, Q^h) is harmonic. But, by Theorem 4.38, (P, P', Q, Q') is harmonic. Thus, $Q^h = Q'$ (by Lemma 4.3). □

Corollary 5.5
Every involution of $\mathsf{PG}(2, \mathbb{R})$ stabilising a non-degenerate, non-empty conic K is a Frégier involution.

Proof Let τ be an involution of $\mathsf{PG}(2, \mathbb{R})$ stabilising a non-degenerate conic K. Let (A, B) and (C, D) be two pairs of points of τ (so $A^\tau = B$ and $C^\tau = D$). Let P be the intersection of AB and CD, and let f_P be the Frégier involution with Frégier point P. Then P and τ agree on four points of K (namely, A, B, C, D) and hence must be identical (by Theorem 2.5). Alternatively, recall from Theorem 4.35 that τ is a harmonic homology, and it would fix P and its polar line ℓ with respect to K. □

Theorem 5.6
Let P and Q be two points not lying on a non-degenerate conic $\mathcal{K}$. Then the Frégier involutions with Frégier points P and Q commute if and only if P and Q are conjugate with respect to $\mathcal{K}$.

Proof Suppose first that Q is external to $\mathcal{K}$, and let t_X, t_Y be the two tangent lines to $\mathcal{K}$ on Q with points of contact X and Y. Then $Q = t_X \cap t_Y$ and so

$$Q^P = t_X^P \cap t_Y^P = t_{X^P} \cap t_{Y^P}.$$

Therefore, $Q = Q^P$ if and only if $\{X, Y\} = \{X^P, Y^P\}$, or, in other words, P is incident with the line XY. That is, $f_P f_Q = f_Q f_P$ if and only if P lies on the polar line of Q. This argument also works when P is an external point, so we can assume for the rest of the proof that P and Q are both internal points. By Exercise 3.15, P and Q are not conjugate and so we need to show that $f_P f_Q \neq f_Q f_P$. Suppose P and Q are not equal. Then there exists a point X on $\mathcal{K}$ not lying on the line joining P and Q. Let ℓ be the line spanned by X^P and X^{PQ}. Now the line joining X^P and X is incident with P and so X and P lie on the same side of ℓ. Similarly, the line joining X^{PQP} and X^{PQ} is incident with P and so X^{PQP} and P lie on the same side of ℓ. Therefore, X and X^{PQP} lie on the same side of ℓ. Now the line spanning X^{PQP} and X^{PQPQ} passes through Q, and Q is incident with ℓ, so X^{PQP} and X^{PQPQ} lie on different sides of ℓ. It then follows that X and X^{PQPQ} lie on different sides of ℓ, and hence $(f_P f_Q)^2 \neq 1$; that is, $f_P f_Q \neq f_Q f_P$.   □

Theorem 5.7

Let τ be a Frégier involution on the points of a non-degenerate, non-empty conic K of $\mathrm{PG}(2, \mathbb{R})$, and let P be the Frégier point of τ. Then

  (i) P is an external point to K if and only if τ fixes two points of K;
  (ii) P is an internal point to K if and only if τ is fixed-point-free.

Proof First note that τ fixes at most two points, because a line on P can only intersect K in at most two points. Suppose τ fixes one point X of K. So P is necessarily an external point since PX is tangent to K. However, there is another tangent on P meeting K at a point Y. So τ fixes Y as well, and hence two points. If P is internal to K, then for every point X of K, we have that PX is a secant line. So, in this case, τ is clearly fixed-point-free.   □

Theorem 5.8

Let K be a non-degenerate, non-empty conic of $\mathrm{PG}(2, \mathbb{R})$ and let G be the group generated by the Frégier involutions of K. Then G acts 3-transitively on the points of K.

Proof First, it is not difficult to see that G acts transitively on the set O of points of K. Let X and Y be two points of O, and let P be any point of the line XY not equal to X or Y; then $X^P = Y$. So, now, fix a point X of O and consider

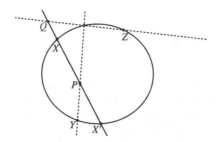

Figure 5.2 Proof of Theorem 5.8.

the remaining points $O \setminus \{X\}$. Take two points Y and Z of $O \setminus \{X\}$, and this time, let P be the intersection of YZ with the tangent t_X to K at X (see Figure 5.2). So $X^P = X$ and $Y^P = Z$, and so G_X acts transitively on $O \setminus \{X\}$, or, in other words, G acts 2-transitively on O. Finally, let X and X' be two distinct points of O and consider the rest, $O \setminus \{X, X'\}$. Take two points Y and Z of $O \setminus \{X\}$. Let P be any point of XX' not equal to X or X', and let Q be the point of intersection of $Y^P Z$ and XX'. Then $X^{PQ} = X'$ and $(X')^{PQ} = X$, and $Y^{PQ} = Z$. Therefore, $G_{X,X'}$ acts transitively on $O \setminus \{X, X'\}$, or, in other words, G acts 3-transitively on O. $\quad\square$

A permutation group G acting on a set Ω is said to be **sharply n-transitive** if it is (i) n-transitive and (ii) the pointwise stabiliser of any n-subset of Ω is the trivial subgroup. Or, in other words, for any two n-tuples of distinct elements of Ω, there is a *unique* element of g mapping one to the other.

Theorem 5.9

Let K be a non-degenerate, non-empty conic of $\mathsf{PG}(2, \mathbb{R})$ *and let G be the group generated by the Frégier involutions of K. Then G acts sharply 3-transitively on the points of K.*

Proof We have from Theorem 5.8 that G acts 3-transitively on the points of K. So suppose $g \in G$ fixes three distinct points X, Y, Z of K. Then g fixes the pole P of XY, which is a point not lying on K. Moreover, since K is non-degenerate, the points X, Y, Z are not collinear, but we also know that the point P does not lie on the lines XY. Moreover, PX and PY are tangent to K, and so Z does not lie on PX or PY. It follows that X, Y, Z, P are four points with no three collinear, and so $g = 1$ (by Theorem 2.5). $\quad\square$

Corollary 5.10

Let K be a non-degenerate conic of $\mathsf{PG}(2, \mathbb{R})$ *and let G be the group generated by the Frégier involutions of K. Then G is the full stabiliser of the points of K in* $\mathsf{PGL}(3, \mathbb{R})$.

Corollary 5.11

The stabiliser of a non-degenerate, non-empty conic in $\mathrm{PGL}(3,\mathbb{R})$ is isomorphic to $\mathrm{PGL}(2,\mathbb{R})$.

So now we have another way to realise the permutational isomorphism between the stabiliser of a non-empty, non-degenerate conic in $\mathrm{PGL}(3,\mathbb{R})$ acting on the conic, and $\mathrm{PGL}(2,\mathbb{R})$ acting on $\mathrm{PG}(1,\mathbb{R})$. To see this, take three points O, I, and P of $\mathrm{PG}(1,\mathbb{R})$. Then the sharply 3-transitive property of $\mathrm{PGL}(2,\mathbb{R})$ means that every element of $\mathrm{PG}(1,\mathbb{R})$, besides O and I, can be written uniquely as P^g where g is an element of $\mathrm{PGL}(2,\mathbb{R})$ fixing O and I. Likewise, the stabiliser G of the irreducible conic K of $\mathrm{PG}(2,\mathbb{R})$ is sharply 3-transitive, so we can label the points of K in precisely the same way. Thus, the permutational isomorphism is readily apparent via this labelling. This, is surprising. The projective line has cross-ratio, so the conic inherits cross-ratio through this permutational isomorphism. Indeed, we explore *cross-ratio on a conic* in Section 5.4.

5.3 Pencils of Conics

Given distinct conics $Q(x,y,z) = 0$ and $Q'(x,y,z) = 0$, the **pencil** of conics they generate is the set of conics of the form $u \cdot Q(x,y,z) + v \cdot Q'(x,y,z) = 0$, where $u, v \in \mathbb{R}$, $(u,v) \neq (0,0)$. A pencil is uniquely determined by two conics belonging to the pencil. Some members of the pencil will be degenerate.

Worked Example 5.12

Let $Q(x,y,z) = y^2 - 2xz$ and let $Q'(x,y,z) = -x^2 + y^2 - z^2$. If we take $(u,v) = (1,-1)$, we see that

$$u \cdot Q(x,y,z) + v \cdot Q'(x,y,z) = -2xz - x^2 + z^2 = (z-x)^2$$

which yields a degenerate conic – a double-line. $\square$

Consider the fundamental quadrangle $(1,0,0)$, $(0,1,0)$, $(0,0,1)$, $(1,1,1)$ and suppose the conic

$$K: Ax^2 + Bxy + Cy^2 + Dxz + Eyz + Fz^2 = 0$$

passes through all four points. Then we have $A = C = F = 0$ and $B+D+E = 0$. Therefore, the equation of such a conic is of the form:

$$B(xy - yz) + D(xz - yz) = 0.$$

That is, the set of conics passing through the fundamental quadrangle is the pencil of conics on $K_1: (x-z)y = 0$ and $K_2: (x-y)z = 0$.

Theorem 5.13

The set of all conics on a quadrangle is a pencil.

Proof The six coefficients of the equation of a conic satisfy four linearly independent linear equations if they pass through a given quadrangle, leaving a solution space of dimension two, which is necessarily a pencil. □

The degenerate members of the pencil of conics on a quadrangle are the pairs of lines consisting of opposite sides of the quadrangle.

Recall that a projectivity from a line ℓ to a line m is a composition of finitely many perspectivities. The dual notion allows us to speak of a **projectivity of pencils**. Instead of considering the points on ℓ and m, we take two points A and B and their pencils of lines pencil(A) and pencil(B), respectively. A perspectivity from pencil(A) to pencil(B) from the line ℓ not in either pencil is the map $a \mapsto \langle (\ell \cap a), B \rangle$. So then a projectivity of pencils is a composition of finitely many perspectivities of pencils.

Lemma 5.14

Let $\mathcal{K}$ be a conic. For a point A of $\mathcal{K}$, define the map $\pi_A \colon \mathcal{K} \to$ pencil(A) by assigning to A the tangent $\pi(A)$ to $\mathcal{K}$ at A, and $\pi_A(X) = AX$, for $X \in \mathcal{K}$, $X \neq A$. Then, for $A, B \in \mathcal{K}$, $\pi_A \pi_B^{-1}$ is a projectivity (of pencils).

Proof Without loss of generality, we may assume that $\mathcal{K}$ is $z^2 = xy$ and $A(1,0,0)$ and $B(0,1,0)$, since $\mathsf{PGL}(3,\mathbb{R})$ is transitive on non-empty, non-degenerate conics (by Theorem 3.13), and the stabiliser of a conic is 2-transitive on points of the conic. So pencil(A) consists of the lines $[0, s, t]$ and pencil(B) consists of the lines $[s', 0, t']$; two lines $[0, s, t]$ and $[s', 0, t']$ correspond if and only if they intersect on $\mathcal{K}$, or, equivalently, $(st', s't, -ss')$ is on $\mathcal{K}$. This is then equivalent to $ss' = tt'$ and hence $\pi_A \pi_B^{-1}$ maps $[0, s, t]$ to $[t, 0, s]$. By Exercise 4.7, the map $[0, s, t] \mapsto [t, 0, s]$ preserves cross-ratio and so is a projectivity between pencil(A) and pencil(B). □

The *Wallace-Simson line theorem* (compare with Theorem 5.15) states that the feet of the perpendiculars from a point on the circumcircle of a triangle to the sides of the triangle are collinear. The following is a generalisation of this result, and the earliest attribution we have for it is to Toshio Seimiya, who apparently discovered Theorem 5.15 at the age of 16. Since Toshio Seimiya was born in Tokyo on 30 March 1910, this means the result was discovered in 1926. (See Totten, 2006 for more.)

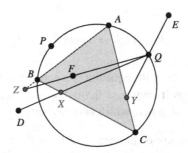

Figure 5.3 Theorem 5.15.

Theorem 5.15 (Seimiya's theorem)
*Let P and Q be points on the circumcircle of triangle ABC, and let D, E, F be
the reflections of P with respect to the lines BC, CA, and AB, respectively (see
Figure 5.3). Let X, Y, and Z be the intersections of QD, QE, and QF with BC,
CA, and AB, respectively. Then X, Y, and Z are collinear.*

We provide the following projective generalisation.

Theorem 5.16 (Projective Seimiya's theorem)
*Let lines p, q, t and point P, Q be fixed (distinct and with no incidences). Define
a projectivity g from p to q by*

$$X^g := (XP \cap t)Q \cap q.$$

Then $QX^g \cap PX$ is on a line ℓ, for all X on p.

Proof Define a projectivity h from the pencil on P to the pencil on Q by

$$x \mapsto Q(x \cap p)^g.$$

Now $(PQ \cap p)^g = (PQ \cap q)$, so $PQ^h = PQ$. By the dual of Corollary 4.9, h is
a perspectivity. Thus, there is a line ℓ with $x^h = Q(x \cap \ell)$. Now $Q(x \cap p)^g =
Q(x \cap \ell)$ for all lines x on P, so $QX^g = Q(PX \cap \ell)$ for all X on p. Thus,
$QX^g \cap PX = PX \cap \ell$ is on ℓ.   □

Steiner's 'oblique Simson theorem' is:

If, from any point D on a conic circumscribed around a given triangle ABC, lines are
drawn to the sides of the triangle, respectively parallel to the diameters passing through
the midpoints of the sides, their feet are collinear.

Proof of Seimiya's Theorem 5.15. Apply Steiner's oblique Simson theorem. Let p and q be the Steiner/Simson lines of P and Q (respectively) with respect to ABC. Let t be the line at infinity. Then Theorem 5.16 gives the result. □

5.4 Cross-Ratio on a Conic

The reader is probably familiar with the proposition in Euclidean geometry that three noncollinear points determine a circle. It is more complicated for parabolae and hyperbolae; three points do not suffice, and the solution for four or five points requires special case analysis. Projective geometry offers a clean and elegant solution: five points, no three collinear lie on a unique conic (Theorem 5.17). First, we see how we can construct a conic on five points P_1, P_2, P_3, P_4, P_5.

Choose two of the points, say P_1 and P_2, as centres of pencils and construct lines P_1P_3, P_1P_4, P_1P_5 and P_2P_3, P_2P_4, P_2P_5. Then the projectivity taking the first triple of lines to the second defines a point conic containing the five points.

Alternatively, the points on the conic give five linear equations in the unknown coefficients of the equation $ax^2 + by^2 + cz^2 + dxy + exz + fyz = 0$ of the conic:

$$(a,b,c,d,e,f)\begin{bmatrix} x_1^2 & x_2^2 & x_3^2 & x_4^2 & x_5^2 \\ y_1^2 & y_2^2 & y_3^2 & y_4^2 & y_5^2 \\ z_1^2 & z_2^2 & z_3^2 & z_4^2 & z_5^2 \\ x_1y_1 & x_2y_2 & x_3y_3 & x_4y_4 & x_5y_5 \\ x_1z_1 & x_2z_2 & x_3z_3 & x_4z_4 & x_5z_5 \\ y_1z_1 & y_2z_2 & y_3z_3 & y_4z_4 & y_5z_5 \end{bmatrix} = (0,0,0,0,0,0).$$

The above matrix has nullity at least 1, for it is a 6×5 matrix, and so there exists a conic on our five points.

Theorem 5.17
Five points, no three collinear lie on a unique conic, and the conic is non-degenerate.

Proof We will directly compute the conic on these five points by using some symmetry. By Theorem 2.5, we may assume that four of our five points are the fundamental quadrangle P: $(1,0,0)$, Q: $(0,1,0)$, R: $(0,0,1)$, S: $(1,1,1)$. This means that the equation of a conic containing these four points has the form $dxy + exz + (-d - e)yz = 0$. Let T be a fifth point such that no three of $\{P, Q, R, S, T\}$ are collinear. This means that T does not lie on any side of the quadrangle $PQRS$, and hence $T = (1, u, v)$ where $u, v \neq 0, 1$ and $u \neq v$. (Note that the sides of the quadrangle have equations $x = 0$, $y = 0$, $z = 0$, $x = y$, $y = z$, and $x = z$.) Since T lies on the conic, we have

$$0 = du + ev + (-d - e)uv = du(1 - v) + ev(1 - u).$$

Since $u(1 - v) \neq 0$, we have

$$d = -\frac{v(1 - u)}{u(1 - v)} e$$

and hence, by dividing out e, the conic has equation

$$v(1 - u) \cdot y(x - z) - u(1 - v) \cdot (x - y)z = 0.$$

Therefore, there is a unique conic passing through P, Q, R, S, and T.   □

An alternative proof of uniqueness, which essentially shows which the 6×5 matrix above has full rank, will be given in Section 20.3.

In *Conics*, Book IV, Apollonius suggests that Theorem 5.17 was first stated by Conon of Samos (c. 280 BC–c. 220 BC). Book IV of Apollonius' *Conics* is largely concerned with questions relating to the possible number of points of intersection of two non-degenerate plane conics. The preface to Book IV makes clear that Conon of Samos first raised this question, and that he asserted that two non-degenerate plane conics could share at most four points without giving a correct proof. He was criticised for omitting a correct proof by Nicoteles of Cyrene. Apollonius seems to assert that his proof is the first he knows of.

Let A, B, C, D be four points of an irreducible conic $\mathcal{K}$. Take a fifth point E of K and define

$$\mathrm{R}(A, B; C, D)_{\mathcal{K}} := \mathrm{R}(EA, EB; EC, ED).$$

By Theorem 5.18, we know this definition to be well-defined and to be independent of the point E.

Theorem 5.18 (Chasles, 1852)
Let $\mathcal{K}$ be a conic, and A_1, A_2, A_3, A_4 be distinct points of $\mathcal{K}$. Then the cross-ratio

$$\mathrm{R}(XA_1, XA_2; XA_3, XA_4)$$

is independent of the point X on $\mathcal{K} \setminus \{A_1, A_2, A_3, A_4\}$.

Proof By Theorem 5.17, five points, no three collinear, determine a conic. Let $X, X' \in \mathcal{K} \setminus \{A_1, A_2, A_3, A_4\}$. Now $\mathcal{K}$ is the unique conic on X, A_1, A_2, A_3, A_4, and is also the unique conic on X', A_1, A_2, A_3, A_4. By Lemma 5.14, there is a projectivity taking pencil(X) to pencil(X') which takes XA_i to $X'A_i$, for $i = 1, 2, 3, 4$. Hence, $R(XA_1, XA_2; XA_3, XA_4) = R(X'A_1, X'A_2; X'A_3, X'A_4)$. □

The following result yields a beautiful way to generate a conic from five points that makes use of the cross-ratio on a conic.

Corollary 5.19

Let A_1, A_2, A_3, A_4, A_5 be five points, no three collinear. Then a point X distinct from these lies on the conic on A_1, A_2, A_3, A_4, A_5 if and only if

$$R(XA_1, XA_2; XA_3, XA_4) = R(A_5A_1, A_5A_2; A_5A_3, A_5A_4).$$

Corollary 5.20

Two ordered quadruples of points on a conic are in the same orbit of the stabiliser of the conic if and only if they have the same cross-ratio with respect to the conic.

So we have an alternative proof of Theorem 5.9. Let g be an element of the stabiliser of a conic $\mathcal{K}$, and suppose g fixes three points A_1, A_2, A_3 of $\mathcal{K}$. Take another point A_4 of $\mathcal{K}$. Then

$$R(A_1, A_2, A_3, A_4)_\mathcal{K} = R(A_1^g, A_2^g, A_3^g, A_4^g)_\mathcal{K} = R(A_1, A_2, A_3, A_4^g)_\mathcal{K}$$

and hence $A_4 = A_4^g$ (by Lemma 4.3). Therefore, $g = 1$.

Corollary 5.21

The stabiliser of a non-empty, non-degenerate conic acts sharply 3-transitively on the points of the conic. Hence, the stabiliser of a non-empty, non-degenerate conic acting on the conic is permutationally isomorphic to $(\mathrm{PG}(1, \mathbb{R}), \mathrm{PGL}(2, \mathbb{R}))$.

Recall from the beginning of Section 4.5 that for four points A, B, C, D lying on a common line, we have $R(A, B; C, D) < 0$ if $\{A, B\}$ *separates* $\{C, D\}$. We have an analogue for cross-ratio on the conic.

Theorem 5.22

Four points A, B, C, D on a conic $\mathcal{K}$ have $R(A, B; C, D)_\mathcal{K} < 0$ if and only if AB and CD meet in an internal point.

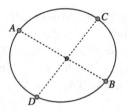

Figure 5.4 $\{A, B\}$ separates $\{C, D\}$ on the conic.

Proof Let $X := AB \cap CD$. Since cross-ratio is invariant under the action of $PGL(3, \mathbb{R})$ (by 4.15), and $PGL(3, \mathbb{R})$ acts transitively on conics (by Theorem 3.13), we may suppose that K is the conic $xz = y^2$. Next, the stabiliser of K acts 3-transitively on the points of K (by Theorem 5.8), and so we can assume $A: (1, 0, 0)$, $B: (0, 0, 1)$, and $C: (1, 1, 1)$. Notice that we can write $D: (t, 1, t^{-1})$ where $t \notin \{0, 1\}$ (as $D \neq A, B, C$). Then $X = (t - 1, 0, t^{-1} - 1)$. Notice that X is an internal point if and only if $t < 0$ (i.e., $0 < Q(t - 1, 0, t^{-1} - 1) = 1 - t - t^{-1}$ where $Q(x, y, z) = xz - y^2$). Now we let $E: (u, 1, u^{-1})$ where $u \notin \{0, 1\}$. Then

$$EA = [0, -1, u] \qquad EC \qquad = [1, -u - 1, u] = EA - EB$$
$$EB = [-1, u, 0] \qquad ED \qquad = [1, -t - u, tu] = tEA - EB.$$

So with EA and EB as reference lines, we have

$$R(EA, EB; EC, ED) = \frac{(-1 - 0)(-1/t - \infty)}{(-1 - \infty)(-1/t - 0)} = t.$$

Therefore, X is an internal point if and only if $R(A, B; C, D)_K < 0$. □

Notice that with the natural total ordering of the points of the conic, we have that AB and CD meet in an internal point when $\{A, B\}$ separates $\{C, D\}$ (see Figure 5.4).

Theorem 5.23
Four points A, B, C, D on a conic K have $(A, B; C, D)$ harmonic on the conic if and only if AB and CD are conjugate with respect to the conic.

Proof By Corollary 5.20, $(A, B; C, D)$ is harmonic on the conic K if and only if there is an element of the subgroup of $PGL(3, \mathbb{R})$ stabilising K taking $(A, B; C, D)$ to $(A, B; D, C)$. Such an element g is a homology with axis AB and centre the pole of AB. But g must then fix CD. So CD passes through the centre of the homology; that is, AB and CD are conjugate with respect to the conic. Conversely, if AB and CD are conjugate with respect to the conic, the the pole of AB lies on CD, so the involutory homology with centre the pole of AB and axis AB (which fixes the conic) interchanges C and D; thus, there is an element of the group of the conic taking $(A, B; C, D)$ to $(A, B; D, C)$, and so $(A, B; C, D)$ is harmonic on the conic. □

The following is the dual of Theorem 4.14.

Theorem 5.24

If g is a collineation, and P a point, then the restriction of g to the pencil of lines on P is a projectivity from the pencil of lines on P to the pencil of lines on P^g.

Theorem 5.25 (Axis of projectivity on a conic (Russell, 1893, chapter XVI; Chasles, 1865, article 234, pp. 151–152; Steiner, 1832, §38, IV, pp. 139))

*Let $g \in \mathrm{PGL}(3, \mathbb{R})$ stabilise a non-degenerate conic $\mathcal{K}$ in $\mathrm{PG}(2, \mathbb{R})$. Then, for $P \neq Q$ in $\mathcal{K}$, $PQ^g \cap P^g Q$ lies on a fixed line ℓ, called the **axis** of g.*

Proof By Theorem 5.24, the restriction of g to the pencil of lines on P is a projectivity with image the pencil of lines on P^g fixing the line PP^g, so is a perspectivity; thus, there is a line ℓ_P such that $PX^g \cap P^g X$ is on ℓ_P for all X in $\mathcal{K}$ with $X \neq P$. Similarly, there is a line ℓ_Q such that $QX^g \cap Q^g X$ is on ℓ_Q for all $X \in \mathcal{K}$ with $X \neq Q$.

Let $I = \ell_P \cap \ell_Q = PQ^g \cap QP^g$. Let $R \in \mathcal{K}$ with $P \neq R \neq Q$. Let $J = PR^g \cap P^g R$. Consider the involution t with centre J interchanging P and $R^g Q$, and P^g and R. In fact, t stabilises $\mathcal{K}$ (by Frégier's theorem (Theorem 5.4). Let $S_1 = Q^t$ and $S = Q^{gt}$. Then, on the conic $\mathcal{K}$,

$$\mathrm{R}(P, Q; R, S)_{\mathcal{K}} = \mathrm{R}(P^g, Q^g; R^g, S^g)_{\mathcal{K}}$$

and $\mathrm{R}(P, Q; R, S)_{\mathcal{K}} = \mathrm{R}(R^g, S_1; P^g, Q^g)_{\mathcal{K}}$ via t. Hence, $\mathrm{R}(P^g, Q^g; R^g, S^g)_{\mathcal{K}} = \mathrm{R}(P^g, Q^g; R^g, S_1)_{\mathcal{K}}$, giving $S_1 = S^g$ (by Lemma 4.3). In other words, $J = QS^g \cap Q^g S$; hence, J is on ℓ_Q. Therefore, $\ell_P = IJ = \ell_Q$. □

The first person to consider projectivities on a conic was Bellavitis (1838).

Corollary 5.26 (Adler's lemma (Adler, 2006))

Suppose points A, B, C, D, E, F, G, H lie on a conic $\mathcal{K}$. Let $X = AF \cap BE$, $Y = BG \cap CF$, $Z = CH \cap DG$. Then $\mathrm{R}(A, B; C, D)_{\mathcal{K}} = \mathrm{R}(E, F; G, H)_{\mathcal{K}}$ if and only if X, Y, Z are collinear.

Proof Now $\mathrm{R}(A, B; C, D)_{\mathcal{K}} = \mathrm{R}(E, F; G, H)_{\mathcal{K}}$ if and only if there exists g in $\mathrm{PGL}(3, \mathbb{R})$ with $A^g = E$, $B^g = F$, $C^g = G$, $D^g = H$. By Theorem 5.25, this implies that X, Y, Z are collinear in a line ℓ. Conversely, suppose X, Y, Z are collinear. Consider the projectivity g stabilising $\mathcal{K}$ with $A^g = E$, $B^g = F$,

$C^g = G$. Then, by Theorem 5.25, the axis of g is ℓ. Hence, $D^g \in \mathcal{K}$ with $CD^g \cap DG$ on ℓ; so $D^g = H$.   □

We have shown that a map between two lines is a projectivity if and only if it preserves cross-ratio, and that cross-ratio can be defined on a conic. Thus, it is reasonable to *define* a projectivity between a line and a conic or between two conics as a map preserving cross-ratio. With this definition, we can prove the following, which we leave as an exercise:

Theorem 5.27
Any projectivity between two conics is induced by a collineation.

Exercises

5.1   Show that if A, B, C, D are on a conic and AB is conjugate to CD with respect to the conic, then $(XA, XB; XC, XD)$ is harmonic for any point X of the conic, distinct from A, B, C, D.

5.2   Let $\mathcal{K}$ be a conic of $PG(2, \mathbb{R})$ and τ be a projectivity stabilising $\mathcal{K}$ with fixed points U and V. Suppose $P^\tau = Q$, $(P')^\tau = Q'$, and that the tangent lines to $\mathcal{K}$ at U, V meet PQ at X, Y, respectively, and meet $P'Q'$ at X', Y', respectively. Show that $R(P, Q; X, Y)_{\mathcal{K}} = R(P', Q'; X', Y')_{\mathcal{K}}$.

5.3   Let $\mathcal{K}$ be a conic of $PG(2, \mathbb{R})$ on the distinct points A, B, C, D.
   (a)   Let P be a variable point on $\mathcal{K}$, distinct from A, B, C, D. Let $M = PB \cap AD$, $N = PC \cap AD$. Prove that the cross-ratio $R(A, D; M, N)$ is independent of the choice of P.
   (b)   Let Q be another point of $\mathcal{K}$, distinct from A, B, C, D. Let $QM \cap \mathcal{K} = \{Q, B'\}$ and $QN \cap \mathcal{K} = \{Q, C'\}$. Prove that $R(A, D; B, C)_{\mathcal{K}} = R(A, D; B', C')_{\mathcal{K}}$.

5.4   Let A, B, C be distinct points on a conic of $PG(2, \mathbb{R})$. Suppose that the tangents at A, B, C meet BC, CA, AB, respectively at A', B', C'. Show that A', B', C' are collinear. (Hint: consider the projectivity on the conic in which A, B, C correspond respectively to B, C, A.)

5.5   Let C, C' be conics of $PG(2, \mathbb{R})$ and suppose that $A, B \in C \cap C'$ and that the tangent to C at A is the tangent to C' at A and the tangent to C at B is the tangent to C' at B. Suppose ℓ is tangent to C at V, $\ell \cap AB = T$, $\ell \cap C' = \{P, Q\}$. Prove that P, Q are harmonic conjugates with respect to V, T.

5.6   (Maclaurin, 1748, posthumous appendix *De Linearum Geometricarum Proprietatibus Generalibus Tractatus*) Let $\mathcal{K}$ be a non-degenerate conic

of PG$(2, \mathbb{R})$ on the points A, B, C, D and let a, b, c, d be the respective tangent lines to $\mathcal{K}$ at A, B, C, D. Show that
  (i) the diagonals of $ABCD$ and $abcd$ are concurrent and harmonic;
  (ii) the points of intersection of the opposite sides of $ABCD$ and $abcd$ are collinear and harmonic;
  (iii) AC and BD are incident with $b \cap c$, $d \cap a$, respectively.

5.7 Show that the cross-ratio of four points on a conic equals the cross-ratio of the four points of intersection of a tangent line with the tangent lines to the conic at those four points.

5.8 Let ρ be a polarity of PG$(2, \mathbb{R})$ defined by a conic K and let A, B, C, D be distinct, collinear points. Show that R$(A, B; C, D) = $ R$(A^\rho, B^\rho; C^\rho, D^\rho)$.

5.9 The points A, B, C and A', B', C' lie on the lines OL and OM, respectively, in PG$(2, \mathbb{R})$ and LM is the Pappus line on $AB' \cap A'B$, $AC' \cap A'C$, $BC' \cap B'C$. Let $X := CC' \cap AA'$, $Y := CC' \cap BB'$, $S := OX \cap A'B$, $Q := OX \cap A'C$, $R := OY \cap B'A$, $P := OY \cap B'C$. Prove that
  (a) $AY \cap BX$, $AC' \cap OX$, $BC' \cap OY$, M are collinear;
  (b) RX, SY, PQ pass through L.

6

Involution

> 'Real' geometry is found to be honeycombed with exceptions and dis-
> tinctions until it has become aesthetically intolerable. Homographies
> may or may not have double elements, straight lines may or may not
> meet conics; just as, in the arithmetic of rationals, $X^2 = a$ may or may
> not have roots. And the process by which these intolerable anomalies
> are banished is simply that of replacing the rational number in one case,
> the 'real point' in the other, by some wider and more complex logical
> construction. In the one case the deus ex machina is the section of Dede-
> kind, in the other the involution.

G. H. Hardy (1920)[1]

As we drift away from Euclidean geometry towards projective geometry,
we gain freedom and symmetry; we lose the shackles that distance, angle, parallelism, and circle placed on us. However, for each Euclidean property we
dispense with, we also strive to retain as much of it as is projectively invariant. Cross-ratio takes the place of distance (though for four collinear points
rather than two), we fix a line to simulate the properties of parallelism, and
we have pole–polar and cross-ratio on a conic instead of circles, parabolas,
and hyperbolas. The remaining vestige of Euclidean geometry that we explore
in this chapter is the remnant of **perpendicularity**. Perpendicularity of lines
of $\mathbb{R}^2$ induces a map that swaps pairs of points at infinity. So the projective
counterpart will be a line equipped with an **involution**.

---

[1] Review of *The Theory of the Imaginary in Geometry*, by J. L. S. Hatton. *The Mathematical Gazette*, **10**(146), 77–9. Reproduced with permission from Cambridge University Press.

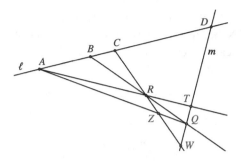

Figure 6.1 The proof of Lemma 6.1.

6.1 Basics on Involutions

A **projective involution**, or simply **involution**, is a projectivity $\ell \to \ell$ of order two, for some line ℓ of PG(2, $\mathbb{R}$). Suppose that A, B, C, D are collinear points. Let R be a point not on their line ℓ, and let m be a line on D distinct from ℓ. Further, let $T := RA \cap m$, $Q := RB \cap m$, $W := RC \cap m$, $Z := AQ \cap RC$. (See Figure 6.1.) Then

$$(A, B, C, D) \overset{Q}{\barwedge} (Z, R, C, W) \overset{A}{\barwedge} (Q, T, D, W) \overset{R}{\barwedge} (B, A, D, C).$$

So we have:

Lemma 6.1
Any four collinear points can be interchanged in pairs by a projectivity.

Theorem 6.2
Any projectivity that interchanges two distinct points of PG(1, $\mathbb{R}$) *is an involution.*

Proof Suppose that the distinct points A and A' are interchanged. Let X be a point on AA', distinct from A and A'. Let X' be the image of X under the projectivity interchanging A and A'. By Lemma 6.1, $(A, A', X, X')\overline{\wedge}(A', A, X', X)$. Since the composition of these two projectivities fixes A, A', and X', by Theorem 4.8, it is the identity. Therefore, the original projectivity maps X' to X. Since X was arbitrary, the result follows. $\qquad\square$

Corollary 6.3
An involution is determined by any two of its pairs.

Corollary 6.3 still holds if the two pairs consist of repeated points, that is, two fixed points. Notice that it makes sense to write an involution as a product of disjoint cycles, each of length of at most 2; and this notation determines the involution (by Corollary 6.3):

$$(A A')(B B')\dots.$$

Corollary 6.4
If an involution fixes one point, then it fixes another.

Proof Suppose $(A)(B C)\dots$ is an involution, and let D be the harmonic conjugate of A with respect to B and C. Then D is also the harmonic conjugate of A with respect to C and B (i.e., with the roles of our two points reversed). However, our involution is a projectivity and hence preserves cross-ratio (by Theorem 4.7) and so D is a second fixed point of our involution.   □

6.2 Imaginary Points

Points at infinity were introduced in order to enable us to state theorems about lines in all their generality, without having to consider cases of exception: two lines always have a point of intersection. But, as we proceed, we meet another set of cases of exception that cannot be dealt with in the same manner. For example, a projectivity of a line might have two fixed points, or might have none. Nevertheless, the nature of these projectivities is not intrinsically different in the two cases. Similarly, a line may meet a circle in two points or in no points. Two tangents may be drawn from a point to a circle or none at all.

Now it would be extremely convenient if these restrictions could be removed and if, by introducing a new set of ideal elements, which have no visual existence, we could state our theorems in a perfectly general manner. We can do this if we extend our coordinates to allow complex numbers. A similar situation arises in linear algebra, where we use complex numbers to obtain eigenvalues and eigenvectors, even for real matrices. The use of complex points begins as a mere matter of convenience. The following quote of Shreeram S. Abhyankar (1976, p. 413) is instructive, and he reframes the introduction of points at infinity, complex points, and double points as *counting properly*:

[T]he peculiar characteristic wisdom of geometric algebraic geometry is the [dictum:] Study of simple cases gives rise to a nice succinct statement. Take it as an axiom that the statement is true most generally. Make it true by the provision that we learn to 'count properly' the intervening quantities.

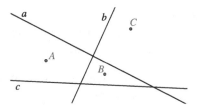

Figure 6.2 Worked Example 6.5.

As an example, consider the following collineation of $PG(1, \mathbb{R})$ represented as a 2×2 matrix with respect to the canonical basis:

$$\begin{bmatrix} 0 & 1 \\ -1 & 0 \end{bmatrix}.$$

What are its fixed points? To work out which points are fixed, we consider the following equation, as we are working with homogeneous coordinates:

$$(x, y)\begin{bmatrix} 0 & 1 \\ -1 & 0 \end{bmatrix} = \lambda(x, y).$$

So the vector (x, y) is an eigenvector of our 2×2 matrix with eigenvalue λ. The problem of finding fixed points is a matter of simple linear algebra, that is, calculating eigenvalues and eigenvectors. We find that the characteristic polynomial of our 2×2 matrix is $X^2 + 1$, which has no real solutions. However, if we were to consider the projective line to be over the complex numbers instead, we would have two fixed points for this collineation: $(i, 1)$ and $(-i, 1)$. These points are called the **imaginary** points of the projective line. Much of the theory we have derived in previous chapters still applies to projective geometry with complex coordinates, but there are some subtle differences, such as the FToPG (Theorem 2.9). The collineation group of $PG(2, \mathbb{C})$ is twice as big as $PGL(3, \mathbb{C})$, since entry-wise complex conjugation induces a collineation.

Worked Example 6.5

Problem   Given three points A, B, C and three lines a, b, c with the given points on none of the given lines (see Figure 6.2), construct a triangle whose vertices lie on the given lines and whose sides pass through the given points.

Solution   Consider the projectivity g from a to a, which is the composition of the perspectivity h_1 with centre C from a to b, the perspectivity h_2 with centre A from b to c, and the perspectivity h_3 with centre B from c to a. Let P be a fixed point of g. Then the desired triangle is $PP^{h_1}P^{h_1h_2}$. (Note that the resulting triangle may be imaginary.)   □

Worked Example 6.6

Problem   Given a conic $\mathcal{K}$ construct a triangle inscribed in $\mathcal{K}$ whose sides pass through three given points A, B, C of the plane not on $\mathcal{K}$.

Solution   Let h_1 be the harmonic homology with centre A and axis the polar line of A, let h_2 be the harmonic homology with centre B and axis the polar line of B, let h_3 be the harmonic homology with centre C and axis the polar line of C. We note that h_1, h_2, h_3 stabilise $\mathcal{K}$. Let $g := h_1 h_2 h_3$. If P is a fixed point of g, the desired triangle is $P P^{h_1} P^{h_1 h_2}$. (Note that the resulting triangle may be imaginary.) This method goes back to Poncelet (1995a).   □

Worked Example 6.7

Problem   Given five points A, B, C, D, E, no three collinear, and a line ℓ, find the points of ℓ on the conic $\mathcal{K}$ through A, B, C, D, E.

Solution   We will see, that by the Chasles–Steiner theorem (Theorem 6.61), $\mathcal{K}$ is defined by the projectivity g taking AC to BC, AD to BD, and AE to BE, and so a point X is on $\mathcal{K}$ if and only if g takes AX to BX. Consider the projectivity h on ℓ, taking $AC \cap \ell$ to $BC \cap \ell$, $AD \cap \ell$ to $BD \cap \ell$, and $AE \cap \ell$ to $BE \cap \ell$. Thus, a point X on ℓ is on $\mathcal{K}$ if and only if X is a fixed point of h. When ℓ is external to $\mathcal{K}$, h has no real fixed points; however, when ℓ is secant to $\mathcal{K}$, h has distinct real fixed points and when ℓ is tangent to $\mathcal{K}$, h has a repeated fixed point. Moreover, when ℓ is external to $\mathcal{K}$, the conjugate pair of imaginary fixed points of h are the zeros of the quadratic polynomial defining $\mathcal{K}$, when restricted to h.   □

We now apply these methods to a problem of Oai Thanh Dao.[2]

Worked Example 6.8 (Dao's two-conic problem)

Problem   Let there be two conics passing through four common points A, B, C, D. Suppose the tangent lines of the first conic at B, D meet the second conic at G, H, respectively (see Figure 6.3). Show that AC, BD, GH are concurrent.

Solution   There is an involution of the plane with centre $AC \cap BD$ stabilising the first conic. Each non-degenerate conic of the pencil of conics on A, B, C and D meets the axis of this involution in a (possibly imaginary) point, which is fixed by the involution, and hence the conic is fixed by the involution. Thus, the second conic is fixed by the involution since five points on it are fixed, and there is a unique conic on these five points (by Theorem 5.17). So G, H, and $AC \cap BD$ are collinear.   □

---

[2]   See Dao's blog: http://oaithanhdao.blogspot.com.

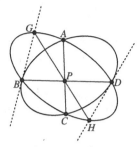

Figure 6.3 Worked Example 6.8.

We began this section with Abhyankar's dictum on extending one's scope so that a geometric property can hold in greater generality, in order to 'count properly'. Alternatively: 'a viewpoint in algebraic geometry is that *a theorem is not something that is true, but it is rather a nice geometric statement that you want to be true. So you adjust your definitions properly*' (Abhyankar, 1990, p. 21).

If we consider objects such as points, lines, conics via their degree, and their basic interaction, we develop the viewpoints that help us navigate from the Euclidean to the projective to the complex world.

- **two lines meet in a point:** so we invent the projective plane to count properly, and we need to also count points at infinity;
- **a line and a conic meet in two points:** so we invent the complex projective plane to count properly, and we need to also count complex points;
- **a tangent line and a conic meet in two points:** we invent *intersection multiplicity* to count properly.

By Exercise 2.6, an involution fixes zero or two points, and is called an **elliptic involution** or a **hyperbolic involution**,[3] respectively. Now if we allow for imaginary points in the complex plane, then every elliptic involution has a conjugate pair of imaginary fixed points. If t is an involution and P, Q are distinct (real or imaginary) fixed points then $(P, Q; X, X^t)$ is harmonic. Conversely, if P, Q are distinct real or distinct conjugate imaginary points then the map interchanging every point with its harmonic conjugate with respect to $\{P, Q\}$ is an involution.

---

[3] These adjectives are common in geometry, owing mainly to the property that an ellipse of $\mathbb{R}^2$ has zero points at infinity, and a hyperbola has two points at infinity.

Theorem 6.9

(a) *Let P, Q be distinct points of* $PG(1, \mathbb{R})$. *The map that fixes P and Q and takes each point X to its harmonic conjugate with respect to P and Q is an involution.*

(b) *Let P, Q be distinct conjugate points of* $PG(1, \mathbb{C})$. *The map taking each point X of* $PG(1, \mathbb{R})$ *to its harmonic conjugate with respect to P and Q is an involution.*

(c) *An involution of* $PG(1, \mathbb{R})$ *has either two real fixed points or a conjugate imaginary pair of fixed points.*

(d) *Every involution of* $PG(1, \mathbb{R})$ *arises in the manner in (a) or in (b).*

Proof Let X be a point, and let X' be its harmonic conjugate with respect to P and Q. Then, also, X is the harmonic conjugate of X' with respect to P and Q, and so the map in question swaps X and X'. So (a) and (b) follow from Corollary 6.3 (when extended to $\mathbb{C}$). We leave (c) for the reader and observe that (d) follows from Theorem 4.35.   □

Theorem 6.10
The map on a pencil of lines through a point not on a conic that interchanges pairs of lines that are conjugate with respect to the conic is an involution.

Proof This follows from Theorem 4.7 and Theorem 6.9.   □

We finish with a second proof of half of the projective Ceva theorem (Theorem 4.33).[4]

Second proof of Theorem 4.33 (compare with Heading, 1958, §3.19P) Let $X := AO \cap \ell$, $Y := BO \cap \ell$, $Z := CO \cap \ell$. By projecting B, C, D, L onto ℓ, firstly from A and secondly from O, we obtain (from Corollary 4.10):

$$R(B, C; D, L) = R(N, M; X, L) = R(Y, Z; X, L).$$

This implies that $R(N, M; X, L) = R(Z, Y; L, X)$ (by Lemma 4.4), showing that $\tau := (LX)(MY)(NZ) \dots$ is an involution. Similarly,

$$R(C, A; E, M) = R(L, N; Y, M) = R(Z, X; Y, M),$$
$$R(A, B; F, N) = R(M, L; Z, N) = R(X, Y; Z, N).$$

---

[4] We restate it here for convenience: Let A, B, C, O be four points in $PG(2, \mathbb{R})$, no three collinear, and ℓ be a line, through none of A, B, C, O. Let $L = BC \cap \ell$, $M = CA \cap \ell$, $N = AB \cap \ell$. If $D = AO \cap BC$, $E = BO \cap CA$, $F = CO \cap BC$, then

$$R(B, C; D, L) \cdot R(C, A; E, M) \cdot R(A, B; F, N) = -1.$$

Now $R(N, X; X, L) = R(N, M; X, Y)R(N, M; Y, L) = R(Z, Y; L, M)R(N, M; Y, L)$ by applying the involution τ, which further equals $R(M, L; Y, Z)R(N, M; Y, L)$. Hence,

$$R(B, C; D, L)R(C, A; E, M)R(A, B; F, N)$$
$$= R(N, M; X, L)R(L, N; Y, M)R(M, L; Z, N)$$
$$= R(M, L; Y, Z)R(N, M; Y, L)R(L, N; Y, M)R(M, L; Z, N)$$
$$= R(M, L; Y, N)R(N, M; Y, L)R(L, N; Y, M)$$
$$= R(Y, N; M, L)R(N, M; Y, L)R(M, Y; N, L) = -1.$$

The converse is similar. Suppose $R(B, C; D, L)R(C, A; E, M)R(A, B; F, N) = -1$. Let $G := AD \cap BE$ and $H := CG \cap AB$. Then, by the other half of the proof, since AD, BE, and CH are concurrent in G, it follows that

$$R(B, C; D, L)R(C, A; E, M)R(A, B; H, N) = -1.$$

Thus, $R(A, B; F, N) = R(A, B; H, N)$, and so $F = H$. Hence, AD, BE, and CF are concurrent. □

Finally, we note that the following result follows directly from Theorem 4.5, once extended to the complex numbers.

Theorem 6.11 (Miquel's theorem (Miquel, 1844))
Given eight collinear points A, B, C, D, P, Q, R, S of $\mathsf{PG}(2, \mathbb{C})$*, if five of the following six cross-ratios are real, so is the sixth:*

$$R(A, B; R, S), R(B, C; P, S), R(C, A; Q, S),$$
$$R(P, Q; C, D), R(Q, R; A, D), R(R, P; B, D).$$

6.3 The Involution Theorems of Pappus and Desargues

A **complete quadrangle** is a set of four points of $\mathsf{PG}(2, \mathbb{R})$, no three are collinear, and the six lines arising from each pair of the four points. The **diagonal points** of a quadrangle $ABCD$ are the points $AB \cap CD$, $AC \cap BD$, and $AD \cap BC$.

Theorem 6.12 (Pappus' involution theorem (c. 340))
The three pairs of opposite sides of a complete quadrangle of $\mathsf{PG}(2, \mathbb{R})$ *meet any line not through a vertex in three pairs of an involution.*

Proof Let P, Q, R, S be a quadrangle, ℓ be a line through no vertex, $A := PS \cap \ell$, $B := QS \cap \ell$, $C := RS \cap \ell$, $D := QR \cap \ell$, $E := PR \cap \ell$, and $F := PQ \cap \ell$ (see Figure 6.4). Then

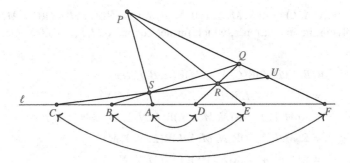

Figure 6.4 Proof of Theorem 6.12.

$$(A, E, C, F) \overset{P}{\overline{\wedge}} (S, R, C, U) \overset{Q}{\overline{\wedge}} (B, D, C, F),$$

where $U := PQ \cap RS$. Combining this with the involution interchanging B and D and C and F gives

$$(A, E, C, F)\overline{\wedge}(D, B, F, C),$$

which is an involution by Theorem 6.2. This involution interchanges the pairs (A, D), (B, E), and (C, F).   □

Theorem 6.13 (Converse of Pappus' involution theorem)
The three pairs of an involution can be seen as the intersections with the pairs of opposite sides of a complete quadrangle of PG$(2, \mathbb{R})$. *Indeed, given an involution* $(A\,D)(B\,E)(C\,F)\ldots$ *on a line* ℓ, *let* P *be a point not on* ℓ, S *be a point on AP, other than A or P,* $Q = BS \cap PF$, $R = CS \cap PE$. *Then* P, Q, R, S *is a quadrangle,* ℓ *is a line through no vertex,* $A = PS \cap \ell$, $B = QS \cap \ell$, $C = RS \cap \ell$, $D = QR \cap \ell$, $E = PR \cap \ell$, *and* $F = PQ \cap \ell$.

Proof Since triangles EFP and BCS are in perspective from A (see Figure 6.5), by Desargues' theorem (Theorem 2.20), the intersections of corresponding sides are collinear; so Q, R, and D are collinear. Now P, Q, R, S is a quadrangle, ℓ is a line through no vertex, $A = PS \cap \ell$, $B = QS \cap \ell$, $C = RS \cap \ell$, $D = QR \cap \ell$, $E = PR \cap \ell$, and $F = PQ \cap \ell$.   □

We remark that Theorem 6.13 still holds if $A = D$; that is, two points are fixed by the involution.

Worked Example 6.14
Pappus' involution theorem and its converse (Theorems 6.12 and 6.13) together imply Pappus' hexagon theorem (Theorem 2.18). The following treatment was taken from Baker (1952, pp. 363–4).

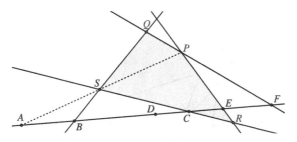

Figure 6.5 The converse of Pappus' involution theorem.

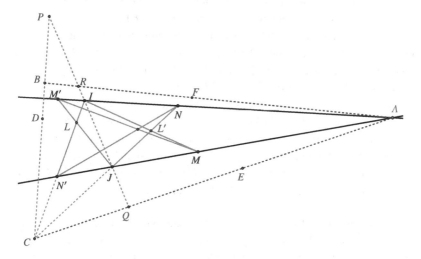

Figure 6.6 Proving Pappus' hexagon theorem.

Let M', I, N be collinear and N', J, M be collinear with none of the points equal to the intersection A of the two lines. Let $L := IN' \cap JM'$, $L' := IM \cap JN$, $B := JN \cap IN'$, $C := IM \cap JM'$. Let IJ meet BC, CA, AB in P, Q, R, respectively; let $D := LL' \cap BC$, $E := MM' \cap CA$, and $F := NN' \cap AB$ (see Figure 6.6).

- Applying Pappus' involution theorem (Theorem 6.12) to the quadrangle $IL'JL$, we see that $(BC)(PD)\ldots$ is an involution.
- Applying Pappus' involution theorem (Theorem 6.12) to the quadrangle $IM'JM$, we see that $(CA)(QE)\ldots$ is an involution.
- Applying Pappus' involution theorem (Theorem 6.12) to the quadrangle $IN'JN$, we see that $(AB)(RF)\ldots$ is an involution.

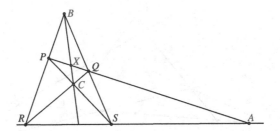

Figure 6.7   A complete quadrangle and a harmonic quadruple.

Since $(A\,C)(Q\,E)\ldots$ and $(A\,B)(R\,F)\ldots$ are involutions, BC, QR, EF are con-current, and so EF is incident with P. Since $(B\,A)(R\,F)\ldots$ and $(B\,C)(P\,D)\ldots$ are involutions, CA, RP, FD are concurrent, and so FD is incident with Q. Since $(C\,B)(P\,D)\ldots$ and $(C\,A)(Q\,E)\ldots$ are involutions, AB, PQ, DE are concurrent, and so DE is incident with R. Now $(LC\,LB)(LP\,LD)\ldots$ is an involution, so, intersecting with IJ, $(J\,I)(P\,P')$ is an involution, where $P' = LD \cap IJ$. Now $(MA\,MC)(MQ\,ME)\ldots$ is an involution, so, intersecting with IJ, $(J\,I)(Q\,Q')$ is an involution, where $Q' = ME \cap IJ$. Now $(NB\,NA)(NR\,NF)\ldots$ is an involution, so, intersecting with IJ, $(J\,I)(R\,R')$ is an involution, where $R' = NF \cap IJ$. Moreover, LL' is on D, MM' is on E, and NN' is on F. Thus, by the converse of Pappus' involution theorem (Theorem 6.13), LL', MM', NN' are concurrent. Hence, $MM' \cap NN'$, $IN' \cap JM$, and $IM \cap JN$ are collinear. $\square$

Let us now relate Pappus' involution theorem to harmonic quadruples. First we recall Theorem 4.31: Let P, Q, R, S be a complete quadrangle in $\mathrm{PG}(2,\mathbb{R})$, and let $A := PQ \cap RS$, $B := PR \cap QS$, $C := PS \cap QR$, and $X := BC \cap PQ$ (see Figure 6.7). Then $(P, Q; A, X)$ is harmonic. If we consider instead the complete quadrangle $RBSC$ and the line $\ell := PQ$, Pappus' involution theorem (Theorem 6.12) says that $(P)(Q)(A\,X)$ is an involution because $BC \cap \ell = Q = BS \cap \ell$ and $BR \cap \ell = P = CS \cap \ell$. Thus, we see that a harmonic quadruple arises as a special case of Pappus' involution theorem when the given line ℓ not through any vertex of the given complete quadrangle passes through two diagonal points of the quadrangle. Indeed, an involution fixing two points is just the correspondence between harmonic conjugates with respect to these two points.

We now use Pappus' involution theorem to prove Apollonius' theorem (Theorem 4.38) but with a subtle variation. By using imaginary points, we can assume that every point not on a conic lies on two tangent lines. We leave it to the reader to assure themselves that Pappus' involution theorem and other results that we proved over the real numbers still hold over the complex numbers.

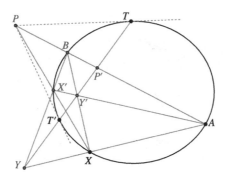

Figure 6.8 A synthetic proof of Theorem 4.38.

Another proof of Theorem 4.38 We will work over the complex numbers. Thus, we may assume that every point not on a conic lies on two tangent lines. Suppose that t and t' are the tangents to $\mathcal{K}$ on P, meeting $\mathcal{K}$ at T, T', respectively, so that $\ell = TT'$ is the polar line to P. Let A be a point on $\mathcal{K}$, other than T, T', $P' := PA \cap \ell$, and suppose $PA \cap \mathcal{K} = \{A, B\}$. It suffices to show that $(P, P'; A, B)$ is harmonic for if Q is a point on PA such that $(P, Q; A, B)$ is harmonic, then $Q = P'$ by Lemma 4.3.

Let X be a point on $\mathcal{K}$, other than T, T', A, B. Let $PX \cap \mathcal{K} = \{X, X'\}$. By Exercise 3.2(a), AX, BX', and TT' are concurrent, in a point Y, say. Let $Y' = AX' \cap BX$ (see Figure 6.8). Then Y' lies on ℓ (again, by Exercise 3.2(a)). By Pappus' involution theorem (Theorem 6.12), applied to the quadrangle $YY'XX'$ and the line PA, $(A)(B)(PP')$ is an involution, and $(P, P'; A, B)$ is harmonic. □

Theorem 6.15 (Hesse, 1840b, p. 17)
If the extremities of each of two diagonals of a complete quadrilateral are conjugate points with respect to a given conic, the extremities of the third diagonal also will be conjugate points with respect to the same conic.

Proof Let $ABXY$ be a complete quadrilateral such that A is conjugate to X, and B to Y, with respect to a given conic K. Let the sides AB, XY meet in C, and the sides AY, BX in Z (see Figure 6.9); we must show that C and Z are conjugate with respect to the conic K. Suppose the polars of the points A, B, C (with respect to K) cut the line AB in A', B', C', respectively. Now $(AA')(BB')(CC')\ldots$ is an involution, so XA', XB', XC' meet in a point Q, by the converse of Pappus' involution theorem (Theorem 6.13). Now since XA is the polar of A and YB the polar of B with respect to K, their point of intersection Q is the pole of AB. Since C is a point on AB and is conjugate to C' on AB, its polar will be QC', but QC' is on Z. Therefore, C and Z are conjugate points. □

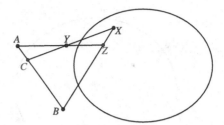

Figure 6.9 Theorem 6.15: a quadrilateral *ABXY* where *A* is conjugate to *X*, and
B is conjugate to *Y*.

The following is the 'involution' version of the projective Ceva theorem
(Theorem 4.33), with perhaps one of the most beautiful proofs of a classical
result.

Theorem 6.16
Let ABC be a triangle in PG(2, ℝ), *and ℓ be a line, through none of A, B, C.*
Let L := BC ∩ ℓ, M := CA ∩ ℓ, N := AB ∩ ℓ. Let L′, M′, N′ be points on ℓ.
Then (L L′)(M M′)(N N′) . . . is an involution if and only if AL′, BM′, CM′ are
concurrent.

Proof If *AL′*, *BM′*, *CM′* are concurrent in *O*, then by Pappus' involution
theorem (Theorem 6.12) applied to *ABCO*, $(L L')(M M')(N N') \ldots$ is an involution. Conversely, if $(L L')(M M')(N N') \ldots$ is an involution, by the converse
to Pappus' involution theorem (Theorem 6.13), there exists *O* with *AO∩ℓ = L′*,
BO ∩ ℓ = M′, *CO ∩ ℓ = N′*, so *AL′*, *BM′*, *CM′* are concurrent.   □

The following uses projective elements, such as Pappus' involution theorem,
in proving a Euclidean theorem. We will explore more of these types of result
in Section 8.

Theorem 6.17 (Oblique Simson line theorem (Poncelet, 1817, XI, p. 10))
If lines are drawn from a point on the circumcircle of a triangle that make any
fixed (directed) angle with the sides, then their feet will be collinear.

Proof This proof is due to Chasles (1852, para. 395). If *ABC* is the triangle, *P* the point on the circumcircle, *A′* the foot on *BC*, *B′* the foot on
CA, *C′* the foot on *AB*, then, by the equal inclination to the sides assumption,
angle ∠*A′PC′* is congruent to an angle supplementary to ∠*ABC*, and similarly
for the angles ∠*C′PB′* and ∠*B′PA′*. Thus, the two angles ∠*APA′* and ∠*CPC′*
have the same bisector. Similarly, ∠*APA′* and ∠*BPB′* have the same bisector.
Thus, $(P A)(P A')(P B)(P B')(P C)(P C') \ldots$ is an involution. So, by the dual

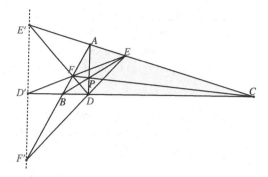

Figure 6.10 The trilinear polar theorem.

of the converse of Pappus' involution theorem (Theorem 6.13), A', B', C' are collinear.   □

In 1865, J.-J.-A. Mathieu (1865, p. 399) introduced the **trilinear polar** with respect to a triangle. (It is also in Poncelet, 1995b, p. 34.)

Theorem 6.18 (Trilinear polar theorem)
*If ABC is a triangle, and P a point on no side of ABC, then, with $D := PA \cap BC$,
$E := PB \cap CA$, $F := PC \cap AB$, if EF, FD, DE intersect the sides BC, CA, AB
of triangle ABC in the points D', E', F', respectively, D', E', F' are collinear.*

Proof By Pappus' involution theorem (Theorem 6.12), $(AD \cap BC, EP \cap BC; AP \cap BC, DE \cap BC)$ is harmonic, and, hence, $(B, F; C, DE \cap BC)$ is harmonic. Therefore, $DE \cap BC = F'$ (see Figure 6.10). Similarly, $EF \cap CA = E'$ and $FD \cap AB = D'$. Now triangles ABC and DEF are in perspective from P, and so D', E', F' are collinear, by Desargues' theorem (Theorem 2.20).   □

The line $D'E'F'$ in Theorem 6.18 is the **trilinear polar** of the point P. Note that, despite the name, this correspondence does not define a polarity of the plane; concurrent lines are not sent to collinear points.

The converse of the trilinear polar theorem is just the existence of the centroid in Euclidean geometry. It says that if ABC is a triangle and ℓ is a line on no vertex, then for the harmonic conjugate D of $\ell \cap BC$ with respect to $\{B, C\}$, the harmonic conjugate E of $\ell \cap CA$ with respect to $\{C, A\}$, and the harmonic conjugate F of $\ell \cap AB$ with respect to $\{A, B\}$, AD, BE, and CF are concurrent, Taking ℓ at infinity, we have that D is the midpoint of $\overline{BC}$, E is the midpoint of $\overline{CA}$, and F is the midpoint of $\overline{AB}$.

Having proved the converse of the trilinear polar theorem from the existence of the centroid, let's prove the trilinear polar theorem itself. Take $D'E'$ to be the line at infinity. Then D is the midpoint of $\overline{BC}$, E is the midpoint of $\overline{CA}$. So $P = AD \cap BE$ is the centroid of ABC, and $F = CP \cap AB$ is the midpoint of $\overline{AB}$. Hence, F' is on the line at infinity. We saw in the discussion following Theorem 4.32, the following property that relates Ceva's theorem to Menelaus' theorem:

If ABC is a triangle, ℓ a line on no vertex of ABC, $L := BC \cap \ell$, $M := CA \cap \ell$, $N := AB \cap \ell$, D is on BC, E is on CA, F is on AB, and F' is the harmonic conjugate of F with respect to $\{A, B\}$, then AD, BE, CF are concurrent if and only if D, E and F', are collinear.

(For a simple proof, use that $R(N, F; A, B) = R(N, F'; A, B)R(F', F; A, B) = -R(N, F'; A, B)$, and apply the projective Menelaus and projective Ceva theorems (Theorems 4.33 and 4.34).) This yields another proof of the trilinear polar theorem.

Another proof of Theorem 6.18 Apply the property above three times. If E' is the harmonic conjugate of E with respect to $\{A, C\}$ and D' is the harmonic conjugate of D with respect to $\{B, C\}$, then

$$AD, BE, CF \text{ are concurrent} \iff D, E, F' \text{ are collinear}$$
$$\iff AD, BE', CF' \text{ are concurrent}$$
$$\iff D', E', F' \text{ are collinear.} \qquad \square$$

Therefore, Theorems 4.33 and 4.34 imply Theorem 6.18. We can also use Theorem 6.18 to show that Theorems 4.33 and 4.34 are equivalent to one another. An alternative statement of the trilinear polar theorem (Theorem 6.18) is the following:

Let ABC be a triangle, and P a point on no side of ABC. Let $D := PA \cap BC$, $E := PB \cap CA$, $F := PC \cap AB$, and let D', E', F' be the **harmonic conjugates** of D, E, F with respect to $\{B, C\}$, $\{C, A\}$, $\{A, B\}$, respectively. Then D', E', F' are collinear.

Let us look again at the cocycle identity:

$$R(X, Y; Z, T) = R(X, W; Z, T)R(W, Y; Z, T)$$

where X, Y, Z, W, T are five collinear points. Now suppose that W is the harmonic conjugate of X with respect to $\{Z, T\}$. Then $R(X, W; Z, T) = -1$ and hence $R(X, Y; Z, T) = -R(W, Y; Z, T)$. Let us now consider a triangle ABC and three points D, E, F lying on the sides BC, CA, AB of the triangle. Let P be the intersection of AD and BE, and let ℓ be the trilinear polar of P. Let $G := PC \cap AB$. Let D', E', G' be the harmonic conjugates of D, E, G with respect to $\{B, C\}$, $\{C, A\}$, $\{A, B\}$, respectively.

Then by the trilinear polar theorem (Theorem 6.18), D', E', G' are collinear. Hence, by the projective Menelaus theorem (Theorem 4.34),

$$R(B, C; D, D')R(C, A; E, E')R(A, B; G, G') = 1.$$

Now A, B, G, G', F lie on the same side of the triangle, and so we can use the above consequence of the cocycle identity: $R(G, G'; A, B) = -R(G, F; A, B)$. So, Ceva's condition appears since the sign on the right-hand side changes:

$$R(B, C; D, D')R(C, A; E, E')R(A, B; G, F) = -1.$$

Hence, CF will pass through P and the lines AD, BE, CF will be concurrent when Ceva's condition is satisfied. So we see that the trilinear polar theorem (Theorem 6.18) implies that different signs occur in projective Ceva and projective Menelaus on the right-hand sides.

Sometimes the trilinear polar theorem (Theorem 6.18) is known as the *threefold degenerate Desargues' theorem* (see Pickert, 1975). To see why, consider Desargues' configuration, where we have two triangles ABC and DEF in perspective from a point P, but with the extra conditions that D lies on BC, E lies on CA, and F lies on AB. Then, indeed, $D = PA \cap BC$, $E = PB \cap CA$, $F = PC \cap AB$ as in Theorem 6.18, with the common consequence (with Desargues' theorem) that $BC \cap EF$, $CA \cap FD$, $AB \cap DE$ are collinear.

$$* * *$$

We now come to the 'second' of the main involution theorems in projective geometry; *Desargues' involution theorem*. We remark that the 'two points' of intersection supposed here could be equal, and then the given line is tangent to the conic.

Theorem 6.19 (Desargues' involution theorem (1639) (compare with Desargues, 1951))
The three pairs of opposite sides of a complete quadrangle of $PG(2, \mathbb{R})$ *meet any line not through a vertex in three pairs of an involution, and any conic on the four vertices that meets that line in two points meets it in another pair of the same involution.*

Proof The first part is Theorem 6.12. Let P, Q, R, S be a quadrangle, ℓ be a line through no vertex, $A := PS \cap \ell$, $B := QS \cap \ell$, $C := RS \cap \ell$, $D := QR \cap \ell$,

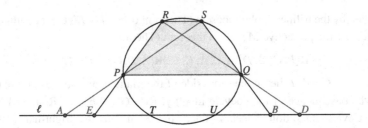

Figure 6.11 Desargues' involution theorem.

$E := PR \cap \ell$, and $F := PQ \cap \ell$ (see Figure 6.11). Let $\mathcal{K}$ be a conic on $PQRS$ meeting ℓ in T and U. Then by Chasles' theorem on cross-ratio on a conic (Theorem 5.18) and Theorem 4.7,

$$(PS, PR, PT, PU)\overline{\wedge}(QS, QR, QT, QU).$$

So, since $PS \cap \ell = A$, $PR \cap \ell = E$, $PT \cap \ell = T$, $PU \cap \ell = U$, $QS \cap \ell = B$, $QR \cap \ell = D$, $QT \cap \ell = T$, and $QU \cap \ell = U$, it follows that

$$(A, E, T, U)\overline{\wedge}(B, D, T, U)\overline{\wedge}(D, B, U, T),$$

where the second projectivity comes from Lemma 6.1. Hence, $(T\,U)$ is a pair of the involution $(A\,D)(B\,E)$ given by Theorem 6.12, as asserted.   □

The above theorem can be used to find a conic on four points and tangent to a given line.

Worked Example 6.20

Problem   Determine a conic through four points A, B, C, D, no three collinear, and tangent to a line ℓ, not on any of the four points.

Solution   By Pappus' involution theorem (Theorem 6.12), the quadrangle $ABCD$ determines an involution t on ℓ. Let E, F be the fixed points of t. By Desargues' involution theorem (Theorem 6.19), the conic on A, B, C, D, and E meets ℓ twice at E, so ℓ is tangent. Similarly, the conic on A, B, C, D, and E meets ℓ twice at F, so ℓ is tangent. Thus, there are two solutions (which may be both real or both imaginary).   □

Theorem 6.21 (Sturm, 1826–7, Theorem on p. 180)

If two conics that are inscribed in a given quadrilateral pass through a given point, their tangents at this point are conjugate lines with respect to any conic inscribed in the quadrilateral. Moreover, the lines joining the point to a pair of opposite vertices are conjugate.

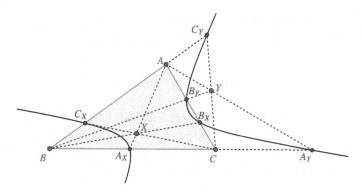

Figure 6.12 Theorem 6.22.

Proof This is simply the dual of Desargues' involution theorem (Theorem 6.19). ∎

Theorem 6.22 (Carnot, 1803, art. 236, p. 293)
Let ABC be a triangle and let X and Y be two points. Project the vertices of ABC to their opposite sides via the two points X and Y to create six new points A_X, B_X, C_X, A_Y, B_Y, C_Y (where the subscripts indicate the point of projection; see Figure 6.12). Then these six points lie on a conic.

Proof The involution induced on BC by the quadrangle AXB_XC_X by Pappus' involution theorem (Theorem 6.12) interchanges A_X and $A'_X = B_XC_X \cap BC$ and fixes B and C. The involution induced on BC by the quadrangle AYB_YC_Y by Pappus' involution theorem (Theorem 6.12) interchanges A_Y and $A'_Y = B_YC_Y \cap BC$ and fixes B and C. Moreover, these two involutions are equal (by Corollary 6.3). Call this involution t. The involution t' induced on BC by the quadrangle $B_XB_YC_XC_Y$, by Pappus' involution theorem (Theorem 6.12), interchanges B and C and A'_X and A'_Y. Now $t'tt'$ is an involution (as it is the conjugate of t by t') and it fixes B and C. Therefore, $t'tt' = t$ (again by Corollary 6.3). Hence,

$$A_X^{t'} = (A'_X)^{tt'} = (A'_X)^{t't} = (A'_Y)^t = A_Y.$$

So t' interchanges A_X and A_Y. Thus, by Desargues' involution theorem (Theorem 6.19), the conic on the five points B_X, C_X, B_Y, C_Y, and A_X also lies on A_Y. ∎

Worked Example 6.23 (ADVANCED: This is a different proof of Theorem 6.22, for the learned reader that has some knowledge of cubic curves.)
Pappus' theorem (Theorem 2.18) applied to the hexagon $AA_XB_YBB_XA_Y$ shows that $R := A_XB_Y \cap B_XA_Y$ lies on XY. Similarly, $S := B_XC_Y \cap B_YC_X$ and $T :=$

$C_X A_Y \cap C_Y A_X$ lie on XY. Consider the cubics $\mathcal{K} := (A_X B_Y)(B_X C_Y)(C_X A_Y)$ and $\mathcal{K}' = (A_Y B_X)(B_Y C_X)(C_Y A_X)$. Now the intersection divisor $\mathcal{K} \cdot (XY)$ is $R + S + T$ and

$$\mathcal{K} \cdot \mathcal{K}' = R + S + T + A_X + A_Y + B_X + B_Y + C_X + C_Y.$$

Choose a point U on XY different from R, S, T and consider the unique cubic $\mathcal{K}''$ in the pencil of cubics generated by $\mathcal{K}$ and $\mathcal{K}'$ passing through U. Since $\mathcal{K}''$ has four points on XY, it follows that it is reducible as $\mathcal{K}' = (XY)\mathcal{K}$. So $\mathcal{K}$ is the conic passing through A_X, A_Y, B_X, B_Y, C_X, and C_Y. □

The following theorem is very similar to Theorem 6.22, but without a conic in the statement. Indeed, Akopyan and Zaslavsky (2007, p. 75) remark: 'Projective properties of conics may be useful for proving results seemingly unrelated to conics.'

Theorem 6.24 (Akopyan and Zaslavsky, 2007, Theorem 3.9)
Suppose we are given a triangle ABC and points P and Q, and suppose that the lines AP, BP, and CP intersect the respective sides of the triangle at points A_1, B_1, C_1 and that the lines AQ, BQ, CQ intersect the respective sides at points A_2, B_2, C_2. Let C_3, C_4 be the intersections of the lines CC_1 and A_2B_2, CC_2 and A_1B_1, respectively; the points A_3, A_4, B_3, B_4 are defined similarly. Then the lines A_1A_4, A_2A_3, B_1B_4, B_2B_3, C_1C_4, C_2C_3 are concurrent.

Proof See Exercise 6.2. □

Theorem 6.25 (Lamé's theorem (Lamé, 1816–17, p. 233; 1818 p. 34; Poncelet, 1822, p. 213 of 1st ed., p. 206 of 2nd ed.))
The polars of a point with respect to a pencil of conics are concurrent.

Proof If two conics inscribed in a quadrilateral pass through a point, then their tangents at this point are conjugate lines with respect to any conic inscribed in this quadrilateral, by the dual of Desargues' involution theorem (Theorem 6.19). There is a unique parabola inscribed in a given quadrilateral, for five tangent lines determine a conic. There is a second conic inscribed in the quadrilateral that passes through the point at infinity on this parabola, for four tangent lines and a point determine two conics. The tangent line to this latter conic on this point at infinity is conjugate to the line at infinity for all conics inscribed in the quadrilateral, and therefore contains all of their centres. By the dual of Desargues' involution theorem (Theorem 6.19), this pair of lines harmonically separates the lines joining this point at infinity to a pair of opposite vertices of the quadrilateral, so the line of centres bisects each diagonal of the quadrilateral, and is therefore its Newton line.   □

With the terminology we have established, we can rephrase Monge's theorem on centres of similitude.

Theorem 6.26
The six centres of similitude of three circles with distinct centres and radii, taken in pairs, are the vertices of a complete quadrangle with diagonal triangle having as its vertices the centres of the three circles. Conversely, given a complete quadrangle, there exist three circles with centres the vertices of the diagonal triangle such that the six vertices of the complete quadrangle are the centres of similitude of the three circles, taken in pairs.

The Newton–Gauss theorem states that the lines joining the midpoints of opposite sides of a quadrilateral meet at a point that bisects the line segment joining the midpoints of the diagonals. This line segment is the **Newton–Gauss line**. The Newton–Gauss theorem follows from the 'converse' component of Theorem 6.26 and another theorem of Monge[5]. See also Court (1943). We also have the following, as a consequence of Lamé's theorem (Theorem 6.25).

Theorem 6.27 (Newton, *Principia*, 1999, Book I, Lemma XXV, Corollary 3)
The centre[6] of a conic inscribed in a quadrilateral lies on the Newton–Gauss line of the quadrilateral.

Proof Let $\mathcal{K}$ be inscribed in a quadrilateral $ABCD$, and let X be the centre of $\mathcal{K}$. Recall from Theorem 5.13, that the set of conics on $ABCD$ is a pencil of conics. By the dual of Lamé's theorem (Theorem 6.25), the centres of the pencil of conics lie on a common line. This line bisects each diagonal of the quadrilateral, and is therefore its Newton–Gauss line.   □

6.4 Pascal's Theorem

In 1639, Blaise Pascal (who was 16 years old) formulated the following generalisation of the theorem of Pappus from pairs of lines to non-degenerate conics.

---

[5] *Monge's theorem:* Let K_1, K_2, K_3 be three circles with distinct radii r_1, r_2, r_3 and with distinct centres. Then the three external centres of similitude of the pairs of circles are collinear.

[6] The **centre** of a conic is a point that bisects every chord passing through it. Alternatively, it is the pole of the line at infinity, with respect to the conic.

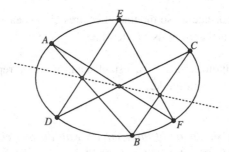

Figure 6.13 Pascal's theorem (Theorem 6.28).

It was published in 1640, but with no proof. The work of Simson (1735, Proposition XLVII) in 1735 is the first publication of Pascal's theorem in a book – and he was apparently unaware of Pascal's work.

Theorem 6.28 (Pascal's theorem (Pascal, 1639))
Let $ABCDEF$ be a hexagon in $\mathrm{PG}(2,\mathbb{R})$ inscribed in a conic $\mathcal{K}$. Then the points of intersection of opposite sides are collinear.

Proof Let $M = BC \cap EF, N = CD \cap AF, L = MN \cap AB, L' = MN \cap DE$ (see Figure 6.13). The theorem says that $L = L'$. Let $MN \cap \mathcal{K} = \{P, P'\}$ and $R = MN \cap CF$. By Desargues' involution theorem (Theorem 6.19) applied to $CDEF$ and the line MN, we have an involution $(PP')(MN)(RL)$. By Desargues' Involution Theorem (6.19) applied to $ABCF$ and the line MN, we have an involution $(PP')(MN)(RL')$. Since these two involutions agree on P and P' and on M and N, by Corollary 6.3 they are equal, and so $L = L'$. $\qquad\square$

The line containing the points of intersection of opposite sides is called the **Pascal line** of the hexagon. We give a second proof of Theorem 6.28.

Another proof of Pascal's theorem (Theorem 6.28) Denote the points of intersection of opposite sides by $X := BC \cap EF$, $Y := CD \cap AF$, $Z := AB \cap DE$. Let $T := CD \cap EF$. Then

$$\begin{aligned}
\mathrm{R}(T, X; E, F) &= \mathrm{R}(CT, CX; CE, CF) \\
&= \mathrm{R}(CD, CB; CE, CF) \\
&= \mathrm{R}(AD, AB; AE, AF) &&\text{(by Theorem 5.18)} \\
&= \mathrm{R}(AD, AZ; AE, A(AF \cap ED)) \\
&= \mathrm{R}(D, Z; E, AF \cap ED) &&\text{(intersection with } DE\text{)}.
\end{aligned}$$

By Corollary 4.10, $DT = CD$, ZX, $(AF \cap ED)F = AF$ are concurrent. So $Y = CD \cap AF$ is incident with ZX. $\qquad\square$

Worked Example 6.29

Before presenting the next beautiful proof of Pascal's theorem, we need to break to discuss how we can envisage a degenerate conic in a pencil of conics generated by one degenerate conic and one non-degenerate conic. Let $\mathcal{K}$ be a non-degenerate conic with equation $s = 0$, where s is a degree 2 polynomial in x, y, z. Let A, B, C, D be four points of $\mathcal{K}$. Then the line pair $(AD)(BC)$ is a degenerate conic. Moreover, if $\alpha = 0$ is a degree 1 equation defining AD (up to a scalar) and $\beta = 0$ defines BC (up to a scalar), then the points of $(AD)(BC)$ are described as the zeros of the equation $\alpha\beta = 0$.

What about the degenerate conic $(AB)(CD)$? It also contains our four points, and so the equation of this degenerate conic should be a combination of the two equations we have already; it lies in the pencil generated by $s = 0$ and $\alpha\beta = 0$. So we can write the equation defining $(AB)(CD)$ as

$$\lambda s + \mu \alpha \beta = 0 \qquad (6.1)$$

where λ and μ are both non-zero. Indeed, since dividing through by λ does not change the set of zeros of this equation, and $-\mu\beta/\lambda = 0$ defines the same line as $\beta = 0$, we can simplify Equation (6.1):

$$s = \alpha\beta.$$

So there is a nice symmetry here in the equations that if $\alpha\beta = 0$ defines the degenerate conic $(AD)(BC)$, then $s = \alpha\beta$ gives us the other degenerate conic in the pencil (namely, $(AB)(CD)$).   □

Alan Robson (1953) gave a spectacularly beautiful proof of Pascal's theorem (Theorem 6.28).

Robson's proof of Pascal's theorem Let $ABCDEF$ be a hexagon on the conic $s = 0$, and suppose AD has equation $\alpha = 0$, BC has equation $\beta = 0$, EF has equation $\gamma = 0$. Then the line pair $(AB)(CD)$ has equation $s = \alpha\beta$ (see Worked Example 6.29), and the line pair $(AF)(ED)$ has equation $s = \alpha\gamma$. Let $X := AB \cap DE$, $Z := BC \cap EF$, $Y := CD \cap FA$. The points X and Y lie on both line pairs, but not on $\alpha = 0$, and so they lie on $\beta = \gamma$, which lies on Z.   □

Given a non-trivial element g of the stabiliser of a non-degenerate conic $\mathcal{K}$, the line joining the fixed points of g is the Pascal line of the hexagon

$$AB^g CA^g BC^g.$$

This line is secant if the fixed points are real and distinct, tangent if there is a repeated fixed point, and external if the fixed points are imaginary.

Theorem 6.30 (Steiner, 1828b)
*Let A, B, C, D, E, F be points on a non-degenerate conic. Then the Pascal
lines of the hexagons ABCDEF, ADCFEB, and AFCBED are concurrent (in
a **Steiner point**).*

Proof Let $L := AB \cap DE$, $M := AB \cap CF$, $N := DE \cap CF$. Now define
the points $L' := CD \cap AF$, $M' := CD \cap BE$, $N' := AF \cap BE$, so that LL'
is the Pascal line of the hexagon $ABCDEF$, MM' is the Pascal line of the
hexagon $ADCFEB$, and NN' is the Pascal line of the hexagon $AFCBED$. Then
$LM \cap L'M' = AB \cap CD$, $LN \cap L'N' = DE \cap AF$, and $MN \cap M'N' = CF \cap BE$. By
Pascal's theorem (Theorem 6.28) applied to the hexagon $AFCDEB$, these three
points are collinear in a line ℓ. By the dual of Desargues' theorem (Theorem
2.20), since LMN and $L'M'N'$ are in perspective from ℓ, it follows that LL',
MM', and NN' are concurrent.   □

Theorem 6.31 (Hopkins, 1950)
*If the diagonals X_1X_2, Y_1Y_2, Z_1Z_2 of a hexagon $X_1Y_2Z_1X_2Y_1Z_2$ are concurrent
in a point I, and the sides X_1Y_2, Y_1X_2 meet in Z, Z_1X_2 and X_1Z_2 in Y, Y_1Z_2
and Z_1Y_2 in X, forming the hexagons $XY_1ZX_1YZ_1$ and $XY_2ZX_2YZ_2$, then the
diagonals XX_1, YY_1, ZZ_1 of the second hexagon are concurrent in a point J,
the diagonals XX_2, YY_2, ZZ_2 of the third are concurrent in a point K, and I, J,
K are collinear.*

Proof (Primrose, 1953) Since the diagonals of $X_1Y_2Z_1X_2Y_1Z_2$ are concurrent, there is a conic inscribed in the hexagon. The hexagons $XY_1ZX_1YZ_1$ and
$XY_2ZX_2YZ_2$ each consist of the same six tangents in a different order and so the
diagonals of each are concurrent. The theorem that the points of concurrence
are collinear is the dual of part of Steiner's theorem (Theorem 6.30).   □

Worked Example 6.32 (A property of six points that lie on a conic)
The following appears on Dao's blog,[7] Problem 10.

Problem Let A, B, C, D, E, F lie on a conic. Let $H = AC \cap BF$, $I = EC \cap BD$,
$K = FD \cap AE$, $O = KI \cap CF$, $N = HI \cap AD$, $P = HK \cap BE$. Show that O, N, P
are collinear.

Solution Consider the hexagon $AECFDB$ on the conic. Let $K := FD \cap AE$,
$I := EC \cap BD$, $O' := FC \cap AB$. So Pascal's theorem (Theorem 6.28) implies
that K, I, O' are collinear. Thus, $O' = KI \cap CF = O$, and we see that O, A, B
are collinear.

---

[7] http://oaithanhdao.blogspot.com

Consider the hexagon $ACDFBE$ on the conic. Let $H := AC \cap BF$, $K := FD \cap AE$, $P' := BE \cap DC$. So Pascal's theorem (Theorem 6.28) implies that H, K, P' are collinear. Thus, $P' = HK \cap BE = P$, and we see that E, F, P are collinear.

Consider the hexagon $ACEFBD$ on the conic. Let $H := AC \cap BF$, $I := EC \cap BD$, $N' := AD \cap EF$. So Pascal's theorem (Theorem 6.28) implies that H, I, N' are collinear. Thus, $N' = HI \cap AD = N$, and we see that C, D, N are collinear.

Consider the hexagon $ABEFCD$ on the conic. Let $O := AB \cap CF$, $N := EF \cap AD$, $P := CD \cap BE$. So Pascal's theorem (Theorem 6.28) implies that N, O, P are collinear. □

A **homography** is a cross-ratio preserving permutation of the points of $PG(2, \mathbb{R})$.

Theorem 6.33

Let ϕ be a homography leaving a conic $\mathcal{K}$ invariant. If A, B, C are distinct points on $\mathcal{K}$ and D, E, F are distinct points on $\mathcal{K}$ then the Pascal lines of the hexagons $AB^\phi CA^\phi BC^\phi$ and $DE^\phi FD^\phi EF^\phi$ are equal. Moreover, the fixed points of ϕ are the intersection of $\mathcal{K}$ with this line.

Proof Let ℓ be the Pascal line of the hexagon $AB^\phi CA^\phi BC^\phi$ and let R, S be the (possibly imaginary) points of ℓ on $\mathcal{K}$. Since ϕ extends to a homography of the conic over the complexes, we may suppose that R and S are real. Let $A' := BC^\phi \cap B^\phi C$, $B' := AC^\phi \cap C^\phi A$, $C' := BA^\phi \cap B^\phi A$, $X := RS \cap AA^\phi$. Then we have two perspectively related pencils (on A^ϕ and A):

$$(A^\phi R, A^\phi A, A^\phi B, A^\phi C, A^\phi S) \overset{\ell}{\overline{\wedge}} (AR, AA^\phi, AB^\phi, AC^\phi, AS).$$

Now

$$\begin{aligned}
R(R, A^\phi; B^\phi, C^\phi) &= R(AR, AA^\phi; AB^\phi, AC^\phi) \\
&= R(A^\phi R, A^\phi A; A^\phi B, A^\phi C) \\
&= R(R, A; B, C) \\
&= R(R^\phi, A^\phi; B^\phi, C^\phi)
\end{aligned}$$

and so $R^\phi = R$ (by Lemma 4.3). Likewise, $S^\phi = S$. So ϕ fixes the points of $\mathcal{K}$ on the Pascal line of $AB^\phi CA^\phi BC^\phi$. Similarly, ϕ fixes the points of $\mathcal{K}$ on the Pascal line of $DE^\phi FD^\phi EF^\phi$. Since ϕ has at most two real (or imaginary) fixed points, this latter line must also be ℓ. □

Theorem 6.34

Let ϕ be a homography preserving a conic $\mathcal{K}$. If A, B, C are distinct points on $\mathcal{K}$ such that AA^ϕ, BB^ϕ, CC^ϕ are concurrent, then ϕ is an involution.

Proof There is an involution τ on $\mathcal{K}$ interchanging A and A^ϕ, B and B^ϕ. By Frégier's theorem (Theorem 5.4), this involution interchanges the points of $\mathcal{K}$ on any line on $AA^\phi \cap BB^\phi$ meeting $\mathcal{K}$, and so it interchanges C and C^ϕ. So $\phi\tau$ fixes A, B, and C, giving $\phi\tau = 1$ (by Theorem 5.21). Therefore, $\phi = \tau$ is an involution.   □

Lemma 6.35

If two involutions have a common pair (X, Y), then the mates of two of the points also form an involution with X, Y; that is, if $(AC)(BD)(XY)\ldots$ is an involution, as is $(AC')(BD')(XY)\ldots$, then $(CD')(C'D)(XY)\ldots$ is an involution.

Proof Observe:

$$
\begin{aligned}
R(C, D; X, Y) &= R(A, B; Y, X), && \text{from the first involution,} \\
&= R(C', D', X, Y), && \text{from the second involution,} \\
&= R(D', C'; Y, X),
\end{aligned}
$$

which proves the lemma since it shows that there is a projectivity taking C to D' and D to C' and interchanging X and Y, which, since it interchanges a pair of points, is an involution (by Theorem 6.2).   □

This lemma applies both to the projective line and to a non-degenerate conic, using cross-ratio on the conic in the proof. Moreover, the lemma does not require that $A \neq C$ or $B \neq D$. We also need the extension of the lemma to the case $X = Y$ (below).

Lemma 6.36

If two involutions have a common fixed point X, then the mates of two of the points also form an involution with fixed point X ; that is, if $(AC)(BD)(X)\ldots$ is an involution, as is $(AC')(BD')(X)\ldots$, then $(CD')(C'D)(X)\ldots$ is an involution.

Proof This can be proved by a calculation in $\mathsf{AGL}(1, \mathbb{R})$. The involutions are maps of the form $x \mapsto -x + c$. Now we write $C = -A + c$, $D = -B + c$, $C' = -A+d$, $D' = -B+d$. So $C-C' = c-d = D-D'$, and, thus, $D'+C = D+C' = e$, say, and $x \mapsto -x + e$ interchanges C and D', and C' and D.   □

The above lemmas lead to a short proof of Pascal's theorem.

Another proof of Pascal's theorem (Theorem 6.28) Let P_1P_2, P_4P_5 meet at L; P_2P_3, P_5P_6 at M; P_3P_4, P_6P_1 at N, where $P_1, \ldots, P_6$ are points on the conic. Let LM meet the conic at E and F, possibly imaginary and possibly equal. Then

$$(P_2\,P_1)(P_5\,P_4)(E\,F)\ldots$$

is an involution since their joins concur at L, as is $(P_2\,P_3)(P_5\,P_6)(E\,F)$ since their joins concur at M. Hence, by Lemmas 6.35 and 6.36, so is

$$(P_1\,P_6)(P_4\,P_3)(E\,F)\ldots,$$

that is, N lies on EF. (We have repeatedly applied Frégier's theorem (Theorem 5.4).) □

Replacing P_1P_2 by the tangent at P_1 proves the five-point Pascal theorem (Theorem 6.66) with $P_1 = P_2$. Similarly, other replacements give the other degenerations of Pascal. For $P_1P_1P_3P_3P_5P_6$, replace P_1P_2 by the tangent at P_1 and P_3P_4 by the tangent at P_3. For $P_1P_1P_3P_4P_4P_6$, replace P_1P_2 by the tangent at P_1 and P_4P_5 by the tangent at P_4. These prove both variants of the four-point Pascal theorem. Finally, for $P_1P_1P_3P_3P_5P_5$, replace P_1P_2 by the tangent at P_1, P_3P_4 by the tangent at P_3, and P_5P_6 by the tangent at P_5.

* * *

We continue with interesting results that are related in some way to Pascal's theorem (Theorem 6.28). The following result is closely related to the theorems of Pappus (Theorem 2.18) and Pascal (Theorem 6.28); however, as Semple and Kneebone (1952, p. 90) observed: 'The Cross Axis Theorem is a key theorem in plane projective geometry, for it enables us actually to construct, by means of the straight-edge, the homographic correspondence between two given lines that is determined by three assigned pairs.'

Theorem 6.37 (Cross-axis theorem (Chasles, 1852, Section 111, pp. 68–9))
Let ϕ be a projectivity between distinct lines ℓ and m. Then

$$\{P^\phi Q \cap Q^\phi P \colon P, Q \in \ell, P \neq Q\}$$

*is a line, called the **cross-axis** of the projectivity.*

Proof Fix three points $X, Y, Z \in \ell$. Consider the hexagram $XY^\phi ZX^\phi YZ^\phi$. The diagonal points of this hexagram are $M := XY^\phi \cap YX^\phi$, $N := YZ^\phi \cap ZY^\phi$, and $O := ZX^\phi \cap XZ^\phi$, which lie on a common line p by Pappus' theorem (Theorem 2.18).

Now consider a point P of ℓ not in $\{X, Y, Z\}$. Let $G := XX^\phi \cap p$, let $F := PX^\phi \cap p$, and let $X' := XF \cap m$. Then,

$$(X, Y, Z, P) \overset{X^\phi}{\barwedge} (G, M, O, F) \overset{X}{\barwedge} (X^\phi, Y^\phi, Z^\phi, X').$$

Now a projectivity is defined by the images of three distinct points, and so $X' = P^\phi$. In other words, $XP^\phi \cap PX^\phi$ lies on p, for all points P of ℓ not equal to X. Therefore, for any two points P and Q of ℓ, the point $P^\phi Q \cap Q^\phi P$ lies on p, and hence $\{P^\phi Q \cap Q^\phi P : P, Q \in \mathcal{K}, P \neq Q\}$ is just the line p. □

Theorem 6.38

Let ℓ, m be lines meeting in O, and let P be a point on neither ℓ nor m. Then the cross-axis n of the perspectivity with centre P from ℓ to m has $(\ell, m; PO, n)$ harmonic.

Proof Let $X \neq Y$ on ℓ, $X' = PX \cap m$, $Y' = PY \cap m$. The result follows from applying the dual of Pappus' involution theorem (Theorem 6.12) to XX', XY', YX', YY', and O. □

Theorem 6.39

Let A, B, C, D, P be five points, no three collinear. Let ℓ be the cross-axis of the perspectivity with centre P from AB to CD, let m be the cross-axis of the perspectivity with centre P from AC to BD, and let n be the cross-axis of the perspectivity with centre P from AD to BC. Then ℓ, m, n are concurrent.

Proof Let $P' := \ell \cap m$. Let $X := AB \cap CD$, $Y := AC \cap BD$, $Z := AD \cap BC$. Notice that X is on ℓ, Y is on m, Z is on n. So by Theorem 6.38, $(AB, CD; PX, \ell)$ is harmonic, as is $(AC, BD; PY, m)$. Thus, $(AB \cap PP' \ CD \cap PP')(P)(P')$ is an involution, as is $(AC \cap PP' \ BD \cap PP')(P)(P')$; necessarily the same involution. Now apply Pappus' involution theorem (Theorem 6.12) to $ABCD$ and PP'. Again

$$(AB \cap PP' \ CD \cap PP')(AC \cap PP' \ BD \cap PP')(AD \cap PP' \ BC \cap PP')$$

is the same involution, so fixes P and P'; thus, $(AD \cap PP', BC \cap PP'; P, P')$ is harmonic, so $(AD, BC; PZ, P'Z)$ is harmonic. Hence, $P'Z = n$ (by Lemma 4.3), and so ℓ, m, and n are concurrent (in P'). □

The point of currency in Theorem 6.39 is called the **conjugate** of P with respect to the quadrangle $ABCD$. The terminology is that of Baker (1943, pp. 55–56).

Theorem 6.40

Let A, B, C, D, P be five points, no three collinear, and let $\mathcal{K}$ be a conic on A, B, C, and D. Then P and its conjugate P′ with respect to ABCD are also conjugate with respect to $\mathcal{K}$.

Proof This follows from Desargues' involution theorem (Theorem 6.19). □

The dual of Lamé's theorem (Theorem 6.25) states that 'the poles of a line with respect to a pencil of conics are collinear'. Notice that this result also defines P' (by applying Theorem 6.40).

The following theorem is due to Gergonne (1826–27), and in the same issue there is a result of Sturm (Gergonne, 1826–27, p. 176) that to the authors' knowledge seems to be a special case of Theorem 6.41. It also appeared in a treatise of Castelnuovo (1904, p. 391, exercise 29) and in the first volume of the classic text by Veblen and Young (1965, p. 140, exercise 29, crediting Castelnuovo), before being rediscovered by Evelyn et al. (1974, Section 2.2)[8].

Theorem 6.41 (Three-conics theorem (Gergonne, 1826–27))

Let S_1, S_2, S_3 be (possibly degenerate) conics of $\mathrm{PG}(2, \mathbb{R})$ such that $S_1 \cap S_2 \cap S_3 = \{I, J\}$ and $S_2 \cap S_3 = \{I, J, P_1, Q_1\}$, $S_1 \cap S_3 = \{I, J, P_2, Q_2\}$, $S_1 \cap S_2 = \{I, J, P_3, Q_3\}$, where the points $I, J, P_1, Q_1, P_2, Q_2, P_3, Q_3$ are distinct. Then $P_1 Q_1, P_2 Q_2, P_3 Q_3$ are concurrent.

Proof Write the equations for S_1, IJ, $P_2 Q_2$ and $P_3 Q_3$ accordingly: $S_1: \Sigma_1 = 0, IJ: \Lambda_1 = 0, P_2 Q_2: \Lambda_2 = 0, P_3 Q_3: \Lambda_3 = 0$. Now S_2 lies in the pencil of conics determined by S_1 and the degenerate conic given by the pair of concurrent lines IJ and $P_3 Q_3$. So S_2 has an equation of the form $\Sigma_1 + \lambda \Lambda_1 \Lambda_3 = 0$ for some fixed constant λ. Similarly, S_3 has an equation of the form $\Sigma_1 + \mu \Lambda_1 \Lambda_2 = 0$ for some constant μ. The equation

$$(\Sigma_1 + \lambda \Lambda_1 \Lambda_3) - (\Sigma_1 + \mu \Lambda_1 \Lambda_2) = \Lambda_1 (\lambda \Lambda_3 - \mu \Lambda_2) = 0$$

yields a conic $\mathcal{K}$ of the pencil determined by Σ_2 and Σ_3; a conic passing through I, J, P_1, and Q_1. Since $\mathcal{K}$ consists of two lines, namely $\Lambda_1 = 0$ (i.e., IJ) and $\lambda \Lambda_3 - \mu \Lambda_2 = 0$, it follows that the second line is $P_1 Q_1$. So the equation for $P_1 Q_1$ is a linear combination of the equations for $P_2 Q_2$ and $P_3 Q_3$, and, hence, $P_1 Q_1$ belongs to the pencil of lines determined by $P_2 Q_2$ and $P_3 Q_3$. The result then follows. □

---

[8] The three-conics theorem (Theorem 6.41) also appears in Graustein (1930, ch. XVI (p. 296), theorem 2), Horadam (1970, ch. 6, exercise 6), and Rosenbaum and Rosenbaum (1949).

It should be remarked that if all three conics are degenerate, then this is essentially Pappus' theorem (Theorem 2.18), while if two are degenerate it is essentially Pascal's theorem (Theorem 6.28).

Theorem 6.42 (Takasu, 1931)
Suppose A, B, C, A', B', C', P are on a conic $\mathcal{K}$ and that $PA' \cap BC$, $PB' \cap CA$, $PC' \cap AB$ are collinear in a line ℓ. Then AA', BB', CC' are concurrent in a point of ℓ.

Proof We apply the three-conics theorem (Theorem 6.41). Note that a conic can be degenerate, for example the product of two lines. The conics $\ell(AA')$, $\ell(BB')$, $\mathcal{K}$ share the two points of $\ell \cap \mathcal{K}$ (over $\mathbb{C}$ if needs be), so ℓ, AA', BB' are concurrent. The conics $\ell(AA')$, $\ell(CC')$, $\mathcal{K}$ share the two points of $\ell \cap \mathcal{K}$ (over $\mathbb{C}$ if needs be), so ℓ, AA', CC' are concurrent.   □

The following theorem, the converse to Pascal's theorem (Theorem 6.28), was the subject of a dispute between Braikenridge and Maclaurin. See Mills (1984). It was published by Maclaurin in 1735 (who claimed to have been teaching the theorem since 1725) and by Braikenridge in 1733 (who claimed to have communicated it to Maclaurin in 1727).

Theorem 6.43 (Braikenridge–Maclaurin theorem)
Let ABCDEF be a hexagon in $\mathrm{PG}(2, \mathbb{R})$ such that the points of intersection of opposite sides are collinear. Then ABCDEF is inscribed in a (possibly degenerate) conic.

Proof Let $M = BC \cap EF, N = CD \cap AF, L = DE \cap AB$. Then L, M, N are collinear. Take the (possibly degenerate) conic $\mathcal{K}$ on $ACDEF$. Let $P := AB \cap CD$, $P' := AB \cap EF$, $Q := AB \cap CE$, $Q' := AB \cap DF$, $R := AB \cap CF$, $R' := AB \cap DE$, and $AB \cap \mathcal{K} = \{A, B'\}$. If $\mathcal{K}$ is non-degenerate, then by Desargues' involution theorem (Theorem 6.19),

$$(P\,P')(Q\,Q')(R\,R')(A\,B')$$

is an involution. By Pascal's theorem (Theorem 6.28), $M' := B'C \cap EF, N := CD \cap AF, L' := DE \cap AB'$ are collinear. But B' lies on AB, and so $AB = AB'$; and hence $L = L'$. Thus, $M' = EF \cap AB = M$, giving $B' = CM \cap AL = B$. Thus, $ABCDEF$ is inscribed in $\mathcal{K}$. If $\mathcal{K}$ is degenerate, say A, C, E are collinear, then B, D, M are collinear by Pappus' theorem (Theorem 2.18) applied to the hexagon $AMCLNE$. So $\mathcal{K}$ is the pair of lines AC and DF, and both lie on B.   □

It seems that the Braikenridge–Maclaurin theorem (Theorem 6.43) was known for the case of an ellipse by Jean-Charles della Faille (1597–1652). It was in an unpublished manuscript of his (*Tratado de las secciones cónicas*) without proof. Of course, in the case of an ellipse, there are no points at infinity, and so no parallel lines appearing in the statement. (See Meskens, 2021.)

Theorem 6.44 (Brianchon, 1806)
If the sides of a hexagon are tangent to a conic, then the diagonals are concurrent.

Proof This follows from Pascal's theorem (Theorem 6.28) and the Principle of Duality. It is necessary to note that the tangent lines to a conic form a conic in the dual plane (a fact that follows from the polarity defined by the conic): that is, the line coordinates of the tangent lines to a conic satisfy a non-degenerate homogeneous quadratic equation. □

The following is the 'heptagonal' version of Theorem 6.43.

Theorem 6.45 (Steiner, 1898, p. 130)
Suppose $A_1A_2A_3A_4A_5A_6A_7$ circumscribes a conic. Then the heptagon with consecutive sides

$$(A_1A_4)(A_2A_5)(A_3A_6)(A_4A_7)(A_5A_1)(A_6A_2)(A_7A_3)$$

is inscribed in a conic.

Proof Let $B_1 := A_1A_4 \cap A_2A_5$, $B_2 := A_2A_5 \cap A_3A_6$, $B_3 := A_3A_6 \cap A_4A_7$, $B_4 := A_4A_7 \cap A_5A_1$, $B_5 := A_5A_1 \cap A_6A_2$, $B_6 := A_6A_2 \cap A_7A_3$, $B_7 := A_7A_3 \cap A_1A_4$. By Brianchon's theorem (Theorem 6.44) applied to $A_2A_3A_4A_5A_6(A_6A_7 \cap A_1A_2)$, we have that $A_3A_6 \cap A_2A_5$ is on the line joining A_4 and $A_6A_7 \cap A_1A_2$. Now $A_7A_1A_4$ is in perspective with $A_6A_2(A_2A_5 \cap A_3A_6)$ (from $A_6A_7 \cap A_1A_2$), and so by Desargues' theorem (Theorem 2.20), $A_7A_1 \cap A_6A_2$, $B_1 = A_1A_4 \cap A_2A_5$, $B_3 = A_4A_7 \cap A_6A_3$ are collinear. Note that $A_6A_2 = B_5B_6$, and so $A_7A_1 \cap B_5B_6$, B_1, B_3 are collinear. Hence, A_7, A_1, $B_5B_6 \cap B_1B_3$ are collinear. Now $B_3B_4 \cap B_6B_7 = A_7$ and $B_4B_5 \cap B_7B_1 = A_1$. So $B_4B_5 \cap B_7B_1$, $B_3B_4 \cap B_6B_7$, and $B_5B_6 \cap B_1B_3$ are collinear. By the Braikenridge–Maclaurin theorem (Theorem 6.43), it follows that B_3, B_4, B_5, B_6, B_7, B_1 lie on a conic. Similarly, B_4, B_5, B_6, B_7, B_1, B_2 lie on a conic. Since these two conics share the five points B_4, B_5, B_6, B_7, B_1, they are identical. □

Theorem 6.46 (Wilkinson, 1872, p. 72, proved on p. 88)

Let ABCDEFGH be an octagon inscribed in a conic; then either

(i) *the points* $P := AB \cap DE$, $Q := DE \cap GH$, $R := GH \cap BC$, *and* $S := BC \cap EF$, $T := EF \cap AH$, $U := AH \cap CD$, $V := CD \cap FG$, $W := FG \cap AB$ *lie on a conic, or*

(ii) *P, R, T, V are collinear and Q, S, U, W are collinear.*

Proof Let $N = BC \cap EH$. By Pascal's theorem (Theorem 6.28) applied to the hexagon $ABCDEH$, N, U, and P are collinear. By the Braikenridge–Maclaurin theorem (Theorem 6.43) applied to the hexagon $PQRSTU$, since E, H, and N are collinear, either there is a conic containing the hexagon $PQRSTU$, or both P, R, T and Q, S, U are collinear. Similarly, either $QRSTUV$ is on a conic, or both Q, S, U and R, T, V are collinear, and either $RSTUVW$ is on a conic or both R, T, V and S, U, V are collinear. Hence, since the conic through Q, R, S, T, and U is unique, either P, Q, R, S, T, U, V, W lie on a conic or P, R, T, V are collinear and Q, S, U, W are collinear. □

Evans and Rigby (2002) consider extensions of this result, for example when the points $ABCDEFGH$ are not necessarily distinct. Suppose $(A, C, E, G) = (B, D, F, H)$. Then the conic in Theorem 6.46 is degenerate when $R(A, C; E, G) = -R(A, G; C, E)$. To see this, let $x := R(A, C; E, G)$. Then $R(A, G; C, E) = (x - 1)/x$, so that $R(A, C; E, G) = -R(A, G; C, E)$ if and only if $x = (1 - x)/x$. We can rewrite this equation as $x^2 + x - 1 = 0$; so x is the golden ratio or its algebraic conjugate. Thus, by Evans and Rigby (2002, Theorem 3.1), P, R, T, and V are collinear. Conversely, by Evans and Rigby (2002, Theorem 3.1), if P, R, T, and V are collinear then $x^2 + x - 1 = 0$. We conjecture that the more general case might be true: the conic of Theorem 6.46 is degenerate if and only if $R(A, C; E, G) = -R(B, H; D, F)$.

THOMAS TURNER WILKINSON was born on 17 March 1815 at Abbot House, Mellor, near Blackburn. He purchased his first mathematical periodical, the *Lady's Diary*, in 1835, and 'the contents so interested me that I began to collect all the English periodicals which contained mathematics (Abram, 1876, p. 88).' He sent his first contribution to the *Lady's Diary* in 1837, and was awarded the mathematical prize in 1852. He contributed many solutions and mathematical essays to that journal (editor W. S. B. Woolhouse). Around the late 1830s, he discovered the York Courant (editors Thomas Tate, then William Tomlinson), and he and several of his pupils contributed to its mathematical and philosophical section. He wrote a series of 29 articles on the English mathematical periodicals in the *Mechanics' Magazine* during 1848, which extended through various volumes of that work up to 1854. These papers were much valued, and led to him being elected a Fellow of the Royal Astronomical Society in December 1850. He died on 6 February 1875.

Another way to state the Braikenridge–Maclaurin theorem (Theorem 6.43) is:

Six distinct points $A_1, A_2, A_3, A_4, A_5, A_6$ lie on a conic or have A_1, A_3, A_5 collinear and A_2, A_4, A_6 collinear if and only if $A_1A_2 \cap A_4A_5$, $A_2A_3 \cap A_5A_6$, $A_3A_4 \cap A_6A_1$ are collinear.

We present a similar result, but in the context of involutions on a conic.

Theorem 6.47
Six distinct points $A_1, A_2, A_3, A_4, A_5, A_6$, no three collinear, lie on a conic and have $(A_1 A_4)(A_2 A_5)(A_3 A_6)$ in involution if and only if $A_1A_2 \cap A_4A_5$, $A_2A_3 \cap A_5A_6$, $A_3A_4 \cap A_6A_1$, $A_1A_3 \cap A_4A_6$ are collinear (in ℓ).

Proof Let h be the projectivity of the conic on $A_1, A_2, A_3, A_4, A_5, A_6$ taking A_1 to A_4, A_3 to A_6, and A_5 to A_2. Then ℓ is the cross-axis of h. Now

$$(A_1 A_4)(A_2 A_5)(A_3 A_6) \ldots$$

is in involution if and only if h is an involution. Suppose $A_1A_3 \cap A_4A_6$ lies on ℓ. Suppose $X^h = A_3$. Then $A_1X^h \cap A_1^h X$ is on ℓ, so $A_1A_3 \cap A_4X$ is on ℓ. So $X = A_6$. Thus, $A_6^h = A_3$, and h is an involution. (Conversely, if h is an involution, then $A_1A_3 \cap A_4A_6 = A_1A_6^h \cap A_1^hA_6$ is on ℓ.) $\square$

In the *Educational Times* in 1897, Augustus De Morgan posed an open problem concerning an octagon inscribed in a conic (see De Morgan, 1867, p. xiv, unsolved problems 2555):

The following is a theorem of which an elementary proof is desired. It was known before I gave it in a totally different form in a communication (April, 1867) to the Mathematical Society on the conic octagram; and the present form is as distinct from the other two as they are from one another. If I, II, III, IV be the consecutive chord-lines of one tetragon inscribed in a conic, and 1, 2, 3, 4 of another; the eight points of intersection of I with 2 and 4, II with 1 and 3, III with 2 and 4, IV with 1 and 3, lie in one conic section. A proof is especially asked for when the first conic is a pair of straight lines. There is, of course, another set of eight points in another conic, when the pairs 13, 24 are interchanged in the enunciation.

The question was repeated a number of times over the years, even after De Morgan's death. For example, two solutions appeared in volume LI in 1889:

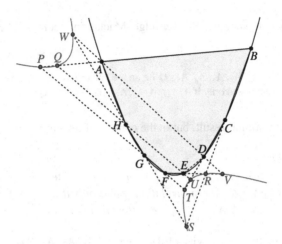

Figure 6.14 De Morgan's theorem.

the first (p. 70), by W. S. Foster, an analytic proof over a full page using a lot of elimination, and the second (pp. 154–7), anonymous, using residuation and so not an elementary proof. The proof below seems to be the first synthetic proof.

Theorem 6.48 (De Morgan, 1867, unsolved problems 2555)
Let ABCDEFGH be an octagon inscribed in a conic; then $P := AB \cap FG$, $Q := AB \cap HE$, $R := BC \cap EF$, $S := BC \cap GH$, $T := CD \cap FG$, $U := CD \cap HE$, $V := DA \cap EF$, $W := DA \cap GH$ lie on a (possibly degenerate) conic.

Proof The reader may want to keep Figure 6.14 in mind for the following argument. By Pascal's theorem (Theorem 6.28) applied to the hexagon *ABCFGH*, *P*, *S*, and $N := AH \cap CF$ are collinear. By Pascal's theorem (Theorem 6.28) applied to the hexagon *ADCFEH*, *V*, *U*, and *N* are collinear. By the Braikenridge–Maclaurin theorem (Theorem 6.43) applied to the hexagon *PQUVWS*, since *A*, *H*, and $N = PS \cap UV$ are collinear, there is a (possibly degenerate) conic $\mathcal{K}_1$ containing *PQUVWS*. (If $\mathcal{K}_1$ is degenerate, *P*, *U*, *W* are collinear and *Q*, *V*, *S* are collinear.)

By Pascal's theorem (Theorem 6.28) applied to the hexagon *BCDGHE*, *S*, *U*, and $M := DG \cap BE$ are collinear. By Pascal's theorem (Theorem 6.28) applied to the hexagon *BADGFE*, *P*, *V*, and *M* are collinear. By the Braikenridge–Maclaurin theorem (Theorem 6.43) applied to the hexagon *RSUQPV*, since *B*, *E*, and *M* are collinear, there is a (possibly degenerate) conic $\mathcal{K}_2$ containing *RSUQPV*. (If $\mathcal{K}_2$ is degenerate, *R*, *U*, *P* are collinear and *Q*, *V*, *S* are collinear.) By the Braikenridge–Maclaurin theorem (Theorem 6.43) applied to the hexagon *PTUVRS*, since *F*, *C*, and *N* are collinear, there is a (possibly degenerate) conic $\mathcal{K}_3$ containing $PTUVRS$. (If $\mathcal{K}_3$ is degenerate, R, U, P are collinear and T, V, S are collinear.)

If R, U, P are not collinear, then $\mathcal{K}_2$ and $\mathcal{K}_3$ are non-degenerate, and since $\mathcal{K}_1 \cap \mathcal{K}_2$ contains $\{P, Q, R.S, U\}$, $\mathcal{K}_1$ is non-degenerate and $\mathcal{K}_2 \cap \mathcal{K}_3$ contains $\{P, R, S, U, V\}$, it follows that $\mathcal{K}_1 = \mathcal{K}_2 = \mathcal{K}_3$ is a conic on P, Q, R, S, T, U, V, W. If R, U, P are collinear, then $\mathcal{K}_2$ and $\mathcal{K}_3$ are degenerate, and since $\mathcal{K}_1 \cap \mathcal{K}_2$ contains $\{P, Q, R.S, U\}$, $\mathcal{K}_1$ is degenerate, and $\mathcal{K}_2 \cap \mathcal{K}_3$ contains $\{P, R, S, U, V\}$, it follows that $\mathcal{K}_1 = \mathcal{K}_2 = \mathcal{K}_3$ is a conic on P, Q, R, S, T, U, V, W and P, R, U, W are collinear and Q, S, T, V are collinear. □

There's a similar theorem due to von Staudt (the dual theorem).

Theorem 6.49 (von Staudt, 1847, p. 293)
If two complete quadrangles have the same diagonal points, their eight vertices lie on a possibly degenerate conic.

Proof Suppose $AB \cap CD = A'B' \cap C'D' = G$, $CA \cap BC = C'A' \cap B'C' = F$, and $BC \cap AD = B'C' \cap A'D' = E$. If three of the vertices, say A, B, A', are collinear, then since G is on both AB and $A'B'$, it follows that B' is also on AB, and since $(GE, GF; AB, CD)$ and $(GE, GF; A'B', C'D')$ are harmonic and $AB = A'B'$, it follows that $CD = C'D'$, and the eight points lie on the two lines AB and CD.

If, on the other hand, no three of the vertices are collinear, take the non-degenerate conic $\mathcal{K}$ on A, B, C, D, A'. Since E, F, G are the diagonal points of the inscribed quadrangle $ABCD$, it follows that G is the pole of EF, and therefore G and $EF \cap A'B'$ are harmonic conjugates with respect to the points $\mathcal{K} \cap A'B'$. One of the points of $\mathcal{K} \cap A'B'$ is A' therefore; the other is B', since A' and B' are harmonically conjugate with respect to G and $EF \cap A'B'$, as E, F, G are the diagonal points of $A'B'C'D'$. Thus, B' lies on $\mathcal{K}$. Similarly, C' and D' lie on $\mathcal{K}$. □

The following is a nice supplement to De Morgan's theorem (Theorem 6.48).

Theorem 6.50
Let $ABCDEFGH$ be an octagon inscribed in a conic $\mathcal{K}$; then $P := AB \cap FG$, $Q := AB \cap HE$, $R := BC \cap EF$, $S := BC \cap GH$, $T := CD \cap FG$, $U := CD \cap HE$, $V := DA \cap EF$, $W := DA \cap GH$ lie on a conic (by Theorem 6.48). Then the conic on P, Q, R, S, T, U, V, W is degenerate if and only if $R_\mathcal{K}(A, F; C, H) = R_\mathcal{K}(G, B; E, D)$.

Proof By Adler's lemma (Corollary 5.26), P, R, and U are collinear if and only if $R_{\mathcal{K}}(A, F; C, H) = R_{\mathcal{K}}(G, B; E, D)$. Now this condition is equivalent to, $R_{\mathcal{K}}(F, C; H, A) = R_{\mathcal{K}}(B, E; D, G)$, which, by Adler's lemma, holds if and only if R, U, and W are collinear. Our proof of De Morgan's theorem (Theorem 6.48) shows that the conic on P, Q, R, S, T, U, V, W is degenerate if and only if P, R, U, and W are collinear.   □

Theorem 6.51 (Möbius, 1827, section/art. 278)
If two triangles are in perspective, the points of intersection of the sides of one with the non-corresponding sides of the other lie on a conic, and the lines joining the vertices of one to the non-corresponding vertices of the other touch another conic.

Proof Let the triangles be ABC, $A'B'C'$, with V, A, A' collinear, V, B, B' collinear, and V, C, C' collinear. Let $L := AB \cap B'C'$, $M := AB \cap C'A'$, $N := AC \cap B'C'$, $O := AC \cap A'B'$, $P := BC \cap A'B'$, $Q := BC \cap C'A'$. By Desargues' theorem (Theorem 2.20), $AB \cap A'B'$, $CA \cap C'A'$, and $BC \cap B'C'$ are collinear. But $AB = LM$, $A'B' = OP$, $CA = NO$, $C'A' = MQ$, $BC = PQ$, and $B'C' = LN$. So $LM \cap OP$, $NO \cap MQ$, and $PQ \cap LN$ are collinear. Thus, by the Braikenridge–Maclaurin theorem (Theorem 6.43) applied to the hexagon $LMQPON$, it follows that L, M, N, O, P, Q lie on a conic. The dual argument finishes the proof.   □

Worked Example 6.52
Problem Let six points be given on the sides of triangle ABC in $\mathrm{PG}(2, R)$: A_1 and A_2 on BC, B_1 and B_2 on CA, and C_1 and C_2 on AB (see Figure 6.15). Let K denote the intersection of A_1B_2 and C_1A_2, L the intersection of B_1C_2 and A_1B_2, and M the intersection of C_1A_2 and B_1C_2. Let T, U, and V be the intersections $A_1B_2 \cap B_1A_2$, $B_1C_2 \cap B_2C_1$, and $C_1A_2 \cap C_2A_1$, respectively. Prove that lines AK, BL, and CM are concurrent if and only if points T, U, and V are collinear.

Solution Applying Desargues' theorem (Theorem 2.20) to ABC and KLM, we have: AK, BL, CM are concurrent if and only if $AB \cap KL$, $BC \cap LM$, $CA \cap MK$ are collinear. Now $\{AB \cap KL, BC \cap LM, CA \cap MK\} = \{C_1C_2 \cap A_1B_2, A_1A_2 \cap B_1C_2, B_1B_2 \cap C_1A_2\}$, so by Pascal's theorem (Theorem 6.28), Pappus' theorem (Theorem 2.18) and the Braikenridge–Maclaurin theorem (Theorem 6.43) these points are collinear if and only if C_1, C_2, A_1, A_2, B_1, B_2 lie on a (possibly degenerate) conic. Similarly, $\{T, U, V\} = \{A_1B_2 \cap B_1A_2, B_1C_2 \cap B_2C_1, C_1A_2 \cap C_2A_1\}$, so again by Pascal's theorem (Theorem 6.28), Pappus' theorem (Theorem 2.18) and the Braikenridge–Maclaurin theorem (Theorem 6.43), these

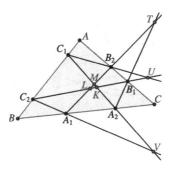

Figure 6.15 Worked Example 6.52.

are collinear if and only if A_1, B_2, B_1, C_2, C_1, A_2 lie on a (possibly degenerate) conic. Thus, AK, BL, CM are concurrent if and only if T, U, V are collinear. □

The upcoming theorem (Theorem 6.53), which we may think of as a projective Simson line theorem, has been rediscovered on several occasions in various different forms (see Quadling, 2012). Johann Lambert ([1761] 2009) proved that if three straight lines touch a parabola, a circle described through their intersections shall pass through the focus of the parabola. We then have Wallace's 1799 Simson line theorem, which was misattributed to Simson by Servois in 1813 (Servois, 1813/14), and repeated by Poncelet in 1822 (for the oblique version, see Poncelet, 1822). Gergonne generalised the Simson line theorem, and provides a converse (Gergonne and Querret, 1824–5, p. 88). This was then followed by Steiner's oblique and affine Simson line theorems (Steiner, 1828–9), Chasles' proof by involution in 1847 (Chasles, 2010), and McKenzie's converse to the oblique Simson line theorem (McKenzie, 1881).

What we call the projective Simson line theorem was posed as a problem in May 1864 in the *Educational Times* as question 1501 by F. D. Thomson and solved by the Reverend R. Townsend the same year and published in October 1864. (Actually, what was proved was the direction showing that the points are collinear.) Thomson was motivated by the following theorem: If a triangle has its sides tangent to a parabola then its orthocentre lies on the directrix.

Aubert (1889, p. 531) rediscovers this theorem, which was again rediscovered by Clawson (1919). Variants on the Simson line theorem were

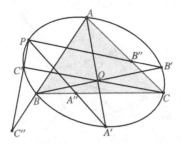

Figure 6.16 The projective Simson line theorem.

given by Turner (1925) (for a pair of inverse points with respect to the circumcircle) and by Seimiya (1926; see Totten, 2006, p. 129) (for two points on the circumcircle). In 1931, Takasu proves the converse of Aubert's theorem (see Theorem 6.42). The story continues, and in recent times, the same theme has prevailed. Riesinger (2004) writes on the projective oblique Simson theorem and generalisations, Oai Thanh Dao (2013) rediscovers the Aubert–Takasu theorem; in fact, Tran Hoang Son (2014) publishes a proof of the Aubert–Takasu theorem, which is attributed to Dao. Oai Thanh Dao (2014) discovers a generalisation of the Aubert–Takasu theorem, which we pursue in Section 10.2. Geoff Smith (2015) publishes a proof of his rediscovery of the Aubert–Takasu theorem, and Giang Ngoc Nguyen (2015) publishes a proof of Dao's generalisation. Tran Minh Ngoc (2018) gives a synthetic proof of Dao's generalisation. It also appears as Starr (1961, p. 246, exercise 14).

Theorem 6.53 (Projective Simson line theorem (Townsend, 1864; Aubert, 1889, p. 531; Smith, 2015; Takasu, 1931))
Let ABC be a triangle and let P and Q be other points in the plane of this triangle. Let K be a conic that passes through A, B that and C. Let AQ meet K, again at A′, and let PA′ meet BC at A″. Points B′, B″, C′, and C″ are similarly defined. Then A″, B″, C″, Q are collinear if and only if P lies on K.

Proof When the six points A, A', P, C', C, and B lie on the conic K, by Pascal's theorem (Theorem 6.28), AA' meets CC' at Q, $A'P$ meets BC at A'', and finally $C'P$ meets AB at C'', so A'', C'', and Q are collinear (see Figure 6.16). Similarly, $B'P$ meets AC at B'', $C'P$ meets AB at C'', and $C'C$ meets $B'B$ at Q, so, under the same hypothesis, by Pascal's theorem (Theorem 6.28), B'', C'', and Q are collinear. By the Braikenridge–Maclaurin theorem (Theorem 6.43), if A'', C'', Q are collinear, then A, A', P, C', C, and B lie on the conic K. □

We now give a converse to the oblique Simson line theorem (Theorem 6.17). (Note that the angles ∠*PLC*, ∠*QMA*, ∠*RNB* are congruent, so that this is the converse of Theorem 6.17.)

Theorem 6.54 (McKenzie, 1881)
The three sides BC, CA, AB of a triangle are cut by a straight line in L, M, N; and lines drawn through A, B, and C, parallel to LMN, cut the circumscribing circle of the triangle ABC in P, Q, and R. Then the lines PL, QM, RN all cut the circle ABC in the same point.

Proof This is Takasu's converse (Theorem 6.42) of the projective Simson line theorem (Theorem 6.53) for a circle with the point of concurrency at infinity. □

6.5 The Chasles–Steiner Theorem

Coxeter (1949, para. 5.61) attributes the following theorem to Chasles (Chasles, 1865, p. 98, art. 135). It also appears in (von Staudt, 1847, p. 136, art. 243) and was known to Ceva in 1678 for an ellipse (Ceva, 1678).

Theorem 6.55 (Chasles, 1828–9, p. 75; Reye, 1886, pp. 220–48)
If two distinct triangles are mutually conjugate with respect to a conic, then they are in perspective from a point.

Proof Let PQR be a triangle. Let the polars of P, Q, R be p, q, r, respectively. Let $P_1 := QR \cap p$, $Q_1 := RP \cap q$, $R_1 := PQ \cap r$. Let r_1 be the line on $p \cap q$ and R. This is the polar of R_1. Let $P' := PQ \cap q$, $R' := QR \cap q$, and p' be the line on $p \cap q$ and R; this is the polar of P'. By application of the the polarity arising from the conic,

$$(R_1, P, P', Q) \,\overline{\wedge}\, (P, R_1, Q, P') \,\overline{\wedge}\, (p, r_1, q, p') \overset{p \cap q}{\overline{\overline{\wedge}}} (P_1, R, R', Q).$$

By Theorem 4.9, $(R_1, P, P') \,\overline{\overline{\wedge}}\, (P_1, R, R')$. The centre $PR \cap P'R' = Q$ of the perspectivity must lie on the line $R_1 P_1$. Hence, P_1, Q_1, R_1 are collinear: the two triangles are in perspective from a line. So, by the dual of Desargues' theorem (Theorem 2.20), the two triangles are in perspective from a point. □

The converse of Theorem 6.55 is due to von Staudt (1847, p. 135, art. 241), and we leave its proof as an exercise (Exercise 6.6):

If two triangles are in perspective, there is a non-degenerate conic for which each is the polar triangle of the other.

Corollary 6.56

If three points are on a non-degenerate conic, then the triangle they form and the triangle formed by the tangents to the conic at those points are in perspective.

The above result (Corollary 6.56) appeared in Chasles' *Apercu* in 1837 (Chasles, 1837), but it first appeared in *A Letter from Mr. Colin Mac Laurin*, Math. Prof. Edinburg. F. R. S. to Mr. John Machin, Astr. Prof. Gresh. & Secr. R. S. concerning the *Description of Curved Lines* (communicated to the Royal Society on 21 December 1732; see also *Philosophical Transactions of the Royal Society* 39(1735):143–65). So it is due to Colin Maclaurin. The special case where one of the vertices of the triangle is at infinity is in Newton's *Principia*, Book I, Lemma 20 (see Newton, 1999). Three more remarks:

1. In Newton's *Principia*, Book I, Section 5, Newton solves the problem of determining a conic when n points and $5 - n$ tangents are given. When $n = 3$ and the two tangents are on two of the points, he must be perilously close to proving this theorem. Indeed, his organic construction of a conic is for the $n = 3$ case, with one of the points at infinity, and he then projects.
2. With the conic a circle, the point of concurrency is the *Gergonne point* (see Theorem 6.58) of the triangle with sides the three tangents.
3. With the conic a circle, the point of concurrency is the Lemoine point of the triangle with vertices on the circle (Mathieu, 1865).

Corollary 6.57 (Bosse, 1672; de La Hire, 1672, 1673; Blondel, 1673)
A conic is determined by three points on it and two tangent lines at these points.

Proof Let A, B, C be three points, let ℓ be the tangent at A, and let m be the tangent at B. Let $D = \ell \cap m$. We may assume that $A(1, 0, 0)$, $B(0, 1, 0)$, $C(0, 0, 1)$, $D(1, 1, 1)$, since $\{A, B, C, D\}$ is a quadrangle. Now a conic on A, B, C has an equation of the form $dxy + exz + fyz = 0$, and this is tangent to AD at A if and only if $(d + e)xy + fy^2 = 0$, which implies $y = 0$; that is, if and only if $d = -e$. Similarly, we have tangency to B at B if and only if $(d + f)xy + ex^2 = 0$, and this implies $x = 0$; that is, if and only if $d = -f$. So the conic is $-xy + xz + yz = 0$.   □

The dual of Corollary 6.57 is that a conic is determined by three tangent lines to it and two points on it and these lines. Abraham Bosse (1672) asked for

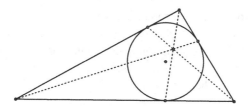

Figure 6.17 The Gergonne point.

a proof, and Philippe de La Hire (1672, 1673) and Nicolas-François Blondel (1673) gave proofs.

Theorem 6.58 (Gergonne point)
*The lines joining each vertex of a triangle to the point of contact of the incircle with the opposite side are concurrent, in a point called the **Gergonne point** (see Figure 6.17).*

Proof The original triangle and the triangle with vertices the points of contact of the incircles of the sides are conjugate, and hence, by Theorem 6.55, in perspective from a point.   □

The above theorem was first published by Giovanni Ceva in 1678 (Ceva, 1678, p. 38).

Theorem 6.59 (Chasles, 1828b; Steiner, 1832; Chasles, 1837, p. 335, n. XV, also p. 340)
Given two pencils of lines of $\mathrm{PG}(2, \mathbb{R})$ *that are in projective but not perspective correspondence, the intersections of the corresponding lines form a conic on the carrier points of both pencils.*

Proof Let P, Q be points and suppose that ϕ: pencil(P) → pencil(Q) is a projectivity that is not a perspectivity. Then the line $(PQ)^\phi \neq PQ$, by Corollary 4.9. Let $\ell := (PQ)^{\phi^{-1}}$ and $m := (PQ)^\phi$. By Theorem 6.57, there is a conic $\mathcal{K}$ tangent to ℓ at P, tangent to m at Q, and passing through $n \cap n^\phi$, for some line n on P with $n \neq \ell, PQ$. By Theorem 5.18, $\mathcal{K}$ defines a projectivity pencil(P) → pencil(Q) that agrees with ϕ at ℓ, n, PQ; thus, this projectivity equals ϕ and $\mathcal{K}$ is the set of intersections of the corresponding lines. Finally, $\ell \cap \ell^\phi = P$ and $PQ \cap (PQ)^\phi = Q$.   □

Recall from von Staudt's theorem that if ρ is a polarity of $\mathsf{PG}(2,\mathbb{R})$ with absolute points, then the set of absolute points of ρ is a conic of $\mathsf{PG}(2,\mathbb{R})$. The following result shows that 'von Staudt conics' are also 'Chasles–Steiner conics', and it bypasses our 'analytic' definition of a conic, and instead uses Seydewitz's theorem (Theorem 4.40).

> **Chasles–Steiner's definition of conic**: the set of intersections of two pencils of lines that are projectively, but not perspectively, related.
> **van Staudt's definition of conic**: the set of absolute points of a polarity (when it has them).

We can recast Seydewitz's theorem in terms of 'von Staudt conics' and the polarity of the conic. That is, if we have absolute points P, Q, R, then any line conjugate to one side of the triangle PQR meets the other two sides in conjugate points. This treatment was taken from Coxeter (1993, §6.5) (which we guess was inspired by Veblen and Young, 1965, §100).

Theorem 6.60
The set of absolute points of a polarity, which has absolute points, is a conic in the sense of Chasles and Steiner (compare with Theorem 6.59).

Proof Let ρ be a polarity having absolute points O, and fix two absolute points P and Q. These will be the points upon which we will base our two pencils. Let p and q be the images of P and Q under ρ, and let $D := p \cap q$. So D is the image of PQ under ρ. Let c be a fixed line through D, but not through P or Q. Let $x \in \text{pencil}(P)$ and $y \in \text{pencil}(Q)$ such that $x \cap y$ is an absolute point R. Define $B := x \cap c$ and $A := y \cap c$. By Seydewitz's theorem (Theorem 4.40), $\{A, B\}$ is a pair of the involution of conjugate points of c (see also Lemma 4.18). Hence, the map $x \mapsto B \mapsto A \mapsto y$ is a projectivity (see Exercise 4.1). $\qquad\square$

In the proof of the following fundamental result – a converse to Chasles' theorem (Theorem 5.18) – we will repeatedly use the dual of Corollary 4.10: Let a, b, c be lines of $\text{pencil}(P)$ and let a', b', c' be lines of $\text{pencil}(P')$. Then $\text{R}(PP',a;b,c) = \text{R}(PP',a';b',c')$ if and only if $a \cap a', b \cap b', c \cap c'$ are collinear (see Figure 6.18).

Theorem 6.61 (Chasles–Steiner theorem (Chasles, 1828a, 1829))
Let A_1, A_2, A_3, A_4 be a quadrangle of $\mathsf{PG}(2,\mathbb{R})$, X, X' be points. Then

$$\text{R}(XA_1, XA_2; XA_3, XA_4) = \text{R}(X'A_1, X'A_2; X'A_3, X'A_4)$$

if and only if $A_1, A_2, A_3, A_4, X, X'$ lie on a conic.

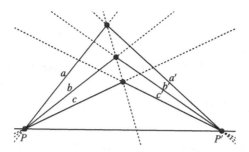

Figure 6.18 The converse of Corollary 4.10.

Proof Suppose $R(XA_1, XA_2; XA_3, XA_4) = R(X'A_1, X'A_2; X'A_3, X'A_4)$. Then, by the dual of Theorem 4.7,

$$(XA_1, XA_2, XA_3, XA_4) \overline{\wedge} (X'A_1, X'A_2, X'A_3, X'A_4).$$

Now (XA_1, XA_2, XA_3, XA_4) and $(X'A_1, X'A_2, X'A_3, X'A_4)$ are not in perspective, as then, by the dual of Corollary 4.10, A_1, A_2, A_3, A_4 would be collinear. So, by Theorem 6.59, $A_1, A_2, A_3, A_4, X, X'$ lie on a conic. Conversely, suppose $A_1, A_2, A_3, A_4, X, X'$ lie on a conic. Then, by Theorem 5.18,

$$R(XA_1, XA_2; XA_3, XA_4) = R(X'A_1, X'A_2; X'A_3, X'A_4). \qquad \square$$

We remark that the 'only if' direction in Theorem 6.61, that is, if A_1, A_2, A_3, A_4, X, X' lie on a conic, then

$$R(XA_1, XA_2; XA_3, XA_4) = R(X'A_1, X'A_2; X'A_3, X'A_4),$$

appears in Pascal's *Essay pour les coniques* (Pascal, 1639) for the unsigned cross-ratio. We can also prove the reverse direction of Theorem 6.61 from the forward direction.

Proof: forward direction $\implies$ *reverse direction, of Theorem 6.61* Suppose

$$R(X'A_1, X'A_2; X'A_3, X'A_4) = R(XA_1, XA_2; XA_3, XA_4),$$

let the conic on X, A_1, A_2, A_3, A_4 be $\mathcal{K}$, and suppose X' is not on $\mathcal{K}$. Let $X'A_1 \cap \mathcal{K} = \{A_1, P\}$. Then by the forward direction of Theorem 6.61,

$$R(PA_1, PA_2; PA_3, PA_4) = R(XA_1, XA_2; XA_3, XA_4)$$
$$= R(X'A_1, X'A_2; X'A_3, X'A_4).$$

By the dual of Corollary 4.10, since $PA_1 = XA_1$, it follows that A_2, A_3, A_4 are collinear, contrary to them being distinct points on the conic $\mathcal{K}$. $\qquad \square$

That four fixed tangents to a conic meet any fifth in four points with a fixed cross-ratio was known to Brianchon (1817, p. 28, Section XXVIII), and it also appears in Poncelet (1822, p. 115). It's unclear to the authors if Brianchon has the full theorem or a special case; however, Poncelet has the full statement of the theorem.

We also remark that there is a classic result that has some striking similarities with the Chasles–Steiner theorem (Theorem 6.61) that can be found in the ancient work of Pappus of Alexandria (1986, Book VII, Proposition 36):

Given four lines in a plane, and given four angles, take a variable point C. Consider now the distances from C to the various given lines, where the distances are measured along lines making the given angles with the given lines. A further condition on C is that the four distances $d(C, D)$, $d(C, F)$, $d(C, B)$, and $d(C, H)$ satisfy

$$\frac{d(C, D) \cdot d(C, F)}{d(C, B) \cdot d(C, H)} = \epsilon$$

where ϵ is a fixed constant. Then C traces a conic.

This thread is also explored in the chapter on 'The three-line and four-line locus' in Heath's book (de Perga and Heath, 2013), and we have this quote from Kurt Vogel (n.d.): 'As for [the] content[] [of Aristaeus' Five Books of the Elements of Conic Sections], it can be determined from the passages by Pappus and Apollonius that the "locus with respect to ... four lines" was treated by Aristaeus.'

In the following, the line XX denotes the tangent at X. This notation will also be used later in this section when we consider degenerate versions of Pascal's theorem (Theorem 6.28).

Corollary 6.62
Let $\mathcal{K}$ be a non-degenerate conic on the distinct points P and Q. Define the map g: pencil(P) $\to$ pencil(Q) by $(PX)^g := QX$. Then g is a projectivity (that is not a perspectivity).

Corollary 6.63 (A generalisation of Newton, *Principia*, 1999, Book I, Section 5, Corollary 1 to Lemma XXV (compare with Newton, 1999))
Let $\mathcal{K}$ be a non-degenerate conic, T, U be distinct points of $\mathcal{K}$, t be the tangent to $\mathcal{K}$ at T, and u be the tangent to $\mathcal{K}$ at U. Define h: $t \to u$ by mapping the intersection P of the tangent to a variable point of $\mathcal{K}$ with t to its intersection P^h with u. Then h is a projectivity (that is not a perspectivity).

Newton's statement is closer to: given two tangents to a fixed conic, the product of the intercepts upon them between the diameter parallel to their chord of contact and any third tangent is constant. He uses a fixed parallelogram circumscribing the conic and a variable tangent.

Worked Example 6.64 (A problem of Oai Thanh Dao[9])
Problem Let two conics pass through four common points A, B, C, D. The tangent lines t_B, t_D to the first conic K_1 at B, D meet the second conic K_2 at G, H, respectively. Show that AC, BD, GH are concurrent.

Solution We repeatedly apply the Chasles–Steiner theorem (Theorem 6.61):

$$R_{K_2}(A, G; C, D) = R(BA, BG; BC, BD) = R(BA, t_B; BC, BD)$$
$$= R_{K_1}(A, B; C, D) = R(DA, DB; DC, t_D)$$
$$= R(DA, DB; DC, DH) = R_{K_2}(A, B; C, H)$$
$$= R_{K_2}(C, H; A, B).$$

So there is a projectivity g of K_2 with $A^g = C$, $G^g = H$, $C^g = A$, $D^g = B$. Since g interchanges A and C, g is an involution. By Frégier's theorem (Theorem 5.5), AC, BD, GH are concurrent. □

We may extend the definition of affine ratio to points that need not be distinct, by defining $\mathsf{ratio}(A, B, B) = 1$, $\mathsf{ratio}(A, B, A) = 0$, and $\mathsf{ratio}(A, A, C) = \infty$ ($\mathsf{ratio}(A, A, A)$ is left undefined). This gives an extension of cross-ratio to points that need not be distinct, by the relationship

$$R(A, B; C, D) = \mathsf{ratio}(C, A, B)/\mathsf{ratio}(D, A, B).$$

Thus,

$$R(A, A; C, D) = 1, \qquad R(A, B; B, D) = 0, \qquad R(A, B; A, D) = \infty,$$
$$R(A, B; C, C) = 1, \qquad R(A, B; C, A) = 0, \qquad R(A, B; C, B) = \infty.$$

Now consider a quadrangle $ABCD$ and the pencil of conics on that quadrangle. There are three degenerate conics in that pencil:

$$(AD)(BC), \quad (AB)(CD), \quad (AC)(BD).$$

For any point E not on one of these three degenerate conics, there is a unique non-degenerate conic $\mathcal{K}$ in the pencil, and, by the Chasles–Steiner theorem (Theorem 6.61), for every point X of $\mathcal{K} \setminus \{A, B, C, D\}$, $R(XA, XB; XC, XD)$ is a constant. With the extension of the values of cross-ratio above,

---

[9] http://oaithanhdao.blogspot.com/2014/11/81-two-conic-problem.html

- for every point X of $(AD)(BC) \setminus \{A, B, C, D\}$, we have

$$\mathrm{R}(XA, XB; XC, XD) = 0;$$

- for every point X of $(AB)(CD) \setminus \{A, B, C, D\}$, we have

$$\mathrm{R}(XA, XB; XC, XD) = 1;$$

- for every point X of $(AC)(BD) \setminus \{A, B, C, D\}$, we have

$$\mathrm{R}(XA, XB; XC, XD) = \infty.$$

Thus, if AB has equation $F = 0$, CD has equation $G = 0$, AC has equation $H = 0$, and BD has equation $K = 0$, then the pencil consists of $HK = 0$ and $FG + \mu HK = 0$, $\mu \in \mathbb{R}$, and the cross-ratio of the conic $FG + \mu HK = 0$ is μ.

Now that we have the Chasles–Steiner theorem (Theorem 6.61), we can give another proof of one direction of Apollonius' theorem (Theorem 4.38).

Another proof of (one direction of) Theorem 4.38 Let P be a point outside a non-degenerate conic $\mathcal{K}$; the two tangents drawn from P to the circle meet $\mathcal{K}$ at R and S. A line through P meets $\mathcal{K}$ at A and B; let the common point of AB and RS be Q. Note that RS is the polar line of P (with respect to $\mathcal{K}$) and so Q is conjugate to P. So by the Chasles–Steiner theorem (Theorem 6.61),

$$\begin{aligned}
\mathrm{R}(A, B; P, Q) &= \mathrm{R}(RA, RB; RP, RQ) \\
&= \mathrm{R}(A, B; R, S) \\
&= \mathrm{R}(SA, SB; SR, SP) \\
&= \mathrm{R}(A, B; Q, P).
\end{aligned}$$

and hence $\mathrm{R}(A, B; P, Q) = \mathrm{R}(A, B; P, Q)^{-1}$. So $\mathrm{R}(A, B; P, Q)^2 = -1$ and hence $\mathrm{R}(A, B; P, Q) = -1$ (as cross-ratio is never equal to 1). $\qquad\square$

Below we give another proof of Pascal's theorem (Theorem 6.28).

Another proof of Pascal's theorem (Theorem 6.28). By the Chasles–Steiner theorem (Theorem 6.59),

$$\mathrm{R}(AD, AB; AF, AE) = \mathrm{R}(CD, CB; CF, CE).$$

Let $S := AF \cap ED$ and $T := CD \cap EF$. Intersecting AD, AB, AF, AE with ED gives D, P, S, E, respectively. Intersecting CD, CB, CF, CE with EF gives T, Q, F, E, respectively. Thus, $\mathrm{R}(D, P; S, E) = \mathrm{R}(T, Q; F, E)$. Hence,

$$\mathrm{R}(RD, RP; RA, RE) = \mathrm{R}(RC, RQ; RF, RE)$$

because $RS = RA$ and $RT = RC$. Now $RD = RC$, $RA = RF$, so there are three lines in common. Since their cross-ratios are equal, the lines RP and RQ are identical (by Lemma 4.3), as required. □

The following proof is from Kaplansky (2003, p. 118).

Another proof of Pascal's theorem (Theorem 6.28). Given six points A, B, C, D, E, F on a non-degenerate conic, let $X := EC \cap BF$, $Y := DC \cap AF$, $Z := DB \cap AE$. We will show that X, Y, and Z are collinear. Let $X' := ZY \cap BF$, $T := DC \cap BF$, $S := DB \cap AF$. Then

$$
\begin{aligned}
R(B,T;X,F) &= R(CB,CD;CE,CF) \\
&= R(AB,AD,AE,AF) \quad &\text{(by Chasles-Steiner (Theorem 6.61))} \\
&= R(B,D;Z,S) \quad &\text{(intersecting with DB)} \\
&= R(YB,YD;YZ,YS) \\
&= R(B,T;X',F) \quad &\text{(intersecting with BR).}
\end{aligned}
$$

So $X = X'$ (by Lemma 4.3). □

We now extend the Chasles–Steiner theorem so that two of the five points can merge so that the line between them can be interpreted as a tangent line. The reader could use the notation AA for the tangent at A, and then Theorem 6.65 would have the elegant formulation '$R(EA, EB; EC, ED) = R(AA, AB; AC, AD)$'.

Theorem 6.65 (Extension to the Chasles-Steiner theorem)
If A, B, C, D, E are on a conic, then $R(EA, EB; EC, ED) = R(t_A, AB; AC, AD)$, where t_A is the tangent to the conic at A.

Proof The cross-ratios $R(EF, EB; EC, ED) = R(AF, AB; AC, AD)$ for F on the conic, distinct from A, B, C, D, E, and these take all but two values, the omitted values being

$$
\{R(EA, EB; EC, ED), R(t_E, EB; EC, ED)\} =
$$
$$
\{R(AE, AB; AC, AD), R(t_A, AB; AC, AD)\},
$$

where t_E is the tangent to the conic at E. Since $R(A, B; C, D) \neq R(E, B; C, D)$ and $R(EA, EB; EC, ED) \neq R(AE, AB; AC, AD)$, we have

$$
R(EA, EB; EC, ED) = R(t_A, AB; AC, AD). \qquad \square
$$

Theorem 6.66 (Five-point Pascal theorem)
Given five points A, C, D, E, F on a non-degenerate conic, the points $CD \cap FA$, $EF \cap AC$, and the intersection of DE and the tangent t_A to the conic at A are collinear.

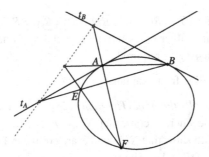

Figure 6.19 The four-point Pascal theorem.

Proof Set $X := CD \cap FE$, $Y := DE \cap FA$, $Z := AC \cap EF$, and $T := DE \cap t_A$. Then

$$R(Z, X; E, F) = R(CA, CD; CE, CF)$$
$$= R(t_A, AD; AE, AF)$$
$$= R(T, D; E, Y)$$

and so ZT, XD, and FY are concurrent (in $CD \cap t_A$). (The second equality is by Theorem 6.65, and the conclusion uses Corollary 4.10.)   □

The above theorem first appeared in 1707 (l'Hôpital, 1707, Corollaire IV, p. 140, para. 208) in the following attractive affine form: if a pentagon inscribed in a conic has two pairs of parallel sides, then the remaining side is parallel to the tangent at the remaining vertex. It also appeared in exactly this form in unpublished work of Newton from c. 1668.

Theorem 6.67 (Four-point Pascal theorem (Newton, 1968, p. 191))
Consider four points A, B, E, F in an irreducible conic, and let t_A and t_B be the tangent lines at A and B (respectively) to the conic. Then the points $t_B \cap FA$, $EF \cap AB$, and $t_A \cap BE$ are collinear (see Figure 6.19).

Proof Set $X := t_B \cap FE$, $Y := BE \cap FA$, $Z := AB \cap EF$, and $T := BE \cap t_A$. Then $R(Z, X; E, F) = R(BA, t_B; BE, BF)$ by joining Z, X, E, F to the point B. If we apply Theorem 6.65 via A and B of our conic, we see that

$$R(t_A, AB; AE, AF) = R(BA, t_B; BE, BF).$$

Furthermore, $R(t_A, AB; AE, AF) = R(T, B; E, Y)$ once we consider the projection from the point A to the line BE. Hence, $R(Z, X; E, F) = R(T, B; E, Y)$ and so ZT, XB, and FY are concurrent (in $t_B \cap FA$), by Corollary 4.10 (and projecting from the point E).   □

Here is another proof of the Chasles–Steiner theorem that uses the four-point Pascal theorem (Theorem 6.67) many times!

Second proof of Theorem 6.61 Let t_X be the tangent to $\mathcal{K}$ at X, $t_{X'}$ be the tangent to $\mathcal{K}$ at X', and $Y := t_X \cap t'_X$. Let $A_{12} := A_1X' \cap A_2X$, $A_{13} = A_1X' \cap A_3X$, $A_{14} = A_1X' \cap A_4X$, $A_{21} := A_2X' \cap A_1X$, $A_{31} := A_3X' \cap A_1X$, $A_{41} := A_4X' \cap A_1X$. By Theorem 6.67 applied to the hexagon $XXA_2X'X'A_1$, we have Y, A_{12}, A_{21} are collinear. By Theorem 6.67 applied to the hexagon $XXA_3X'X'A_1$, we have Y, A_{13}, A_{31} are collinear. Again, by Theorem 6.67 applied to the hexagon $XXA_4X'X'A_1$, we have Y, A_{14}, A_{41} are collinear. Thus, the perspectivity with centre X taking A_1X' to A_1X maps A_{12} to A_{21}, A_{13} to A_{31}, A_{14} to A_{41}, and fixes A_1. Hence, $R(XA_1, XA_2; XA_3, XA_4) = R(X'A_1, X'A_2; X'A_3, X'A_4)$.

Conversely, with all the hypotheses except A_4 on $\mathcal{K}$, if

$$R(XA_1, XA_2; XA_3, XA_4) = R(X'A_1, X'A_2; X'A_3, X'A_4),$$

then $R(A_1, A_{12}; A_{13}, A_{14}) = R(A_1, A_{21}; A_{31}, A_{41})$. Hence, $A_{12}A_{21}$, $A_{13}A_{31}$, $A_{14}A_{41}$ are concurrent, by Corollary 4.10. By Theorem 6.67 applied to the hexagon $XXA_2X'X'A_1$, we have Y, A_{12}, A_{21} are collinear. By Theorem 6.67 applied to the hexagon $XXA_3X'X'A_1$, we have Y, A_{13}, A'_{13} are collinear. So Y, A_{14}, and A_{41} are collinear. By Theorem 6.67 applied to the hexagon $XXA_4X'X'A_1$, and the conic C' on X, X', A_1, A_4 and tangent at X to t'_X, we have $t_{X'} = YX'$ is tangent to C'. So $\mathcal{K}' = \mathcal{K}$; that is, A_4 lies on $\mathcal{K}$. □

Worked Example 6.68

Here is how we can derive 'another' Four-Point Pascal Theorem: Consider four points A, C, D, F in an irreducible conic, and let t_A and t_D be the tangent lines at A and D (respectively) to the conic. Then the points $t_A \cap t_D$, $AC \cap DF$, $CD \cap FA$ are collinear (see Figure 6.20).

Proof The polar ℓ of $O := AD \cap CF$ contains the pole $t_A \cap t_D$ of AD. By Desargues' involution theorem (Theorem 6.19) the line m joining O to $N := AC \cap DF$ meeting the conic in P and Q has $(P, Q; O, N)$ harmonic. Therefore, N lies on ℓ, by Theorem 4.38. Similarly, $CD \cap FA$ lies on ℓ. □

And here is another solution using properties of the cross-ratio.

Second solution

$$R(AF \cap CD, t_A \cap CD; D, C) = R(AF, t_A; AD, AC) \qquad \text{(by joining to } A)$$
$$= R(DF, AD; t_D, DC) \qquad \text{(by Theorem 6.61)}$$
$$= R(AC \cap DF, A; AC \cap t_D, C) \qquad \text{(intersecting with } AC),$$

so $(AF \cap CD)(AC \cap DF)$, t_A, and t_D are concurrent. □

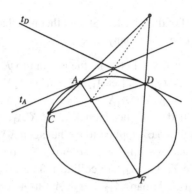

Figure 6.20 Another four-point Pascal theorem.

An elegant way to think of the four-point Pascal theorem (Theorem 6.67) and the five-point Pascal theorem (Theorem 6.66) is to interpret them as degenerate versions of Pascal's theorem (Theorem 6.28) where doubling a point, for example AA, refers to the tangent line at A. So a 'degenerate' hexagon $ABCDEF$ can have adjacent points identified:

Six-point Pascal $ABCDEF$ $AB \cap DE$, $BC \cap EF$, $CD \cap FA$ are collinear
Five-point Pascal $AACDEF$ $AA \cap DE$, $AC \cap EF$, $CD \cap FA$ are collinear
Four-point Pascal $AABBEF$ $AA \cap BE$, $AB \cap EF$, $BB \cap FA$ are collinear
Four-point Pascal $AACDDF$ $AA \cap DD$, $AC \cap DF$, $CD \cap FA$ are collinear.
Three-point Pascal $AABBCC$ $AA \cap BC$, $BB \cap CA$, $CC \cap AB$ are collinear.

The last of the *degenerate* Pascal theorems, the *three-point Pascal theorem*, is due to Lazare Carnot (1803, p. 453, Theoreme XLVI).

Corollary 6.69 (Extended four-point Pascal)
Let $ABCD$ be a quadrilateral inscribed in a conic K and let $M := AD \cap BC$ and $N := AB \cap CD$. Let P be the intersection of the tangents t_A and t_C to K at A and C (respectively). Let Q be the intersection of the tangents t_B and t_D to K at B and D (respectively). Then the points M, N, P, and Q are collinear.

Proof If we apply Theorem 6.67 to the degenerate hexagon $AABCCD$, we see that the points M, N, and P are collinear. Then if we apply Theorem 6.67 to $ABBCDD$, the points M, N, and Q are collinear.   □

The above result first appeared in 1735 in the first edition of Robert Simson's *Sectionum conicarum libri V* (Simson, 1735, Proposition VII). It also appears

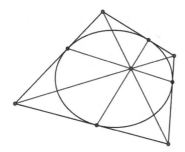

Figure 6.21 A four-line Brianchon theorem.

in Maclaurin's posthumous *De Linearum Geometricarum Proprietatibus Generalibus Tractatus* appendix to his *Algebra* (Maclaurin, 1748). On the other hand, the dual of the above result is one of the results known as a *four-line Brianchon theorem*, appearing in Chasles (1865, p. 89, art. 121).

Theorem 6.70 (Four-line Brianchon theorem)
The diagonals and the lines joining opposite points of tangency of a quadrilateral circumscribed by a conic are concurrent.

Newton's *Principia* (Newton, 1999, Corollary 2 to Lemma XXIV) proves this result for the special case where a pair of sides of the quadrilateral are parallel. Note that it is also related to Theorem 4.39 in which the diagonal triangle of a complete quadrilateral circumscribing a conic is self-polar with respect to that conic.

Theorem 6.71 (Another four-line Brianchon theorem)
Suppose a quadrilateral ABCD circumscribes a conic, with AB touching the conic at E and BC touching the conic at F. Then CE, FA, and BD are concurrent (see Figure 6.21).

Worked Example 6.72
(This was taken from Maclaurin, 1748, Section 38.)

Problem   Given four points A, B, C, D on a conic and the tangent t_A at A, determine the tangent t_B at B.

Solution   Let AD, BC meet in Q and let AC, BD meet in R (see Figure 6.22). Now let QR meet t_A in Z; then ZB is the tangent t_B at B. (Because, by Theorem 6.67, $R = AC \cap BD$, $Q = AD \cap BC$, and $t_A \cap t_B$ are collinear.)   □

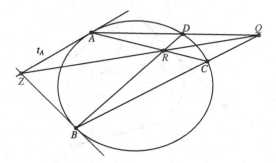

Figure 6.22 Maclaurin's construction of a tangent line.

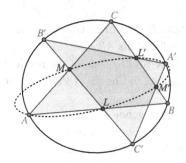

Figure 6.23 The second conic in Theorem 6.73.

Theorem 6.73 (Brianchon, 1817, XXXVII, XXXVIII; Steiner, 1832)
If the vertices of two triangles lie on a conic, then their sides touch a conic, and conversely.

Proof Let the triangles be ABC and $A'B'C$. Define points $L := AB \cap B'C'$, $M := AC \cap B'C'$, $L' := A'B' \cap BC$, and $M' := A'C' \cap BC$ (see Figure 6.23). Now

$$R(C', L; M, B') = R(AC', AB; AC, AB')$$
$$= R(A'C', A'B; A'C, A'B')$$
$$= R(M', B; C, L').$$

So (C', L, M, B') is projectively, but not perspectively, related to (M', B, C, L) on BC hence, by the dual of Theorem 6.59, the six lines $C'M'$, LB, MC, $B'L'$, $B'C'$, BC touch a conic.   □

Theorem 6.74 (Hesse, 1840a; Pascal, 1639)
Two triangles are polar with respect to a conic of $\mathsf{PG}(2, \mathbb{R})$ *and have no three of their vertices collinear if and only if their vertices lie on another conic.*

Proof Suppose ABC and DEF are self-polar triangles with respect to a conic $\mathcal{K}$ and have no three of their vertices collinear. Then, by the Chasles–Steiner Theorem 6.61, $(AB, AC, AE, AF) \barwedge (DC, DB, DF, DE)$, since AB, AC, AE, AF are respectively conjugate to DC, DB, DF, DE with respect to $\mathcal{K}$. Thus,

$$(AB, AC, AE, AF) \barwedge (DB, DC, DF, DE),$$

and this is not a perspective mapping, since no three of the vertices are collinear, so, by the Chasles–Steiner theorem (Theorem 6.61) again, A, B, C, D, E, F lie on a conic. $\qquad\square$

Corollary 6.75
If two triangles are self-polar with respect to a conic, then their vertices lie on another conic.

Worked Example 6.76
If ABC is a triangle, and D, E are distinct points on no side of the triangle, $A' := AD \cap BC$, $B' := BD \cap CA$, $C' = CD \cap AB$, $A'' = AE \cap BC$, $B'' = BE \cap CA$, $C'' = CE \cap AB$, then A', B', C', A'', B'', C'' lie on a conic.

Proof Consider the conic $\mathcal{K}$ on A, B, C, D, E (compare with Theorem 5.17). Then $A'B'C'$ is the diagonal triangle of $ABCD$, so self-polar with respect to $\mathcal{K}$, by Theorem 4.39. Similarly, $A''B''C''$ is the diagonal triangle of $ABCE$, so self-polar with respect to $\mathcal{K}$, by Theorem 4.39. Thus, A', B', C', A'', B'', C'' lie on a conic, by Corollary 6.75. $\qquad\square$

Theorem 6.77 (Todd and Eves, 1942)
If the trilinear polars of the vertices of a triangle PQR with respect to a triangle ABC are concurrent, then the trilinear poles of the sides of PQR (with respect to ABC) are collinear.

Proof Let p, q, r, a, b, c denote the sides of the two triangles, and let P', Q', R', p', q', r' be the trilinear poles and polars of p, q, r, P, Q, R with respect to ABC. Let P_a, Q_a, R_a, P'_a, Q'_a, R'_a, A_p, A_q, A_r, $A_{p'}$, $A_{q'}$, $A_{r'}$ denote the points where a meets the respective lines AP, AQ, AR, AP', AQ', AR', p, q, r, p', q', r'. Similarly, let $P_c, \ldots, C_p, \ldots$ be the points where c meets $CP, \ldots, p, \ldots$.

By the definition of trilinear polarity, the point pairs $\{P_a, A_{p'}\}$, $\{Q_a, A_{q'}\}$, $\{R_a, A_{r'}\}$, $\{P'_a, A_p\}$, $\{Q'_a, A_q\}$, $\{R'_a, A_r\}$ are harmonic conjugates with respect to B and C. Thus, the projectivity $(P_a, Q_a, R_a) \mapsto (A_{p'}, A_{q'}, A_{r'})$ leaves B and C invariant. Hence, if p', q', r' are concurrent, we have

$$(B, P_a, Q_a, R_a) \barwedge (B, A_{p'}, A_{q'}, A_{r'}) \barwedge (B, C_{p'}, C_{q'}, C_{r'}) \barwedge (B, P_c, Q_c, R_c).$$

Hence, $(AB, AP, AQ, AR) \barwedge (CB, CP, CQ, CR)$. Hence, A, B, C, P, Q, R lie on a conic; and a, b, c, p, q, r touch a conic, by Theorem 6.73. The dual of the Chasles–Steiner theorem (Theorem 6.61) now supplies the middle link in the following chain of projectivities:

$$(a \cap b, a \cap p, a \cap q, a \cap r) \barwedge (c \cap b, c \cap p, c \cap q, c \cap r)$$
$$\barwedge (AC, AP', AQ', AR') \barwedge (C, P_{a'}, Q_{a'}, R_{a'})$$
$$\barwedge (C, A_p, A_q, A_r) \barwedge (A, C_p, C_q, C_r)$$
$$\barwedge (A, P_{c'}, Q_{c'}, R_{c'}) \barwedge (CA, CP', CQ', CR').$$

Therefore, P', Q', R' are collinear.   □

Theorem 6.78

Given a line ℓ and a pencil of conics on a quadrangle, such that ℓ is on no points of the quadrangle, the locus of poles of ℓ with respect to the conics of the pencil is itself a conic that passes through the diagonal points of the quadrangle.

Proof By Theorem 4.39, the diagonal points R, S, T of the quadrangle form a self-polar triangle with respect to any of the conics of the pencil. Let P, Q be the (possibly imaginary) fixed points of the involution cut on ℓ by the pencil of conics, by Desargues' involution theorem (Theorem 6.19). Let K be the conic on P, Q, R, S, and T. Let X be a point of ℓ, K' be the conic of the pencil on X, and X' be the other point of that conic on ℓ. Let Y be the pole of ℓ with respect to K'. Then RST, PQY are self-polar triangles with respect to K', so, by Theorem 6.74, there is a conic on R, S, T, P, Q, Y; that is, Y is on K.   □

The following proof uses Desargues' theorem (Theorem 2.20) and Brianchon's other theorem (Theorem 6.44).

Second proof of Theorem 6.73 Suppose the vertices of triangles $A_1A_2A_3$, $A_4A_5A_6$ lie on a conic. Then, by Pascal's theorem (Theorem 6.28), $A_1A_2 \cap A_4A_5$, $A_2A_3 \cap A_5A_6$, and $A_3A_4 \cap A_6A_1$ are collinear in a line m. So the triangles with sides $(A_1A_2)(A_3A_4)(A_5A_6)$ and $(A_4A_5)(A_6A_1)(A_2A_3)$ are in perspective from m. So, by Desargues' theorem (Theorem 2.20), the triangles with sides $(A_1A_2)(A_6A_1)(A_5A_6)$ and $(A_4A_5)(A_3A_4)(A_2A_3)$ are in perspective from a point P. Thus, the hexagon

$$H = A_4A_6(A_5A_6 \cap A_1A_2)A_1A_3(A_2A_3 \cap A_4A_5)$$

has the lines joining opposite vertices concurrent. By Theorem 6.44, the sides of H touch a conic. The sides of H are A_6A_4, A_5A_6, A_1A_2, A_1A_3, A_2A_3, A_4A_5. Thus, the sides of the triangles $A_1A_2A_3$, $A_4A_5A_6$ touch a conic.   □

By the dual of the Braikenridge–Maclaurin theorem (Theorem 6.43), the direct converse of Brianchon's theorem (Theorem 6.44) holds. That is, if the diagonals of a hexagon $ABCDEF$ are concurrent (a so-called *Brianchon hexagon*), then the sides of the hexagon are tangent to a conic. There is another 'converse' to Brianchon's theorem when we instead consider the hexagon derived from the points of intersection of the tangents.

Theorem 6.79 (A converse to Brianchon's theorem (Theorem 6.44))
Let A, B, C, D, E, F be six distinct points in the plane. Let $U := BC \cap DE$, $V := CA \cap EF$, $W := AB \cap FD$, $X := AB \cap EF$, $Y := BC \cap FD$, $Z := CA \cap DE$. Then the lines UX, VY, WZ meet in a point if and only if the points A, B, C, D, E, F lie on a conic.

Proof Apply Brianchon's theorem (Theorem 6.73) on two triangles inscribed in a conic to ABC, DEF: the points A, B, C, D, E, F lie on a conic if and only if AB, BC, CA, DE, EF, FD touch a conic. This is equivalent to the hexagon $WXVZUY$ being a Brianchon hexagon. Moreover, by Brianchon's theorem (Theorem 6.44) and the dual of Theorem 6.43, $WXVZUY$ is a Brianchon hexagon if and only if the lines UX, VY, WZ meet in a point. $\square$

Theorem 6.80 (Braikenridge, 1733, p. 68)
When the n sides of a polygon pass one by one through n fixed points, and $n-1$ of the vertices lie on fixed lines, the remaining vertex lies either on a line or on a conic.

Proof Let the sides be $a_1, \ldots, a_n$, the respective fixed points on the sides $P_1, \ldots, P_n$, and suppose that $a_i \cap a_{i+1}$ lies on m_i, for $i = 1, \ldots, n-1$. Let g_i be the perspectivity from the pencil on P_i to the pencil on P_{i+1} via m_i, for $i = 1, \ldots, n-1$. Then $g = g_1 \ldots g_{n-1}$ is a projectivity from the pencil on P_1 to the pencil on P_n taking a_1 to a_n. So $S = \{\ell \cap \ell^g : \ell \mathrm{I} P_1\}$ is either a line (if g is a perspectivity) or a conic (by Theorem 6.59) and $a_1 \cap a_n$ lies on S. $\square$

We provide another way to develop the Pascal theorems that makes elegant use of the notation for a pencil of conics. Let $ABCDEF$ be a hexagon inscribed in a conic $\mathcal{K}$. If the conic is a line pair, also suppose that none of the vertices are the intersection of the pair of lines, and that the vertices lie alternatingly on the two lines. Let the equations of: AB be $L = 0$; BC be $M = 0$; CD be $N = 0$; DE be $L' = 0$; EF be $M' = 0$; FA be $N' = 0$; AD be $T = 0$.

Then the pencil of conics on $ABCD$ is given by

$$LN + rMT = 0, \quad MT = 0.$$

Thus, there exists r such that $\mathcal{K}$ has equation $LN + rMT = 0$. Also, the pencil of conics on $CDEF$ is given by $L'N' + sM'T = 0$, $M'T = 0$. Thus, there exists s such that $\mathcal{K}$ has equation $L'N' + sM'T = 0$. Hence, there exists t such that $LN + rMT = t(L'N' + sM'T)$, so that

$$LN - tL'N' = (sM' - rM)T.$$

Now $sM' - rM = 0$ is a line ℓ passing through $BC \cap EF$. Thus, $LN - tL'N' = 0$ is a degenerate conic passing through $AB \cap DE$ and $CD \cap FA$. Since $T = 0$ is the line AD, it doesn't pass through $AB \cap DE$ or $CD \cap FA$. So ℓ passes through $AB \cap DE$ and $CD \cap FA$.

Example 6.81 (Four-point Pascal of type $AACCEF$)
Let the equations of the tangent t_A at A be $L = 0$; AC be $M = 0$; the tangent t_C at C be $N = 0$; CE be $L' = 0$; EF be $M' = 0$; FA be $N' = 0$; AD be $T = 0$. Then the pencil of conics on AC tangent to t_A and t_C is given by $LN + rMT = 0$ (and $MT = 0$). Thus, there exists r such that $\mathcal{K}$ has equation $LN + rMT = 0$. Also, the pencil of conics on CEF tangent to t_C is given by $L'N' + sM'T = 0$ (and $M'T = 0$). Thus, there exists s such that $\mathcal{K}$ has equation $L'N' + sM'T = 0$. Hence, there exists t such that $LN + rMT = t(L'N' + sM'T)$, so that $LN - tL'N' = (sM' - rM)T$. Now $sM' - rM = 0$ is a line ℓ passing through $AC \cap EF$. Thus, $LN - tL'N' = 0$ is a degenerate conic passing through $t_A \cap CE$ and $t_C \cap FA$. Since $T = 0$ is the line AC, it doesn't pass through $t_A \cap CE$ or $t_C \cap FA$. So ℓ passes through $t_A \cap CE$ and $t_C \cap FA$.

The other degenerations of Pascal's theorem may be handled similarly.

We continue on the theme of variations of Pascal's theorem.

Theorem 6.82 (Möbius, 1848)
If three sides of a quadrangle inscribed in a conic of $\mathsf{PG}(2, \mathbb{R})$ pass through fixed points on a given line, then the fourth side also passes through a fixed point on that line.

Proof Let quadrangles $ABCD$ and $A'B'C'D'$ be inscribed in a conic S, with three pairs of corresponding sides intersecting at points R_1, R_2, and R_3 of a given line ℓ. Let R_4 be the point of intersection of CD and ℓ; it must be shown that $C'D'$ also meets ℓ at R_4. Consider the hexagon $ABCA'B'C'$. By Pascal's theorem (Theorem 6.28), since opposite sides AB and $A'B'$ meet on ℓ at R_2, and BC and $B'C'$ meet on ℓ at R_3, it follows that $A'C$ and AC' also meet on ℓ, say at P. Now consider the hexagon $ADCA'D'C'$. Again by Pascal's theorem (Theorem 6.28), since opposite sides AD and $A'D'$ meet on ℓ at R_1, and $A'C$ and AC' meet on ℓ at P, it follows that CD and $C'D'$ meet on ℓ. $\qquad\square$

Theorem 6.83 (Möbius, 1848)

Let an n-gon, where n = 2k, be inscribed in a conic of $PG(2, \mathbb{R})$ *and let n − 1 of its sides meet a given line at fixed points. Then the n-th side also meets that line at a fixed point.*

Proof Our proof will be by induction on k. The base case $k = 2$ is Theorem 6.81. Suppose now that our result holds for some integer k; it must be shown to hold for $k + 1$. Let an inscribed polygon have consecutive vertices $P_1, P_2, \ldots, P_n, P_{m-1}, P_m$ where $m = 2(k + 1)$. It is given that $m − 1$ sides of the polygon meet a given line ℓ at fixed points; without loss of generality, let P_1P_2 through $P_{m-1}P_m$ be the sides so given. By supposition, the theorem holds for a polygon with $n = 2k$ sides; specifically, since the sides P_1P_2 through $P_{n-1}P_n$ of the polygon $P_1P_2 \ldots P_n$ pass through fixed points on ℓ, then the side P_1P_n also meets ℓ at a fixed point. Now, P_1P_n, P_nP_{m-1}, $P_{m-1}P_m$, and P_mP_1 are the sides of an inscribed quadrangle, of which sides three are given or have been shown to meet ℓ at fixed points; hence, P_mP_1, the fourth side of the quadrangle and the m-th side of the polygon, also meets ℓ at a fixed point, by Theorem 6.81. □

Theorem 6.84

Theorem 6.81 is equivalent to Pascal's theorem (Theorem 6.28).

Proof It was shown in the proof of Theorem 6.81 that Pascal's theorem implies Theorem 6.81; the reverse implication remains to be shown. Let the hexagon *ABCDEF* be inscribed in a conic, let *AB* meet *DE* at *P*, *BC* meet *EF* at *Q*, *CD* meet *PQ* at *R*, and let *AD* meet *PQ* at *T*. We discern that the quadrangles *ABCD* and *DEFA* possess three pairs of corresponding sides, those which pass through the points *P*, *Q*, and *T* on *PQ*, and hence the fourth pair of corresponding sides, *CD* and *AF*, meet *PQ* at the point *R*. □

Finally, we finish this theme of variations on Pascal's theorem by giving the cross-axis theorem on a conic. The proof is left as an exercise, but we provide a hint that it is very similar to the proof of Theorem 6.37, except that we use Pascal's theorem (Theorem 6.28) instead of Pappus' theorem (Theorem 2.18).

Theorem 6.85 (Cross-axis theorem on a conic)

Let ϕ be a projectivity from a non-degenerate conic $\mathcal{K}$ to itself. Then

$$\{P^\phi Q \cap Q^\phi P : P, Q \in \mathcal{K}, P \neq Q\}$$

*is a line, called the **cross-axis** of the projectivity.*

Proof See Exercise 6.30. □

6.6 Quadrangular Sets

A set of six distinct points on a line ℓ is said to be a **quadrangular set** if it is the set of points of intersection of ℓ with the sides of a complete quadrangle, no vertex of which is on ℓ. We have seen that five points, no three collinear, determine a conic (Theorem 5.17). We now see that five collinear points determine a quadrangular set.

Theorem 6.86 (Fundamental theorem on quadrangular sets (von Staudt, 1847, p. 42, para. 91))
If two quadrangular sets coincide in five of their corresponding points, then they are identical.

Proof Let A, B, C, D, E, F and A, B, C, D, E, F' be quadrangular sets lying on the line ℓ. Suppose further that $\{Q_1, Q_2, Q_3, Q_4\}$, $\{Q'_1, Q'_2, Q'_3, Q'_4\}$ are quadrangles with $A = Q_1Q_2 \cap \ell = Q'_1Q'_2 \cap \ell, B = Q_1Q_3 \cap \ell = Q'_1Q'_3 \cap \ell, C = Q_1Q_4 \cap \ell = Q'_1Q'_4 \cap \ell, D = Q_3Q_4 \cap \ell = Q'_3Q'_4 \cap \ell, E = Q_2Q_4 \cap \ell = Q'_2Q'_4 \cap \ell, F = Q_2Q_3 \cap \ell, F' = Q'_2Q'_3 \cap \ell$. Let $V = Q_1Q_2 \cap Q'_1Q'_2$. Then $Q_1Q_2Q_4$ and $Q'_1Q'_2Q'_4$ are in perspective from ℓ, and so, by the dual of Desargues' theorem (Theorem 2.20), they are in perspective from V. Similarly, $Q_1Q_3Q_4$ and $Q'_1Q'_3Q'_4$ are in perspective from ℓ, and so, by the dual of Desargues' theorem (Theorem 2.20), they are in perspective from $Q_1Q_4 \cap Q'_1Q'_4 = V$. Therefore, $Q_1Q_2Q_3$ and $Q'_1Q'_2Q'_3$ are in perspective from V, and so, by Desargues' theorem (Theorem 2.20), they are in perspective from ℓ. Therefore, $Q_2Q_3 \cap \ell = Q'_2Q'_3 \cap \ell$; that is, $F = F'$. □

Theorem 6.85 is the converse of Pappus' involution theorem (Theorem 6.13), and is likely the first time this was proved. Pappus did prove the converse (Pappus of Alexandria, 1986, Proposition 130); he was even careful to cover the cases where one of the points is at infinity (Pappus of Alexandria, 1986, Propositions 127, 128).

Theorem 6.87
Let $ABCD$ be a quadrangle and ℓ be a line on no vertex. Let $P_1 := AB \cap \ell$, $P_2 := AC \cap \ell, P_3 := BC \cap \ell, Q_1 := DC \cap \ell, Q_2 := DB \cap \ell, Q_3 := DA \cap \ell$. Then $R(Q_3, P_2; Q_1, P_1) = R(Q_2, P_3; Q_1, P_1)$. Conversely, if $ABCD$ is a quadrangle and ℓ is a line on no vertex, with $P_1 := AB \cap \ell, P_2 = AC \cap \ell, P_3 = BC \cap \ell, Q_1 = DC \cap \ell, Q_2 = DB \cap \ell$, and Q_3 on ℓ has $R(Q_3, P_2; Q_1, P_1) = R(Q_2, P_3; Q_1, P_1)$ then $Q_3 = DA \cap \ell$.

Proof Projecting $DC \to \ell$ from B shows that

$$R(D, C; Q_1, DC \cap AB) = R(Q_2, P_3; Q_1, P_1),$$

and projecting $DC \to \ell$ from A shows that

$$R(D, C; Q_1, DC \cap AB) = R(Q_3, P_2; Q_1, P_1).$$

Conversely, let $Q_3' = DA \cap \ell$. Projecting ℓ to DC from B shows that

$$R(D, C; Q_1, DC \cap AB) = R(Q_2, P_3; Q_1, P_1),$$

and projecting ℓ to DC from A shows that

$$R(D, C; Q_1, DC \cap AB) = R(Q_3', P_2; Q_1, P_1).$$

Hence, $Q_3 = Q_3'$. □

This is close to the form of what Pappus proved, but not at all close to his proof. Now $R(Q_3, P_2; Q_1, P_1) = R(Q_2, P_3; Q_1, P_1)$ if and only if

$$R(Q_1, P_2; Q_3, P_1) = R(Q_1, P_3; Q_2, P_1). \tag{6.2}$$

Pappus proved

$$d(Q_1, Q_3)d(Q_2, P_3)d(P_1, P_3) = d(Q_1, Q_2)d(Q_3, P_2)d(P_1, P_2),$$

which is equivalent to Equation (6.2) (for unsigned cross-ratio). Pappus drew eight diagrams for Proposition 130 for various cases. This is perhaps how he finessed the sign problems for the converse; that is, he assumed that one of the eight diagrams represented the implicit ordering assumptions, and, thus, unsigned cross-ratio being equal implied that signed cross-ratio was equal.

Corollary 6.88
If $(P_3, P_2; Q_1, P_1)$ is harmonic, then $R(P_3, P_2; Q_1, P_1) = -1$.

Proof Put $P_2 = Q_2$, $P_3 = Q_3$. Then

$$R(P_3, P_2; Q_1, P_1) = R(P_2, P_3; Q_1, P_1) = 1/R(P_3, P_2; Q_1, P_1). \qquad □$$

Theorem 6.86 shows that six collinear points $P_1, P_2, P_3, Q_1, Q_2, Q_3$ form a quadrangular set if $R(Q_3, P_2; Q_1, P_1) = R(Q_2, P_3; Q_1, P_1)$. We now see that a quadrangular set can be thought of as supporting elements of an involution.

Theorem 6.89
Let $P_1, P_2, P_3, Q_1, Q_2, Q_3$ be six points on a line of $\mathsf{PG}(2, \mathbb{R})$. Then

$$(P_1 \, Q_1)(P_2 \, Q_2)(P_3 \, Q_3) \ldots$$

is an involution if and only if $R(Q_3, P_2; Q_1, P_1) = R(Q_2, P_3; Q_1, P_1)$.

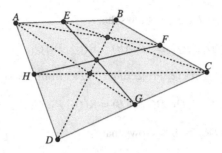

Figure 6.24 van Kempen's theorem.

Proof Note that $R(Q_3, P_2; Q_1, P_1) = R(Q_2, P_3; Q_1, P_1)$ if and only if

$$R(P_3, Q_2; P_1, Q_1) = R(Q_2, P_3; Q_1, P_1).$$

By Theorem 4.7, this is equivalent to there being a projectivity taking P_3 to Q_2, Q_2 to P_2, and interchanging P_1 and Q_1. This is in turn the same as

$$(P_1\, Q_1)(P_2\, Q_2)(P_3\, Q_3)\ldots$$

being an involution, by Theorem 6.2.   □

We have two proofs of a recent result of van Kempen on quadrilaterals: the first proof is an elegant use of the Chasles–Steiner theorem (Theorem 6.61) and Pascal's theorem (Theorem 6.28) the second proof; utilises quadrangular sets.

Theorem 6.90 (van Kempen, 2006, Theorem 2)
If in the quadrilateral $ABCD$ of $\mathrm{PG}(2, \mathbb{R})$, the points E, F, G, and H lie on AB, BC, CD, and DA, respectively, such that AG, CH, and BD are concurrent and AF, CE, and BD are concurrent, then EG, FH, and BD are concurrent (see Figure 6.24).

Proof Let $S := AG \cap CH$ and $T = AF \cap CE$. Now

$$\begin{aligned}
R(AH, AG; AF, AE) &= R(D, S; T, B) &&\text{(by intersection with } DB)\\
&= R(S, D; B, T)\\
&= R(CH, CG; CF, CE) &&\text{(by joining to } C),
\end{aligned}$$

so, by the Chasles–Steiner theorem (Theorem 6.61), A, E, F, C, G, H lie on a conic. Applying Pascal's theorem (Theorem 6.28) to the hexagon $AFHCEG$, $T = AF \cap CE$, $S = AG \cap CH$, and $FH \cap EG$ are collinear, so EG, FH, and $ST = BD$ are concurrent.   □

Another proof of Theorem 6.89 The quadrangular set on the line BD given by the quadrangle $CGAE$ is $\{D, S, P, EG \cap BD, T, B\}$ where $P = CA \cap BD$. That given by the quadrangle $HFAC$ is $\{D, S, P, FH \cap BD, T, B\}$. So, by the converse of Pappus' involution theorem (Theorem 6.13), EG, FH, and BD are concurrent. □

Theorem 6.91
If five sides of a hexagon inscribed in a conic meet a given line at fixed points, then its Pascal line meets that line at a fixed point.

Proof With reference to the construction and notation used in the proof of Theorem 6.81, let AB, CD, DE, EF, and AF be the sides meeting a given line ℓ at fixed points. Application of Theorem 6.81 to the quadrangle $ADEF$ yields that AD meets ℓ at a fixed point, and application of the fundamental theorem on quadrangular sets (Theorem 6.85) to the complete quadrangle $PADR$ yields that PR, the Pascal line of $ABCDEF$, meets ℓ at a fixed point. □

Exercises

6.1 Prove that a conic consisting of just one point is degenerate.

6.2 Prove Theorem 6.24. (Hint: The five points A, B, C, P, Q determine a conic by Theorem 5.17, and the point of concurrency will be the pole of PQ.)

6.3 Let C be a conic of $\mathsf{PG}(2, \mathbb{R})$, P, P', Q, Q' be distinct points of C, $T = PP' \cap QQ'$, ℓ be the tangent to C at A. Prove that $(AP\,AP')(AQ\,AQ')(\ell\,AT)\ldots$ is an involution.

6.4 Let A_i, $i = 0, 1, \ldots, 5$ denote the vertices of a hexagon inscribed in a conic of $\mathsf{PG}(2, \mathbb{R})$ and let B_i denote the intersection of the lines A_iA_{i+2} and $A_{i+1}A_{i+3}$, for $i = 0, 1, \ldots, 5$ (indices modulo 6). Show that the lines B_iB_{i+3}, $i = 0, 1, 2$ are concurrent.

6.5 (Zhao, 2009) Let A, B, C, D be four points in the affine plane. Let lines AC and BD meet at P, lines AB and CD meet at Q, and lines BC and DA meet at R. Let the line through P parallel to QR meet lines AB and CD at X and Z. Show that P is the midpoint of $\overline{XZ}$.

6.6 Use the Braikenridge–Maclaurin theorem (Theorem 6.43) to prove the converse of Theorem 6.55.

6.7 Write down the dual of the cross-axis theorem (Theorem 6.37) and use it to prove Brianchon's theorem (Theorem 6.44).

6.8 Let A, B, C be three points of a conic and suppose the tangents at B and C meet in A'. The points B' and C' are similarly defined. Prove that the lines AA', BB', CC' are concurrent. Deduce 'Euclidean' theorems for a hyperbola and a parabola by taking the line at infinity (i) as the line BC, (ii) as the line $B'C'$.

6.9 If $ABCD$ is a complete quadrangle of PG$(2, \mathbb{R})$ whose sides AB, AC, AD, BC, BD, CD are cut by a line ℓ in points P, Q, R, S, T, U and if E, F, G, H, I, J are the harmonic conjugates of these points with respect to the pair of vertices on the side, show that a conic through the diagonal points of $ABCD$ lies on E, F, G, H, I, J.

6.10 Suppose $ABCD$ is a quadrangle, $X := AD \cap BC$, and $Y := AC \cap BD$. Prove that the midpoints of AB, DC, and XY are collinear.

6.11 A variable conic $\mathcal{K}$ has one focus at a fixed point P and touches two distinct fixed lines ℓ and m (which do not pass through P). Show that the locus of the centre of $\mathcal{K}$ is a straight line.

6.12 Show that if the lines joining the points X, Y on the respective sides AB, AC to the opposite vertices of the triangle ABC meet on the median through A, then XY is parallel to BC.

6.13 Let ABC be a triangle, and let P be a point on no side of the triangle. Through the midpoints of BC, CA, AB, lines parallel to PA, PB, PC respectively are constructed. Show that these lines are concurrent.

6.14 Prove that the parallels to the sides of a triangle through any point on no side of the triangle, and not on the lines through a vertex parallel to the opposite side, intersect the sides in six points that lie on a conic.

6.15 A line meets the sides BC, CA, CB of the triangle ABC in L, M, N, respectively, and $MANP$, $NBLQ$, $LCMR$ are parallelograms. Show that P, Q, R are collinear.

6.16 Two triangles ABC, $A'B'C'$ are such that the lines through A, B, C parallel to $B'C'$, $C'A'$, $A'B'$, respectively, are concurrent. Prove that the same is true of the lines through A', B', C' parallel to BC, CA, AB, respectively.

6.17 A triangle is self-polar with respect to a parabola. Prove that the lines joining the midpoints of the sides are tangent to the parabola.

6.18 Suppose PR and QS are parallel chords of a conic, and T is the pole of PQ. Prove that PS and QR intersect on the chord through T parallel to the given chords, and that this chord is bisected at the point of intersection.

6.19 Two parabolas touch the sides of a triangle ABC and intersect one another in P, Q, R, S. Prove that the line joining any two of the points P, Q, R, S passes through one of the vertices of the triangle formed by the lines through the vertices of ABC parallel to the opposite sides.

6.20 Given any acute-angled triangle ABC and one altitude AH, select any point D on AH, then draw BD and extend it until it intersects AC in E.

Draw CD and extend it until it intersects AB in F. Prove that $\angle AHE$ is congruent to $\angle AHF$.

6.21 Show that if K is a conic in $\mathsf{PG}(2,\mathbb{R})$ with self-polar triangle $P_1 P_2 P_3$ and C is a point not on K, then the lines that join C to the points in which the polar line of C meets the sides of the triangle and the lines that join C to the vertices of the triangle form three pairs of an involution whose fixed lines are the tangents t_1, t_2 to K on C. Deduce that if a triangle is self-polar with respect to a circle, the centre of the circle is the orthocentre of the triangle.

6.22 Five points A, B, H, K, P are given on a line ℓ of $\mathsf{PG}(2,\mathbb{R})$. Lines a, b, p are given through A, B, P, respectively; $X := p \cap a$, $Y := p \cap B$. A conic $\mathcal{K}$ on X, Y, H, K is taken, meeting a, b again in points U, V, respectively. Show that UV meet ℓ in a fixed point determined completely by A, B, H, K, P and independent of the choices of a, b, p, and $\mathcal{K}$. (Hint: Use Desargues' involution theorem.)

6.23 (i) Prove the following:

Given five points A_0, A_1, A_2, A_3, A_4, no three collinear, let $P := A_0 A_1 \cap A_2 A_3$, Y be the harmonic conjugate of P with respect to $\{A_0, A_1\}$, Z be the harmonic conjugate of P with respect to $\{A_2, A_3\}$, and A_5 be the harmonic conjugate of A_4 with respect to $\{P, YZ \cap A_4 P\}$. Then A_5 lies on the conic $\mathcal{K}$ on A_0, A_1, A_2, A_3, A_4.

(ii) Use the result of (i) to show that a point X not equal to A_0, A_1, A_2, A_3, A_4, A_5 lies on $\mathcal{K}$ if and only if the pairs of lines

$$(XA_0, XA_1), \ (XA_2, XA_3), \ (XA_4, XA_5)$$

are in involution.

6.24 Let $A_1 A_2 A_3 A_4 A_5 A_6$ be a hexagon circumscribed about a conic in $\mathsf{PG}(2,\mathbb{R})$, and form the intersections $P_i = A_i A_{i+2} \cap A_{i+1} A_{i+3}$ $(i = 1, \ldots, 6$, all indices modulo 6). Show that the P_i are the vertices of a hexagon inscribed in a conic.

6.25 Let ABC be a triangle. Let D, D', E, E', F, F' be three pairs of points on each side BC, AC, and AB, respectively, such that AD, BE, CF and AD', BE', CF' are concurrent. Prove that $FE \cap F'E'$, $FD \cap F'D'$, $ED' \cap E'D$, and C are collinear.

6.26 We will use the same notation and context as in Theorem 6.22. Let $B_X C_X$ and $B_Y C_Y$ meet in P, $C_X A_X$ and $C_Y A_Y$ meet in Q, and $A_X B_X$ and $A_Y B_Y$ meet in R. Show that AP, BQ, and CR are concurrent in the pole of XY with respect to the conic A_X, A_Y, B_X, B_Y, C_X, and C_Y.

6.27 Suppose there are two conics $\mathcal{K}_1$ and $\mathcal{K}_2$ of $\mathsf{PG}(2,\mathbb{R})$ that touch at S and intersect at A, B. Also suppose that the common tangent at P and at Q

meets the tangent at S in the point R. Let $T := AB \cap PQ$. Show that $R(R, T; P, Q) = -1$.

6.28 Quadrilateral $ABCD$ is circumscribed about a conic. The conic touches the sides AB, BC, CD, DA at points E, F, G, H, respectively.

  (a) Show that AC, BD, EG, and FH are concurrent.

  (b) Also show that lines AC, EF, GH are concurrent at the pole of BD.

6.29 Prove the following result of Carnot (see Carnot, 1803, art. 400):

Let A, B, C, D be distinct points on a conic $\mathcal{K}$, with respective tangents t_A, t_B, t_C, t_D. Let $A^* := t_A \cap t_B$, $B^* = t_B \cap t_C$, $C^* = t_B \cap t_C$, and $D^* = t_C \cap t_D$. Then AC, BD, A^*C^*, B^*D^* are concurrent.

6.30 Prove Theorem 6.85.

7

Real Affine Plane Geometry from a Projective Perspective

That's done, as near as the extremest ends of parallels.

William Shakespeare, *Troilus and Cressida*, Act I, Scene iii

To recover the affine plane $\mathsf{AG}(2,\mathbb{R})$ from $\mathsf{PG}(2,\mathbb{R})$, we remove a line, and it does not matter which line we choose (by Theorem 2.1). This line is now the *line at infinity* ℓ_∞. Two lines ℓ and m (not equal to ℓ_∞) are then parallel in the affine plane, if their point of intersection in $\mathsf{PG}(2,\mathbb{R})$ lies on ℓ_∞. If A, B, C are three collinear affine points, then B is the midpoint of A and C if in $\mathsf{PG}(2,\mathbb{R})$ we have that $(A,B;C,D)$ is harmonic, where D is the point at infinity of the line AC (see Theorem 7.1). Moreover, we can recover affine ratio from cross-ratio by stipulating that the second coordinate is a point at infinity (see Worked Example 4.2). With these methods in hand, we can simulate all that we know of the real affine plane within a projective setting.

Theorem 7.1
Given collinear points B, C, D of $\mathsf{AG}(2,\mathbb{R})$, let A be the point of BC on the line at infinity. Then B is the midpoint of CD if and only if $(A,B;C,D)$ is harmonic.

Proof Recall that

$$\mathsf{ratio}(B,C,D) = |\mathsf{R}(B,A;C,D)|.$$

Now B is the midpoint of CD if and only if $d(B,C) = d(B,D)$, or, in other words, $\mathsf{ratio}(B,C,D) = 1$. This is equivalent to $|\mathsf{R}(B,A;C,D)| = 1$, and hence $\mathsf{R}(B,A;C,D) = -1$. □

The following treatments use the Braikenridge–Maclaurin theorem, and give 'projective' proofs of affine theorems.

Theorem 7.2 (Veblen and Young, 1918, p. 281, Section 108, Exercise 7)
The lines joining the midpoints of the opposite sides of a hexagon, which has all pairs of opposite sides parallel, are concurrent.

Proof By the Braikenridge–Maclaurin theorem (Theorem 6.43), the vertices of the hexagon lie on a conic. By Apollonius' theorem (Theorem 4.38), the polar line of the parallel class given by a pair of opposite sides is the line joining the midpoints. Hence, the three lines joining the midpoints of the opposite sides are concurrent, since their poles are collinear. □

The following theorem, sometimes known as Weiss' theorem (Theorem 7.3), was discovered before her by Boon (1978–9).

Theorem 7.3 (Asia Ivić Weiss c. 1981, unpublished, see Snapper, 1981)
Let P be any point of $\mathsf{AG}(2, \mathbb{R})$. Let ℓ_A be the line through the vertex A of an arbitrary triangle ABC that is parallel to the line through P and the midpoint A' of the opposite side BC. Let the lines ℓ_B, ℓ_C be constructed similarly. Then the three lines ℓ_A, ℓ_B, ℓ_C have a point Q in common, the points P, Q, and G (the centroid of ABC) are collinear, and $\mathsf{ratio}(G, Q, P) = -2$. Let A'', B'', C'' be the respective midpoints of AQ, BQ, CQ, and let D, E, F be the respective intersections $\ell_A \cap BC, \ell_B \cap CA, \ell_C \cap AB$. Then $A', B', C', A'', B'', C'', D, E, F$ lie on a conic with centre the midpoint of PQ.

Proof The first part of this theorem is Snapper's theorem.[1] We will consider four Pascal hexagons: $A'B''C''A''B'C''$, $A'B'C'B''A''D$, $B'C'A'C''B''E$, and $C'A'B'C''B''F$. By the Braikenridge–Maclaurin theorem (Theorem 6.43), the conic on A', B'', C', A'', B' is on C'', D, E, and F. That the centre of this conic is the midpoint of PQ is left as an exercise. □

If P is the circumcentre of ABC, then this conic is the **nine-point circle** of ABC.

Theorem 7.4 (Butterfly theorem for quadrilaterals (Kung, 2005))
Through the intersection I of the diagonals AC, BD of a convex quadrilateral $ABCD$, draw two lines EF and HG that meet the sides of $ABCD$ at E, F, G, H. If $M = EG \cap AC$, $N = HF \cap AC$, then

$$\mathsf{ratio}(M, A, I) \cdot \mathsf{ratio}(N, I, C) = \mathsf{ratio}(I, A, C).$$

---

[1] Snapper's theorem: Let P be any point of $\mathsf{AG}(2, \mathbb{R})$. Let ℓ_A be the line through the vertex A of an arbitrary triangle ABC that is parallel to the line through P and the midpoint A' of the opposite side BC. Let the lines ℓ_B, ℓ_C be constructed similarly. Then the three lines ℓ_A, ℓ_B, ℓ_C have a point Q in common, the points P, Q, and G (the centroid of ABC) are collinear, and $\mathsf{ratio}(G, Q, P) = -2$.

Table 7.1 *Central collineations viewed in the affine plane.*

Case	$a = \ell_\infty$	$a \neq \ell_\infty$
$C\,I\,a$	Translation	Shear
$C\,\cancel{I}\,a$	Dilatation	Strain

Proof By the Braikenridge–Maclaurin theorem (Theorem 6.43), A, C, E, F, G, H lie on a conic. By Chasles' theorem (Theorem 5.18),

$$R(GA, GE; GH, GC) = R(FA, FE; FH, FC).$$

Intersecting with AC, we have $R(A, M; I, C) = R(A, I; N, C)$. Hence,

$$\text{ratio}(M, A, I) \cdot \text{ratio}(N, I, C) = \text{ratio}(I, A, C). \qquad \square$$

Recall from Chapter 1 that we initially constructed the real projective plane as the geometry arising from the parallel classes of lines and planes of Euclidean 3-space $\mathbb{R}^3$. We will show how we can develop the affine plane from this structure. Delete a parallel class $[\pi_\infty]$ of planes and all lines parallel to it to obtain a smaller incidence structure $\mathcal{A}$. Choose a plane π of $[\pi_\infty]$, and a point P not on π. Consider the map that takes

- a parallel class $[\ell]$ of lines to the intersection of its unique member on P with π, and
- a parallel class $[\sigma]$ of planes to the intersection of its unique member on P with π.

Since the parallel classes have parallel representatives if and only if the unique members on P are contained in one another, this is if and only if the images are contained in one another. So $\mathcal{A}$ is isomorphic to the affine plane π. Moreover, the deleted objects (i.e., $[\pi_\infty]$ and all parallel lines) thus naturally give the points at infinity of π as lines on P parallel to π, and a line at infinity as the plane on P parallel to π.

7.1 Affinities

Recall from Theorem 2.4 that the group of affine transformations/affinities $\text{AGL}(2, \mathbb{R})$ can be identified with a stabiliser of a line ℓ_∞ (of $\text{PG}(2, \mathbb{R})$) in $\text{PGL}(3, \mathbb{R})$. A central collineation ϕ of $\text{PG}(2, \mathbb{R})$ can have its centre C and axis a located in different ways with respect to ℓ_∞. The various configurations of C and a with respect to ℓ_∞ give rise to the elementary affinities of $\text{AG}(2, \mathbb{R})$. Table 7.1 shows how we obtain a translation, dilatation, shear, and strain via these configurations.

Recall that Thales' side-splitter theorem tells us that if we have a triangle in $AG(2, \mathbb{R})$, then a line intersecting two sides of the triangle and parallel to the other side yields a division of the two sides in a common ratio of lengths. We can synthesise this property in $PG(2, \mathbb{R})$ by considering what happens to a central projection between lines that intersect at infinity (i.e., parallel lines).

Theorem 7.5 (Thales' side-splitter theorem)
Central projection g between parallel lines of $AG(2, \mathbb{R})$ involves a scale factor $\lambda = d(A^g, B^g)/d(A, B)$, which does not depend on the choice of the (distinct) points A, B on the line that is the domain.

Proof Consider the central projection g via P from ℓ to m, where $\ell \cap m$ is a point at infinity. The dilatation δ with centre P taking ℓ to m has its restriction to ℓ equal to g, so $d(A^g, B^g)/d(A, B) = d(A^\delta, B^\delta)/d(A, B)$, which is independent of the choice of the (distinct) points A, B on ℓ.   □

Recall that a **median** of a triangle is a line joining a vertex to the midpoint of its opposite side.

Theorem 7.6
*The three medians of a triangle meet in a common point, namely, the **centroid** of the triangle.*

Proof Let ABC be a triangle, A' be the midpoint of $\overline{BC}$, B' be the midpoint of $\overline{CA}$, C' be the midpoint of $\overline{AB}$. By Thales' side-splitter theorem (Theorem 7.5), AB is parallel to $A'B'$, BC is parallel to $B'C'$, and CA is parallel to $C'A'$. So AB and $A'B'$, BC and $B'C'$, CA and $C'A'$ all meet on the line at infinity. By the dual of Desargues' theorem (Theorem 2.20), AA', BB', and CC' are concurrent.   □

Worked Example 7.7 (2011 Indian Regional Mathematical Olympiad Advanced Problem 5)

Problem Let $ABCD$ be a convex quadrilateral. Let E, F, G, H be the midpoints of $\overline{AB}, \overline{BC}, \overline{CD}, \overline{DA}$, respectively. If AC, BD, EG, FH concur at a point O, prove that $ABCD$ is a parallelogram.

Solution Central projection g with centre O from AB to CD has $(AEB)^g = CGD$. Since there is an affinity h with $(AB)^h = CD$ and hence $E^h = G$, we have $g = h$, as g and h agree at three points (by Exercise 2.6). It follows that

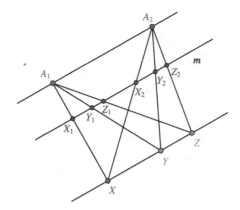

Figure 7.1 The Fairfax–Wallis theorem (Corollary 7.9).

$AB \parallel CD$ (as h, and so g, maps the point at infinity of AB to the point at infinity of CD). Similarly, $BC \parallel AD$. □

We will use the following result on central projections to prove the Fairfax–Wallis theorem.

Theorem 7.8

The restriction to a line m of the composition of two central projections $g\colon m \to m'$ with centre P and $h\colon m' \to m$, with centre Q where PQ is parallel to m, is an affinity (and hence preserves affine ratio).

Proof First, gh is a projectivity of m. Moreover, each affine point of m has an affine image under gh, so gh fixes the point at infinity of m. So the restriction of gh to m is an affinity (and hence preserves affine ratio). □

Corollary 7.9 (Fairfax–Wallis theorem (Wallis, 1693))

If A_1 and A_2 are two points not on a line m with A_1A_2 parallel to m and X, Y, Z are collinear points on a line n not parallel to m and the projections of X, Y, Z from A_1 onto m are X_1, Y_1, Z_1 and those of X, Y, Z onto m from A_2 are X_2, Y_2, Z_2, then ratio$(X_1, Y_1, Z_1) =$ ratio(X_2, Y_2, Z_2) (see Figure 7.1).

Proof Let $g\colon m \to n$ be the central projection with centre A_1, and let $h\colon n \to m$ be the central projection with centre A_2. By Theorem 7.8, the restriction of gh to m is an affinity. Therefore,

$$\mathrm{ratio}(X_1, Y_1, Z_1) = \mathrm{ratio}(X_1^{gh}, Y_1^{gh}, Z_1^{gh})$$
$$= \mathrm{ratio}(X^h, Y^h, Z^h)$$
$$= \mathrm{ratio}(X_2, Y_2, Z_2).$$ □

Theorem 7.10 (Sachse, 1882)

Let ABCD be a quadrilateral, $O := AC \cap BD$, E be the intersection of the parallel to AB on O with CD, F be the intersection of the parallel to BC on O with DA, G be the intersection of the parallel to CD on O with AB, and H be the intersection of the parallel to DA on O with BC. Then E, F, G, H are collinear.

Proof We work in the projective completion with ℓ as the line at infinity. Consider the unique projectivity g interchanging both A and C, and B and D. Then g fixes O, AC, BD and the line joining O to $AB \cap CD$, so fixes O line-wise. Let $W := AB \cap \ell, X := BC \cap \ell, Y := CD \cap \ell, Z := DA \cap \ell$. Then

$$W^g = (AB)^g \cap OW = CD \cap OW = E,$$
$$X^g = (BC)^g \cap OX = AD \cap OX = F,$$
$$Y^g = (CD)^g \cap OY = AB \cap OY = G,$$
$$Z^g = (DA)^g \cap OZ = BC \cap OZ = H.$$

Hence, E, F, G, H all lie on ℓ^g.   □

7.2 Parallel Projection

Let π_1, π_2 be distinct planes of $\mathbb{R}^3$. Let $[\ell]$ be a parallel class of lines that are not parallel to π_1 or to π_2. Define a map $\pi_1 \to \pi_2$ by mapping a point X of π_1 to the point $m \cap \pi_2$ of π_2, where m is the unique member of $[\ell]$ incident with X. Recall that this mapping gives us parallel projection from π_1 to π_2 defined by $[\ell]$. Now each plane π_1 and π_2 carries its own affine structure; we have two copies of AG(2, $\mathbb{R}$). It turns out that parallel projection induces an affine map on AG(2, $\mathbb{R}$) (upon identification). However, in the projective space PG(3, $\mathbb{R}$), $[\ell]$ is the set of lines on a point P lying on the plane at infinity.[2] Therefore, parallel projection is a special case of central projection where the focal point P lies on the plane at infinity.

This model also allows us to view the two-dimensional case of projection between lines in the plane. Let k_1 and k_2 be lines of AG(2, $\mathbb{R}$) not parallel to a line ℓ of AG(2, $\mathbb{R}$). Define a map $k_1 \to k_2$ by mapping a point X of k_1 to the point $m \cap k_2$ of k_2 where m is the unique member of $[\ell]$ incident with X. Again, parallel projection is a special case of central projection where the focal point P lies on the plane at infinity. Indeed, consider our AG(2, $\mathbb{R}$) here to be another

---

[2] We will visit this in Chapter 13.

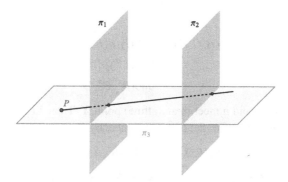

Figure 7.2 Two-dimensional projection in 3-space.

plane π_3 of $\mathbb{R}^3$ containing the line ℓ that we had before, and note that π_3 meets π_1 and π_2 in lines, which we take to be our lines k_1 and k_2. (See Figure 7.2.)

Given points C and D of $\mathsf{AG}(2, \mathbb{R})$, let A be the point of CD on the line at infinity. Then for any point B of CD, we say that B is **between** C and D if and only if $\{A, B\}$ separates $\{C, D\}$ (that is, $R(A, B; C, D) < 0$). The **line segment** with **endpoints** C and D is the set of all points B on the line CD that lie between C and D.

Theorem 7.11 (Darboux, 1880)
Parallel projection from a line m to a line n of $\mathsf{AG}(2, \mathbb{R})$ preserves betweenness and ratios of line segments.

Proof Suppose we have a point B between two points C and D, on m. Let A be the point at infinity on m. Then $R(A, B; C, D) < 0$. Parallel projection π from m to n is simply a central projection via a point P at infinity. Therefore, $(A, B, C, D) \overset{P}{\overline{\wedge}} (A, B^\pi, C^\pi, D^\pi)$, and so, by Theorem 4.7, $R(A, B^\pi; C^\pi, D^\pi) < 0$. Hence, B^π is between C^π and D^π.

Finally, we will show that $\mathsf{ratio}(B, C, D) = \mathsf{ratio}(B^\pi, C^\pi, D^\pi)$. Suppose m and n are parallel. Now we can complete $\mathsf{AG}(2, \mathbb{R})$ to $\mathsf{PG}(2, \mathbb{R})$ by a line ℓ_∞ at infinity, which contains a point P lying on each line BB^π, CC^π, and DD^π, but ℓ_∞ also contains the point A for which m and n meet. But now we take a new line ℓ'_∞ on A, but not containing P, so that parallel projection becomes central projection via P in the affine plane formed by removing ℓ'_∞. So now, in this new copy of $\mathsf{AG}(2, \mathbb{R})$, we can apply Theorem 7.5 to obtain

$$d(B^\pi, C^\pi)/d(B, C) = d(B^\pi, D^\pi)/d(B, D).$$

Hence,

$$\text{ratio}(B, C, D) = d(B, C)/d(B, D)$$
$$= d(B^\pi, C^\pi)/d(B^\pi, D^\pi)$$
$$= \text{ratio}(B^\pi, C^\pi, D^\pi).$$

Now suppose m and n meet in an (affine) point X. Then XDD^π is a triangle in $\text{AG}(2, \mathbb{R})$ and BB^π is parallel to DD^π. Hence, by Theorem 7.5,

$$\text{ratio}(B, X, D) = \text{ratio}(B^\pi, X, D^\pi).$$

Similarly, $\text{ratio}(B, X, C) = \text{ratio}(B^\pi, X, C^\pi)$. Therefore,

$$\text{ratio}(B, C, D) = \text{ratio}(B, X, D)/\text{ratio}(B, X, C)$$
$$= \text{ratio}(B^\pi, X, D^\pi)/\text{ratio}(B^\pi, X, C^\pi)$$
$$= \text{ratio}(B^\pi, C^\pi, D^\pi). \qquad \square$$

7.3 The Theorems of Ceva and Menelaus

From the projective Ceva theorem (Theorem 4.33), we immediately have a 'projective' version of the fact that the medians of a triangle of $\text{AG}(2, \mathbb{R})$ are concurrent (in the centroid). Part (ii) of the following result will allow us to study the **medial triangle** in the projective setting (see Richter-Gebert, 2011, remark 19.2).

Theorem 7.12
Let ABC be a triangle in $\text{PG}(2, \mathbb{R})$ and let ℓ be a line not incident with any vertex of this triangle. Furthermore, let Z', X', Y' be the intersections of the lines AB, BC, CA with ℓ and let X, Y, Z be such that $(A, B; Z, Z')$, $(B, C; X, X')$, and $(C, A; Y, Y')$ are harmonic. Then:

(i) the lines AX, BY, and CZ are concurrent.
(ii) each of the triples (X, Y, Z'), (X, Y', Z), (X', Y, Z) is a set of collinear points.

Proof Observe that each of $\text{R}(A, B; Z, Z')$, $\text{R}(B, C; X, X')$, and $\text{R}(C, A; Y, Y')$ is equal to -1, and so their product is equal to -1. By Theorem 4.33, part (i) of the result holds. Part (ii) can be proved by noting, for example, that Z' is the point we would arrive at if we were to find the harmonic conjugate of the point $XY \cap CZ$ (see Figure 7.3) with respect to X and Y; and similarly for the other two triples of points. $\qquad \square$

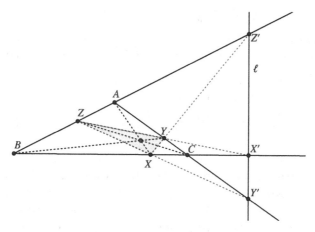

Figure 7.3 The projective generalisation of the medians meeting in the centroid.

By Theorem 7.12, if ℓ is the line at infinity, then XYZ is the medial triangle of ABC.

Corollary 7.13
Let ABC be a triangle of $\mathbb{R}^2$, and let X, Y, Z be the midpoints of the sides BC, CA, AB, respectively. Then the centroid of ABC is also the centroid of XYZ.

Proof We use Theorem 7.12 where ℓ is the line at infinity. Since $(B, Z; A, Z')$ is harmonic, Z is the midpoint of $\overline{AB}$ (by Theorem 7.1). Now

$$(B, X, C, X') \overset{A}{\overline{\overline{\wedge}}} (Z, ZY \cap AX, Y, X')$$

and so $(Z, ZY \cap AX, Y, X')$ is also harmonic (by Theorem 4.7). Therefore, $ZY \cap AX$ is the midpoint of ZY (by Theorem 7.1). Likewise, BY subdivides ZX, and CZ subdivides XY. Therefore, the centroid of ABC is also the centroid of XYZ. □

Theorem 7.14 (Ceva's theorem)
If the sides AB, BC, CA of a triangle are divided by points F, D, E in the ratios $1 : \lambda$, $1 : \mu$, $1 : \nu$ then the three lines CF, AD, BE are concurrent if and only if $\lambda\mu\nu = 1$.

Proof Take ℓ to be the line at infinity in the projective Ceva theorem (Theorem 4.33). □

Theorem 7.15 (Menelaus' theorem)
Suppose points D, E, F are respectively on lines BC, AC, AB and each is distinct from the vertices of the triangle ABC. Then D, E, F are collinear if and only if

$$\text{ratio}(F, A, B)\text{ratio}(D, B, C)\text{ratio}(E, C, A) = 1.$$

Proof Take the line dual to O at infinity in the dual of the projective Ceva theorem.   □

Theorem 7.16 (Carnot's theorem (Carnot, 1806, Chapitre: Essai sur la théorie des transversales))
Let ABC be a triangle of $\mathsf{AG}(2, \mathbb{R})$ *whose sides intersect a (possibly degenerate) conic* $\mathcal{K}$: $AB \cap \mathcal{K} = \{P, P'\}$, $BC \cap \mathcal{K} = \{Q, Q'\}$, $CA \cap \mathcal{K} = \{R, R'\}$. *Then*

$$\text{ratio}(A, B, P)\text{ratio}(A, B, P')\text{ratio}(B, C, Q)\cdot$$
$$\text{ratio}(B, C, Q')\text{ratio}(C, A, R)\text{ratio}(C, A, R') = 1.$$

Conversely, if

$$\text{ratio}(A, B, P)\text{ratio}(A, B, P')\text{ratio}(B, C, Q)\cdot$$
$$\text{ratio}(B, C, Q')\text{ratio}(C, A, R)\text{ratio}(C, A, R') = 1$$

then there is a (possibly degenerate) conic passing through the six points

$$\{P, P', Q, Q', R, R'\}.$$

Proof If $\mathcal{K}$ is degenerate, say P, Q, R are collinear and P', Q', R' are collinear, then by applying Menelaus' theorem (Theorem 7.15) twice:

$$\text{ratio}(A, B, P)\text{ratio}(B, C, Q)\text{ratio}(C, A, R) = -1,$$
$$\text{ratio}(A, B, P')\text{ratio}(B, C, Q')\text{ratio}(C, A, R') = -1,$$

so

$$\text{ratio}(A, B, P)\text{ratio}(A, B, P')\text{ratio}(B, C, Q)\text{ratio}(B, C, Q')\text{ratio}(C, A, R)$$
$$\text{ratio}(C, A, R') = 1.$$

Now consider any non-degenerate conic through Q, Q', R, R' meeting the line AB in P_1, P_1'. By Desargues' involution theorem (Theorem 6.19), $(A\,B)(P\,P')(P_1\,P_1')$ is an involution. Hence,

$$(P, P_1', A, B)\overline{\wedge}(P', P_1, B, A)\overline{\wedge}(P_1, P', A, B),$$

giving $\text{ratio}(P, A, B)\text{ratio}(P', A, B) = \text{ratio}(P_1, A, B)\text{ratio}(P_1', A, B)$. Since

$$\text{ratio}(A, B, P)\text{ratio}(A, B, P')\text{ratio}(B, C, Q)\text{ratio}(B, C, Q')\text{ratio}(C, A, R)$$
$$\text{ratio}(C, A, R') = 1$$

holds in the degenerate case,

$$\text{ratio}(A, B, P_1)\text{ratio}(A, B, P_1')\text{ratio}(B, C, Q)\text{ratio}(B, C, Q')\text{ratio}(C, A, R)$$
$$\text{ratio}(C, A, R') = 1$$

holds in the non-degenerate case.

Conversely, if

$$\text{ratio}(A, B, P)\text{ratio}(A, B, P')\text{ratio}(B, C, Q)\text{ratio}(B, C, Q')\text{ratio}(C, A, R)$$
$$\text{ratio}(C, A, R') = 1,$$

then let $S := BC \cap R'P$, $T := AC \cap QP'$, and $U := AB \cap Q'R$. Since S, R', P are collinear, by Menelaus' theorem (Theorem 7.15),

$$\text{ratio}(A, B, P)\text{ratio}(B, C, S)\text{ratio}(C, A, R') = -1.$$

Similarly,

$$\text{ratio}(A, B, P')\text{ratio}(B, C, Q)\text{ratio}(C, A, T) = -1,$$
$$\text{ratio}(A, B, U)\text{ratio}(B, C, Q')\text{ratio}(C, Q, R) = -1,$$

and hence $\text{ratio}(A, B, U)\text{ratio}(B, C, R)\text{ratio}(C, A, S) = -1$. By Menelaus's theorem (Theorem 7.15) again, S, T, U are collinear. By the Braikenridge–Maclaurin theorem (Theorem 6.43), there is a (possibly degenerate) conic passing through the six points $\{P, P', Q, Q', R, R'\}$. $\square$

Theorem 7.17
Let P, Q, R, S, T be five points, no three collinear. Then the six points $A = QR \cap PS$, $B = RP \cap QS$, $C = PQ \cap RS$, $A' = QR \cap PT$, $B' = RP \cap QT$, $C' = PQ \cap RT$ all lie on a conic.

Proof Note that A, B, C are the Cevians for S on triangle PQR and A', B', C' are the Cevians for T on triangle PQR. Let ℓ be a line on none of A, B, C, A', B', C', P, Q, R, S, T and let $L = \ell \cap QR$, $M = \ell \cap PR$, $N = \ell \cap PQ$. By the projective Ceva theorem (Theorem 4.33),

$$R(Q, R; A, L)R(R, P; B, M)R(P, Q; C, N) = -1$$

and

$$R(Q, R; A', L)R(R, P; B', M)R(P, Q; C', N) = -1.$$

So

$$R(Q, R; A, L)R(R, P; B, M)R(P, Q; C, N)R(Q, R; A', L)R(R, P; B', M)$$
$$R(P, Q; C', N) = 1.$$

By the projective Carnot theorem (Theorem 7.16), A, B, C, A', B', C' lie on a conic.   □

Theorem 7.17 had appeared as a question on the Mathematical Tripos at Cambridge before 1958, when it was published as problem C51 in Heading's *Methods of pure projective geometry* (labelled as from the Tripos; see Heading, 1958) and a variant as C26, also labelled as from the Tripos, whereas Coxeter didn't attribute it to Seymour Schuster until 1963 (Coxeter, 1974, p. 89), so it's likely the result pre-dates Schuster's discovery.

7.4 Affine Conics: Ellipse, Hyperbola, Parabola

We now categorise *affine* conics with respect to a fixed line at infinity. A conic of $PG(2, \mathbb{R})$ is an **ellipse** if it is disjoint from the line at infinity, a **parabola** if the line at infinity is tangent, and a **hyperbola** if the line at infinity is secant. Hence, the stabiliser $AGL(2, \mathbb{R})$ of the line at infinity has at least three orbits on conics of $PG(2, \mathbb{R})$. Indeed, there are only three orbits, and three *kinds* of affine conic.

Theorem 7.18
$AGL(2, \mathbb{R})$ *is transitive on each of the following: ellipses, parabolas, hyperbolas.*

Proof By Theorems 3.13 and 5.1, $PGL(3, \mathbb{R})$ has three orbits on conic–line pairs, accordingly as the line and the conic meet in 0, 1, or 2 points. Hence, $PGL(3, \mathbb{R})_{\ell_\infty}$, thought of as $AGL(2, \mathbb{R})$ (compare with Theorem 2.4), has three orbits on affine conics: ellipses, parabolas, and hyperbolas.   □

Theorem 7.19 (Group law on an affine conic)
Let $\mathcal{K}$ be a non-empty, non-degenerate conic in $AG(2, \mathbb{R})$. Distinguish a point O on $\mathcal{K}$. Define $+\colon \mathcal{K} \times \mathcal{K} \to \mathcal{K}$ by setting $A + B$ to be the second point of $\mathcal{K}$ on the line ℓ incident with O parallel to AB, if ℓ is not tangent to $\mathcal{K}$, and to be O if ℓ is tangent to $\mathcal{K}$. Then $\mathcal{K}$ under this operation is an Abelian group $\mathcal{K}^+$.

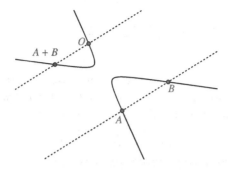

Proof Suppose we have three points A, B, C. In this proof, we will use the convention that the line UU is the tangent line at the point U. The line $O(A+B)$ is parallel to AB and the line $O(B + C)$ is parallel to BC. Let $X := (A + B) + C$ and let $Y := A + (B+C)$. To show that $X = Y$, it suffices to show that $A(B+C)$ is parallel to $(A + B)C$. Consider the hexagram $ABC(A + B)O(B + C)$. Let $(AB)_\infty$ be the point at infinity on AB, and let $(BC)_\infty$ be the point at infinity on BC. By Pascal's theorem (Theorem 6.28), or its degenerate versions when some of the adjacent points of the hexagram are equal, we find that the diagonal point $(A + B)C \cap A(B + C)$ lies on the line joining $(AB)_\infty$ and $(BC)_\infty$; which is the line at infinity. In other words, $(A + B)C$ is parallel to $A(B + C)$, as required. Therefore, $+$ is an associative binary operation. Next, note that the point O is the identity element for this operation, and if A is a point, then the other point of $\mathcal{K}$ on AO is the additive inverse of A. So, indeed, $\mathcal{K}$ forms a group. Finally, it is straightforward to see that $+$ is a commutative operation simply because the line spanned by A and B is also the line spanned by B and A.   □

Corollary 7.20

   (i) For a hyperbola, $\mathcal{K}^+$ is isomorphic to $\mathbb{R}^*$ under multiplication.

  (ii) For a parabola, $\mathcal{K}^+$ is isomorphic to $\mathbb{R}$ under addition.

 (iii) For an ellipse, $\mathcal{K}^+$ is isomorphic to the complex numbers of absolute value 1 under multiplication.

In each case, the permutation action on the conic is regular.

Proof We will demonstrate how (i) is done, and leave the rest to the reader. Without loss of generality (by Theorem 7.18), we can assume that the hyperbola has homogeneous equation $xy = z^2$, and let $O := (1, 1, 1)$ be the fixed point of the hyperbola. Two points $A: (1, a^2, a)$ and $B: (1, b^2, b)$ (where $a, b \neq 0$) span the line with homogeneous coordinates $[a^2b - ab^2, a - b, b^2 - a^2]$, and so the point at infinity on AB is $X := (1, -ab, 0)$. Let $T := (1, t^2, t)$ be a point on the hyperbola collinear with O and X. (So T is $A + B$.) Then, by Theorem 1.5,

$$\begin{vmatrix} 1 & 1 & 1 \\ 1 & -ab & 0 \\ 1 & t^2 & t \end{vmatrix} = 0,$$

which implies that $t = 1$ or $t = ab$. In the former case, $T = O$ and OT is tangent to the hyperbola. Otherwise, we find that $A + B = (1, a^2b^2, ab)$. Therefore, the map $\mathcal{K}^+ \to \mathbb{R}^*$ defined by $(1, x^2, x) \mapsto x$ is an isomorphism of groups.   □

The **centre** of an ellipse or hyperbola is the pole of the line at infinity, and these conics are known as **central conics**. The centre of an ellipse is an internal point and the centre of a hyperbola is an external point. The tangent lines to the conic on the centre of a hyperbola are the **asymptotes** of the hyperbola. Their points of contact lie on the line at infinity, and they decompose the hyperbola into two **branches**. In general, an **asymptote to a curve** is a tangent line to the projective completion of the curve at a point at infinity of the curve.

The following result is from Archimedes, for parabolas only, and the general theorem is from Apollonius de Perge (2010), *Conics*, Book II, Proposition 44.

Theorem 7.21 (Archimedes *Quadrature of the parabola* (third century BC), Proposition 1; see Dijksterhuis, 1987)
*The midpoints of parallel chords to a conic all lie on a line called a **diameter**. Conversely, a line on the centre of a central conic is a diameter.*

Proof The join of two midpoints of parallel chords is the polar line with respect to the conic of the point at infinity on both lines, by Theorem 4.38. Hence, it is incident with the pole of the line at infinity, which is the centre of the conic. Conversely, the chords parallel to the tangent at the extremity of a diameter ℓ have a common point at infinity, which has polar line ℓ since it is the intersection of the tangents at the extremities of ℓ, and so, by Theorem 4.38, ℓ contains the midpoints of each of these chords.   □

Theorem 7.22 (Hamilton, 1758, liber 1, prop LV)
If A, N are points on a hyperbola K and m is a line on the point B of K parallel to an asymptote a of K, ratio$(B, AD \cap m, ND \cap m)$ is independent of the choice of D on K ($\neq A, N$ and not on m).

Proof Let X be the point at infinity of the asymptote a. Then

$$R(DX, DB; DA, DN) = \text{ratio}(B, AD \cap m, ND \cap m),$$
$$R(D'X, D'B; D'A, D'N) = \text{ratio}(B, AD' \cap m, ND' \cap m),$$
$$R(DX, DB; DA, DN) = R(X, B; A, N)_K = R(D'X, D'B; D'A, D'N)$$

(by the Chasles–Steiner theorem (Theorem 6.59) for all D, D' on K (not A, B, X, N). □

Hamilton mentions a letter to Digby from Fermat and Wallis' demonstration, which we will see later in this section.

Theorem 7.23 (de la Hire, 1685)
Any pair of conjugate diameters of a hyperbola form a harmonic quadruple with the asymptotes.

Proof Applying the polarity, we must show that a conjugate pair of points at infinity forms a harmonic quadruple with the pair of points at infinity on the conic. But this is the content of Theorem 4.38. □

The following result demonstrates the usefulness of the language of harmonic separation.

Theorem 7.24
Given five points A_1, A_2, A_3, A_4, A_5 on a central conic, take any three points $\{A_1, A_2, A_3, A_4, A_5\} \setminus \{A_i, A_j\}$, and through the centroid of these three points draw a line ℓ_{ij} parallel to the line through the centre of the conic and the pole of the line through the remaining two points. Then these 10 lines are concurrent.

Proof Let $\{X, Y\}$ be the points at infinity on the conic. Then

$$(X, Y; \ell_{ij} \cap \ell_\infty, A_i A_j \cap \ell_\infty)$$

is harmonic. Hence, ℓ_{ij}, ℓ_{ik}, ℓ_{jk} are concurrent. □

The forward direction of the following result is due to Apollonius, and the backward direction is due to ibn al Haytham, *Completion of the Conics* (written before 1021) P4, Proposition 10 (see translation by Hogendijk, 1985).

Theorem 7.25 (Apollonius de Perge, 2010, III.41; Hogendijk, 1985, pp. 17–18, 186–90)
Let m, m', a, b, c be five lines in $AG(2, \mathbb{R})$, no three concurrent. Let $A := m \cap a$, $B := m \cap b$, $C := m \cap c$, $A' := m' \cap a$, $B' := m' \cap b$, $C' := m' \cap c$. Then there is a parabola tangent to m, m', a, b, and c if and only if ratio$(A, B, C) =$ ratio(A', B', C').

Proof Let ℓ_∞ be the line at infinity $M := \ell_\infty \cap m$, $M' := \ell_\infty \cap m'$. Then there is a parabola of $AG(2, \mathbb{R})$ tangent to m, m', a, b, and c if and only if there is a conic of $PG(2, \mathbb{R})$ tangent to ℓ_∞, m, m', a, b, and c. By the Chasles–Steiner theorem

(Theorem 6.61), this is equivalent to $R(M, A; B, C) = R(M', A'; B', C')$, which is in turn equivalent to $\text{ratio}(A, B, C) = \text{ratio}(A', B', C')$.   □

Theorem 7.26 (Fermat, 19 June 1658 letter to Kenelm Digby)
If A, N are points on a parabola K and m is a line on the point B of K parallel to the axis of K, then $\text{ratio}(B, AD \cap m, ND \cap m)$ is independent of the choice of D on K ($\neq A, N$ and not on m).

Proof Let X be the point at infinity of K. Then, for all D, D' on K (not A, B, X, N),

$$R(DX, DB; DA, DN) = \text{ratio}(B, AD \cap m, ND \cap m),$$

$$R(D'X, D'B; D'A, D'N) = \text{ratio}(B, AD' \cap m, ND' \cap m),$$

and

$$R(DX, DB; DA, DN) = R(X, B; A, N)_K = R(D'X, D'B; D'A, D'N)$$

(by the Chasles–Steiner theorem (Theorem 6.61)).   □

Theorem 7.26 was first published in Wallis (1658) (in Letter XLVII, Fermat-Kenelm Digby, 19 June 1658, p. 188). It is also in de Fermat (1679, *Porismatum Euclideorum*, pp. 116–19, *Porisma Secundum*, p. 117), Tannery and Henry (1894, p. 407), Wallis (2003, Letter 165: Pierre de Fermat to Kenelm Digby June 1658, pp. 491–6 (in English))) and Fermat's posthumously printed tract on porisms *Porismatum Euclideorum* (Fermat, 1999, pp. 76–84 as Porisma Secundum, p. 79).

Theorem 7.27
Let K be a conic of $\text{AG}(2, \mathbb{R})$ and P be a point, not on K, and on no asymptote of K. Then

$$\left\{ M : PM \cap K = \{Q, R\}, \quad M \text{ is the midpoint of } \overline{QR} \right\}$$

is a conic. In other words, the midpoints of the chords on P lie on a conic.

Proof The map g: $\text{pencil}(P) \to \ell_\infty, n \mapsto n \cap \ell_\infty$ is a projectivity. If K is central with centre C, let $S := \text{pencil}(P)$. If K is a parabola, let S be the parallel class of the axis of K. The map h: $\ell_\infty \to S$, $X \mapsto$ the polar line of X with respect to K, is a projectivity. So gh is a projectivity.

Suppose gh is a perspectivity. Then the line n of S on P has $n^{gh} = n$, so n^g is the pole of n and on n and ℓ_∞; so n is an asymptote of K. So gh is not a perspectivity. By Theorem 6.59, $\{n \cap n^{gh} : n \text{ on } P\}$ is a conic. Let n be a line on P, $M = n \cap n^{gh}$, and $n \cap K = \{Q, R\}$. Since M is on the polar line of $n \cap \ell_\infty$, by

Apollonius' theorem (Theorem 4.38), $R(n \cap \ell_\infty, M; Q, R) = -1$. So M is the midpoint of Q and R. $\square$

Theorem 7.28 (Steiner, 1846)
The midpoints of the six sides of a quadrangle of $AG(2, \mathbb{R})$ *and the three diagonal points all lie on a conic.*

Proof We argue in $PG(2, \mathbb{R})$. Let $ABCD$ be the given quadrangle. Let AB cut the line at infinity at K. Then the midpoint K' of $\overline{AB}$ has (A, B, K, K') harmonic. Similarly, let CD, AC, BD, AD, BC, respectively, cut the line at infinity at L, M, N, O, P and let L', M', N', O', P' be the respective midpoints. Now

$$(A, B, K, K') \overset{W}{\barwedge} (A, C, M, M'),$$

so BC, KM, and $K'M'$ are concurrent, by Corollary 4.10; that is, $K'M'$ passes through P. Similarly, $K'N'$, $K'O'$, and $K'P'$ pass through O, N, and M, respectively. Hence,

$$R(K'M', K'N'; K'O', K'P') = R(P, O; N, M) = R(M, N; O, P).$$

Similarly,

$$R(L'M', L'N'; L'O', L'P') = R(M, N; O, P).$$

Therefore, by Theorem 6.61, K', L', M', N', O', P' lie on a conic $\mathcal{K}$, which is uniquely determined by Theorem 5.17. Let $U := AB \cap CD$, $V := AC \cap BD$, $W := AD \cap BC$. Now $(A, C; M, M')$ is harmonic, so $(UA, UC; UM, UM')$ is harmonic. Similarly, $(B, D; N, N')$ is harmonic, so $(UA, UC; UN, UN')$ is harmonic; $(A, D; O, O')$ is harmonic so $(UA, UC; UO, UO')$ is harmonic; and $(B, C; P, P')$ is harmonic, so $(UA, UC; UP, UP')$ is harmonic. Hence,

$$R(UM', UN'; UO', UP') = R(M, N; O, P) = R(K'M', K'N'; K'O', K'P'),$$

so U also lies on $\mathcal{K}$. Similarly for V and W. $\square$

In 1814, Reverend Thomas Scurr from Hexham (a small town in the north of England), who was Master of Hexham Grammar School, posed as question 1029 in *The Gentleman's Diary* the first appearance of the butterfly theorem.

Theorem 7.29 (Butterfly theorem)
Let A, B, C, D, E, F be points on a conic $\mathcal{K}$, let O be the midpoint of $\overline{AB}$, and suppose CF and DE pass through O. Then O is the midpoint of the line segment with endpoints $CE \cap AB$ and $DF \cap AB$ (see Figure 7.4).

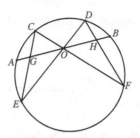

Figure 7.4 Butterfly theorem.

We give five proofs: the first is a beautifully simple application of cross-ratio, the second uses projectivities and harmonic conjugacy, the third uses the pole–polar properties of a conic, the fourth uses Desargues' involution theorem (Theorem 6.19), and the fifth utilises Möbius' theorem (Theorem 6.81).

First proof Here we present a beautiful proof by John Casey (1892). By Chasles' theorem (Theorem 5.18), R($EA, EC; ED, EB$) = R($FA, FC; FD, FB$), and so (intersecting with AB), R($A, G; O, B$) = R($A, O; H, B$) by Corollary 4.10. Take the reflection σ_ℓ about the perpendicular line ℓ to AB, through the point O. Since σ_ℓ is induced by a collineation of PG(2, $\mathbb{R}$), and hence preserves cross-ratio (by Theorem 4.15), we have

$$R(A, O; H, B) = R(A, G; O, B)$$
$$= R(A^{\sigma_\ell}, G^{\sigma_\ell}; O^{\sigma_\ell}, B^{\sigma_\ell})$$
$$= R(B, G^{\sigma_\ell}; O, A)$$
$$= R(A, O; G^{\sigma_\ell}, B).$$

Since three of the points are equal now on both sides, and their cross-ratios are the same (by Lemma 4.3), the fourth points H and G^{σ_ℓ} are equal. By definition of a reflection, O is the midpoint of $\overline{GH}$. □

Second proof Let $G = CE \cap AB$, $H = DF \cap AB$. Let I be the point at infinity on AB. Now, by the Chasles–Steiner theorem (Theorem 6.61) (and Theorem 4.7) we have $(DA, DF, DE, DB) \barwedge (CA, CF, CE, CB)$. Noting the points at which these two pencils meet the line AB, we have $AHOB \barwedge AOGB \barwedge BGOA$. Since the projectivity $AHOB \barwedge BGOA$ interchanges A and B, it must be an involution. Since it has one fixed point, O, it must have another which is the harmonic conjugate of O with respect to A and B, namely, I. Since H and G are another pair of this involution, these likewise must be harmonic conjugates with respect to O and I. Hence, O is the midpoint of $\overline{GH}$. □

Third proof Let $G = CE \cap AB$, $H = DF \cap AB$. Let CE meet DF in P, and let CD meet EF in Q. Then PQ is the polar of O with respect to the conic. But any line through O intersects the conic and PQ in points which, with O, form a harmonic quadruple; therefore, A, O, B, and the intersection $I := AB \cap PQ$ are harmonic. But $\mathrm{ratio}(A, O, B) = 1$: thus I must be at infinity. That is, AB and PQ are parallel. Furthermore, for the complete quadrangle $CDEF$, we have that the lines PC, PO, PD, PQ are a harmonic quadruple, by Theorem 4.31. These lines meet AB in points G, O, H, I, which are thus a harmonic quadruple; but I is at infinity, so O is the midpoint of $\overline{GH}$. □

Fourth proof This is an example of Desargues' involution theorem (Theorem 6.19) concerning conics passing through the vertices of a complete quadrangle. Now O is a fixed point of the involution determined by $CDFE$ on AB. Since A, B form a pair of corresponding points in this involution, the other fixed point is at infinity, and the points G, H (where CE, DF meet AB) have O as their midpoint. □

Fifth proof The butterfly theorem is also a consequence of Theorem 6.81. First the *double butterfly theorem* states that if two re-entrant quadrangles $ABCD$ and $A'B'C'D'$ inscribed in a circle have their respective sides meeting a chord of the circle at R_1, R_2, R_3, R_4 and R'_1, R'_2, R'_3, R'_4, and if the chord's midpoint M bisects the segments $\overline{R_1 R'_1}$, $\overline{R_2 R'_2}$, and $\overline{R_3 R'_3}$, then M bisects $\overline{R_4 R'_4}$. Theorem 6.81 generalises this by removing the midpoint and its metric aspects, by allowing all quadrangles, by removing the restriction that the given line intersect the circle, and by including conics other than the circle. Indeed, the proof of the double butterfly theorem becomes a simple matter of using Theorem 6.81: map $ABCD$ to $A''B''C''D''$ by reflection in the midpoint M. By hypothesis, therefore, three corresponding sides of $A'B'C'D'$ and $A''B''C''D''$ pass through R'_1, R'_2, and R_3. By Theorem 6.81, the fourth side $B'C'$ and $B''C''$ pass through R_4, and since $B''C''$ is a reflection of BC in M, we have that M is the midpoint of $\overline{R4R4'}$. To prove the classical butterfly theorem (Theorem 7.29) using Theorem 6.81, align $ABCD$ and $A'B'C'D'$ so that $R_1 = R'_4$ and $R_2 = R'_2 = R_3 = R'_3 = M$, assume that R_4 is distinct from R', and apply the above argument. □

For more, see Jones (1980). Here is a similar 'butterfly' result.

Theorem 7.30 (Fritsch, 2015, Theorem 3.4)
Let the quadrangle $ABCD$ be inscribed in a conic K, g be a line not through any vertex of the quadrangle and not a tangent to the conic, L be a point on the line g, k be the polar of L with respect to K, $M = g \cap k$, $T = AB \cap g$, $Y = BC \cap g$, $U = CD \cap g$, $X = DA \cap g$, $W = BC \cap g$, $V = AC \cap g$. If one of the

quadruples (T, U, L, M), (V, W, L, M), (X, Y, L, M) *is harmonic, so are the other two.*

Proof We may assume that (T, U, L, M) is harmonic. By Pappus' involution theorem (Theorem 6.12), $ABCD$ defines the involution $t = (TU)(VW)(XY)\ldots$ on g. By Desargues' involution theorem (Theorem 6.19), the conic K also meets the line g in two points P and Q, (that possibly coincide and) that are also exchanged by t. Since the points L and M are conjugate with respect to the conic K, the quadruple (P, Q, L, M) is harmonic. Since (T, U, L, M) is harmonic, the unique involution fixing L and M must interchange both T and U, and P and Q, and hence equals t. Thus, (V, W, L, M) and (X, Y, L, M) are harmonic. □

Exercises

7.1   Let $ABCD$ be a parallelogram, P on AD, Q on BC, R on CD, S on AB, $PQ \parallel AB$, $RS \parallel AD$, $T := PQ \cap RS$. Show that BP, CT, DS are concurrent.

7.2   Point M lies on diagonal BD of parallelogram $ABCD$. Line AM intersects side CD and line BC at points K and N, respectively. Let K_1 be the circle with centre M and radius $\overline{MA}$, and K_2 be the circumcircle of triangle KCN. Suppose K_1 and K_2 intersect at P and Q. Prove that MP, MQ are tangent to K_2.

7.3   Show that if a parallelogram is inscribed in a conic, the conic is central, and the diagonals intersect in the centre.

7.4   Let S_1, S_2, S_3 be the midpoints of three concurrent Cevians of triangle ABC in $\mathsf{AG}(2, \mathbb{R})$. Let $S_2 S_3$, $S_3 S_1$, $S_1 S_2$ meet the sides BC, CA, AB in $A_1, B_1, C_1; A_2, B_2, C_2; A_3, B_3, C_3$, respectively. Show that
       (a) A_1, B_2, C_3 are collinear;
       (b) $A_2, A_3, B_3, B_1, C, C_2$ lie on a conic.

7.5   (Siebeck, 1865; Bôcher, 1892/3; Grace, 1902) Given a triangle in $\mathsf{AG}(2, \mathbb{R})$, show that there is a unique ellipse inscribed in the triangle which touches the midpoints of the sides of the triangle.

7.6   Let $\mathcal{K}$ be a conic of $\mathsf{AG}(2, \mathbb{R})$, O be a point not on $\mathcal{K}$. For X on $\mathcal{K}$, let $OX \cap C = \{X, X'\}$. Show that the locus of the midpoint of $\overline{XX'}$ as X varies on $\mathcal{K}$ is a conic on O and on the centre of $\mathcal{K}$.

7.7   A variable tangent to a parabola of $\mathsf{AG}(2, \mathbb{R})$ meets two fixed tangents at X, Y. Show that the locus of the midpoints of $\overline{XY}$ as the tangent varies is a line.

7.8 The triangle ABC is self-polar with respect to a parabola Σ of $\mathsf{AG}(2, \mathbb{R})$; L, M, N are the midpoints of $\overline{BC}$, $\overline{CA}$, $\overline{AB}$, respectively. Show that the lines MN, NL, LM are tangent to Σ.

7.9 A point P is taken on the diagonal BD of the parallelogram $ABCD$, and a line through P meets AD in V and AB in W. The line through W parallel to AD meets CD in N; the line through V parallel to AB meets BC in M.

   (a) Prove that M, N, P are collinear.

   (b) The lines BV, DW meet in U; the lines BN, DM meet in L. Prove that CU and AL pass through the point of intersection of WN, VM.

8

Euclidean Plane Geometry from a Projective Perspective

> [I]t is the glory of geometry that from so few principles, brought from without, it is able to produce so many things.

Isaac Newton (Newton and Motte, 1848, p. 67)

8.1 Perpendicularity of Lines and Triangle Theorems

To derive the real affine plane from the real projective plane we singled out for special treatment a line to be the line at infinity; $\ell_\infty : z = 0$. Similarly, Euclidean geometry can be derived from affine geometry by picking out an involution on the line at infinity, that allows us to say when two lines are perpendicular. According to a theorem of Carathéodory (see Theorem 8.50), every property of (similarity) Euclidean geometry follows from this involution, including measurement of angle and eccentricity of ellipses. Recall from secondary school that two lines $ax + by + c = 0$ and $a'x + b'y + c' = 0$ of the Euclidean plane are perpendicular if $a'b' = -ab$. Now the points of intersection of these two lines with $\ell_\infty : z = 0$ are

$$(b, -a, 0) \quad \text{and} \quad (b', -a', 0).$$

So we see that the two lines are perpendicular if the points at infinity can be interchanged by an involution. This (elliptic) involution is

$$*: (x, y, 0) \mapsto (-y, x, 0).$$

Two lines ℓ and m of the Euclidean plane $\mathbb{R}^2$ are **perpendicular** if and only if $(\ell \cap \ell_\infty)^* = m \cap \ell_\infty$. The following 'Euclidean' results have projective proofs, which essentially only utilise the involution at infinity.

Theorem 8.1

(i) Two lines of $\mathbb{R}^2$, perpendicular to the same line, are parallel.

(ii) If ℓ, m, n are lines of $\mathbb{R}^2$ such that $\ell \parallel m$ and $\ell \perp n$, then $m \perp n$.

Proof

(i) Let m be a line and suppose ℓ, ℓ' are two lines perpendicular to m. Then, by definition,

$$(\ell \cap \ell_\infty)^* = m \cap \ell_\infty = (\ell' \cap \ell_\infty)^*$$

and hence $\ell \cap \ell_\infty = \ell' \cap \ell_\infty$; that is, ℓ and ℓ' are parallel.

(ii) Suppose $\ell \parallel m$ and $\ell \perp n$. Then $\ell \cap \ell_\infty = m \cap \ell_\infty$ and $\ell \cap \ell_\infty = (n \cap \ell_\infty)^*$. Therefore, $(n \cap \ell_\infty)^* = m \cap \ell_\infty$ and hence $m \perp n$. □

Theorem 8.2 (Desargues (1639), see Desargues, 1951 and Field and Gray, 1987, p. 52)

Given four collinear points A, B, C, D and a point O not on AB, any two of the following statements imply the third.

(i) $(OA, OB; OC, OD)$ is harmonic.

(ii) OB bisects angle $\angle AOC$ internally.

(iii) OB and OD are perpendicular to each other.

Proof We show that (i) and (iii) imply (ii) and leave the rest to the reader. Draw $EF \parallel AO$ such that E is on OC and F is on OD. Now, $AO \parallel EF$ implies $\frac{d(A,C)}{d(C,B)} = \frac{d(A,O)}{d(E,B)}$, and since $AO \parallel EF$, we have $\frac{d(A,D)}{d(B,D)} = \frac{d(A,O)}{d(B,F)}$. Now $(OA, OB; OC, OD)$ is harmonic, so $(A, B; C, D)$ is harmonic, so $\frac{d(A,C)}{d(C,B)} = \frac{d(A,D)}{d(B,D)}$, giving $d(E, B) = d(B, F)$, that is, B is the midpoint of $\overline{EF}$. Now, we already have that angles $\angle OBF$ and $\angle BOA$ are right angles, and $\overline{EB}$ is congruent to $\overline{BF}$, so triangles OBE and OBF are congruent. Therefore, OB bisects $\angle EOF$, but since this is perpendicular to OA, we have that OA also bisects $\angle EOF$. Since R$(A, B; C, D)$ is negative, $\{C, D\}$ separates $\{A, B\}$, so OB is the internal bisector and OA the external bisector (see Figure 8.1). □

Corollary 8.3

If a line ℓ bisects the (internal) angle given by lines a and b, then the perpendicular to ℓ at the point of intersection also bisects the (external) angle between a and b and $(\ell, m; a, b)$ is harmonic.

Proof We simply assign O to $a \cap b$, place A on a, B on ℓ, and C on b, and then we have (i) and (ii) of Theorem 8.2 if D lies on the perpendicular to ℓ.

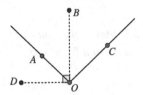

Figure 8.1 Theorem 8.2.

Since (iii) subsequently holds, it follows that OD bisects the (external) angle between a and b. □

Corollary 8.4 (Desargues (1639))
Given four lines ℓ, m, a, b through a point such that $(\ell, m; a, b)$ is harmonic, if ℓ and m are perpendicular, then they bisect the angles between a and b.

Proof This is just (i) and (iii) imply (ii) in Theorem 8.2. □

Corollary 8.5
Given an angle $\angle APB$, let ℓ be the internal angle bisector and m be the external angle bisector of PB and ℓ at P. Let $C := AB \cap \ell$, $D := AB \cap m$. Then $(A, B; C, D)$ is harmonic.

Proof By Theorem 8.2, $(PA, PB; \ell, m)$ is harmonic, or, in other words, the 4-tuple $(PA, PB; PC, PD)$ is harmonic. So, by Theorem 4.7, $(A, B; C, D)$ is harmonic. □

Worked Example 8.6
This example is from Dobos (2011).

Problem In a convex quadrilateral $ABCD$, with $O := AC \cap BD$, the internal bisectors of $\angle AOB$, $\angle BOC$, $\angle COD$, $\angle DOA$ intersect AB, BC, CD, DA at M, N, P, Q respectively. Prove that MQ, NP, BD are concurrent.

Solution Let $AC \cap MQ = S$ and $MQ \cap BD = L$. The reflection in OM interchanges OA and OB and also fixes OQ and OM (as they are perpendicular, being the internal and external angle bisectors of $\angle AOB$). So $(OQ, OA; OM, OB)$ is harmonic. Intersecting with QM, we see that $(Q, S; M, L)$ is harmonic. By using a perspectivity from A taking QM to DB gives $(D, O; B, L)$ harmonic. Similarly, $(ON, OC; OP, OD)$ is harmonic. Intersecting with NP, we see that $(N, AC \cap NP; PNP \cap BD)$ is harmonic. By using a perspectivity from C taking NP to DB gives $(B, O; B, NP \cap BD)$ harmonic. So $L = NP \cap BD$. □

Theorem 8.7

*The altitudes of a triangle are concurrent (in the **orthocentre**).*

Proof Let ABC be a triangle in $\mathbb{R}^2$. If ABC is right-angled, then it has only two altitudes, so they meet in a point. So suppose that ABC is not right-angled, and let ℓ be the altitude through A and m be the altitude through B. Since BC and AC meet at C, ℓ and m are not parallel. Let $D := \ell \cap m$. Since ABC is not right-angled, $ABCD$ is a quadrangle. Apply Pappus' involution theorem (Theorem 6.12) to the quadrangle $ABCD$ and the line at infinity ℓ_∞ to obtain an involution τ on ℓ_∞ interchanging $BC \cap \ell_\infty$ and $AD \cap \ell_\infty$, $AC \cap \ell_\infty$ and $BD \cap \ell_\infty$. Since BC is perpendicular to AD and AC is perpendicular to BD, τ agrees with the perpendicularity involution on ℓ_∞ at four points, and therefore equals it. Hence, CD is perpendicular to AB. □

Theorem 8.8 (Dobos, 2011)

Let ABC be a triangle and let the angle bisector of angle $\angle BAC$ meet BC at A'. Let X be a point of AA' between A and A'. Let $BX \cap AC = B'$, $CX \cap AB = C'$, $A'B' \cap CC' = P$, $A'C' \cap BB' = Q$. Then angles $\angle PAC$ and $\angle QAB$ are congruent.

Proof Let σ be the reflection in AA'. Then $(\angle PAC)^\sigma = \angle P^\sigma AB$ and so we are required to show that σ maps AP to AQ. In projective language, we need only show that $R(AB, AA'; AQ, AC) = R(AC, AA'; AP, AB)$. Intersecting with CP, we see that

$$R(AB, AA'; AQ, AC) = R(C', X; Q', C) = R(QC', QX; QA, QC)$$
$$= R(A', B; BC \cap QA, C) = R(AX, AB; AQ, AC).$$

By the dual of Pappus' involution theorem (Theorem 6.12), $R(AX, AB; AQ, AC) = -1$, and, similarly, $R(AC, AA'; AP, AB) = -1$. Therefore, $R(AB, AA'; AQ, AC)$ and $R(AC, AA'; AP, AB)$ are equal. □

Lemma 8.9 (Richter-Gebert, 2011, Theorem 8.2)

Let A, B, C be three distinct points on the line at infinity ℓ_∞ of $\mathsf{PG}(2, \mathbb{R})$. Then $\{A, B, C, A^*, B^*, C^*\}$ is a quadrangular set.

The above result holds more generally if we instead have an involutory collineation on ℓ_∞ in place of $*$.

Another proof of Theorem 8.7 Let P, Q, R be a triangle, where none of the vertices lie on ℓ_∞. Take the point S that lies on two altitudes of the triangle; those which emanate from Q and R. Let $A = PQ \cap \ell$, $B = PR \cap \ell$, $C = PS \cap \ell$, $D = RS \cap \ell$, $E = QS \cap \ell$, $F = QR \cap \ell$. Since $PQRS$ is a complete quadrilateral,

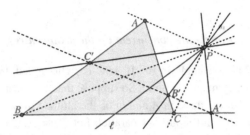

Figure 8.2 Theorem 8.10.

the set $\{A, B, C, D, E, F\}$ is a quadrangular set. By definition of altitude, we have $A^* = D$ and $B^* = E$. By Lemma 8.9, $\{A, B, C, D, E, C^*\}$ is a quadrangular set. Therefore, by Theorem 6.86, $C^* = F$ and hence $PS \perp QR$ as required. □

$$* * *$$

In what follows, we will need to dualise the converse to Pappus' involution theorem (Theorem 6.13).

Let ℓ, m, n be lines on a point P, and ABC be a triangle. If $(PA\,\ell)(PB\,m)(PC\,n)\ldots$ is an involution, then $BC \cap \ell$, $CA \cap m$, $AB \cap n$ are collinear.

Theorem 8.10 (USA Mathematical Olympiad 2012, problem 5)
Let P be a point in the plane of a given triangle ABC. If A', B', C' are the points where the reflections of lines PA, PB, PC in a given line ℓ through P meet the sides BC, CA, AB, respectively, then A', B', and C' are collinear (see Figure 8.2).

Proof Note that $(PA\,PA')(PB\,PB')(PC\,PC')\ldots$ is an involution (induced by the reflection) on the pencil of lines through P, so by the dual of the converse of Pappus' involution theorem, A', B', and C' are collinear. □

Let P be a point in the plane of a triangle $\triangle ABC$. Reflect the lines PA, PB, and PC in the bisectors of angles at A, B, and C, respectively. The lines thus obtained are concurrent in the **isogonal conjugate** P' of P (see Figure 8.3). The existence of isogonal conjugates follows from Ceva's theorem, but we also prove it (with projective methods) in Theorem 10.49. When P happens to lie on the circumcircle of $\triangle ABC$, the reflections of PA, PB, and PC in the angle bisectors become parallel, and so the isogonal conjugate of P lies on the line at infinity. This mapping of a point to its isogonal conjugate, taking the line at infinity to the circumcircle of the triangle, is an example of a **quadratic transformation**. We will encounter more of these kinds of maps in later sections.

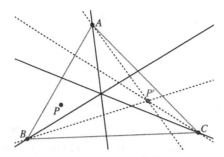

Figure 8.3 The isogonal conjugate P' of a point P.

For a line a' on the vertex A, its **isogonal conjugate** is the line obtained by reflecting a' in the angle bisector at A (and likewise for lines emanating from the other vertices).

We define the isogonal conjugate of a point X on a side of the triangle in the following way. Suppose, without loss of generality, that X lies on AB. Take the reflection ℓ of CX in the angle bisector at C, and let X' be the point of intersection of ℓ with AB. Then X' is the isogonal conjugate of X. The reader can verify the following property of isogonal conjugacy (see Exercise 8.11):

Let I be the point at infinity on AB. Then $R(X, A; B, I) \cdot R(X', A; B, I) = 2$ if X and X' lie on AB and are isogonal conjugates with respect to $\triangle ABC$.

Using this simple cross-ratio formula, and the projective Menelaus' theorem (Theorem 4.34), we have the following well-known property of isogonal conjugacy.

Corollary 8.11
Three points, one of each side of a triangle, are collinear if and only if their isogonal conjugates are collinear.

Proof Let the triangle in question be ABC, and the three points, D, E, F (on sides BC, CA, AB, respectively). Let ℓ_∞ be the line at infinity. Let D', E', F' be the isogonal conjugates of D, E, F, respectively. By the above cross-ratio formula,

$$R(D, B; C, BC \cap \ell_\infty) \cdot R(D', B; C, BC \cap \ell_\infty) = 1$$
$$R(E, C; A, CA \cap \ell_\infty) \cdot R(E', C; A, CA \cap \ell_\infty) = 1$$
$$R(F, A; B, AB \cap \ell_\infty) \cdot R(F', A; B, AB \cap \ell_\infty) = 1.$$

By Theorem 4.34, D, E, F are collinear if and only if

$$R(B, C; D, BC \cap \ell_\infty) \cdot R(C, A; E, CA \cap \ell_\infty) \cdot R(A, B; F, AB \cap \ell_\infty) = 1,$$

which is equivalent to

$$R(B, C; D', BC \cap \ell_\infty) \cdot R(C, A; E', CA \cap \ell_\infty) \cdot R(A, B; F', AB \cap \ell_\infty) = 1.$$

This in turn is equivalent to D', E', F' being collinear.   □

The following was posed as a problem by Arnold Droz-Farny in 1899 (Droz-Farny, 1899). The Droz-Farny theorem was first proved by Albert Noyer (1893) 6 years before Droz-Farny posed the problem, and 10 years before by M. W. Mantel (1889).

Theorem 8.12 (Droz-Farny theorem)
If two perpendicular straight lines are drawn through the orthocentre of a triangle, they intercept a segment on each of the sidelines. The midpoints of these three segments are collinear.

Proof Apply Theorem 8.10 using either of the angle bisectors of the pair of perpendicular lines. The resulting triple of collinear points are precisely the midpoints of the three segments. For if the perpendicular lines m and n meet the side BC of triangle ABC in A', A'', respectively, then the altitude at A for triangle ABC is also an altitude for triangle $A'HA''$, where H is the orthocentre of ABC. Hence, its reflection in the angle bisector ℓ at H of triangle $A'HA''$ passes through the circumcentre of triangle $A'HA''$ (as the orthocentre and the circumcentre are isogonal conjugates, by Exercise 8.36), so the intersection of the reflection in ℓ of AH with BC is the midpoint of $\overline{A'A''}$, and we may argue similarly for the other cases.   □

Here is another application of Pappus' involution theorem (Theorem 6.12) to a result in Euclidean geometry.

Theorem 8.13 (Blaikie, 1905)
Let ABC be a triangle and let O be another point, and let any straight line ℓ through O meet BC in P, CA in Q, AB in R. Let P' be the reflection of P in O, Q' be the reflection of Q in O, and R' be the reflection of P in O. Then AP', BQ', CR' are concurrent (see Figure 8.4).

Proof Let $D := BQ' \cap CR'$. Apply Pappus' involution theorem (Theorem 6.12), to $ABCD$ and ℓ: so $(P\,AD \cap \ell)(Q\,Q')(R\,R')\ldots$ is an involution t. The reflection in O is an involution u of ℓ: $(P\,P')(Q\,Q')(R\,R')\ldots$ Since t and u share the pairs $(Q\,Q')(R\,R')$, they are equal: so $P' = AD \cap \ell$. Thus, AP', BQ', CR' are concurrent in D.   □

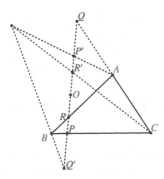

Figure 8.4 Theorem 8.13.

The following results of Maxwell and Steiner can also be found in Pedoe (1967), where these proofs were taken.

Theorem 8.14 (Maxwell, 1864, p. 258)
If ABCV is a quadrangle and A'B'C' is a triangle with VA ∥ B'C', VB ∥ C'A', VC ∥ A'B' then the lines through A' parallel to BC, through B' parallel to CA, through C' parallel to AB are concurrent.

Proof By Pappus' involution theorem (Theorem 6.12), if the point at infinity on *BC* is *X*, the point at infinity on *AV* is *X'*, the point at infinity on *CA* is *Y*, the point at infinity on *BV* is *Y'*, the point at infinity on *AB* is *Z*, the point at infinity on *CV* is *Z'*, then $(XX')(YY')(ZZ')\ldots$ is an involution. Now *B'C'* is on *X'*, *C'A'* is on *Y'*, and *A'B'* is on *Z'*. Let $V' = A'X \cap B'Y$. Then *A'B'C'V'* is a quadrangle giving the involution $(XX')(YY')$ on the line at infinity and so this involution interchanges the point *Z'* at infinity on *A'B'* and *Z*, which must therefore be the point at infinity on *C'V'*. Thus, *A'X*, *B'Y*, and *C'Z* are concurrent. □

Here is a restatement of the above theorem: *If the parallels on the vertices of one triangle to the sides of another are concurrent, then the corresponding parallels on the vertices of the latter to the sides of the former are concurrent.*

Theorem 8.15 (Steiner, 1827b)
If the perpendiculars from the vertices of one triangle to the sides of another are concurrent, then the corresponding perpendiculars from the vertices of the latter to the sides of the former are concurrent.

Proof Rotate the second triangle by a right angle and apply Theorem 8.14. □

Theorem 8.16 (Thébault's problem)
Through the vertex A of a triangle ABC, a straight line AD is drawn, cutting the side BC at D. Let I be the centre of the incircle of ABC. Let P be the centre of the circle which touches DC, DA at E, F, and the circumcircle of ABC, and let Q be the centre of a further circle which touches DB, DA in G, H and the circumcircle of ABC. Then P, I, and Q are collinear.

Although Theorem 8.16 was proposed in the 'Problems and solutions' section of the *American Mathematical Monthly* in 1938 by Victor Thébault (1882–1960), it was not solved until 1973, when three independent solutions were published in *Nieuw Tijdschrift voor Wiskunde*, volume 61. However, a more general problem had been posed and solved in the *Monthly* in 1905 by Y. Sawayama, an instructor at the central military school in Tokyo. In 2003, a purely synthetic proof of this theorem was given by Ayme, based on Sawayama's work.

First we will need a lemma.

Lemma 8.17
Through the vertex A of a triangle ABC, a straight line AD is drawn, cutting the side BC at D. Let P be the centre of the circle C_1 which touches DC, DA at E, F and the circumcircle C_2 of ABC at K. Then the chord of contact EF passes through the incentre I of triangle ABC.

Unfortunately, the proof of Lemma 8.17 is involved. It uses Miquel's pivot theorem and radical axes. It also refers to two theorems in F. G.-M. (1912) (here, 'F. G.-M.' is Frère Gabriel-Marie of the Institut des Frères des Écoles Chrétiennes, who published the book. Frère Gabriel-Marie (1835–1916) was appointed Superior General of his Order in 1897. The fifth edition was the final one produced in F. G.-M.'s lifetime).

In the following, the notation $QG//PE$ means that the points Q and G separate the points P and E, on a given conic.

Proof of Theorem 8.16 According to the hypothesis, $QG \perp BC$, $BC \perp PE$; so $QG//PE$. By Lemma 8.17, GH and EF pass through I. Triangles DHG and QGH being isosceles in D and Q, respectively, DQ is (1) the perpendicular bisector of GH, and (2) the D-internal angle bisector of triangle DHG. Mutatis mutandis, DP is (1) the perpendicular bisector of EF, (2) the D-internal angle bisector of triangle DEF. As the bisectors of two adjacent

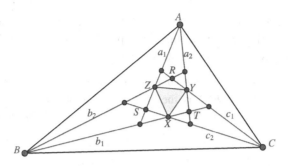

Figure 8.5 Morley's theorem (Theorem 8.18).

and supplementary angles are perpendicular, we have $DQ \perp DP$. Therefore, $GH // DP$ and $DQ // EF$. Now, by the Braikenridge–Maclaurin theorem (Theorem 6.43) applied to the hexagon $PEIGQD$, P, I, and Q are collinear. □

Theorem 8.18 (Morley's theorem c. 1899, (Morley and Morley, 1933, §140)) *Given a triangle ABC, the pairwise intersections of the adjacent angle trisectors form an equilateral triangle (see Figure 8.5).*

Proof This proof is due to Marchand (1931). We will first prove that the hexagon $RZXSTY$ circumscribes a conic, and so the diagonals RX, SY, and TZ are concurrent (by Brianchon's theorem (Theorem 6.44)). For then simple calculations of angles will show that these lines form between them angles of $2\pi/3$ and are respectively perpendicular to the sides YZ, ZX, and XZ of the triangle XYZ. It will then follow that XYZ is equilateral.

Let a, b, c be the (internal) angle bisectors at A, B, C, respectively, and let σ_a, σ_b, σ_c be the reflections about a, b, c, respectively. Then the isogonal conjugate of R is

$$(BR)^{\sigma_b} \cap (CR)^{\sigma_c} = b_2^{\sigma_b} \cap c_1^{\sigma_c} = b_1 \cap c_2 = X.$$

Now R and X are isogonally conjugate with respect to the triangle ABC, as well as B and C.

The dual pencil of conics that have b_1, b_2, c_1, and c_2 as common tangents contains the degenerate conics given by the lines BC and RX. This dual pencil of conics gives an involution τ on the pencils of lines on A, by Desargues' involution theorem (Theorem 6.19), with pairs the pairs of tangent lines on A to each conic of the dual pencil. Note that τ interchanges AB and AC because

$$(AB)^\tau = A(b_1 \cap b_2)^\tau = A(c_1 \cap c_2) = AC.$$

We also know that τ interchanges AR and AX because

$$(AR)^\tau = A(b_2 \cap c_1)^\tau = A(c_2 \cap b_1) = AX.$$

The reflection σ_a also interchanges AB and AC, and AR and AX, and so τ is induced by σ_a. Hence, each member of a pair of τ is equally inclined to a. Therefore, $RZSXTY$ is a Brianchon hexagon.   □

Robson's proof of Morley's theorem (Robson, 1923) Now AC'' is the angle bisector of $\angle BAC'$ and BC'' is the angle bisector of $\angle ABC'$, so C'' is the incentre of ABC', and thus $C'C''$ is the angle bisector of $\angle AC'B$; thus,

$$\angle AC''C' = \pi/2 + (\angle ABC)/3$$

(by consideration of the angles of triangle $AC''C'$). Similarly, B'' is the

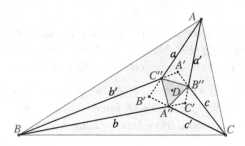

incentre of $AB'C$, so $\angle C''B'B''$ is $\pi/2 - (\angle BAC + \angle ACB)/3$. The difference of these two angles is $\angle C''OB'$ (by consideration of the angles of triangle $C''OB'$), where $O := B''B' \cap C''C'$, and therefore $\angle C''OB' = \pi/3$. Similarly, angles $\angle B'OA''$, $\angle A''OC'$, $\angle C'OB''$, $\angle B''OA'$, $\angle A'OC''$ are all $\pi/3$. Since they have a common base and congruent angles at each of their extremities, triangles $OC''A'$ and $OB''A'$ are congruent, as are triangles $A''C''A'$ and $A''B''A'$. So $A''C''$ and $A''B''$ are congruent. Similarly, $A''B''$ and $B''C''$ are congruent. So $A''B''C''$ is equilateral.   □

Gorjian et alia's proof (Gorjian et al., 2015) Consider a triangle $\triangle ABC$ with $\angle A = 3\alpha$, $\angle B = 3\beta$, and $\angle C = 3\gamma$. We shall show that one can construct this triangle and its angle trisectors backward from its Morley triangle.

Let us put $x = \pi/3 + \alpha = \alpha^*$; $y = \pi/3 + \beta = \beta^*$; and $z = \pi/3 + \gamma = \gamma^*$. Construct the triangle $\triangle AZB$ on the side AB of the given triangle $\triangle ABC$ with the base angles equal to α, β at A, B, respectively. Now construct the triangles $\triangle AZY$, $\triangle BZX$, on the sides AZ and BZ, with the base angles as we have in Figure 8.5. We will try to ignore C for the moment, since we are going to show that it can be built from this configuration.

We will be using Exercise 8.18. This result shows that the triangle $\triangle XYZ$ is equilateral (note $\angle YZX = \pi/3$ and $ZY = ZX$ by Exercise 8.18). Next, from the points Y and B, we draw two lines that make angles equal to x and b with the segments YX and BX, and we let these lines intersect at the point C'. Using Exercise 8.18 again, we infer that X is on the bisector of the angle that is formed by the intersection of BZ and $C'Y$. But X is on the bisector of the angle $\angle C'BZ$; hence, X must also be on the bisector of the angle $\angle BC'Y$. Consequently, we may put $\angle YC'X = \angle XC'B = \theta$.

Now, by considering the summation of the interior angles of the quadrilateral $BZYC'$, we see that $\theta = \gamma$. We note that in the triangle $C'XY$ we have $\angle C'XY = y$. So far, we have constructed the unique triangle $\triangle C'XY$ on XY with the given base angles. Similarly, if instead of the points Y and B, we take the points X and A and apply the same argument as above (i.e., drawing lines from X and A to make the angles equal to y and α with the segments XY and AY), and consider the summation of the interior angles of the quadrilateral $AZXC'$, we obtain the same triangle $\triangle C'XY$ with $\angle XC'Y = \angle YC'A = \gamma$, and we are finished. □

* * *

We now give a projective proof of the famous 'Euler line' theorem.

Theorem 8.19
The orthocentre, circumcentre, and centroid of a triangle are collinear (in the **Euler line***).*

Proof This proof uses Desargues' theorem (Theorem 2.20). Let ABC be the given triangle, and let M_{AB} be the midpoint of $\overline{AB}$, and let M_{AC} be the midpoint of $\overline{AC}$. Let G, H, O be the centroid, orthocentre, and circumcentre, respectively, of ABC. Consider the two triangles $M_{AB}M_{AC}O$ and BCH. The line $M_{AB}M_{AC}$ is parallel to BC, by Theorem 7.5. Moreover, the perpendicular bisectors $M_{AB}O$ and $M_{AC}O$ are parallel to the altitudes BH and CH (respectively). Therefore, the sides of $M_{AB}M_{AC}O$ are parallel to the sides of BCH, and so the the parallel pairs of sides meet in three points at infinity. Therefore, by the (dual of) Desargues' theorem, $M_{AB}M_{AC}O$ and BCH are in perspective from a point. Therefore, the lines OH, BM_{AC}, and CM_{AB} are concurrent. Now the latter two lines meet in G, and, therefore, G lies on OH. □

Let ABC be any given triangle. Let the medians through the vertices A, B, C meet the circumcircle of triangle ABC at A', B', and C', respectively. Let DEF be the triangle formed by the tangents at A, B, and C to the circumcircle of triangle ABC. (Let D be the vertex opposite to the side formed by the tangent at the vertex A, E be the vertex opposite to the side formed by the tangent

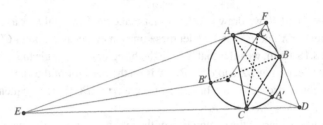

Figure 8.6 The Exeter point.

at the vertex B, and F be the vertex opposite to the side formed by the tangent at the vertex C.) Then the lines through DA', EB', and FC' turn out to be concurrent (see the proof of Theorem 8.20). The point of concurrence is the **Exeter point** (see Figure 8.6) of triangle ABC.

Theorem 8.20
The Exeter point lies on the Euler line.

Proof In the above notation, the medial triangle and the circum-medial triangle $A'B'C'$ are in perspective from the centroid. The medial triangle and the tangential triangle DEF are in perspective from the circumcentre. The two perspectivities have the same axis. Therefore, the circum-medial triangle and the tangential triangle are in perspective. The centre of perspectivity is the Exeter point (by definition). The three centres of perspectivity are collinear, since all three perspectivities fix the line joining the first two centres.   □

The following is a posthumously published theorem of Dan Barbilian (a.k.a. Ion Barbu).

Theorem 8.21 (Barbilian and Vodă, 1984)
Let ABC be a triangle of $\mathbb{R}^2$, A' be the midpoint of $\overline{BC}$, B' be the midpoint of $\overline{CA}$, and C' be the midpoint of $\overline{AB}$. Let A_1 be the foot of the altitude from A to BC, B_1 be the foot of the altitude from B to CA, and C_1 be the foot of the altitude from C to AB. Then $A_1B' \cap A'B_1$, $B_1C' \cap B'C_1$, $C_1A' \cap C'A_1$ lie on the Euler line of ABC.

Proof Apply Pappus' theorem (Theorem 2.18) to the points A, B_1, B' of AC and the points B, A_1, A' of BC. Then $AA_1 \cap BB_1$ is the orthocentre of ABC, and $AA' \cap BB'$ is the centroid of ABC, so $B_1A' \cap A_1B'$ lies on the Euler line of ABC. Similarly, $B_1C' \cap B'C_1$, $C_1A' \cap C'A_1$ lie on the Euler line of ABC.   □

8.2 Circular Points and Euclidean Conics

Embed $PG(2, \mathbb{R})$ into $PG(2, \mathbb{C})$ and consider the **circular points**

$$I: (1, i, 0), \quad J: (1, -i, 0)$$

introduced by Poncelet (1822) (Poncelet, 1995b). Note that these points are fixed by the involution $*$.

Theorem 8.22

A non-empty, non-degenerate conic of $PG(2, \mathbb{R})$ is a circle of $\mathbb{R}^2$ if and only if it passes through the circular points.

Proof Moving from the non-homogeneous equation $aX^2 + bY^2 + cXY + dX + eY + f = 0$ to the homogeneous equation $ax^2 + by^2 + cxy + dxz + eyz + fz^2$ by putting $X = x/z$ and $Y = Y/z$ and multiplying by z^2, the intersection with $z = 0$ is given by $\{(x, y, 0): ax^2 + by^2 + cxy = 0\}$, so this is $\{(1, i, 0), (1, -i, 0)\}$ if and only if $a = b$ and $c = 0$, in which case the original conic had equation $aX^2 + aY^2 + dX + eY + f = 0$, which can be rewritten as

$$\left(X - \frac{d}{2a}\right)^2 + \left(Y - \frac{e}{2a}\right)^2 = \left(\frac{d}{2a}\right)^2 + \left(\frac{e}{2a}\right)^2 - \frac{f}{a},$$

which is (i) empty if the right-hand side is negative, (ii) degenerate if the right-hand side is 0, or (iii) the circle with centre $(\frac{d}{2a}, \frac{e}{2a})$ and radius $\sqrt{(\frac{d}{2a})^2 + (\frac{e}{2a})^2 - \frac{f}{a}}$ if the right-hand side is positive. Conversely, a circle with centre (s, t) and radius r has equation $(X - c)^2 + (Y - d)^2 = r^2$, which homogenises to $(x - cz)^2 + (y - dz)^2 = r^2 z^2$. This circle meets the line with equation $z = 0$ in $\{(x, y, 0): x^2 + y^2 = 0\} = \{(1, i, 0), (1, -i, 0)\}$. $\square$

Note that the above theorem implies that such a conic has no real points on the line $z = 0$ at infinity, and so is an ellipse of $AG(2, \mathbb{R})$.

Theorem 8.23

Let K be a circle of $\mathbb{R}^2$.

(a) Two diameters of K are perpendicular if and only if they are conjugate with respect to K.

(b) Let ℓ, m be lines not on the centre O of K. Let L be the pole of ℓ with respect to K and M be the pole of m with respect to K. Then ℓ and m are perpendicular if and only if OL and OM are perpendicular.

Proof Every tangent is perpendicular to the diameter it lies on. Hence, perpendicular diameters have as poles with respect to K points at infinity that correspond to perpendicular parallel classes. Thus, two diameters of a circle are perpendicular if and only if they are conjugate.

The pole of OL is a point L' at infinity. And by (a), L' is on OM if and only if OL and OM are perpendicular. Let M_∞ be the point at infinity of m. Now M_∞ is on m so its polar line is OM. Thus, OL and OM are perpendicular if and only if L' is on the polar line of M_∞, which is equivalent to OL is on M_∞. Moreover, OL is on M_∞ if and only if $OL \parallel m$, which is the same as $OM \parallel \ell$. Finally, this last condition is equivalent to ℓ and m being perpendicular.   □

Thus, two lines that are not diameters of K are perpendicular if and only if the joins of their poles with respect to K with the centre of K are perpendicular.

Corollary 8.24
Let K be a circle of $\mathbb{R}^2$ with centre O.

(a) *A line ℓ not on O is conjugate to a line m on O if and only if m is perpendicular to ℓ.*

(b) *The orthocentre of a self-polar triangle for K is O.*

Proof

(a) By Theorem 8.23(a), the pole of m is the point P at infinity on the diameter n of K perpendicular to m. So ℓ and m are conjugate if and only if P lies on ℓ, which is the same as n being parallel to ℓ. Furthermore, this is equivalent to m being perpendicular to ℓ.

(b) By (a), the lines joining the vertices of the triangle to O are perpendicular to the opposite sides of the triangle.   □

Theorem 8.25
If a conic has two distinct pairs of perpendicular diameters, it is a circle.

Proof See Exercise 8.1.   □

Recall from Theorem 6.10 that if K is a conic and P is a point not on K, then the map on pencil(P) that interchanges pairs of conjugate lines (with respect to K) is an involution. By taking the conic to be a circle and P to be the centre of the circle, we have:

Corollary 8.26
The map on a pencil of lines through a point that interchanges pairs of perpendicular lines is an involution.

Similarly, we have the following consequence of Theorem 6.10:

Corollary 8.27

Every circle induces the same involution on the line at infinity by interchanging conjugate points.

Proof If points are interchanged then the lines joining them to the centre of the circle are perpendicular. If two lines are perpendicular, a line parallel to the first and a line parallel to the second are taken, then the new lines will also be perpendicular. □

Corollary 8.28

There is a conjugate pair of imaginary points on the line at infinity that lie on every circle. Conversely, every conic through this pair of points is a circle.

Proof Take the fixed points of the involution of Corollary 8.27. Since no real line is perpendicular to itself, these are a conjugate imaginary pair. Conversely, if a conic passes through these points then the involution it induces via conjugation has the same fixed points as the one induced by any circle, and so equals that involution. Hence, it maps every line on the centre of the conic to a perpendicular line. Since the conic has more than one pair of perpendicular diameters, it is a circle, by Theorem 8.25. □

Worked Example 8.29

Let K be a circle with centre O. Take two points A and B of K so that AB does not pass through O. Then the angle bisector of the tangents t_A and t_B at A and B (respectively) is the perpendicular bisector of the chord $\overline{AB}$, and passes through O. To see this, we use the polarity ρ given by K. Let g be the reflection in the angle bisector of t_A and t_B (so $t_A^g = t_B$). Then g fixes the point X of intersection of t_A and t_B. Moreover, OX is the angle bisector and so g fixes O. It follows that g fixes K and, hence, g commutes with ρ. The image of X under ρ is the line AB. Therefore,

$$A^g = A^{\rho g \rho} = t_A^{g\rho} = t_B^{\rho} = B.$$

So, g is the reflection in the perpendicular bisector of $\overline{AB}$. □

The following is equivalent to the fact that a quadrilateral is circumscriptable if and only if the product of its angle bisectors in cyclic order is 1.

Theorem 8.30 (King and Schattschneider, 1997, pp. 29–32)

A quadrilateral is cyclic if and only if the product of the reflections in its perpendicular bisectors in cyclic order is the identity.

Proof Let $ABCD$ be a quadrilateral. Suppose that $ABCD$ is cyclic (with this cyclic ordering), and so lies on a circle K. Let $g_{AB}, g_{BC}, g_{CD}, g_{DA}$ be the reflections in the perpendicular bisectors of the respective sides of the quadrilateral, and let ρ be the polarity arising from K. By Example 8.29, each of these reflections is the (internal) angle bisector for each pair of corresponding pairs of tangent lines. For instance, g_{AB} is the angle bisector of t_A and t_B. Therefore, the product of these reflections (in cyclic order) is the identity. The converse also follows through by applying the observation of Example 8.29.   □

Theorem 8.31 (Thales' theorem (Euclid's *Elements*, Book III, Proposition 31)) *A diameter of a circle subtends a right angle.*

Proof Let $\mathcal{K}$ be a non-degenerate conic of $\mathsf{PG}(2, \mathbb{R})$ passing through the circular points I and J; that is, $\mathcal{K}$ is a circle of the Euclidean plane. The centre O of $\mathcal{K}$ is the pole of the line at infinity (containing I and J), and a diameter ℓ of $\mathcal{K}$ is a line through the centre of $\mathcal{K}$. Let A and B be the points of ℓ on $\mathcal{K}$. Let X be a point of $\mathcal{K}$ not incident with ℓ. We will show that $\angle AXB$ is a right angle. Let $a := AX$, $b := XB$, $L := a \cap \ell_\infty$, and $M := b \cap \ell_\infty$. We want to show that $(L, M; I, J)$ is harmonic. Let Z be a point of $\mathcal{K}$ on the perpendicular bisector of $\overline{AB}$. Now $\mathrm{R}(L, M; I, J) = \mathrm{R}(XL, XM; XI, XJ) = \mathrm{R}(XA, XB; XI, XJ)$, and, by Theorem 5.18, $\mathrm{R}(XA, XB; XI, XJ) = \mathrm{R}(ZA, ZB; ZI, ZJ)$. Since we know that OZ is perpendicular to AB, it follows that ZA is perpendicular to ZB; that is, $\mathrm{R}(ZA, ZB; ZI, ZJ) = -1$. So $\mathrm{R}(L, M; I, J) = -1$ as required.   □

In Exercise 8.12, one can essentially reverse the argument in the proof above to obtain a converse to Thales' theorem:

Given distinct points A, B, C, if angle $\angle ACB$ is a right angle then C lies on the circle with diameter $\overline{AB}$.

Theorem 8.32
Suppose the perpendicular bisector of $\overline{BC}$ meets the angle bisector of angle $\angle BAC$ at Y. Then Y lies on the circumcircle of ABC.

Proof Since Y lies on the perpendicular bisector of $\overline{BC}$, $d(Y, B) = d(Y, C)$, triangle YBC is isosceles; thus, angles $\angle YBC$ and $\angle YCB$ are congruent. By Theorem 8.31, angles $\angle BAY$ and $\angle YCB$ are congruent, as are angles $\angle CAY$ and $\angle YBC$; hence, angles $\angle BAY$ and $\angle CAY$ are congruent. So AY is the internal angle bisector of angle $\angle BAC$.   □

Theorem 8.33
In every circumscriptable quadrilateral, the line connecting the incentre with the intersection point of its diagonals is perpendicular to the line connecting the intersections of its opposite sides.

Proof Let A, B, C, D lie on circle K with centre O, $I := AC \cap BD$, $M :=$ $AB \cap CD$, $N := AD \cap BC$. Let $L := AB \cap K$, $X := BC \cap K$, $J := CD \cap K$, $P := DA \cap K$. Then M is the pole of JL with respect to K, N is the pole of PX with respect to K. So $JL \cap PX$ is the pole of MN with respect to K. By the four-point Pascal theorem (Theorem 6.67), JL, PX, AC, BD are concurrent at I, so I is the pole of MN. So, by the converse to Thales' theorem, OI and MN are perpendicular. □

A circumscriptable cyclic quadrilateral is called **bicentric**.

Corollary 8.34 (Durrande, 1824–5, p. 139, Theorem XI)
The circumcentre, incentre, and intersection of the diagonals of a bicentric quadrilateral are collinear.

Recall the Wallace–Simson theorem (compare with Theorem 5.15): *The feet of the perpendiculars from a point on the circumcircle of a triangle to the sides of the triangle are collinear.* If we apply the polarity of the incircle of the triangle to all of the points and lines in question, we obtain what seems to be an entirely new theorem of Euclidean geometry.

Theorem 8.35 (Pătraşcu, 2010)
Let ABC be a triangle and let $\mathcal{K}$ be the incircle of ABC, with centre I. Let a, b, c be the perpendiculars to AI, BI, CI, passing through I. Let m be a tangent line to I, and let $A_1 := a \cap m$, $B_1 := b \cap m$, $C_1 := c \cap m$. Then the lines AA_1, BB_1, CC_1 are concurrent.

Proof We will simply take the image of each point and line under the polarity ρ associated with $\mathcal{K}$. Define $M := m^\rho$, which is a point of $\mathcal{K}$. Then the images under ρ of the sides of the triangle are the points of tangency A', B', C' of $\mathcal{K}$ with BC, CA, AB, respectively. So now $\mathcal{K}$ is the circumcircle of the triangle $A'B'C'$. Also, I^ρ is the line at infinity ℓ_∞, and so the images of AI, BI, and CI are points at infinity incident with the tangent lines to the intersections of AI, BI, CI with $\mathcal{K}$. Since the lines a, b, c are perpendicular to AI, BI, CI, respectively, we see that a, b, c are parallel to the aforementioned tangent lines, and so

$$(AI)^\rho = a \cap \ell_\infty, \quad (BI)^\rho = a \cap \ell_\infty, \quad (CI)^\rho = a \cap \ell_\infty.$$

Therefore, A_1^ρ is the line through M parallel to AI, and so forth (for B_1^ρ and C_1^ρ). So $(AA_1)^\rho$, $(BB_1)^\rho$, $(CC_1)^\rho$ are precisely the feet of the perpendiculars from M to the sides of the triangle $A'B'C'$. By the Wallace–Simson theorem, these points are collinear. □

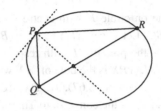

Figure 8.7 Theorem 8.36.

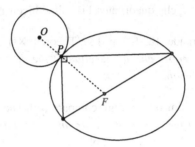

Figure 8.8 Another proof of Theorem 8.36, using Euclid III.31.

One of the most well-known of Frégier's results is the following. Suppose we have a non-degenerate conic $\mathcal{K}$ of the Euclidean plane and a point P on $\mathcal{K}$. If we take all right-angled triangles inscribed in a non-degenerate conic, and having the vertex at the right angle fixed at a point P, then all hypotenuses pass through a common point (see Figure 8.7). We give a beautiful proof using Frégier involutions.

Theorem 8.36 (Frégier, 1815–16)
If from a point P on a non-degenerate conic $\mathcal{K}$ any two perpendicular lines are drawn cutting the conic in points Q and R, then the line QR meets the normal at P at a point P' independent of the choice of the pair of perpendicular lines.

Proof The map fixing P and interchanging Q and R (for every such pair of perpendicular lines) is an involution on the conic and P' is the Frégier point of this involution (compare with Exercise 6.3).   □

We give a proof of Theorem 8.36 that connects to Euclid III.31 (Theorem 8.31).

Another proof of Theorem 8.36 (Schröcker, 2017) Take an arbitrary circle $\mathcal{K}'$, tangent to $\mathcal{K}$ at P (see Figure 8.8). There exists a homology h with centre P that maps $\mathcal{K}'$ to $\mathcal{K}$. (Its axis is the Desargues axis of two triangles that correspond in h and are inscribed into $\mathcal{K}'$ and $\mathcal{K}$, respectively.) By Theorem 8.31, the

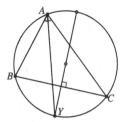

Figure 8.9 Theorem 8.37.

Frégier point of P is then $F := O^h$ where O is the centre of the circle $\mathcal{K}'$. Moreover, if we apply Euclid's III.18, the perpendicular on P to the tangent at P passes through the centre of the circle. Thus, the Frégier point of P lies on the normal at P. □

Now the map $\mathcal{K} \to \mathcal{K}$ taking X to Y where PX is at right angles to PY is the restriction to $\mathcal{K}$ of the map of the pencil of lines on P that maps each line to its perpendicular at P. Hence, it is an involution.

We can also uncover Corollary 5.5 with this approach. Let f be an involution on a conic $\mathcal{K}$. Then there are two cases: f has 0 or 2 fixed points. If f has no fixed points, then we apply the 'projective' version of Theorem 8.36 and we see that f is a Frégier involution. If f has two fixed points, then we use the 'projective' version of the Euclidean theorem that the reflection in a diameter fixes a circle. See also Theorem 5.7.

Now, in the notation above, take a circle $\mathcal{K}'$ with centre P. Apply the polarity with respect to $\mathcal{K}'$. The polar image of a conic is a dual conic (of lines). Since P is on $\mathcal{K}$, the line at infinity is tangent to the polar image of $\mathcal{K}$, so we have the tangent lines to a parabola $\mathcal{P}$. The other vertices of an inscribed triangle with right angle at P have polar images that are tangent lines to $\mathcal{P}$, and the hypotenuse has polar image the point of intersection of these two tangent lines. Since these vertices subtend a right angle at P, their polar images meet at right angles. Finally, the polar image of the Frégier point is a line. So we can deduce another Euclidean theorem, namely, Theorem 8.63 (later in this section) that the tangents to a parabola that intersect at right angles meet on a fixed line (the directrix). We will see more results of this type, where a Euclidean result derives another via a projective viewpoint, in Chapter 10.

We have another application that begins with Theorem 8.31.

Theorem 8.37
Let BC be a chord of a circle K, and let Y be a point of K on the diameter perpendicular to BC. For any point A on $K \setminus \{B, C, Y\}$, the line AY bisects the angle $\angle BAC$ (see Figure 8.9).

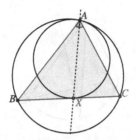

Figure 8.10 Corollary 8.38.

Proof Let x be the perpendicular to AY at A, and let Y' be the other point of K on the diameter d perpendicular to BC. Note that $x = AY'$ by Thales' theorem (Theorem 8.31) applied to the diameter $d = YY'$ and the point A.

Consider the case that $A = Y'$. Let M be the midpoint of $\overline{BC}$. The diameter to a circle perpendicular to a chord bisects the chord,[1] and so M is the intersection $BC \cap d$. Since AY is equal to d in this case, we have that x is parallel to BC, and hence $R(AB, AC; AY, AY') = R(B, C; M, BC \cap \ell_\infty) = -1$ where ℓ_∞ is the line at infinity. Therefore, AY bisects the angle $\angle BAC$.

Now we suppose the generic case, where A is some point of $K \setminus \{B, C, Y\}$. By Chasles' theorem (Theorem 5.18), varying A does not change $R(AB, AC; AY, AY')$ and hence AY bisects the angle $\angle BAC$.   □

Corollary 8.38

Let ABC be a triangle, K be the circumcircle, and K' be the circle tangent to K at A and tangent to BC at X. Then X lies on the angle bisector of angle $\angle BAC$ (see Figure 8.10).

Proof Consider the dilatation δ mapping K' to K, with centre A. Then X is mapped to the other point Y of AX on K. The chord BC is mapped to the tangent t to K at Y, and t is parallel to BC. Let d be the diameter of K passing through Y. Then d is the unique common perpendicular to BC and t. By Theorem 8.37, AX is the angle bisector of $\angle BAC$.   □

Theorem 8.39

Let $\mathcal{K}$ be a circle, A, B be distinct points of $\mathcal{K}$. For X on $\mathcal{K} \setminus \{A, B\}$, the oriented angle between AX and BX equals the oriented angle between AB and the tangent t to $\mathcal{K}$ at B and AB.

---

[1] To see this, consider the reflection in the diameter, and note that this reflection leaves the circle invariant (Euclid III.8).

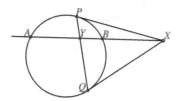

Figure 8.11 Theorem 8.40.

Proof The projectivity taking *XI* to *BI*, *XJ* to *BJ*, and *XA* to *BA* defines $\mathcal{K}$, since this uniquely determines a projectivity τ, and $\mathcal{K}$ is on *I*, *J*, *A*, *X*, and *B*. This is induced by the product of a rotation about *X* and a translation taking *X* to *B*, which is orientation-preserving (since *I* and *J* are fixed). Since *B* is on $\mathcal{K}$, *BX* is mapped to *t* by τ. So the oriented angle between *AX* and *BX* equals the oriented angle between *AB* and *t*.　　　　□

Theorem 8.40 (Dobos, 2011)
Let X be a point outside circle K, with the two tangents drawn from X to the circle meeting K at P and Q. A line through X meets K at A and B. Then X and Y := AB ∩ PQ are harmonic conjugates with respect to A and B (see Figure 8.11).

Proof Using the notation *PP* for the tangent at *P*, we simply compute:

$$R(A, B; X, Y) = R(PA, PB; PX, PY) = R(PA, PB; PP, PQ)$$
$$= R(QA, QB; QP, QQ) \qquad \text{by Theorem 6.65}$$
$$= R(QA, QB; QY, QX)$$
$$= R(A, B; Y, X).$$

Therefore, $R(A, B, X, Y) = -1$.　　　　□

Theorem 8.41 (Dobos, 2011)
Let A be a point outside circle K. Draw tangents from A to K; the points of tangency are B and C. A line through B parallel to AC meets K at D. Suppose DA meets K at E, and BE meets AC at F. Then F is the midpoint of AC (see Figure 8.12).

Proof Let *I* be the point at infinity of *AC*, *T* be the intersection of *BC* and *AD*. Let *M* be the midpoint of $\overline{AC}$. Then $R(A, C; M, I) = -1$. Now $(A, T; E, D)$ is harmonic (by Theorem 4.38), and so

$$R(A, C; F, I) = R(BA, BC; BF, BI) = R(BA, BT; BE, BD)$$
$$= R(A, T; E, D) = -1.$$

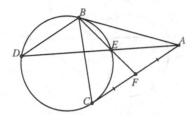

Figure 8.12 Theorem 8.41.

Therefore, $F = M$ (by Lemma 4.3).   □

The following example comes from Primrose (1960). First, we make a simple observation about two conics meeting in four distinct points:

Theorem 8.42

Given two conics meeting at four distinct points A, B, C, D, let $V := AC \cap BD$ and $R(A, C; V, U) = -1$, $R(A, C; V, W) = -1$. Then the harmonic homology with centre V and axis UW stabilises both conics.

Proof Note that the line UW is the polar of V with respect to both conics.   □

Corollary 8.43

Given two conics S, S' meeting at four distinct points A, B, C, D, if the pole of AB for S' is the pole of CD for S, then the pole of CD for S' is the pole of AB for S.

Corollary 8.44

Given two conics S, S' meeting at four distinct points A, B, C, D, if the pole of AB for S' lies on S, then the pole of CD for S' lies on S.

Euclidean Interpretation of Corollary 8.43.

Suppose we have two circles S, S' meeting at distinct points C, D (see Figure 8.13). The following are equivalent:

(i) The centre of S' is the intersection of the tangents at C and D to S.
(ii) S meets S' orthogonally.
(iii) The centre of S is the intersection of the tangents at C and D to S'.

Euclidean Interpretation of Corollary 8.44.

Given two circles S, S' meeting at distinct points C, D (see Figure 8.14). If the centre of S' lies on S, then the tangents to S' at C and D meet on S.

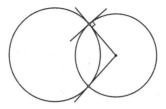

Figure 8.13 Euclidean interpretation of Corollary 8.43.

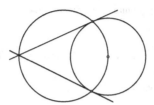

Figure 8.14 Euclidean interpretation of Corollary 8.44.

* * *

Theorem 8.45

Let K be a circle and P be a point outside K. Let ℓ be a line on P meeting K in A, B, with A between P and B. Let C, D be the points of K so that PC and PD are tangents, and let $Q := CD \cap \ell$. Then $R(C, D; A, B)_K = -1$. Conversely, given a cyclic quadrilateral $ACBD$, let P be the intersection of the tangents at C and at D, and $Q := CD \cap AB$. Then A, B, P, Q are collinear and $R(P, Q; A, B) = -1$.

Proof Now CD is the polar line to P with respect to K, by Theorem 4.38, $R(P, Q; A, B) = -1$. By Theorem 6.65, $R(C, D; A, B)_K = R(m, CD; CA, CB)$, where m is the tangent to K at C. Intersecting with ℓ, we see that this last cross-ratio is $R(P, Q; A, B)$. The converse follows from Theorem 4.38. □

Such cyclic quadrilaterals $ABCD$ are called **harmonic quadrilaterals**.

Worked Example 8.46 (Asia-Pacific Math. Olympiad 2013, Problem 5)

Problem Let $ABCD$ be a quadrilateral inscribed in a circle K, and let P be a point on AC such that PB and PD are tangent to K. The tangent at C intersects PD at Q and the line AD at R. Let E be the second point of intersection of AQ and K. Prove that B, E, R are collinear.

Solution Set $T := BD \cap CR$, $L := AC \cap BD$, $Z := AB \cap CR$ and let E' be the second intersection of BR with K. Since $ABCD$ is harmonic, we have T, L, B, D collinear and therefore

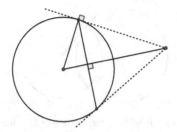

Figure 8.15 Theorem 8.47.

$$-1 = R(T, L; B, D) = R(AT, AC; AZ, AR) = R(BD, BC; BA, BE').$$

But *DACE* is harmonic, so $E = E'$.   □

Theorem 8.47
The polar of a point with respect to a circle, other than the centre, is perpendicular to the diameter through the point. Hence, a tangent line to a circle is perpendicular to the diameter through its point of contact (see Figure 8.15).

Proof The reflection in the diameter through the point fixes the point, the diameter, and the circle and, therefore, commutes with the polarity defined by the circle. Thus, this reflection fixes the polar line to the point, which, therefore, must be perpendicular to the diameter through the point.   □

Theorem 8.48 (Brocard's theorem (Brocard, 1877))
The orthocentre of the diagonal triangle of a cyclic quadrilateral is the circumcentre of the quadrilateral.

Proof By Theorem 4.39, the diagonal triangle is self-polar with respect to the circle. By Theorem 8.47, the line joining a diagonal point to the circumcentre of the quadrilateral is perpendicular to the lines joining the other two diagonal points. Hence, the orthocentre of the diagonal triangle of a cyclic quadrilateral is the circumcentre of the quadrilateral.   □

Theorem 8.49 (Steiner, 1846, §XII)
The centre of a circle is the orthocentre of the triangle of diagonal points of every inscribed quadrilateral.

Proof Suppose that the quadrilateral *ABCD* is inscribed in the circle K with centre *O*. Let $E = AB \cap CD, F = AD \cap BC, G = AC \cap BD$. We will prove that *EG* is the polar of *F*. Let $X = EG \cap BC$ and $Y = EG \cap AD$. Then

$$ADYF \stackrel{E}{\overline{\wedge}} BCXF \stackrel{G}{\overline{\wedge}} DAYF.$$

So $(A, D; Y, F)$ and $(B, C; X, F)$ are harmonic. Thus, the points X and Y lie on the polar line of the point F. Hence, EG is the polar line of the point F. It follows that EG is perpendicular to OF. Similarly, FG is perpendicular to OE, and so O is the orthocentre of the triangle EFG. □

Second proof of Theorem 8.49 First, apply the four-point Pascal theorem (Theorem 6.67) to $AABCCD$ and $ABBCDD$. Then apply Theorem 8.2, to deduce that OG is perpendicular to EF. The argument is similar for the other sides. □

The first proof of Theorem 8.49 reproves Theorem 4.39 in its opening seven sentences. The second proof also gives an alternate proof of Theorem 4.39. The four-point Pascal theorem (Theorem 6.67) applied to $AABCCD$ shows that (in the notation above) E, F, and the pole of AC are collinear, and the four-point Pascal theorem (Theorem 6.67) applied to $ABBCDD$ shows that E, F, and the pole of BD are collinear, and, thus, PQ is the polar of $AC \cap BD = G$. The argument is similar for the other sides.

$$* * *$$

Carathéodory's theorem states that a bijection $g: \mathbb{R}^2 \to \mathbb{R}^2$ takes circles to circles if and only if g is a similarity. A consequence of this result is that we can characterise affinities that are similarities as those that preserve perpendicularity.

Theorem 8.50
An affinity that preserves perpendicularity is a similarity.

Proof An affinity maps circles to ellipses and diameters of ellipses to diameters of ellipses. An ellipse is a circle if and only if it has more than one pair of perpendicular diameters (by Theorem 8.25). Therefore, an affinity that preserves perpendicularity is a similarity, by Carathéodory's theorem. □

Theorem 8.51
Given an affinity g of the plane, if there is a circle K such that K^g is a circle, then g is a similarity.

Proof Work in the complexification of the projective completion. Now g fixes ℓ_∞ and so $(K \cap \ell_\infty)^g = (K \cap \ell_\infty)$, so g fixes the set $\{I, J\}$ of circular points at infinity. Hence, g maps conics on $\{I, J\}$ to conics on $\{I, J\}$, that is, circles to circles, so, by Carathéodory's theorem, g is a similarity. □

Theorem 8.52 (Crannell et al., 2017, Corollary 2)
Fix a line ℓ in $PG(2, \mathbb{R})$ *to play the role of the line at infinity and fix complex conjugate points I, J on ℓ to play the role of the circular points. Every collineation g of* $PG(2, \mathbb{R})$ *is the product of a homology and a similarity.*

Proof Choose coordinates so that $\ell: z = 0$, $I: (1, i, 0)$, $J: (1, -i, 0)$. We have two cases:

g **fixes** ℓ. Let $K = I^{g^{-1}}$. Then there exist $e, f \in \mathbb{R}$ with $K(1, e + if, 0)$ and $f \neq 0$. Now $h: (x, y, z) \mapsto (x - ey/f, y/f, z)$ takes K to I. So $h^{-1}g$ fixes I. Hence, $h^{-1}g$ fixes J (as $h^{-1}g \in PGL(3, \mathbb{R})$). So $h^{-1}g$ is a similarity. Now h is a homology with centre $(-e, 1, 0)$ and axis $y = 0$.

g **does not fix** ℓ. Let $m = \ell^{g^{-1}}$ and $K = I^{g^{-1}}$. By Exercise 8.7, the (complex) line KI has a real point C. There is a homology h with centre C taking m to ℓ. Now h fixes KI (as it is on the centre) and so $K^h = I$. Since $h \in PGL(3, \mathbb{R})$, $h^{-1}g$ fixes J. So $h^{-1}g$ is a similarity. □

A hyperbola is **rectangular** if its asymptotes are perpendicular.

Worked Example 8.53
Let $k \in \mathbb{R}^*$. The strain $(x, y) \mapsto (x, ky)$ and the strain $(x, y) \mapsto (x/k, y)$ commute. Their composition $h_k: (x, y) \mapsto (x/k, ky)$ is called a **hyperbolic rotation**. Now the hyperbolic rotations $G = \{h_k: k \in \mathbb{R}^*\}$ form a group, whose orbits on $\mathbb{R}^2$ are $\{(0, 0)\}$, $\{(x, 0): x \in \mathbb{R}^*\}$, $\{(0, y): y \in \mathbb{R}^*\}$ and the hyperbolas $xy = a$, $a \in \mathbb{R}^*$. Moreover, $\{xy = a: a \in \mathbb{R}\}$ is a pencil of conics of $\mathbb{R}^2$. Each hyperbola $xy = a$ has $x = 0$ and $y = 0$ as asymptotes. Thus, they are rectangular hyperbolas: the asymptotes meet at right angles.

More generally, any element of $AGL(2, \mathbb{R})$, conjugate to h_k, is called a hyperbolic rotation. The set of all hyperbolic rotations fixing a pair of intersecting lines forms a group, with orbits the point of intersection, each line minus the point of intersection, and a pencil of hyperbolas.

If we move to $AGL(2, \mathbb{C})$, h_k is conjugate to a rotation, and G is conjugate to the group $SO(2)$ in $Isom(\mathbb{R}^2)$ of all rotations fixing the origin O. The orbits of $SO(2)$ on $\mathbb{R}^2$ consist of $\{O\}$ and the circles with centre O. Each circle passes through the circular points I, J at infinity, and OI, OJ are perpendicular lines; each rotation fixes I and J. Viewed in $\mathbb{C}^2$, each circle is a hyperbola. The eigenvalues of a rotation about O anticlockwise by angle θ are zeros of

$$\lambda^2 - 2(\cos\theta)\lambda + 1,$$

namely, $\cos\theta + i\sin\theta$, $\cos\theta - i\sin\theta$. Since the constant term of this polynomial is 1, these eigenvalues are mutual reciprocals of one another. Moreover, the composition of the strain in OI with factor $k = \cos\theta + i\sin\theta$ and the strain in OJ with factor $1/k$ is the rotation about O anticlockwise by angle θ. □

Theorem 8.54 (Brianchon and Poncelet, 1821)
Every non-degenerate conic through the vertices and the orthocentre of a triangle that is not right-angled is a rectangular hyperbola.

Proof Let ABC be a triangle that is not right-angled and let $\mathcal{K}$ be a conic through the vertices and the orthocentre O of ABC. So $ABCO$ is a quadrangle. By Theorem 5.13, the set of conics on $ABCO$ form a pencil $\mathcal{P}$. Recall that the degenerate members of $\mathcal{P}$ are the pairs of lines consisting of opposite sides of the quadrangle $ABCO$. Now three of these pairs of opposite sides are perpendicular, by definition of an orthocentre. So we have three rectangular hyperbolae in the pencil $\mathcal{P}$. Two pairs of the involution defined by $\mathcal{P}$ on the line at infinity harmonically separate the absolute points, and so all pairs of the involution harmonically separate the circular points I and J. Therefore, every conic of $\mathcal{P}$ is a rectangular hyperbola. □

Another proof of Theorem 8.54 By Desargues' involution theorem (Theorem 6.19), the conics of the pencil induce the same involution on the line at infinity by conjugation with respect to the conic. The degenerate conics induce the perpendicularity involution. Hence, every line on the centre is a diameter. So a non-degenerate conic of the pencil either is a circle or has perpendicular real asymptotes (i.e., is a rectangular hyperbola). But the circumcircle of a triangle does not pass through its orthocentre, so the conic is a rectangular hyperbola. □

Theorem 8.55 (Kiepert, 1869)
Directly similar isosceles triangles ABC', BCA', CAB' erected externally or internally on the sides of a triangle ABC have AA', BB', CC' concurrent in the points of a rectangular hyperbola on A, B, C, the orthocentre H, and the centroid G.

Proof Let g be a direct similarity taking C to B and taking A to C. Then angle $\angle CAB'$ is congruent to $\angle CA(B')^g$ and $\angle AB'C$ is congruent to $\angle AB'^g C$, so $(B')^g = A'$. Let ℓ be the perpendicular bisector of $\overline{AC}$ and m be the perpendicular bisector of $\overline{BC}$. Then $\ell^g = m$. Hence, the restriction ϕ of g to ℓ is a projectivity that is not a perspectivity. Thus, $\psi \colon \text{pencil}(B) \to \text{pencil}(A)$ with $n^\psi = A(n \cap \ell)^\phi$ is a projectivity that is not a perspectivity. Hence,

$$\mathcal{K} = \{n \cap n^\psi : n \in \text{pencil}(B)\}$$

Table 8.1 *Some points on the Kiepert hyperbola.*

Point	Angle-Parameter
orthocentre	$\pi/2$
centroid	0
outer and inner Fermat points	$\pm\pi/3$
outer and inner Napoleon points	$\pm\pi/6$
outer and inner Vecten points	$\pm\pi/4$
third Brocard point	$-\omega$
Tarry point	$\pi/2 - \omega$

is a non-degenerate conic, containing A, B, and $BB' \cap AA'$, for all values θ of the directed $\angle BCB'$, with $\theta = -\angle BCA$ giving C, $\theta = 0$ giving G, and $\theta = \infty$ giving H. By the preceding theorem, $\mathcal{K}$ is a rectangular hyperbola. Permuting the roles of B and C gives another non-degenerate conic $\mathcal{K}' = \{n \cap n^{\psi'} : n \in \text{pencil}(C)\}$ on A, B, C, H, G, and $CC' \cap AA'$. Since five points (no three collinear) determine a conic, $\mathcal{K}' = \mathcal{K}$, and so AA', BB', CC' are concurrent.   □

This conic is called the **Kiepert hyperbola** of ABC. Each isosceles triangle involved has two equal *base angles*, and we let the positively oriented one be the **angle-parameter** of the Kiepert hyperbola. Table 8.1 shows a list of points on the Kiepert hyperbola according to their angle-parameter, where ω is the Brocard angle:

We add also that the centre of the Spieker circle has the more complicated angle-parameter $\arctan\left(\tan(\frac{1}{2}\angle BAC)\tan(\frac{1}{2}\angle ACB)\tan(\frac{1}{2}\angle BAC)\right)$. We refer the reader to Eddy and Fritsch (1994) and Casey (1885) for more.

Theorem 8.56 (Dobos, 2011)
Let ABC be a triangle such that $d(A, B) = d(A, C)$. Let ABC', BCA', CAB' be similar triangles drawn externally on the sides of ABC. Then AA', BC', and CB' are concurrent.

Proof Let us denote the intersection of AA' and BC by D. From the conditions of the problem, we know that angles $\angle A'BC$ and $\angle B'CA$ are congruent, as are angles $\angle CBA$ and $\angle ACB$, and angles $\angle ABC'$ and $\angle BCA'$. So the following four lines through B and C have the same cross-ratio: $R(A'B, CB; AB, C'B) = R(B'C, AC; BC, A'C)$. Since $R(B'C, AC; BC, A'C) = R(A'C, BC; AC, B'C)$, it follows that $R(A'B, CB; AB, C'B) = R(A'C, BC; AC, B'C)$. These four lines through B and C meet AA' in such a way that the first three of the lines pass through A', D, and A. Since the cross-ratio is the same and it uniquely determines the place of the fourth (by Lemma 4.3), it follows that $C'B$ and $B'C$ meet AA' at the same point.   □

$$* * *$$

Theorem 8.57 (The reflection property; Lipót Fejér, c. 1899–1900 (unpublished, see Rademacher and Toeplitz, 1957, p. 30))
A tangent to a conic forms congruent angles with the lines that join its point of contact to its foci.

Proof Suppose now F_1 and F_2 are the two foci of a conic (in the case of the parabola, one of the foci lies at infinity). Any two conjugate lines SP_1 and SP_2 which are perpendicular to each other are harmonically separated by the points F_1 and F_2 and, hence, by the lines SF_1 and SF_2; they therefore bisect the angles between SF_1 and SF_2. If S is a point of the conic on one of the lines SP_1, SP_2 is tangent to the conic; if S is external to the conic, then SP_1 and SP_2 also bisect the angles between the two tangents which can be drawn from S to the curve, since these are also harmonically separated by SP_1 and SP_2, by the dual of Theorem 4.33.   □

Corollary 8.58

(a) *The reflection of a line through one focus of an ellipse in the tangent line at the point of intersection passes through the other focus.*
(b) *The reflection of a line parallel to the axis of a parabola in the tangent line at the point of intersection passes through the focus.*

Worked Example 8.59
The following comes from Prasolov and Tikhomirov (2001, exercise 4.17).

Problem Parallel beams of light, reflected from a plane curve C, converge to point F. Show that the curve C is a parabola.

Solution We may assume that the beams are represented vertically, that the curve passes through the origin, and that the focus is $(0, 1/4)$. Then the curve can be written as $y = f(x)$, and f is differentiable since the curve has tangents at every point. We know that $y = f'(a)x + f(a) - f'(a)a$ is the tangent at $(a, f(a))$. The reflection of $x = a$ in the tangent is

$$y = -(1 - f'(a)^2)x/2 \cdot f'(a) + f(a) + (1 - f'(a)^2)x/2 \cdot f'(a)$$

and, since it passes through $(0, 1/4)$, it follows that

$$4f(a)f'(a) + 2a - 2af'(a)^2 - f'(a) = 0.$$

The only solution to this differential equation, with boundary condition $f(0) = 0$, is $f(x) = x^2$. Thus, C is a parabola.   □

Let K be any central conic of $\mathbb{R}^2$, not a circle; then I, J are not on K. Let m_1, m_2 be the tangents to K on I, and m_3, m_4 be the tangents to K on J. One of $m_1 \cap m_3$, $m_1 \cap m_4$ is real; label the tangents so that it is $F_1 = m_1 \cap m_3$. Let $F_2 = m_2 \cap m_4$, $F_1' = m_1 \cap m_4$, $F_2' = m_2 \cap m_3$. By the dual of Desargues' involution theorem (Theorem 6.19), if P is a point on no side of the diagonal triangle of the complete quadrilateral $IJF_1F_2F_1'F_2'$, then the pairs of tangents to the conics tangent to m_1, m_2, m_3, m_4 form an involution

$$t = (PI\,PJ)(PF_1\,PF_2)(PF_1'\,PF_2')\ldots.$$

If P is on K, the tangent v to K on P is fixed by t. If u is also fixed by t, then $(PI, PJ; v, u)$ is harmonic, so v and u are perpendicular. Also $(PF_1, PF_2; v, u)$ is harmonic, so the tangent and the normal to K at a point P of K are the bisectors of $\angle F_1PF_2$. This is the reflection property of a central conic.

Now let K be a parabola of $\mathbb{R}^2$ with axis a and focus F. Let $a \cap \ell_\infty = A$. By the dual of Desargues' involution theorem (Theorem 6.19), if P is a point, then the pairs of tangents to the conics tangent to FI, FJ, and ℓ_∞ at A (all of which are parabolas with focus F and axis a) form an involution $t = (PI\,PJ)(PF\,PA)\ldots.$ If P is on K, the tangent v to K on P is fixed by t. If u is also fixed by t, then $(PI, PJ; v, u)$ is harmonic, so v and u are perpendicular. Also $(PF, PA; v, u)$ is harmonic, so the tangent and the normal to K at a point P of K are the bisectors of $\angle FPA$. This is the reflection property of a parabola.

* * *

A **directrix** is the polar of a focus. The **vertex** of a parabola is the unique point of the parabola lying on the perpendicular to the directrix on the focus.

Theorem 8.60
Two tangents PA and PB to a parabola are perpendicular to each other if their point of intersection P is on the directrix.

Proof The polar of P passes through the focus F and contains the points of contact A and B of the two tangents, and each of these tangents forms the same angle with AB as with an arbitrary diameter. Consequently, in the triangle APB, the sum of the angles A and B equals the sum of the angles which PA and PB make with the diameter PC passing through P, that is, equal to the angle APB; and, since the angles A, B, and P together make up two right angles, then APB must be a right angle.   □

Theorem 8.61

The points of intersection of all tangents to a parabola with the perpendiculars to them from the focus F are incident with the tangent at the vertex of the parabola.

Proof In order to prove this, draw through N, the point of intersection of an arbitrary tangent (at the point A) with the tangent at the vertex S, a line NF_1 parallel to the axis, and draw NF where F is the focus. Then the angles $\angle FNA$ and $\angle SNF_1$ are congruent, since NF_1 passes through the second focus (at infinity); and, since SNF_1 is a right angle, FNA must be one also.   □

Corollary 8.62

(a) *(Construction of a parabola with four given lines, no three concurrent as tangents:) From the four tangent triangles that can be formed from the four given tangents, we choose two and draw the circumcircle for each. The point of intersection of the two circumcircles is the focus. We then find the reflections of the focus in two of the tangents and in this way obtain two points of the directrix, which gives us the directrix. Now we can use the focus-directrix construction.*

(b) *A parabola touching four given lines, no three concurrent, has its focus lying on the circumcircle of every triangle formed by any three of the said lines. Hence, the circumcircles of the four triangles formed by any four lines, no three concurrent, meet in a point.*

(c) *The directrix of the parabola touching four given lines, no three concurrent, passes through the orthocentres of the four triangles formed by those lines. Hence, the orthocentres of the four triangles formed by any four straight lines, no three concurrent, are collinear.*

Newton in the *Principia* in 1687 was the first to solve the problem of constructing a parabola that passes through four given points, no three collinear. The theorem in question is a special case of Brianchon's theorem (Theorem 6.44).

Theorem 8.63

Perpendicular tangents to a parabola meet on the directrix.

Proof This proof is from Coxeter (1949, Section 9.81). Let a and a' be the tangents to a parabola from a point U on the directrix f, and let these meet the tangent at the vertex in T and T'. Since the focus F is conjugate to U, the dual of Seydewitz's theorem (Theorem 4.40) shows that FT and FT' are conjugate; therefore, they are perpendicular, since they lie on the focus. Now FT is perpendicular to a, and FT' is perpendicular to a'. Hence, $FTUT'$ is a rectangle, and a is perpendicular to a'. Since any other tangent perpendicular to a would be parallel to a' (which is impossible, since then the point at infinity on those tangents would lie on three tangent lines to the conic – one at infinity), this completes the proof.   □

Theorem 8.64 (Steiner, 1827a, p. 191, Lehrsatz [Theorem] 29)
Let $\mathcal{K}$ be a parabola and a, b, c be tangents to $\mathcal{K}$. Then the orthocentre of the triangle with sides a, b, and c lies on the directrix of $\mathcal{K}$.

Proof This proof is due to John C. Moore, and published by George Salmon (1954, p. 247, art. 268). Let $A := c \cap b$, $B := a \cap c$, $C := b \cap a$. Let b', c' be tangent to $\mathcal{K}$ and perpendicular to b, c, respectively. By the preceding theorem, $P := b \cap b'$ and $Q := c \cap c'$ lie on the directrix. If H is the orthocentre, then BH is perpendicular to AC, and so BH is parallel to b' meeting in a point U at infinity. Similarly, CH meets c' at V on the line ℓ at infinity. By Brianchon's theorem (Theorem 6.44) applied to the six tangents ℓ, a, b, c, b', c' to $\mathcal{K}$, the lines PQ, BU, and CV are concurrent; that is, the altitudes BH and CH meet on the directrix.   □

Recall the theorem of the Wallace–Simson line (compare with Theorem 5.15):

The feet of the perpendiculars from a point on the circumcircle of a triangle to the sides of the triangle are collinear.

Theorem 8.64 can be deduced from this result:

- The focus F of the parabola lies on the circumcircle of the triangle.
- The orthogonal projections of F onto the sides of the triangle lie on the tangent t at the vertex of the parabola.
- By the Wallace–Simson line theorem, t is a Simson line of the triangle.
- Let H be the orthocentre of the triangle. Then $\overline{HF}$ is bisected by t (Casey, 1892, Book III, Prop. 14).
- Therefore, H lies on the directrix of the parabola (see Figure 8.16).

Conversely, in any given triangle, and with any point on its circumscribed circle as focus, inscribe a parabola. Then, since the sides of the triangle are tangents to the parabola, the feet of the three focal perpendiculars upon them must lie on the tangent at the vertex. Hence, if from any point on the circumscribed circle of a triangle perpendiculars are dropped to its sides, the feet of the three perpendiculars will be collinear (in the Wallace–Simson line).

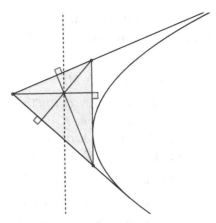

Figure 8.16 A parabola inscibed in a triangle has its directrix on the orthocentre of the triangle.

Worked Example 8.65

The following is from McGregor et al. (1922, p. 184).

Problem Two parabolas have parallel axes. Prove that their common chord bisects their common tangent.

Solution Let the two conics C_1 and C_2 be tangent to the line at infinity of $\mathbb{R}^2$ at Q_∞. Then C_1 and C_2 are parabolas whose axes are parallel. Let a common tangent be L_1L_2, tangent to C_1 at L_1 and to C_2 at L_2, and intersecting the line at infinity in L_∞. Let the common chord be AB, intersecting L_1L_2 in M. Let the intersections of AQ_∞ and BQ_∞ with L_1L_2 be R and S, respectively.

Since the line L_1L_2 is tangent to C_1 at L_1, we have that L_1 is a fixed point of the involution determined on L_1L_2 by the pencil of conics on R, S, M, and L_∞ by Desargues' involution theorem (Theorem 6.19). Similarly, L_2 is the other fixed point of the same involution. Hence, M is the harmonic conjugate of L_∞ with respect to L_1 and L_2. Therefore, M is the midpoint of L_1L_2. $\square$

$$* \ * \ *$$

Theorem 8.66 (Nine-point circle)

If ABC is a triangle, then the vertices of its medial triangle, together with the orthocentre H of ABC, the feet of the altitudes of ABC, and the midpoints of segments from each vertex of ABC to the orthocentre, lie on a circle.

Proof This elegant *projective* proof is due to Pickford (1909, §111).

Let $K := AB \cap \ell_\infty$, $L := CH \cap \ell_\infty$, $M := AC \cap \ell_\infty$, $N := BH \cap \ell_\infty$, $O := AH \cap \ell_\infty$, $P := BC \cap \ell_\infty$. First note that $R(A, B; K, F) = R(A, C; M, E)$ (as they are both equal to -1) and so, by Corollary 4.10, BC, $KM = \ell_\infty$, EF are concurrent. This means that EF passes through P. Similarly, FU, FV, FD pass through O, N, M, respectively. Hence,

$$R(FE, FV; FU, FD) = R(P, O; N, M) = R(M, N; O, P).$$

Similarly, $R(WE, WV; WU, WD) = R(M, N; O, P)$. Therefore, by Theorem 6.61, F, W, E, V, U, D lie on a common conic $\mathcal{K}$.

On the other hand, RE, RV, RU, RD are harmonic conjugates of RM, RN, RO, RP with respect to RA and RC. So, by Corollary 4.10,

$$R(RE, RV; RU, RD) = R(M, N; O, P) = R(FE, FV; FU, FD).$$

Therefore, R lies on $\mathcal{K}$ (by Theorem 6.61), and, by a similar argument, V and W also lie on $\mathcal{K}$. We have shown that the nine points D, E, F, P, Q, R, U, V, W lie on a conic $\mathcal{K}$: we have not yet used the fact that AH, BH, CH are perpendicular to BC, AC, AB. In this case, $K^* = L$, $M^* = N$, and $O^* = P$, and $\mathcal{K}$ passes through I and J. That is, $\mathcal{K}$ is a circle.   □

Theorem 8.67 (Eleven-point conic)
The conic K of the last theorem passes through the diagonal triangle, the fixed points of the involution on ℓ determined by the pencil of conics on the quadrangle, and the six harmonic conjugates of the intersections of ℓ with the sides of the complete quadrangle with respect to the vertices on that side.

Proof First, the points I and J of the last proof are the fixed points of the involution on ℓ determined by the pencil of conics on the quadrangle, by Desargues' involution theorem (Theorem 4.7). Let $A_0A_1A_2A_3$ be the quadrangle, $X_1 := \ell \cap A_0A_1$ and Y_1 be the harmonic conjugate of X_1 with respect to $\{A_0, A_1\}$. Let $P := \ell \cap A_2Y_1$ and Q be the harmonic conjugate of A_2 with respect to $\{P, Y_1\}$. Consider the conic C of the pencil on $A_0A_1A_2A_3$ which is on Q. Since P is the harmonic conjugate of Y_1 with respect to $\{A_2, Q\}$ and X_1 is the harmonic conjugate of Y_1 with respect to $\{A_1, A_0\}$, it follows that $PX_1 = \ell$ is the polar of Y_1 with respect to C. Hence, Y_1 is on K, and, likewise, the other points lie on K.   □

Second proof of Theorem 8.67 The intersections of ℓ with A_2Y_1, A_0A_1; A_0Y_3, A_1A_2; A_1Y_2, A_0A_1 are in involution, so by the converse to Pappus' involution theorem (Theorem 6.13), A_2Y_1, A_0Y_3, A_1Y_2 are concurrent in a point. Let $X_2' := \ell \cap A_1A_3$, and $X_3' := \ell \cap A_0A_3$. Now triangles $A_0Y_3X_3'$ and $A_1Y_2X_2'$ are in perspective from X_1, so this point lies on the line joining A_3 and $Y_3X_3' \cap Y_2X_2'$. The last point is the pole of ℓ with respect to the conic on A_0, A_1, A_2, I, and J

(see Theorem 5.17). (The polar of X_3 with respect to the conic on A_0, A_1, A_2, I, and J is $Y_3 X_3'$ because $(A_1, A_2; X_3, Y_3)$ is harmonic and $(X_3 X_3')$ are in involution with respect to this conic by Desargues' involution theorem (Theorem 4.7), and the polar of X_2 with respect to the conic on A_0, A_1, A_2, I, and J is $Y_2 X_2'$ because $(A_0, A_2; X_2, Y_2)$ is harmonic and $(X_2 X_2')$ are in involution with respect to this conic by Desargues' involution theorem (Theorem 4.7).) □

Theorem 8.67 goes back to two papers of Beltrami (1862, 1863)). Its statement can also be found in Wilkinson (1858).

By applying Snapper's Theorem, we see that we obtain four Euler lines: one is given as follows. Let $X_2 := \ell \cap A_0 A_2$ and Y_2 be the harmonic conjugate of X_2 with respect to A_0, A_2. Let $X_3 := \ell \cap A_1 A_2$ and Y_3 be the harmonic conjugate of X_3 with respect to $\{A_1, A_2\}$. By Snapper's theorem, the lines $A_2 Y_1$, $A_0 Y_3$, $A_1 Y_2$ are concurrent in a point lying on the line joining A_3 and the pole of ℓ with respect to the conic on A_0, A_1, A_2, I, and J. These four Euler lines are concurrent in the centre of the nine-point conic (see Theorem 8.66).

Theorem 8.68 (Brianchon and Poncelet, 1821)
The centre of a rectangular hyperbola circumscribing a triangle lies on its nine-point circle.

Proof Let ABC be a triangle with orthocentre H. By Lamé's theorem (Theorem 6.25), the locus of the centres of the pencil of conics on A, B, C, H is a conic. Since it is the isogonal conjugate of the line at infinity, it is the circumcircle. □

A quadrilateral is **orthocentric** if one of its points is the orthocentre of the other three. So the nine-point circles of the triangles obtained by omitting one point of a quadrilateral are distinct if and only if the quadrilateral is not orthocentric. The following follows from Theorem 8.68.

Corollary 8.69 (Brianchon and Poncelet, 1821)
The four nine-point circles of the triangles obtained by omitting one point of a quadrilateral which is not orthocentric have a common point, which is the centre of the rectangular hyperbola circumscribing the quadrilateral.

The following was taken from O'Hara and Ward (1937, pp. 133–5, §6.3).

Theorem 8.70

Given a quadrangle and a line ℓ on no vertex and no diagonal point of the quadrangle, the poles of ℓ with respect to the non-degenerate conics of the pencil on the quadrangle all lie on a conic K.

Proof Let I and J be the (possibly imaginary) points of tangency on ℓ to conics of the pencil. Let C be a non-degenerate conic of the pencil not tangent to ℓ. Then C meets ℓ in (possibly imaginary) points L, L' and $(I, J; L, L')$ is harmonic, by Desargues' involution theorem (Theorem 4.7). Let the tangents to C at L and L' meet at T. Then IJ has pole T with respect to C, TI has pole J with respect to C, and TJ has pole I with respect to C. So TIJ is self-polar with respect to C. By theorem (Theorem 4.39), the diagonal triangle of the quadrangle is self-polar with respect to C. By theorem (Theorem 6.74), there is a conic containing the diagonal triangle and TIJ. Since, by Theorem 6.74, this conic is the unique such conic on the diagonal triangle and TIJ, it doesn't depend on C and so contains the poles of ℓ with respect to all the non-degenerate conics of the pencil on the quadrangle.   □

8.3 Axes, Diameters, Foci

An **axis** of a conic is a diameter that is perpendicular to the chords it bisects.

Theorem 8.71

For a central conic, a diameter is an axis if and only if it is perpendicular to its conjugate diameter.

Proof The pole of a diameter is the point at infinity corresponding to the parallel class of the chords it bisects.   □

Theorem 8.72

A central conic, other than a circle, has exactly two axes.

Proof Let A be a symmetric matrix such that the conic has equation $xAx^\top = 0$, having changed coordinates so that the conic has centre $P(0, 0, 1)$. Consider the map $\phi: X \mapsto X^\rho P \cap \ell_\infty$ on $\ell_\infty: z = 0$, where ρ is the polarity defined by the conic. Composing ϕ with $*$ gives a projectivity ψ of ℓ_∞ with matrix the 2 x 2 top-left corner B of A^{-1}, which is symmetric. Since symmetric matrices have real eigenvalues, it follows that there is a real eigenvector, and the point Q spanned by this eigenvector is fixed by ψ. Thus, the line PQ is mapped to a line perpendicular to PQ by ϕ. Hence, PQ is perpendicular to its conjugate diameter. If the eigenvalues of B are distinct, then there are exactly two axes.

If there are more than two axes, then A^{-1} (and so A) acts as a scalar matrix on ℓ_∞, so the conic has equation $x^2 + y^2 + cz^2 = 0$; since the conic is non-empty and non-degenerate, $c < 0$, so $c = -r^2$ for some $r > 0$. Thus, the conic is the circle with centre the origin and radius r for some $r > 0$ if there are more than two axes. $\square$

Theorem 8.73

*A parabola has one and only one diameter which is perpendicular to the chords which it bisects. This is called the **principal diameter**.*

Proof All the diameters are parallel and pass through a fixed point P at infinity. Each diameter d determines a set of parallel chords which it bisects, and these meet the line at infinity in the same point Q. As the line d alters its position, so does Q, and there will be one position of d for which P, Q are harmonic conjugates with respect to the circular points. $\square$

A **focus** of a conic is a point F such that a pair of lines on F is perpendicular if and only if they are conjugate with respect to the conic.[2]

Corollary 8.74

A point is a focus of a central conic if and only if the involution induced by the conic on the pencil of lines through the point by the dual of Desargues' involution Theorem 6.19 is the same as the involution taking each line of the pencil to its perpendicular in the pencil.

Proof This follows from the details of the proof of Theorem 8.72, noting that the involution given by Theorem 6.19 is $\ell \mapsto P(\ell \cap \ell_\infty)^\phi$. $\square$

Corollary 8.75

An affine point is a focus of a conic if and only if the lines joining it to the circular points are tangent to the conic.

Proof This follows for central conics from Corollary 6.3 since both involutions fix the lines on the circular points. For a parabola, let F be the intersection of the tangent lines on the circular points, distinct from the line at infinity. The join ℓ of $F(a, b, c)$ and $I(1, i, 0)$ is $[c, ci, a + ib]$; the join m of F and $J(1, -i, 0)$ is $[c, -ci, a - ib]$; and ℓ and m are perpendicular since $[d, e, f]$ is perpendicular to $[d', e', f']$ if and only if $dd' + ee' = 0$. Now the involution induced by the

---

[2] In Book III, Propositions 45–52 of the *Conics* (c. 200 BC), Apollonius deals with certain properties of the foci of ellipse and hyperbola. He has nothing on the focus of a parabola. Diocles' *On burning mirrors* (c. 190–180 BC, Propositions 4 and 5) is the earliest surviving work with the focus of a parabola.

dual of Desargues' theorem by the conic on the lines through F fixes the joins of F with I and J so two lines on F are conjugate with respect to the conic if and only if they are perpendicular. Conversely, if F is a focus of the conic, then the involution induced on the pencil of lines on F by perpendicularity fixes the joins of F to I and J, so these lines are self-conjugate with respect to the conic, which means that they are tangent lines.   □

The proof above also gives an alternative proof for Corollary 8.74, with a few modifications.

Theorem 8.76
If there are two foci, then their join is an axis.

Proof Let F, F' be foci. The perpendiculars to FF' at F and F' are conjugate to FF'; therefore, the pole of FF' is at infinity. So FF' is a diameter, and, therefore, an axis.   □

Theorem 8.77
The foci of a conic are internal points (with respect to the conic).

Proof Suppose by way of contradiction that a focus F of a conic $\mathcal{K}$ is an external point. Then any tangent at F, being self-conjugate, would have to be self-perpendicular; a contradiction.   □

The centre of a circle is a focus. Two rays which are conjugate with respect to a circle intersect at right angles only if one or both are diameters; thus, a circle has no focus except the centre.

Theorem 8.78 (Pappus of Alexandria, 1986, Propositions 313–18)
   (i) *Every ellipse that is not a circle has exactly two foci (Pappus of Alexandria, 1986, Book VIII, Proposition 14).*
   (ii) *Every parabola has a unique focus (Diocles et al., 1976, Proposition 4).*
   (iii) *Every hyperbola has exactly two foci.*

Proof The following is based on the treatment in Coxeter (1993, Section 9.7). Recall that a conic is an ellipse, a parabola, or a hyperbola according to whether the line ℓ_∞ is external, tangent, or secant to the conic. By Theorem 8.76, any foci that exist must lie on the one axis a of the conic. Moreover, the foci of a conic are interior to the conic (by Theorem 8.77). Let A be a point of the axis lying on the conic; a *vertex* of the conic. We will construct the foci in each case, thereby giving a characterisation of the foci in the process.

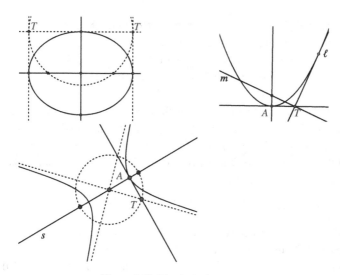

Figure 8.17 The foci of a conic.

(i) To construct the two foci F and F' of an ellipse that is not a circle (see Figure 8.17):

  - Take tangent lines at the vertices of the ellipse, and take their points of intersection.
  - Two of the tangent lines are parallel to the major axis of the ellipse (the longer of the two axes), so let T and T' be the points of intersection on one of these tangent lines.
  - Let C be the circle with diameter $\overline{TT'}$.
  - Then F and F' are the points of intersection of the major axis with C.

    Let's see why this works. Let b be the tangent at one end of the minor axis, and let a and a' be tangents at the ends of the major axis. Let $T := a \cap b$ and let $T' = a \cap b'$, and suppose X is a point on the major axis. We see that the ellipse is inscribed in the triangle with sides a, a', and b; the intersection $a \cap a'$ lying on the line at infinity. Now the dual of Seydewitz's theorem (Theorem 4.40) states that if a triangle is circumscribed about a conic, any point conjugate to one vertex (of the triangle) is joined to the other two vertices (of the triangle) by conjugate lines. So XT and XT' are conjugate lines, and, thus, X is a focus if XT and XT' are perpendicular. Alternatively, X lies on the circle with diameter $\overline{TT'}$ by the converse to Thales' theorem.[3]

---

[3] Given distinct points A, B, C, if angle $\angle ACB$ is a right angle then C lies on the circle with diameter $\overline{AB}$.

(ii) To construct the focus F of a parabola (see Figure 8.17):
- Let ℓ be any tangent to the parabola, not at A.
- Let T be the point of intersection of the tangent a at A and ℓ.
- Let m be the line through T perpendicular to ℓ.
- Then m meets the axis at the focus F.

Just as before, we use the dual of Seydewitz's theorem (Theorem 4.40) with the triangle given by the line ℓ_∞ at infinity and the lines ℓ and a.

(iii) To construct the foci F and F' of a hyperbola (see Figure 8.17):
- Take an asymptote b and a tangent a at the vertex A and take their intersection $T := a \cap b$.
- One of the axes is secant and the other is external. Let s be the secant axis.
- Then the circle centred at the centre of the hyperbola, and passing through T, meets s in F and F'.

In this case, we use the dual of Seydewitz's theorem (Theorem 4.40) with the triangle given by the line ℓ_∞ at infinity and the lines a and a', where a' is the tangent at the other vertex A' of the hyperbola (so that $s = AA'$). □

The following problem was proposed by J. M. Feld in Altshiller-Court et al. (1932, problem 3578):

Two points are isogonal conjugates with respect to a triangle if and only if they are the foci of an inscribed conic.

In 1957, the *American Mathematical Monthly* declared that the problem was unsolved (see *American Mathematical Monthly*, 1957, pp. 65–75). However, the result does appear in the books Gallatly (1910), Morley and Morley (1933), Deaux (1957), Robson (1949), and Coxeter (1993, p. 198), though never with a synthetic proof: Deaux and Morley and Morley use the representation of the real Euclidean plane by complex numbers, Coxeter and Robson analytic geometry, Gallatly trigonometry.

Theorem 8.79
Let ABC be a triangle and P, Q be points on no side of ABC (with Q possibly at infinity). Then P and Q are isogonally conjugate if and only if there is a conic tangent to the sides of ABC with foci P and Q.

Proof Suppose there is a conic tangent to the sides of *ABC* with foci P and Q. Then, by the reflection property (Theorem 8.57), P and Q are isogonally conjugate. Conversely, suppose P and Q are isogonally conjugate. Consider the conic $\mathcal{K}$ tangent to the sides of ABC, PI, and PJ. Then, by Corollary 8.75, P is a focus of $\mathcal{K}$. Now by the reflection property, the other focus of $\mathcal{K}$ is isogonally conjugate to P with respect to triangle ABC, and hence equals Q. □

Corollary 8.80
If the sides of a triangle are tangent to a conic, then there is a focus of this conic on the circumcircle of the triangle if and only if the conic is a parabola.

Note how close this is to Wallace's 1797 problem that began the Simson line episode (earlier proved by Lambert in 1761; see Chapter 6, text under Worked Example 6.52), namely: 'If three straight lines touch a parabola, a circle described through their intersections shall pass through the focus of the parabola.'

Theorem 8.81
Let $\mathcal{K}$ be a conic with foci F_1 and F_2, and P be a point on two tangents to $\mathcal{K}$, with points of tangency X and Y. Then the angles $\angle F_1 PX$ and $\angle F_2 PY$ are congruent.

Proof This proof is due to Hatton (1913). By the dual of Desargues' involution theorem (Theorem 6.19), the conics tangent to four given lines (no three concurrent) give an involution t on the pencil of lines on any point on none of the four lines. A point F is a focus of a conic if and only if FI, FJ are tangent lines; thus, the conics with two given foci give an involution on the pencil of lines on any real point distinct from the foci. Let F and F' be the foci, P the point; then the circumscribing quadrilateral is $FI, FJ, F'I, F'J$. Since PI and PJ are a pair of the involution t, the fixed lines ℓ, m have $(PI, PJ; \ell, m)$ harmonic; hence, ℓ is perpendicular to m. Now PF, PF' is another pair of t, so PF, PF' are at congruent directed angles to ℓ.

Now let $\mathcal{K}$ be a conic on T, T' with PT and PT' tangent to $\mathcal{K}$ and foci F, F'. Then PT, PT' is another pair of t, so PT, PT' are at congruent directed angles to ℓ. Hence, the directed angles PT, PF and PT', PF' are congruent. □

Corollary 8.82 (Poncelet, 1817, p. 5, Théorème II)
The angles between the tangents from a point P to a conic and between the lines joining P to the foci have the same bisector.

The **envelope** of a conic $\mathcal{K}$ is the set of tangent lines to $\mathcal{K}$. Recall (see Example 3.6) that a conic, with associated symmetric matrix A, yields a polarity defined by

Table 8.2 *The result of taking the polar image of a circle with respect to another circle (see Figure 8.18).*

Case	Conic type of $\mathcal{K}$
O lies inside α	Ellipse
O is the centre of α	Circle
O lies on α	Parabola
O lies outside α	Hyperbola

$$(x, y, z) \mapsto [(x, y, z)A],$$
$$[a, b, c] \mapsto (A[a, b, c]^{\mathsf{T}}),$$

where we interchange round with square brackets accordingly. Suppose we have two non-degenerate conics defined by symmetric matrices M and A. We will take the image of the singular points of A under the polarity defined by M, to obtain a new conic A'. To do this, suppose v is a point of the conic defined by A; that is, $vAv^{\mathsf{T}} = 0$. Then the tangent line at v to A is given by $[vA]$, where we have identified v with its homogeneous coordinates. We then apply the polarity defined by M to obtain the point $w := vAM^{\mathsf{T}}$. Now define $A' := M^{-\mathsf{T}}A^{-1}M^{-1}$. Note that A' is invertible and symmetric. Then

$$wA'w^{\mathsf{T}} = \left(vAM^{\mathsf{T}}\right)M^{-\mathsf{T}}A^{-1}M^{-1}\left(vAM^{\mathsf{T}}\right)^{\mathsf{T}}$$
$$= vAM^{\mathsf{T}}M^{-\mathsf{T}}A^{-1}M^{-1}MAv^{\mathsf{T}}$$
$$= vAA^{-1}Av^{\mathsf{T}} = vAv^{\mathsf{T}} = 0.$$

Therefore, w lies on the non-degenerate conic defined by A'.

Theorem 8.83

Every non-empty, non-degenerate conic $\mathcal{K}$ of $\mathbb{R}^2$ can be obtained by taking the polar image of the envelope of a circle α, with respect to another circle ω. Moreover, if O is the centre of ω, then $\mathcal{K}$ has the characteristics described by Table 8.2.

Proof We will exploit the projective setting, that is, we will make use of the fact that $\mathsf{PGL}(3, \mathbb{R})$ acts transitively on non-empty, non-degenerate conics (Theorem 3.13). Let $\mathcal{K}$ be a conic with associated invertible symmetric matrix Z. Without loss of generality, we may suppose that ω is the unit circle with symmetric matrix

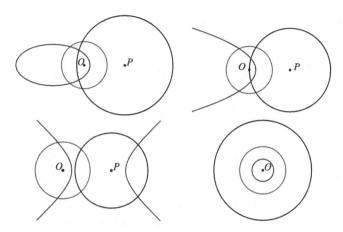

Figure 8.18 Four cases: The polar image of a circle with respect to a circle ω.

$$M := \begin{bmatrix} 1 & 0 & 0 \\ 0 & 1 & 0 \\ 0 & 0 & -1 \end{bmatrix}$$

and we can assume that a focus of $\mathcal{K}$ is the centre O of ω.

Let $A := M^{-\top}Z^{-1}M$. Then A is invertible and symmetric, so by the remarks above, the conic $\mathcal{K}$ is the the polar image of the envelope of the circle α defined by A. To verify that α is indeed a circle, note that I and J lie on α, and that, by Corollary 8.71, Z^{-1} acts trivially on the line at infinity, since O is a focus:

$$(1, \pm i, 0)M^{-\top}Z^{-1}M(1, \pm i, 0)^\top = (1, \pm i, 0)Z^{-1}(1, \pm i, 0)^\top = 0. \qquad \square$$

The **distance** $d(P, \ell)$ from a point P to a line ℓ is the distance to $d(P, \ell \cap h)$, where h is the perpendicular to ℓ on P.

Theorem 8.84
Let C be a conic of $\mathbb{R}^2$ and let F be a focus of C, and let f be the polar (directrix) of F. Then there exists a number e such that $d(P, F) = e \cdot d(P, f)$ for all points P of C.

Proof First, by Theorem 8.83, we will assume that C is the image of the envelope of a circle α with respect to the unit circle ω, and α has radius r and centre A; ω has centre F. Let P be a point of C and let p be the polar line of P with respect to ω. Suppose p meets α at T, meets the line FA at M, and meets the line FP at P'. Now f and the polar of M meet the line FA at A' and M'. Let K be the foot of the perpendicular from P to f. Then

$$\frac{d(P,K)}{d(F,P)} = \frac{d(F,A') - d(F,M')}{d(F,P)} = d(F,P')\left(\frac{1}{d(F,A)} - \frac{1}{d(F,M)}\right)$$

$$= \frac{d(F,P')}{d(F,M)}\left(\frac{d(F,M)}{d(F,A)} - 1\right) = \frac{d(A,T)d(A,M)}{d(A,M)d(F,A)}$$

$$= \frac{r}{d(F,A)}.$$

The result follows upon setting $e := r/d(F,A)$.   □

The converse of the above theorem then gives us a characterisation of conics, commonly known as the *focus-directrix* definition of a conic. (Note that the focus-directrix characterisation of a parabola is due to Diocles, On burning mirrors (c. 190–180 BC; see Diocles et al., 1976, prop. 4).

Theorem 8.85 (Pappus (c. 340) (see Pappus of Alexandria, 1986))
Given a point F and a line f not on F, and a positive real number e, the locus of points P with $d(P,F) = e \cdot d(P,f)$ is a conic.

Proof Let ω be the circle with centre F and tangent to f at a point A. Let α be the circle with centre A and radius $d(F,A)/e$. Then the locus of points P satisfying $d(P,F) = e \cdot d(P,f)$ is the image of the envelope of α with respect to the polarity defined by ω. The result then follows from (the discussion preceding) Theorem 8.83.   □

The number e is the **eccentricity** of the conic: $0 < e < 1$ for an ellipse that is not a circle, $e = 1$ for a parabola, and $e > 1$ for a hyperbola. Since the limit as $e \to 0$ is a circle (with F fixed), a circle is usually defined to have eccentricity 0. Another way of seeing this is to define the semi-length of an axis to be the distance from the centre to a point on the conic and on the axis. Then, since an ellipse that is not a circle has two axes, it has two semi-lengths a and b of axes, where $b < a$. (If $a = b$, then the ellipse is a circle.) Then it turns out that

$$e = \sqrt{1 - \frac{b^2}{a^2}}.$$

All semi-lengths of axes of a circle are equal (to the radius) and so this formula gives eccentricity $e = 0$ for a circle. Of course, in Theorem 8.85, F is a focus of the conic, and f is a directrix.

* * *

The following result relies on Theorem 10.7. It states:

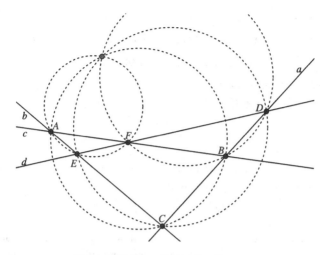

Figure 8.19 Miquel's four-line theorem.

Let K be a circle, A', B' be distinct points of K, and P, Q be points of K on the same side of $A'B'$. Then angles $\angle A'PB'$ and $\angle A'QB'$ are congruent.

Theorem 8.86 (Miquel's four-line theorem (Wallace, 1806, see p. 22, part C (1804) and pp. 169–70, part I (1805)))
Let a, b, c, d be four lines in the plane. Let $A := b \cap c$, $B := c \cap a$, $C := a \cap b$, $D := a \cap d$, $E := b \cap d$, $F := c \cap d$. Then, the circumcircles of triangles EAF, BDF, EDC, and BAC have a common point (see Figure 8.19).

Proof Let M be the point of intersection of the circumcircles of triangles EAF and EDC other than E. Then, angles $\angle MCD$ and $\angle MED$ are congruent, by Theorem 10.7, since M lies on the circumcircle of triangle EDC and angles $\angle MEF$ and $\angle MAF$ are congruent, by Theorem 10.7, since M lies on the circumcircle of triangle EAF. Hence, all the following angles are congruent: $\angle MCB$, $\angle MCD$, $\angle MED$, $\angle MEF$, $\angle MAF$, and $\angle MAB$. Thus, the points M, B, C, A are concircular, by the converse to Theorem 10.7. Equivalently, M lies on the circumcircle of triangle BAC. Similarly, M lies on the circumcircle of triangle BDF. Thus, the circumcircles of triangles EAF, BDF, EDC, and BAC have a common point. □

Recall that in Theorem 7.24 we showed that 5 points on a central conic determine 10 concurrent lines via taking centroids of every triangle made from 3 of the 5 points. We have the following corollary for circles of the Euclidean plane.

Corollary 8.87
Given five points A_1, A_2, A_3, A_4, A_5 on a circle, take any three points

$$\{A_1, A_2, A_3, A_4, A_5\} \setminus \{A_i, A_j\},$$

and through the centroid of these three points draw a line ℓ_{ij} perpendicular to the line A_iA_j. Then these 10 lines are concurrent.

$$* * *$$

In a similar way to when we introduced cross-ratio, we will define the **bracket** of three points P, Q, R of $\mathrm{PG}(2, \mathbb{R})$ to be

$$[P, Q, R] = \det(P \mid Q \mid R),$$

where we consider the points to be written in homogeneous coordinates and placed into the matrix as columns on the right-hand side. The following result can be found in Richter-Gebert (2011, Theorem 18.10), and can be derived from work of Cayley (1859) and Klein (1871, 1873).

Theorem 8.88
The Euclidean distance between two points P and Q can be calculated by

$$\frac{\sqrt{[P, Q, I][P, Q, J]} \cdot [A, I, J][B, I, J]}{\sqrt{[A, B, I][A, B, J]} \cdot [P, I, J][Q, I, J]}$$

where $d(A, B) = 1$ is a reference length.

Proof Write $P = (p_1, p_2, 1)$ and $Q = (q_1, q_2, 1)$ so that we regard P and Q as points of $\mathrm{PG}(2, R)$. By definition of the Euclidean distance, we have

$$
\begin{aligned}
d(P, Q) &= \sqrt{(p_1 - q_1)^2 + (p_2 - q_2)^2} \\
&= \sqrt{(q_1 - p_1 + i(q_2 - p_2)) \cdot (q_1 - p_1 - i(q_2 - p_2))} \\
&= \left(\begin{vmatrix} p_1 & q_1 & -i \\ p_2 & q_2 & 1 \\ 1 & 1 & 0 \end{vmatrix} \begin{vmatrix} p_1 & q_1 & i \\ p_2 & q_2 & 1 \\ 1 & 1 & 0 \end{vmatrix} \right)^{1/2} \\
&= \sqrt{[P, Q, I][P, Q, J]}.
\end{aligned}
$$

In order for this formula to be independent of the choice of homogeneous coordinates for P and Q, we introduce the extra terms so that scaling P and Q leaves the distance unchanged:

$$\frac{\sqrt{[P, Q, I][P, Q, J]} \cdot [A, I, J][B, I, J]}{\sqrt{[A, B, I][A, B, J]} \cdot [P, I, J][Q, I, J]}. \qquad \square$$

The following formula is very useful and allows us to recapture angle measure from cross-ratio. Note that ln in the formula below is the complex logarithm function.

Theorem 8.89 (Laguerre, 1853)
Given two lines ℓ, m of $\mathbb{R}^2$ meeting at P, let n, o be the lines through P and the circular points $I(1, i, 0)$, $J(1, -i, 0)$, respectively. Then the measure of the directed angle from ℓ to m in radians is

$$\frac{1}{2i} \ln(\mathrm{R}(\ell, m; n, o)) \quad (\mathrm{mod} \ \pi).$$

Proof Let ℓ have line coordinates $[l_1, l_2, l_3]$ and m have $[m_1, m_2, m_3]$. Then with $L = \ell \cap \ell_\infty$, $M = m \cap \ell_\infty$, $\mathrm{R}(\ell, m; n, o) = \mathrm{R}(L, M; I, J)$, and we have the coordinates $L(l_2, -l_1, 0)$, $M(m_2, -m_1, 0)$. The vectors $(l_2, -l_1)$, $(m_2, -m_1)$ are normal to the lines L, M, respectively, so the measure of the directed angle from ℓ to m in radians is the directed angle between these vectors (modulo π). Let $z_\ell = l_2 - i l_1 \in \mathbb{C}$ and $z_m = m_2 - i m_1 \in \mathbb{C}$ and write them in polar form: $z_\ell = r_\ell e^{i\psi_\ell}$, $z_m = r_m e^{i\psi_m}$. Now, after some calculation, we obtain

$$\mathrm{R}(L, M; I, J) = \frac{z_\ell \bar{z}_m}{\bar{z}_\ell z_m} = e^{2i(\psi_\ell - \psi_m)},$$

giving $\psi_\ell - \psi_m = \frac{1}{2i} \ln(\mathrm{R}(\ell, m; n, o))$ (mod π). □

Corollary 8.90
Let ℓ, m be lines of $\mathbb{R}^2$, and let $L := \ell \cap \ell_\infty$, $M := m \cap \ell_\infty$. Then ℓ is perpendicular to m if and only if $(L, M; I, J)$ is harmonic, where I and J are the circular points.

Corollary 8.91 (Euclid, Book I, Proposition 32)
The sum of the measures of the angles in a triangle is π radians.

Proof Let ABC be a triangle. By the *cocycle identity* for cross-ratio (see Theorem 4.20 and preceding discussion), we have

$$\mathrm{R}(AB, AC; AI, AJ) \cdot \mathrm{R}(BC, BA; BI, BJ) \cdot \mathrm{R}(CA, CB; CI, CJ) = 1.$$

Apply Theorem 8.89. □

Another proof of Theorem 8.7 Let ℓ be ℓ_∞ in Theorem 6.16, and let L' be the point at infinity on the altitude on A, M' be the point at infinity on the altitude on B, and N' be the point at infinity on the altitude on C. By Corollary 8.90, $(I, J; L, L')$, $(I, J; M, M')$, $(I, J; N, N')$ are harmonic. So $(L L')(M M')(N N')\ldots$ is an involution. Thus, the altitudes of ABC are concurrent. □

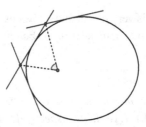

Figure 8.20 Theorem 8.92.

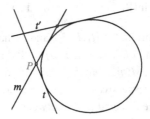

Figure 8.21 Proof of Theorem 8.92.

Theorem 8.92 (de la Hire, 1685, Book VIII, p. 190, Prop. XXIV; Poncelet, 1822, p. 267, art. 464)
The tangents to a conic subtend equal angles at a focus (see Figure 8.20).

Proof Let F be a focus of a conic K. Fix tangents t, t' to conic K and define $h: t \to t'$ as follows. For P on t, let m be the other tangent to K lying on P (apart from t). Then define P^h to be $m \cap t'$ (see Figure 8.21). Theorem 6.61 implies that h is a projectivity.

Define $g: \text{pencil}(F) \to \text{pencil}(F)$ as follows. For n in $\text{pencil}(F)$, let $n^g := F(n \cap t)^h$. Then g is a projectivity on pencils. Now n is fixed by g if and only if n is tangent to K. Since F is a focus of K, we have that FI, FJ are tangent to K. By Theorem 4.36, $R(FI, FJ; n, n^g)$ is independent of n. Let s be a tangent to K other than t, t', and let $M := s \cap t$, $M' := s \cap t'$. Then $M^h = M'$, so $(FM)^g = FM'$, and $R(FI, FJ; FM, FM')$ is independent of s. By Laguerre's formula (Theorem 8.89), $\angle MFM'$ is independent of s. □

We will show how a theorem of Juel yields Euclid's III.20.

Theorem 8.93 (Juel, 1934, CH. VII, Section 2.II, p. 90)
Let $\mathcal{K}$ be a conic, A, B, C, D be points of $\mathcal{K}$, and E be the intersection of the tangent at A and the tangent at B. Then $R(EA, EB; EC, ED) = R_{\mathcal{K}}(A, B; C, D)^2$.

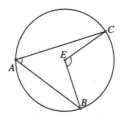

Figure 8.22 Theorem 8.94.

Proof Let t_A be the tangent at A, and let t_B be the tangent at B. Let $A_1 :=$ $t_A \cap CD$, $B_1 := t_B \cap CD$. Let $X := AB \cap CD$. Then

$$R(EA, EB; EC, ED) = R(A_1, B_1; C, D) = R(A_1, X; C, D) \cdot R(X, B_1; C, D).$$

Now,

$$R(A_1, X; C, D) = R(AA_1, AX; AC, AD) = R(t_A, AB; AC, AD) = R_{\mathcal{K}}(A, B; C, D),$$

and, likewise,

$$R(X, B_1; C, D) = R(BX, BB_1; BC, BD) = R(BA, t_B; BC, BD) = R_{\mathcal{K}}(A, B; C, D).$$

Therefore, $R(EA, EB; EC, ED) = R_{\mathcal{K}}(A, B; C, D)^2$. □

Theorem 8.94 (Euclid's *Elements*, Book III, Proposition 20)
In a circle, the angle at the centre is double the angle at the circumference when the angles have the same circumference as their base (see Figure 8.22).

Proof Take A and B to be the circular points in Theorem 8.93. The rest follows from Laguerre's formula (Theorem 8.89). □

Another consequence of Juel's theorem (Theorem 8.93) is an old result of Euclidean geometry, attributed to Heron of Alexandria (in his commentary on Euclid's *Elements*) by al-Faḍl b Ḥātim al-Nayrīzī c. 920 (see Lo Bello, 2009, pp. 110–11). It was rediscovered by Ramus in 1569.

Theorem 8.95 (Heron of Alexandria; Ramus, 1569, lib XV, prop. 15, no. 3)
Let P be an external point to a circle K. Let Q, R be the points of tangency to K of the tangent lines to K on P. Then $\overline{PQ}$ and $\overline{PR}$ are congruent.

Proof It suffices to show that the angles $\angle PRQ$ and $\angle PQR$ have the same measure. To do this, we show that $|R(QP, QR; QI, QJ)| = |R(RP, RQ; RI, RJ)|$ and use Laguerre's formula (Theorem 8.89). We denote the tangents at Q and R by

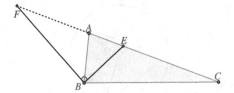

Figure 8.23   Theorem 8.96 and its converse.

QQ and *RR* (respectively), which are also equal to *PQ* and *PR* (respectively). Then

$$R(QP, QR; QI, QJ)^2 = R(QQ, QR; QI, QJ)^2$$
$$= R_{\mathcal{K}}(Q, R; I, J)^2 \qquad \text{by Theorem 6.65}$$
$$= R(PQ, PR; PI, PJ) \quad \text{by Juel's theorem (Theorem 8.93).}$$

Also, $R(RP, RQ; RI, RJ)^2 = R(PR, PQ; PI, PJ)$ and so $R(QP, QR; QI, QJ)^2 = R(RP, RQ; RI, RJ)^2$ as required.   □

Another proof of Theorem 8.95   Let *g* be the harmonic homology with axis the line *m* joining *P* and the centre *O* of *K*, and having centre the pole of *m*. Then *g* fixes *K*, and, since the axis is on the centre of *K*, has centre on the line at infinity. Thus, *g* interchanges *I* and *J*, so is a similarity. Since *g* fixes *O* and *K*, *g* is an isometry. Since *g* fixes *P*, it also fixes the polar line *QR* of *P*; hence, *g* interchanges *Q* and *R*. Thus, $\overline{PQ}$ and $\overline{PR}$ are congruent.   □

Theorem 8.96 (Pappus, 1986, Book VI, Proposition 52)
Let ABC be a triangle, E be the intersection of AC and the internal angle bisector of angle ∠CBA, F on AB with ratio(E, A, C) = ratio(F, A, C). *Then angle ∠EBF is a right angle (see Figure 8.23).*

   Conversely ...

Theorem 8.97 (Commandino, 1588)
Let ABC be a triangle, E on AC, F on AC with angle ∠EBF a right angle, and EB bisecting ∠CBA. Then ratio(E, A, C) = ratio(F, A, C).

Proofs of Theorem 8.96 and Theorem 8.97.   By Theorem 8.89, if *EB* bisects ∠*CBA*, (*BE, BF; BI, BJ*) is harmonic, so, by Corollary 8.90, ∠*EBF* is a right angle. Conversely, if ∠*EBF* is a right angle, and *EB* bisects ∠*CBA*, then, by Theorem 8.89 and Corollary 8.90, (*BE, BF; BI, BJ*) is harmonic, so (*E, F; A, C*) is harmonic and hence ratio(*E, A, C*) = ratio(*F, A, C*).   □

Theorem 8.98 (Pappus of Alexandria, 1986, Proposition 156)
Let A, B be points on circle K, m the diameter of K bisecting A and B, {D, E} = K ∩ m. For any point C on K, if G = BC ∩ m, and F = AC ∩ m, ratio(G, D, E) = ratio(F, D, E).

Proof Since AB is perpendicular to m, the angle $\angle BDE$ will be congruent to the angle $\angle EDA$, but the angle $\angle DAB$ is in fact congruent to the angle $\angle GCE$ in the same segment, while the angle $\angle EDA$ is congruent to the angle $\angle ECF$, which is an exterior angle of the cyclic quadrilateral $DACE$. Therefore, the angle $\angle BCE$ is congruent to the angle $\angle ECF$ and the angle $\angle DCE$ in the semicircle is a right angle. Therefore, by Theorem 8.97, ratio(G, D, E) = ratio(F, D, E). ◻

Theorem 8.99
In every cyclic quadrilateral, the line connecting the circumcentre with the intersection of its diagonals is perpendicular to the line connecting the intersections of its opposite sides.

Proof Let K be a circle with centre O, and let A, B, C, D be four points on K. Let $P := AC \cap BD$, $Q := AB \cap CD$, and $R := BC \cap DA$. By Miquel's four-line theorem (Theorem 8.86), the circumcircles of triangles RCD, QAD, RAB, and QCB have a common point. Since M lies on the circumcircle of triangle RAB, angles $\angle RMB$ and $\angle RAB$ are congruent. Since M lies on the circumcircle of triangle BOD, angles $\angle BMO$ and $\angle BDO$ are congruent. Since O is the centre of the circle K and the points B, D, and A lie on K, by Theorem 8.94, $\angle BDO = 2\angle DAB$. Thus,

$$\angle RMO = \angle RMB + \angle BMO = \angle RAB + \angle BDO$$
$$= \angle DAB + (\pi/2 - \angle DAB)$$
$$= \pi/2,$$

and hence OM is perpendicular to QR. Since the point M lies on OP, OP is perpendicular to QR. ◻

Worked Example 8.100
The following comes from Demir et al. (1992).

Problem Suppose that F and F' are points situated symmetrically to the centre O of a given circle, and that S is a point of the circle not on FF'. Let P and P' be the second points of the circle on FS and $F'S$, respectively. If (see Figure 8.24) the tangents t_P, $t_{P'}$ to the circle at P and P' intersect at T, prove that the perpendicular bisector of $\overline{FF'}$ passes through the midpoint of $\overline{ST}$.

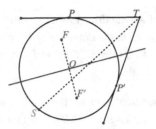

Figure 8.24 Worked Example 8.100.

Solution Let K be the point of the circle diametrically opposite S. Let S' be the other point of the circle on the line ℓ through S parallel to FF' (with $S' = S$ if ℓ is tangent to the circle). Now O is the midpoint of $\overline{FF'}$, so $(\ell, SP; SK, SP')$ is harmonic. By the (extended) Chasles–Steiner theorem (Theorem 6.65), it follows that, for any point X on the circle, $(XS', XP; XK, XP')$ is harmonic, with XX interpreted as the tangent line to the circle at X. Thus, $(PS', PT; PK, PP')$ and $(P'S', P'P; P'K, P'T)$ are harmonic. Hence, S', K, T are collinear. If $S \neq S'$, the diameter SK subtends a right angle at the point S' of the circle, so $S'K$ is perpendicular to SS', which is parallel to FF'. If $S = S'$, the diameter SK is perpendicular to the tangent line ST at S, which is parallel to FF'. Now the perpendiculars to FF' through T, O, and S meet TS in the same ratio as they meet FF'; thus, the perpendicular bisector of $\overline{FF'}$ passes through the midpoint of $\overline{ST}$.   □

Exercises

8.1   Prove Theorem 8.25.

8.2   How is measurement modulo π of angles reflected in the properties of the complex logarithm function?

8.3   Using Laguerre's angle formula, show that the angle between two real lines is real.

8.4   Using Laguerre's angle formula, show that interchanging two lines reverses their angle.

8.5   Using Laguerre's angle formula, show how angles between concurrent lines add.

8.6   Interpret Exercise 5.5 by taking A and B to be the circular points.

8.7   Show that every line of $PG(2, \mathbb{C})$ contains a real point.

8.8   Let P, Q, R be non-collinear points of $PG(2, \mathbb{R})$, embedded in $PG(2, \mathbb{C})$. Let A and B be two more points of $PG(2, \mathbb{R})$, not lying on PQ. Show that there exists a collineation of $\mathsf{PG}(2, \mathbb{C})$ such that $\{A, B\}$ is mapped to the circular points, and such that P, Q, R are mapped to points of $\mathsf{PG}(2, \mathbb{R})$.

8.9 Let $ABCD$ be a rectangle, and let E, F be two points of the diagonal BD, neither of which is the midpoint of $\overline{BD}$, and such that $(E, F; B, D)$ is harmonic. Let K be a point of BD, and let L be the point on BD such that CL is parallel to AK. Show that $(E, F; K, L)$ is harmonic if and only if $\{K, L\} = \{B, D\}$.

8.10 Prove the following: The Wallace–Simson line of a point P on the circumcircle of a triangle ABC of $\mathbb{R}^2$ is the tangent at the vertex of a parabola tangent to the sides of ABC and having its focus at P.

8.11 Let ABC be a triangle, and let I be the point at infinity on the line AB. Let X and X' be a pair of isogonally conjugate points lying on AB, with respect to $\triangle ABC$. Show that $\mathrm{R}(X, A; B, I) \cdot \mathrm{R}(X', A; B, I) = 2$

8.12 Prove the converse to Thales' theorem: Given distinct points A, B, C, if angle $\angle ACB$ is a right angle then C lies on the circle with diameter $\overline{AB}$.,

8.13 Show that the circumcircle of the triangle formed by three tangents to a parabola of $\mathbb{R}^2$ passes through the focus.

8.14 Given a triangle ABC of $\mathbb{R}^2$, prove that the bisectors of the interior and exterior angles at C, the side AB and its perpendicular bisector, and the perpendiculars to AC at A and to BC at B are all tangent to a parabola. Locate its focus.

8.15 Modify the proof of Theorem 8.52 to prove the following theorem.

Theorem 8.101 (Möbius, 1827, para. 230)
Every quadrangle is the perspective image of a square.

8.16 Let ℓ, m be concurrent lines of $\mathbb{R}^2$, and let n, o also be concurrent lines. Show that the directed angle from ℓ to m is congruent to the directed angle from n to o if and only if

$$(\ell \cap \ell_\infty \ n \cap \ell_\infty)(m \cap \ell_\infty \ o \cap \ell_\infty)(I J)\ldots$$

is an involution.

8.17 Show that if $\ell_1, \ell_2, \ell_3, \ell_4$ are concurrent lines of $\mathbb{R}^2$, and m_1, m_2, m_3, m_4 are another set of concurrent lines, such that the directed angle $\angle \ell_i \ell_{i+1}$ is congruent to $\angle m_i m_{i+1}$ for each $i \in \{1, 2, 3\}$, then (in the extended Euclidean plane) $\mathrm{R}(\ell_1, \ell_2; \ell_3, \ell_4) = \mathrm{R}(m_1, m_2; m_3, m_4)$.

8.18 Suppose that A is a point inside the angle $\angle XOY$, and B and C are points on the arms of OX and OY, respectively. Then if any two of the following statements hold, so does the third.

  (i) A lies on the bisector of the angle $\angle XOY$.

  (ii) $AB = AC$.

  (iii) Angles $\angle OBA$ and $\angle OCA$ are either equal or supplementary.

8.19 $AXBZ$ is a jointed rhombus of $\mathbb{R}^2$ connected with a fixed point O by two equal rods OA, OB. $OCZD$ is a jointed rhombus and YC, YD are equal rods. Prove that as Y describes a circle, X describes a conic.

8.20 Given a triangle ABC whose altitudes are AA', BB', CC' with feet A', B', C', respectively, prove that the Euler lines of the triangles $AB'C'$, $A'BC'$, $A'B'C$ are concurrent on the nine-point circle.

8.21 Interpret the result of Exercise 6.9 when ℓ is the line at infinity, AB is perpendicular to CD, AC is perpendicular to BD, and AD is perpendicular to BC.

8.22 ('The problem of Pappus') Fix a quadrangle $ABCD$ in $\mathbb{R}^2$ and a constant $t \in \mathbb{R}$, with $t \neq 0, 1$. Show that the following is a conic:

$\{P: \text{R}(\text{perpendicular from } P \text{ to } AB, \text{perpendicular from } P \text{ to } CD;$

$\text{perpendicular from } P \text{ to } AD, \text{perpendicular from } P \text{ to } BC) = t\}$

$\cup \{A, B, C, D\}.$

8.23 Suppose a, b, c are lines in $\mathbb{R}^2$ and A is a variable point on a. We also have lines through A at right angles to b, c meet b and c at B, C, respectively. Show that BC is tangent to a fixed conic.

8.24 The lines ℓ, ℓ' of $\mathbb{R}^2$ meet at a point A, and let B be another point. A variable circle through A and B meets ℓ, ℓ' again at P, P', respectively. Show that PP' is tangent to a fixed conic.

8.25 Suppose Σ is a circle of $\mathbb{R}^2$ and C is a point of Σ. The lines PC, QC are at right angles and the tangent to Σ at C passes through the midpoint of $\overline{PQ}$. Show that P and Q are conjugate points with respect to Σ.

(a) Apply Frégier's theorem (Theorem 8.36) to prove Frégier's projection theorem:

Let C be a conic, ABC be a triangle inscribed in C, D be a point of C, other than A, B, C, E, F, G be the feet of the perpendiculars of D on the sides of ABC, and $\mathcal{K}$ be the circumcircle of EFG. Then $\mathcal{K}$ passes through the projection of D on the polar of the Frégier point of D.

(b) Apply Frégier's projection theorem to prove the following theorem.

Theorem 8.102 (Bobillier, 1827–8, p. 349)
Let $\mathcal{H}$ be a rectangular hyperbola, ABC be a triangle inscribed in $\mathcal{H}$, D be a point of $\mathcal{H}$, other than A, B, C, E, F, G be the feet of the perpendiculars of D on the sides of ABC, and $\mathcal{K}$ be the circumcircle of EFG. Then $\mathcal{K}$ passes through the centre of $\mathcal{H}$.

8.26 Suppose F and F' are points situated symmetrically with respect to the centre O of a given circle (so that $F^{H_O} = F'$) and not equal to O, and suppose S is a point on the circle not on the line FF'. Let P and P' be the second points of intersection of SF and SF', respectively, with the circle. If the tangents to the circle at P and P' intersect at T, prove that the perpendicular bisector of FF' passes through the midpoint of the line segment $\overline{ST}$.

8.27 Use Menelaus' theorem to show that the tangents at the vertices to the circumcircle of a triangle, provided none of them is parallel to the opposite side, meet the opposite sides in collinear points.

8.28 Show that the centre of a central conic that is not a circle is the midpoint of the line segment with endpoints the two foci of the conic.

8.29 Prove the following theorem.

Theorem 8.103
Given four points A, B, C, D in $\mathbb{R}^2$, the necessary and sufficient condition that AD is perpendicular to BC, AC is perpendicular to BD, and AB is perpendicular to CD is that the midpoints of $\overline{AB}$, $\overline{AC}$, $\overline{AD}$, $\overline{BC}$, $\overline{BD}$, and $\overline{CD}$ are concircular.

8.30 Use Theorem 8.23 to prove the following theorem.

Theorem 8.104 (Bobillier, 1829)
If ABC is a triangle and P is a point, A' is the intersection of BC and the perpendicular to PA on P, B' is the intersection of CA and the perpendicular to PB on P, and C' is the intersection of AB and the perpendicular to PC on P, then A', B', C' are collinear.

8.31 Prove the following theorem. (Hint: Take the reflection g with axis ZC.)

Theorem 8.105 (Witelo *Perspectiva*, proposition 125, c. 1270–8; see Unguru, 1977)
If B, C, D, K are collinear points of $\mathbb{R}^2$ with $d(B, K): d(K, D) = d(B, C): d(C, D)$ and Z has angles $\angle BZC$ and $\angle CZD$ congruent, then $\angle CZK$ is a right angle.

8.32 Prove the following theorem.

Theorem 8.106 (Neël, 1879, art. 6, p. 11)
Let ABC be a triangle and the altitude on B have foot Y on CA, P be a point of $\overline{BY}$. Let $AP \cap BC = X$ and $CP \cap AB = Z$. Then the line BY bisects $\angle XYZ$.

8.33 Let A_1, A_2, B_1, B_2 be distinct points.

  (a) Suppose every circle on A_1, A_2 meets every circle through B_1, B_2.

    i Show that A_1, A_2, B_1, B_2 are collinear or concircular.
    ii Show that the cross-ratio of $(A_1, B_1; A_2, B_2)$ is negative.

  (b) Conversely, show that if A_1, A_2, B_1, B_2 are collinear or concircular and the cross-ratio of $(A_1, B_1; A_2, B_2)$ is negative then every circle through A_1, A_2 meets every circle through B_1, B_2.

8.34 An acute triangle ABC with $d(A, B) \neq d(A, C)$ is given. Let V and D be the feet of the altitude and the angle bisector from A, and let E and F be the intersection points of the circumcircle of $\triangle AVD$ with sides AC and AB, respectively. Prove that AD, BE, and CF have a common point.

8.35 (i) Let P be a variable point on line BC of triangle ABC. The circle with diameter $\overline{BP}$ intersects the circumcircle of ACP again at Q. Prove that PQ passes through a fixed point, as P varies.

  (ii) Furthermore, let H be the orthocentre of ABP, and $M := PQ \cap AC$. Prove that, as P varies, MH passes through a fixed point.

8.36 Use a similar technique to the proof of Corollary 8.11 to show that the orthocentre and the circumcentre of a triangle are isogonally conjugate. (Hint: Take the images of the altitudes of the triangle in the respective angle bisectors. Show that they intersect in the circumcentre.)

9

Transformation Geometry: Klein's Point of View

> The importance of group theory was emphasized very recently when some physicists using group theory predicted the existence of a particle that had never been observed before, and described the properties it should have. Later experiments proved that this particle really exists and has those properties.
>
> Irving Adler (1968, p. 7)

We give a very brief overview of different kinds of geometry from Klein's point of view: Euclidean geometry (in two flavours: congruence and similarity), affine geometry, projective geometry, and topology.

Symmetry was first introduced into geometry by Legendre in 1794 (Hon and Goldstein, 2008) in his influential textbook *Éléments de géométrie* (Legendre, 1794). The genius behind bringing symmetry to the fore in geometry was Christian Felix Klein in his Erlangen Programme (1872) (Klein, 1893). This saw the differences between the types of geometries in terms of their allowable groups of transformations. Building on the development of the concept of group by Galois in the 1830s (Neumann, 2011), and by Jordan in the 1860s (Jordan, 1989), and on the slow acceptance of the non-Euclidean geometries of János Bolyai (Bonola, 1955) and Nikolai Ivanovich Lobachevski (1946–51; Bonola, 1955) (published around 1830), Klein pioneered a perspective under which *geometry became definitively plural*. Projective geometry had been pioneered by Desargues (1951; Bosse, 1648) in the seventeenth century (with some projective results known to Pappus of Alexandria (1986)) and came into full flower at the hands of Poncelet (1995b) and Gergonne (1825/6), early in the nineteenth century, with the development of the principle of duality. The first text in projective geometry was Poncelet's *Treatise* (1822) (Poncelet, 1995b). Affine geometry has its origins in the books of Möbius (1827)

241

Table 9.1 *A tower of groups and the properties they preserve.*

Group	Adjective	Property
Homeo(PG(2, $\mathbb{R}$))	topological	compactness, connectedness
PGL(3, $\mathbb{R}$)	projective	collinearity, cross-ratio, incidence
AGL(2, $\mathbb{R}$)	affine	parallel, ellipse, midpoint, ratios of lengths
Sim($\mathbb{R}^2$)	(similarity) Euclidean	circle, isosceles, angle
Isom($\mathbb{R}^2$)	(congruence) Euclidean	distance, congruence

(Möbius, 1827) and Grassmann (1844) (Grassmann, 1862). (Non-Euclidean geometry developed as an outcome of attempts to prove the controversial parallel postulate of Euclid by, for example, Saccheri (1733) (Saccheri, 2011) and Lambert (1766) (Lambert, 1946, 1948) at the end of a long list of predecessors. Here, again, spherical geometry was an unrecognised precursor.) Before Klein, projective and affine geometry were seen just as ways of doing Euclidean geometry. After Klein, or more properly, by the 1890s, they were seen as geometries in their own right.

9.1 A Tower of Groups

Consider the following groups: the Euclidean group Isom($\mathbb{R}^2$), Sim($\mathbb{R}^2$), AGL(2, $\mathbb{R}$), PGL(3, $\mathbb{R}$), Homeo(PG(2, $\mathbb{R}$)), where the last group is the homeomorphism group of the real projective plane. (A **homeomorphism** is a continuous bijection with continuous inverse.) Each group is isomorphic to a subgroup of the following group in the list.

Klein's point of view on geometry is illustrated by saying that, in each case, there is a corresponding geometry. Respectively, these are: the Euclidean geometry of congruence, the Euclidean geometry of similarity, affine geometry, projective geometry, and topology. (Topology is the abstract study of continuity.) A property of subsets (or relation between subsets) of PG(2, $\mathbb{R}$) is **topological** if it is preserved by Homeo(PG(2, $\mathbb{R}$)), but not by the symmetric group on the points of PG(2, $\mathbb{R}$). So *compactness, connectedness* are topological properties, but *cardinality, collinearity* are not (the first since it is preserved by the symmetric group, the second since it is not preserved by the homeomorphism group). We can continue this way through the tower of groups (see Table 9.1).

A subtlety occurs in moving from the affine group to the projective group: the permutation action changes from being on $\mathbb{R}^2$ to being on PG(2, $\mathbb{R}$). Thus, the abstract groups are subgroups, but the permutation groups are not.[1] This subtlety underlies some of the conceptual difficulties of moving to projective geometry historically: the need for the invention of *points at infinity*.

There are other forms of these geometries: direct Euclidean congruence geometry (associated with $\mathbb{R}^2 \rtimes SO(2)$ and preserving orientation); direct Euclidean congruence geometry (associated with $\mathbb{R}^2 \rtimes \mathbb{R}^+ SO(2)$ and also preserving orientation), though this latter group neither contains nor is contained in the Euclidean congruence group $\mathbb{R}^2 \rtimes O(2)$ but, rather, is off to the side; direct affine geometry (associated with the **equiaffine group** $ASL(2, \mathbb{R})$, also preserving orientation), which is also off to the side.

9.2 Cayley–Klein Geometries

> The analytical geometry of Descartes and the calculus of Newton and Leibniz have expanded into the marvellous mathematical method more daring than anything that the history of philosophy records – of Lobachevsky and Riemann, Gauss and Sylvester. Indeed, mathematics, the indispensable tool of the sciences, defying the senses to follow its splendid flights, is demonstrating today, as it never has been demonstrated before, the supremacy of the pure reason.

> Nicholas Murray Butler (1895, p. 45)

Ascending the chain of subgroups involves forgetting structure, first distance, then circles, then parallels, then collinearity. Descending the chain involves imposing structure. Can the structure be imposed intrinsically? The answer is a resounding yes. Moreover, it brings about a unification of previously different domains: Euclidean geometry and parts of non-Euclidean geometry. Moving from the projective to the affine is easy: just delete a hyperplane.[2] But how do we move from the affine to the Euclidean (or the non-Euclidean)? This is the realm of Cayley–Klein geometries (Cayley, 1859, 1872; Klein, 1873, 1893, 1968). We have a hyperplane at infinity that we add back to our affine space to return to the projective setting. We put extra structure on this hyperplane, namely, a polarity. We then consider, as allowable transformations in our new geometry, only those affine transformations that commute with the polarity. If the polarity has no absolute points, Euclidean

---

[1] To see why $(PG(2, \mathbb{R}), AGL(2, \mathbb{R}))$ is not permutationally isomorphic to $(\mathbb{R}^2, AGL(2, \mathbb{R}))$, note that $AGL(2, \mathbb{R})$ has two orbits on the points of $PG(2, \mathbb{R})$, yet it acts transitively on the points of $\mathbb{R}^2$.

[2] We are referring here to projective, affine, and Euclidean geometries of higher dimensions than we have touched on so far. This is the subject of forthcoming sections.

similarity geometry is the outcome. But if the polarity has absolute points, a (non-classical) non-Euclidean geometry arises. For example, using the polarity arising from the quadric $x^2 + y^2 + z^2 - c^2 t^2$ (where c is the speed of light) gives rise to (a supergroup of) the Poincaré group (Poincaré, 1905) $\mathbb{R}^4 \rtimes L$, where $L = \mathsf{SO}(3, 1)$ is the Lorentz group (Lorentz, 1904): the group of isometries of the quadratic form defining the quadric. We obtain the geometry of Minkowski space-time (Minkowski, 1909) and the physics of special relativity (Einstein, 1905). (The supergroup is $\mathbb{R}^4 \rtimes \mathbb{R}^+ L$, which bears the same relation to the Poincaré group as the Euclidean similarity group does to the Euclidean congruence group.) This is a hyperbolic non-Euclidean geometry.

To return to the Euclidean geometry of the plane, the polarity $(a, b) \mapsto [a, b]$ of $\mathsf{PG}(1, \mathbb{R})$ gives as its centraliser the Euclidean similarity group $\mathbb{R}^2 \rtimes \mathbb{R}^+ \mathsf{O}(2)$. Since 'circle, is a Euclidean similarity concept, we should be able to identify the circles of $\mathbb{R}^2$ with the help of the polarity. Embed $\mathsf{PG}(1, \mathbb{R})$ in $\mathsf{PG}(1, \mathbb{C})$ (and $\mathsf{PG}(2, \mathbb{R})$ in $\mathsf{PG}(2, \mathbb{C})$), and consider the absolute points of the polarity. They are (with the line at infinity being the usual $z = 0$) $(1, i, 0)$ and $(1, -i, 0)$. These are the **circular points**, introduced by Poncelet. A non-empty, non-degenerate conic of $\mathsf{PG}(2, \mathbb{R})$ is a circle of $\mathbb{R}^2$ if and only if it passes through the circular points, by Theorem 8.22.

Recall the three-reflection theorem: every isometry of the Euclidean plane can be written as the product of at most three reflections.[3] Using this theorem, we showed that the isometry group of the Euclidean plane is the semi-direct product of the group of translations and $\mathsf{O}(2)$, and that every similarity is the product of a dilatation and an isometry. By a theorem of Carathéodory (see Theorem 8.50, every bijection mapping circles to circles is a similarity. Thus, we have recovered Euclidean similarity geometry from the polarity on the line at infinity.

If we wish to consider an alternative to the parallel axiom, we may impose an orthogonal polarity on the whole plane. If we choose the polarity $(a, b, c) \mapsto [a, b, c]$ of $\mathsf{PG}(2, \mathbb{R})$ and consider, as allowable transformations in our new geometry, only those projective transformations that commute with the polarity, we obtain the classical non-Euclidean elliptic geometry. If, instead of the polarity $(a, b, c) \mapsto [a, b, c]$ of $\mathsf{PG}(2, \mathbb{R})$, we use the polarity $(a, b, c) \mapsto [a, b, -c]$ of $\mathsf{PG}(2, \mathbb{R})$, and consider, as allowable transformations in our new geometry, only those projective transformations that commute with the polarity and take as our points the points internal to the conic $x^2 + y^2 - z^2 = 0$ of absolute points and lines the lines secant to this conic, we obtain the classical

---

[3] Élie Cartan (1938) generalised this to real orthogonal groups in all dimensions and then Jean Dieudonné (1948) generalised to fields of characteristic not 2: an isometry of n dimensions is the product of at most $n + 1$ reflections.

non-Euclidean hyperbolic plane. This is a surface of constant Gaussian curvature -1, for which the hyperbolic trigonometric functions play the role that the trigonometric functions play for the Euclidean plane, which is itself a surface of zero Gaussian curvature.[4] This is all related to the Gauss–Bonnet theorem (Bonnet, 1848; Gauss, 1900), connecting the curvature of a surface to its Euler characteristic.

9.3 Descending the Klein Tower

Let us begin at the top of the tower. Consider the real projective plane as a topological surface, with the associated group of symmetries being $\mathsf{Homeo}(\mathsf{PG}(2, \mathbb{R}))$. This group does not preserve incidence of points and lines, but once we insist on lines being preserved, we arrive at the subgroup of collineations of $\mathsf{PG}(2, \mathbb{R})$. By the FToPG (Theorem 2.9), we have:

Theorem 9.1
Any collineation of $\mathsf{PG}(2, \mathbb{R})$ *is a homeomorphism.*

At this level of the tower, we have the real projective plane as an incidence geometry. By singling out 'a line at infinity' we create the real affine plane. From the real affine plane we descend to the Euclidean similarity geometry by either insisting on the circular points being preserved or insisting on an involution with no real fixed points on the line at infinity being preserved ($x \mapsto -1/x$ being the canonical choice). From this point of view, we can easily distinguish the orientation-preserving (e.g., dilatations, translations, half-turns) and orientation-reversing similarities (e.g., reflections): orientation-reversing similarities interchange the circular points; orientation-preserving similarities fix each circular point.

Recall from Carathéodory's theorem that a bijection on $\mathbb{R}^2$ maps circles to circles if and only if it is a similarity. The importance of this result cannot be overstated, for it *means* that Euclidean planes are point–circle incidence geometries, and that the circular points are sufficient to establish every facet of the Euclidean plane from the blank slate of the real projective plane. In other words, we can *measure angles*, once we can establish the meaning of *right angle*!

Finally, after stipulating that length be preserved, we step down from similarity geometry to congruence geometry, and as we descend a chain of subgroups, from $\mathsf{Homeo}(\mathsf{PG}(2, \mathbb{R}))$ to $\mathsf{Isom}(\mathbb{R}^2)$, we inherently introduce the fundamental geometric invariants preserved by our symmetries.

---

[4] These are also analogous to the ancient results in spherical trigonometry, related to the fact that a sphere of radius 1 is a surface of constant Gaussian curvature 1.

10

The Power of Projective Thinking

> In modern mathematics the investigation of the symmetries of a given mathematical structure has always yielded the most powerful results.

Emil Artin (1957, p. 54)

In this chapter, we celebrate the projective point of view, and the dividends it brings. Euclidean and affine theorems have not only been realised as derivatives of projective results; surprisingly, there are unnoticed connections between results of ancient times, only visible through the projective lens. Involutions become conspicuous in the projective realm, and we explore how involutions can be regarded as the foundational transformations of projective geometry.

10.1 Euclidean and Affine Connections

Below are some examples that show that seemingly unconnected Euclidean theorems are just different interpretations of one projective theorem, equivalent to each of them. This illustrates the power of projective thinking.

First, we present a theorem of Euclidean geometry about two circles with two common tangents. This example is taken from Horadam (1970, pp. 178, 234–5).

Theorem 10.1
Two circles K_1 and K_2, mutually tangent at S with common tangent ℓ, have m as another common tangent line, with points of tangency P on K_1 and Q on K_2. Let $R := \ell \cap m$. Then R is the midpoint of $\overline{PQ}$ and $\angle PSQ$ is a right angle (see Figure 10.1).

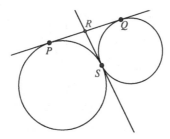

Figure 10.1 Theorem 10.1.

Proof By Theorem 8.95, $d(R, P) = d(R, S)$. By Theorem 8.95 again, $d(R, S) = d(R, Q)$. Hence, R is the midpoint of $\overline{PQ}$, and the circle K with centre R and radius $d(R, P)$ passes through P, Q, and S. Hence, by 'angle-angle-angle', $\angle PSQ$ is a right angle. □

Now we give the 'projective' counterpart to Theorem 10.1. An alternative proof is given in the solution to Exercise 6.27.

Theorem 10.2

Two conics $\mathcal{K}_1$ and $\mathcal{K}_2$ touch at S and intersect at A, B. Then their common tangent at P and at Q meets the tangent at S in the point R. Let $T := AB \cap PQ$. Then $R(R, T; P, Q) = -1$ and $R(SA, SB; SP, SQ) = -1$.

Proof By Exercise 8.8, there exists a collineation g of $\mathsf{PG}(2, \mathbb{C})$ such that $\{A, B\}$ is mapped to the circular points $\{I, J\}$, and such that P, Q, S are mapped to points of $\mathsf{PG}(2, \mathbb{R})$. Therefore, this result follows from the Euclidean theorem above, because g maps our two conics to two circles. □

From the projective counterpart, we obtain a seemingly different theorem in Euclidean geometry, but now for parabolas! This result also appeared in Worked Example 8.65.

Theorem 10.3

The common chord of two parabolas with parallel axes bisects their common tangents (see Figure 10.2).

Proof This follows from the projective theorem, by putting SR as the line at infinity. □

We can take a theorem of Apollonius, and take a route through projective geometry to a theorem of Euclid.

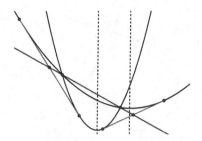

Figure 10.2 Theorem 10.3.

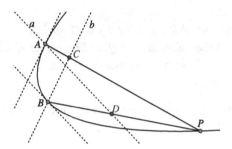

Figure 10.3 Theorem 10.4.

Theorem 10.4 (Apollonius, *Conics*, Book III, Proposition 54 (rephrased))
If A, B are two fixed points on a conic, and parallels a to the tangent at B and on A and b to the tangent at A and on B are taken, and P is a variable point on the conic, and $PA \cap b = C$, $PB \cap a = D$, then $d(A, D) \cdot d(B, C)$ is constant (see Figure 10.3).

Corollary 10.5
The map $a \to b$ taking D to C is a projectivity.

Proof Let Q be the point at infinity on a, R be the point at infinity on b. Then A maps to R and Q maps to B. Let D_1, D_2 be points on a, and let C_1, C_2 be their images on b. We list a series of equivalent statements: $R(A, D_1; D_2, Q) = R(R, C_1; C_2, B)$ if and only if $\text{ratio}(A, D_2, D_1) = \text{ratio}(B, C_1, C_2)$, if and only if $d(A, D_2) \cdot d(B, C_2) = d(A, D_1) \cdot d(B, C_1)$, if and only if $d(A, D) \cdot d(B, C)$ is constant. □

Corollary 10.6
The map g from the pencil on A to the pencil on B taking AP to BP is a projectivity.

Thus, we have one direction of the Chasles–Steiner theorem (Theorem 6.61).

Theorem 10.7 (Euclid *Elements*, Book III, Proposition 21)
Let K be a circle, A', B' be distinct points of K, and P, Q be points of K on the same side of A' B'. Then angles ∠A'PB' and ∠A'QB' are congruent.

Proof Apply Corollary 10.6: R(PA', PB'; PI, PJ) = R(QA', QB'; QI, QJ). So, by Laguerre's angle formula (Theorem 8.89), angles ∠$A'PB'$ and ∠$A'QB'$ are congruent. □

Without projective thinking, who would have seen a connection between Apollonius III.54 and Euclid III.21?

Notice that we have above, as a by-product, proved one direction of the Chasles–Steiner theorem (Theorem 6.61).

Proof Angles subtended by the same segment on a circle are congruent (if on the same side of the segment) or supplementary (if on opposite sides of the segment), and hence have the same sine. By the interpretation of cross-ratio of four concurrent lines in terms of the sines of the angles between them (see Worked Example 4.22), it follows that the forward direction of Theorem 6.61 holds for a circle. By projection, the forward direction of Theorem 6.61 holds for all non-degenerate conics. □

The following example was adapted from Milne (1911, p. 269), and it begins with a well-known Euclidean theorem. Note that all three parts of the theorem follow from the fact that the reflection in *CT* fixes the circle, and therefore interchanges *A* and *B*.

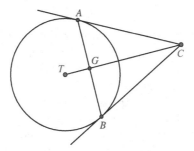

Figure 10.4 Theorem 10.8.

Theorem 10.8 (compare with Euclid's *Elements*, Book III, Proposition 3)
Suppose CA and CB are tangents to a circle at A and B (respectively), with centre T, and suppose AB and CT meet in a point G. Then:

(i) *CT bisects the angle* $\angle ATB$.
(ii) *CT bisects the angle* $\angle ACB$.
(iii) *CT bisects AB at right angles at G (see Figure 10.4).*

The corresponding projective result is:

Theorem 10.9
Suppose IJ and AB are two chords of a conic, with T and C their poles respectively. Then:

(i) *TC is fixed by* $(TI\ TJ)(TA\ TB)$.
(ii) *If CT, AB, AC, BC meet IJ in F, H, K, L, then CT is fixed by*

$$(CI\ CJ)(CK\ CL).$$

(iii) $(A, B; C, H)$ *and* $(I, J; F, H)$ *are harmonic. Hence, TC and TH are fixed by*
$(TI\ TJ)(TA\ TB)(TK\ TL).$

From the above we also obtain the following:

Tangents to a conic subtend equal (or supplementary angles) at the focus,[1] and if T is the focus and C the pole of a chord AB, then AB is divided harmonically by CT and the directrix, the polar of T.

Again, previously unsuspected connections between theorems are discerned by projective thinking.

* * *

The next example shows that we can prove a result that holds in the projective plane using a well-known result in Euclidean geometry.

Theorem 10.10 (Ehrmann and van Lamoen, 2004)
Let ABC be a triangle and E be an inscribed conic. Let ℓ and ℓ' be two lines intersecting at P, tangent to E, and d be a line not passing through P. Let X, Y, Z (respectively X', Y', Z'; X_d, Y_d, Z_d) be the intersections of ℓ (respectively ℓ', d) with the lines BC, CA, AB (see Figure 10.5). If X'_d is the harmonic conjugate of X_d with respect to $\{X, X'\}$, and similarly for Y'_d and Z'_d, then X'_d, Y'_d, Z'_d lie on a common line d' if and only if d touches E.

---

[1] 'Tangents to a conic subtend equal angles at the focus' means that if ZX is tangent to a conic K at X, and ZY is tangent to K at Y, and K has focus F, then FZ bisects the angle $\angle XFY$.

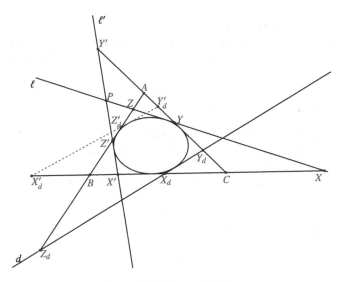

Figure 10.5 Theorem 10.10.

To prove the Ehrmann–van Lamoen theorem (Theorem 10.10), we prove the dual theorem:

Theorem 10.11 (Dual of Theorem 10.10)
Let A, B, C, I, J be a set of five points, no three collinear, and let S be a point not on IJ. The harmonic conjugates of AS with respect to {AI, AJ}; BS with respect to {BI, BJ}; CS with respect to {CI, CJ} concur if and only if A, B, C, I, J, S lie on a conic.

We prove this theorem by taking I and J to be the circular points. Thus, we must show that given a triangle ABC and a point S, the perpendiculars to AS, BS, CS at A, B, C, respectively, concur if and only if S lies on the circumcircle of ABC. But this is the Simson line theorem (Theorem 6.53) and its converse!

Putting d as the line at infinity gives:

Corollary 10.12
The midpoints of XX′, YY′, ZZ′ lie on a same line d′ if and only if ℓ and ℓ′ touch the same inscribed parabola. In this case, if ℓ and ℓ′ touch the parabola at M and M′, d′ is the tangent to the parabola parallel to MM′.

Putting P to be the orthocentre of ABC in the corollary gives the Droz-Farny theorem (Theorem 8.12) (because P lies on the directrix of an inscribed parabola).

Theorem 10.13 (Droz-Farny, 1899)
If two perpendicular straight lines are drawn through the orthocentre of a triangle, they intercept a segment on each of the sidelines. The midpoints of these three segments are collinear.

Proof Note that P lies on the directrix of any inscribed parabola.   □

The next example shows that in passing from an affine to a Euclidean perspective and studying conics, the eccentricity tells all.

Theorem 10.14 (Apollonius de Perge, 2009, Book VI, Proposition 11)
Any pair of parabolas are similar.

Proof By Theorem 8.84, a parabola is determined by its focus and directrix, so we need only show that $\mathrm{Sim}(\mathbb{R}^2)$ is transitive on non-incident point-line pairs, which is straightforward and is the subject of Exercise 2.2.   □

Theorem 10.15
Two conics of $\mathbb{R}^2$ are similar if and only if they have the same eccentricity.

Proof The forward direction is an immediate consequence of Theorem 8.84. Now, for the backward direction, note that all circles are similar, so we may assume that the eccentricity is positive. By Theorem 8.84, a conic is determined by its focus, directrix, and eccentricity, so we need only show that $\mathrm{Sim}(\mathbb{R}^2)$ is transitive on non-incident point-line pairs, which is straightforward.   □

The following example is from Vajda (1946). We begin with a theorem of Euclidean geometry, with a 'Euclidean' proof.

Theorem 10.16 (Carnot, 1801, pp. 101–3, nos. 142–4)
The reflections of the orthocentre of an acute triangle in its sides lie on the circumcircle.

Proof Let ABC be the triangle, H the orthocentre, R be the foot of the altitude on AC, Q be the foot of the altitude on BC, S be the other intersection of AH with the circumcircle of ABC. Angles $\angle CAS$ and $\angle CBS$ are congruent, by Euclid's *Elements*, Book III, Proposition 21. Hence, triangles RAH and QBH are similar. Thus, triangles QBS and QBH are congruent. So S is the reflection of H in BC.   □

Theorem 10.16 can be readily transformed into a result that holds in the real projective plane.

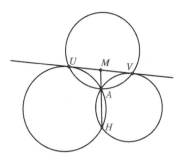

Figure 10.6 Theorem 10.18.

Theorem 10.17 (Projective version of Theorem 10.16)
Let ABCH be a quadrangle, with the pencil of conics on A, B, C, H giving an involution on the line ℓ, on none of A, B, C, H. Consider the conic 𝒦 passing through A, B, C and the fixed points U and V of the involution. Let Q := BC ∩ AH, M := AH ∩ ℓ, {S, A} = AH ∩ 𝒦. Then R(H, S ; Q, M) = −1.

Proof This follows from Theorem 10.16 by projection.   □

We then gain a new Euclidean theorem via the intervening projective result.

Theorem 10.18
Two circles intersect in A and H. Draw one of the common tangents ℓ which touches at U and V, respectively (see Figure 10.6). Draw further the circle K through A, U, and V. Let M := AH ∩ ℓ, {S, A} = AH ∩ 𝒦. Then M is the midpoint of $\overline{HS}$.

Proof Take *B* and *C* to be the circular points in Theorem 10.17.   □

Here is another example where the projective point of view can give insight into the connections between results in Euclidean geometry. First, let us consider a theorem on equilateral triangles which follows from applications of the theorems of Menelaus and Ceva.

Theorem 10.19
If UVW is an equilateral triangle, A, B, C are the respective midpoints of the sides VW, WU, UV; A′ is any point on line VW, B′ any point on line WU, and C′ any point on line UV, P := BC ∩ B′C′, Q := CA ∩ C′A′, R := AB ∩ A′B′, then the lines A′P, B′Q, C′R are concurrent (see Figure 10.7).

Proof Applying Menelaus' theorem,

$$\text{ratio}(P, B', C')\text{ratio}(C, C', U)\text{ratio}(B, U, B') = 1,$$
$$\text{ratio}(Q, C', A')\text{ratio}(A, A', V)\text{ratio}(C, V, C') = 1,$$
$$\text{ratio}(R, A', B')\text{ratio}(B, B', W)\text{ratio}(A, W, A') = 1,$$

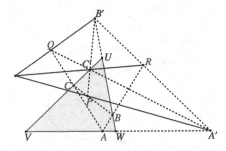

Figure 10.7 Theorem 10.19.

giving $\text{ratio}(P, B', C')\text{ratio}(Q, C', A')\text{ratio}(R, A', B') = -1$, because

$$\text{ratio}(C, C', U)\text{ratio}(B, U, B')\text{ratio}(A, A', V)\cdot$$
$$\text{ratio}(C, V, C')\text{ratio}(B, B', W)\text{ratio}(A, W, A') = -1.$$

Therefore, $A'P$, $B'Q$, $C'R$ are concurrent, by Ceva's theorem (Theorem 7.14). □

From the above theorem, we obtain an *affine* result.

Theorem 10.20
If UVW is a triangle, A, B, C are the respective midpoints of the sides VW, WU, UV; A' is any point on line VW, B' any point on line WU, and C' any point on line UV, P := BC ∩ B'C', Q := CA ∩ C'A', R := AB ∩ A'B', then the lines A'P, B'Q, C'R are concurrent.

Proof Apply an affinity to take UVW to an equilateral triangle and apply Theorem 10.19. □

Next in this sequence of theorems is a *projective* result that appears in part I of the 1909 mathematical tripos.

Theorem 10.21
If UVW is a triangle, A, B, C are points of the sides VW, WU, UV such that UVW and ABC are in perspective; A' is any point on line VW, B' any point on line WU, and C' any point on line UV, P := BC ∩ B'C', Q := CA ∩ C'A', R := AB ∩ A'B', then the lines A'P, B'Q, C'R are concurrent.

Proof Apply a projective transformation to take UVW to an equilateral triangle and the centre of perspectivity of UVW and ABC to the centroid of UVW and apply Theorem 10.19. □

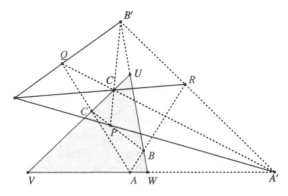

Figure 10.8 Theorem 10.22.

Now we see the benefit of the ascension to the projective point of view; we have first a Euclidean theorem that begins with the altitudes of an arbitrary triangle, and a second theorem, which concerns the incircle of a triangle.

Theorem 10.22

If UVW is a triangle, A, B, C are the feet of the altitudes of the sides VW, WU, UV; A' is any point on line VW, B' any point on line WU, and C' any point on line UV, P := BC ∩ B'C', Q := CA ∩ C'A', R := AB ∩ A'B', then the lines A'P, B'Q, C'R are concurrent (see Figure 10.8).

Proof This follows from the projective theorem and the concurrency of the altitudes (Theorem 8.7). □

Theorem 10.23

If UVW is a triangle, A, B, C are the points of tangency of the incircle with the sides VW, WU, UV; A' is any point on line VW, B' any point on line WU, and C' any point on line UV, P := BC ∩ B'C', Q := CA ∩ C'A', R := AB ∩ A'B', then the lines A'P, B'Q, C'R are concurrent (see Figure 10.9).

Proof This follows from the projective theorem and the concurrency of the internal angle bisectors. □

Thus, from a projective perspective, Theorems 10.19, 10.22, and 10.23 are equivalent!

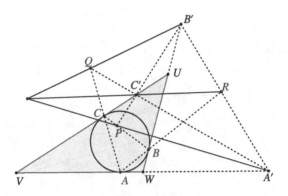

Figure 10.9 Theorem 10.23.

10.2 Advanced Theory Using Involutions

In this section, we explore two threads. The first concerns the results that are closely related to the Lehmer–Daus theorem (Theorem 10.38). There are several interesting results in the last 150 years that follow more or less directly from this theorem, so it is a foundational result on involutions abounding the pencils of lines attached to a triangle. The second thread concerns a result attributed to Oai Thanh Dao, which also has many corollaries, some of which were not known to be corollaries. Remarkably, these two threads are inextricably linked, as we shall see.

We begin by recalling the dual of the converse to Pappus' involution theorem (Theorem 6.13). Much of what we derive in this section begins with this result.

Theorem 10.24 (Dual of Theorem 6.13, restated)
If $(PA\,PA_1)(PB\,PB_1)(PC\,PC_1)\ldots$ is an involution and ABC is a triangle, with $A_1 = BC \cap PA_1$, $B_1 = CA \cap PB_1$, $C_1 = BC \cap PC_1$, then A_1, B_1, C_1 are collinear.

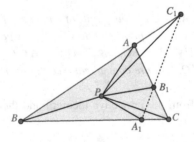

From Theorem 10.24, we can deduce a 'Euclidean' version of a result of Hjelmslev (1907). We use the notation σ_ℓ for the reflection about the line ℓ.

Theorem 10.25 (Euclidean counterpairing theorem)
Let ABC be a triangle and $a', b', c', a^*, b^*, c^*$ be elements of a pencil of lines which correspond in a counterpairing, that is, $\sigma_{a'}\sigma_{b'} = \sigma_{b^*}\sigma_{a^*}$ and $\sigma_{a'}\sigma_{c'} = \sigma_{c^*}\sigma_{a^*}$ and $\sigma_{b'}\sigma_{c'} = \sigma_{c^*}\sigma_{b^*}$, and suppose a' is incident with A, b' is incident with B, and c' is incident with C. Then $BC \cap a^*$, $CA \cap b^*$, $AB \cap c^*$ are collinear.

Proof Let P be the centre of the given pencil. Consider the following map θ_a on pencil(P):

$$\theta_a \colon \ell \mapsto \ell^*, \quad \text{where } \sigma_{\ell^*} := \sigma_{a'}\sigma_\ell\sigma_{a^*}.$$

The three reflections theorem of Euclidean geometry states that the product of three reflections in concurrent lines is a reflection and so θ_a is well-defined. Next, note that θ_a has order 2 on pencil(P) and interchanges a and a^*. In fact, θ_a is induced by a projectivity, as we shall see now. Let w be the angle bisector of $\angle a'a^*$. Then $\sigma a'\sigma_w = \sigma_w\sigma_{a^*}$ and hence

$$\sigma_{a'}\sigma_\ell\sigma_{a^*} = \sigma_w\sigma_{a^*}\sigma_w\sigma_\ell\sigma_{a^*} = \sigma_w\sigma_{a^*}(\sigma_w\sigma_\ell\sigma_{a^*})$$
$$= \sigma_w\sigma_{a^*}(\sigma_{a^*}\sigma_\ell\sigma_w) = \sigma_w\sigma_\ell\sigma_w = \sigma_{\ell^\sigma w},$$

and so the map θ_a is induced by a reflection about the angle bisector w. Therefore, θ_a is a projective involution. We can define θ_b and θ_c similarly.

Consider the product $\theta = \theta_a\theta_b\theta_c$. Then, for each ℓ,

$$\sigma_{c'}\sigma_{b'}\sigma_{a'}\sigma_\ell\sigma_{a^*}\sigma_{b^*}\sigma_{c^*} = \sigma_{a'}\sigma_{a'}\sigma_{c'}\sigma_{b'}\sigma_{a'}\sigma_\ell\sigma_{a^*}\sigma_{b^*}\sigma_{c^*}\sigma_{a^*}\sigma_{a^*}$$
$$= \sigma_{a'}\sigma_{b'}\sigma_{c'}\sigma_\ell\sigma_{c^*}\sigma_{b^*}\sigma_{a^*},$$

and so θ is a projective involution on pencil(P). Moreover,

$$\sigma_{c'}\sigma_{b'}\sigma_{a^*}\sigma_{b^*}\sigma_{c^*} = (\sigma_{b'}\sigma_{c^*}\sigma_{a^*}\sigma_{b^*})\sigma_{c^*} = (\sigma_{a'}\sigma_{c^*})\sigma_{c^*} = \sigma_{a'},$$

and hence θ interchanges a' and a^*. Also,

$$\sigma_{c'}\sigma_{b'}\sigma_{a'}\sigma_{b'}\sigma_{a^*}\sigma_{b^*}\sigma_{c^*} = \sigma_{c'}\sigma_{b'}(\sigma_{a'}\sigma_{b'}\sigma_{a^*}\sigma_{b^*})\sigma_{c^*} = \sigma_{c'}\sigma_{b'}\sigma_{c^*} = \sigma_{b'},$$

and hence θ interchanges b' and b^*. Also,

$$\sigma_{c'}\sigma_{b'}\sigma_{a'}\sigma_{c'}\sigma_{a^*}\sigma_{b^*}\sigma_{c^*} = \sigma_{c'}\sigma_{b'}\sigma_{c^*}\sigma_{b^*}\sigma_{c^*} = \sigma_{c^*},$$

and hence θ interchanges c' and c^*. So θ is the involution $(a'\, a^*)(b'\, b^*)(c'\, c^*) \ldots ..$
Therefore, by Theorem 10.24, $BC \cap a^*$, $CA \cap b^*$, $AB \cap c^*$ are collinear. $\square$

Corollary 10.26 (Goormaghtigh, 1930)
Given a triangle ABC of $\mathbb{R}^2$, a point D distinct from A, B, C and a line ℓ passing through D, suppose that A^*, B^*, C^* belong to BC, CA, AB, respectively, such that DA^*, DB^*, DC^* are the images of DA, DB, DC, respectively, under the reflection in ℓ (see Figure 10.10). Then A^*, B^*, C^* are collinear.

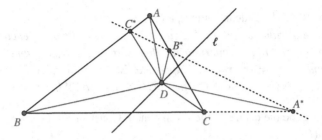

Figure 10.10 Goormaghtigh's theorem (Theorem 10.26).

Table 10.1 *Corollaries of Theorem 10.27.*

Oblique Simson line theorem (Poncelet, 1817)	Theorem 6.17
Goormaghtigh's theorem (Goormaghtigh, 1930)	Theorem 10.26
Nixon's theorem (Nixon, 1891)	Theorem 10.29
Zaslavsky's theorem, unpublished, 2003	Theorem 10.30
Bliss' theorem, unpublished, 1999 (van Lamoen, 2000)	Theorem 10.31
Lalesco's theorem (Lalesco, 1937; Collings, 1973)	Theorem 10.32

Proof Let $a' := AD$, $b' := BD$, $c' := CD$, and let $a^* := A^*D$, $b^* := B^*D$, $c^* := C^*D$. Notice that $A^* = BC \cap A^*D = BC \cap a^*$, and similarly for B^* and C^*. Now $a^* = (a')^{\sigma_\ell}$, $b^* = (b')^{\sigma_\ell}$, and $c^* = (c')^{\sigma_\ell}$, and so

$$\sigma_{b^*}\sigma_{a^*} = \sigma_{b'}^{\sigma_\ell}\sigma_{a'}^{\sigma_\ell} = \sigma_\ell\sigma_{b'}\sigma_{a'}\sigma_\ell = \sigma_{a'}\sigma_{b'},$$

and similarly $\sigma_{a'}\sigma_{c'} = \sigma_{c^*}\sigma_{a^*}$ and $\sigma_{b'}\sigma_{c'} = \sigma_{c^*}\sigma_{b^*}$. Therefore, by Theorem 10.25, A^*, B^*, C^* are collinear. □

We refer to Struve and Struve (2016) for more on counterpairing and Goormaghtigh's theorem.

Second proof of Theorem 8.12 Apply Goormaghtigh's theorem (Theorem 10.26), with ℓ an angle bisector of the pair of perpendicular straight lines. □

The following theorem is an extension of Theorem 6.53 and Takasu's theorem (Theorem 6.42), and a generalisation of a theorem of Dao Thanh Oai (2013) (see also Cohl, 2014 and Ngoc, 2018). The ensuing corollaries are direct consequences of this foundational result, and we summarise them in Table 10.1 as a convenient reference:

Theorem 10.27
Let $\mathcal{K}$ be a conic, A, B, C be distinct points of $\mathcal{K}$, P a point not on $\mathcal{K}$, nor on any side of ABC, $PA \cap \mathcal{K} = \{A, A_1\}$, $PB \cap \mathcal{K} = \{B, B_1\}$, $PC \cap \mathcal{K} = \{C, C_1\}$,

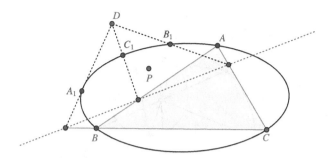

Figure 10.11 Theorem 10.27 (where *D* is on the polar line of *P*).

*D be a point distinct from A, B, C, A$_1$, B$_1$, C$_1$ (see Figure 10.11). Then A$_0$:=
DA$_1$ ∩ BC, B$_0$:= DB$_1$ ∩ CA, C$_0$:= DC$_1$ ∩ AB are collinear if and only if D is
on the given conic or the polar line of P.*

Proof The Frégier involution with centre *P* fixes every point on its axis, leaves
$\mathcal{K}$ invariant, and interchanges *A* and *A$_1$*, *B* and *B$_1$*, *C* and *C$_1$*. Hence on $\mathcal{K}$,
$(A A_1)(B B_1)(C C_1)\ldots$ is an involution. If *D* is on the polar line of *P* or a point
of $\mathcal{K}$, we have that $(DA\, DA_1)(DB\, DB_1)(DC\, DC_1)\ldots$ is an involution. Apply
the converse of the dual of Pappus' involution theorem (Theorem 10.24) to the
quadrilateral with sides *AB*, *BC*, *CA* and $\ell = (DA_1 ∩ BC)(DB_1 ∩ CA)$ and the
point *D*. This gives an involution

$$(DA\, DA_1)(DB\, DB_1)(DC\, D(\ell ∩ AB))\ldots.$$

Hence, $D(\ell ∩ AB) = DC_1$; thus, $DC_1 ∩ AB = \ell ∩ AB$; in other words, $DA_1 ∩ BC$,
$DB_1 ∩ CA$, $DC_1 ∩ AB$ are collinear.

Conversely, given a point *D* with homogeneous coordinates (x, y, z), the condition that $DA_1 ∩ BC$, $DB_1 ∩ CA$, $DC_1 ∩ AB$ are collinear is cubic in (x, y, z).
Since it holds for more than three points of the polar line of *P*, the polar line of
P is a component of this cubic, and the other component is given by a quadratic
condition, so is a conic. Since it holds for more than four points of the given
conic, it equals the given conic. That proves the converse. □

Corollary 10.28
*The points A$_0$, B$_0$, C$_0$, P defined in Theorem 10.27 are collinear if and only if
D lies on the conic $\mathcal{K}$.*

Proof If *D* lies on $\mathcal{K}$, then by applying Pascal's theorem (Theorem 6.28) to
the hexagon $AA'DB'BC$, we see that *P*, *B$_0$*, and *A$_0$* are collinear. Therefore, *P*
lies on ℓ. Conversely, suppose *P* lies on ℓ and consider the Frégier involutions

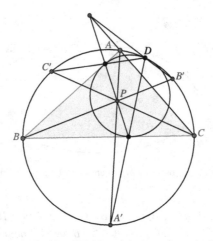

Figure 10.12 Corollary 10.29, and the ingredients for its proof.

t_P, t_{A_0}, t_{B_0} with centres P, A_0, B_0, respectively. Since these three points are collinear, we have $(t_P t_{A_0} t_{B_0})^2 = 1$. So, in particular,

$$(A')^{t_{A_0}} = A^{t_P t_{A_0}} = A^{t_{B_0} t_{A_0} t_P t_{B_0}} = C^{t_{A_0} t_P t_{B_0}} = B^{t_P t_{B_0}} = (B')^{t_{B_0}}.$$

Now $D = A'A_0 \cap B'B_0$, and so $D = (A')^{t_{A_0}} = (B')^{t_{B_0}}$ and D lies on $\mathcal{K}$.   □

Corollary 10.29 (Nixon, 1891, p. 107)
If a circle is tangent internally to the circumcircle of a triangle and to two sides of that triangle, then the line joining the points of contact of the sides passes through the incentre.

Proof Suppose we have a triangle ABC with circumcircle S, and another circle S' which is tangent to S, tangent to AB, and tangent to BC (see Figure 10.12). Let P be the incentre of ABC, and let D be the point of contact of S' with S. Let A', B', C' be the images of A, B, C under the Frégier involution with centre P. Let $A_0 := DA' \cap BC$, $B_0 := DB' \cap AC$; $C_0 := DC' \cap AB$. By Corollary 10.28, A_0, B_0, C_0, P are collinear.

We will show now that A_0 is the point of contact of BC with S'. It will follow also that C_0 is the point of contact of AB with S' (by taking the reflection in the line BB'). By definition of incentre, A' lies on the angle bisector of $\angle BAC$. By Theorem 8.32, A' is the intersection of the perpendicular bisector m of $\overline{BC}$ with S. Let A_0' be the point of contact of S' with BC. By Corollary 8.38, A_0' lies on the angle bisector n of $\angle BDC$. By Theorem 8.32 (applied to BDC),

$m \cap n$ lies on S, and hence $m \cap n = A'$. So A' lies on $n = A_0'D$ and hence $A_0' = DA' \cap BC = A_0$. $\qquad\qquad\qquad\qquad\qquad\qquad\qquad\qquad\qquad\qquad\qquad\square$

Another proof of Goormaghtigh's theorem (Theorem 10.26) Let A', B', C' be the images of A, B, C (respectively) under the reflection in ℓ. Then, by the Braikenridge–Maclaurin theorem (Theorem 6.43), A, B, C, A', B', C' lie on a conic $\mathcal{K}$ (because the diagonal points of $AA'BB'CC'$ lie on ℓ), and $AA' \parallel BB' \parallel CC'$. So Theorem 10.27 applies with P being the point at infinity for this parallel class. (Note that D lies on the polar of P because ℓ is perpendicular to the lines of this parallel class.) Therefore, A^*, B^*, C^* are collinear. $\qquad\square$

Another proof of the oblique Simson line theorem (Theorem 6.17) The result follows from Theorem 10.27 when we let the conic $\mathcal{K}$ be the circumcircle of the given triangle, D is on the conic, and P is at infinity. $\qquad\qquad\square$

Corollary 10.30 (Alexey Zaslavsky, unpublished, 2003)
Given a triangle ABC, a point P, and reflection $A'B'C'$ of ABC in P. Let three parallel lines through A', B', and C' intersect BC, AC, and AB in X, Y, Z, respectively. Then X, Y, Z are collinear.

Proof This result follows from Theorem 10.27 when P is the centre of the conic. $\qquad\qquad\qquad\qquad\qquad\qquad\qquad\qquad\qquad\qquad\qquad\qquad\qquad\qquad\square$

Corollary 10.31 (Panakis, undated c. 1965, vol. 2, p. 654, art. 582; Adam Bliss, unpublished, 1999; van Lamoen, 2000)
The reflections of the parallels to a line ℓ through the midpoints of the sides of triangle ABC in the side whose midpoint it goes through are concurrent in a point of the nine-point circle.

Proof Apply Theorem 10.27 with the nine-point circle of ABC as the conic, the point of concurrence at infinity, with D, E, F and the other points of intersection of the three parallel lines with the nine-point circle. $\qquad\qquad\square$

Corollary 10.32 (Lalesco, 1937, p. 99; Collings, 1973)
The reflections of a line ℓ in the sides of triangle ABC are concurrent if and only if ℓ passes through the orthocentre. In this case, the intersection is a point on the circumcircle.

Proof If ℓ is on the orthocentre of triangle ABC, and A', B', C' are the respective images of H under the reflections in BC, CA, AB, then A, B, C, A', B', C' all lie on the circumcircle of ABC; since AA', BB', CC' concur at H, and D,

Table 10.2 *Consequences of Theorem 10.34.*

Jeřábek's theorem (Jeřábek, 1886)	Theorem 10.33
Dao's theorem	Theorem 10.27
Theorem 10.35	
Papelier's theorem (Papelier, 1927)	Theorem 10.36

E, F are points on BC, CA, AB, respectively, with D, E, F, H collinear, then Theorem 10.27 says that DA' (the reflection of ℓ in BC), EB' (the reflection of ℓ in CA), and FC' (the reflection of ℓ in AB) are concurrent. (The converse follows from Theorem 8.86.)   □

The following was posed as a question by Jeřábek in 1886, and was solved on the conic in *Mathesis Tome* 8 in 1888 by Laurens (pp. 204–6), Beyens (p. 206), Servais (pp. 206–7) and Jeřábek (pp. 207–8). It also appears in Neuberg (1888) in the same issue. Interestingly, Steiner in 1867 gives a passing demonstration of this result (see Steiner, 1867, pp. 330–1). Baker states one direction of it (Baker, 1943, Chapter VII, Section 4, p. 57, example 18) where he considers nine points on the projective line and certain involutions involving these points. It also appears in Baker (1943, p. 307, example 155).

Theorem 10.33 (Jeřábek, 1886, p. 23; Steiner, 1867, pp. 330–1)
Given a triangle ABC inscribed in a circle and two points D_1, D_2 not on the circle, suppose (A, C_2, A_1), (B, C_2, B_1), (C, C_2, C_1), (A, D_2, A_2), (B, D_2, B_2), (C, D_2, C_2) are collinear triples of points. Then the triangle ABC and the triangle with the sides A_1A_2, B_1B_2, C_1C_2 are in perspective.

The proof of Theorem 10.33 will follow from the next result, which is a generalisation of Dao's theorem (Theorem 10.27) and Theorem 10.33. In Table 10.2, we extend our list of corollaries to Theorem 10.27, with consequences from Theorem 10.34 (see Table 10.2):

Later we will see that Theorem 10.35 leads to the Lehmer–Daus theorem (Theorem 10.38) and its many corollaries.

Theorem 10.34
Let ABC be a triangle inscribed in a conic S and A_1, A_2, B_1, B_2, C_1, C_2 be six points on S such that AA_1, BB_1, CC_1 are concurrent (see Figure 10.13). All six points are distinct except that it is possible for A_1, B_1, C_1 to coincide. Then AA_2, BB_2, CC_2 are concurrent if and only if $A_3 := A_1A_2 \cap BC$, $B_3 := B_1B_2 \cap CA$, and $C_3 := C_1C_2 \cap AB$ are collinear.

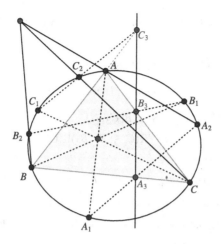

Figure 10.13 A generalisation of Dao's theorem.

Proof We first need to project all the points onto the sides of the triangle ABC. Let $A_4 := AA_1 \cap BC$, $A_5 := AA_2 \cap BC$, $B_4 := BB_1 \cap CA$, $B_5 := BB_2 \cap CA$, $C_4 := CC_1 \cap AB$, $C_5 := CC_2 \cap AB$. Let t_A be the tangent to S at A and $A_6 := BC \cap t_A$; let t_B be the tangent to S at B and $B_6 := CA \cap t_B$; let t_C be the tangent to S at C and $C_6 := AB \cap t_C$.

First we show

$$R(B, C; A_3, A_4) = R(B, C; A_5, A_6) \tag{10.1}$$

$$R(C, A; B_3, B_4) = R(C, A; B_5, B_6) \tag{10.2}$$

$$R(A, B; C_3, C_4) = R(A, B; C_5, C_6). \tag{10.3}$$

For Equation (10.1), we have

$$
\begin{aligned}
R(B, C; A_3, A_4) &= R(A_1 B, A_1 C; A_1 A_3, A_1 A_4) \\
&= R(A_1 B, A_1 C; A_1 A_2, A_1 A) \\
&= R(AB, AC; AA_2, t_A) && \text{(by Theorem 6.65),} \\
&= R(B, C; A_5, A_6) && \text{(intersecting with } BC).
\end{aligned}
$$

A similar argument validates Equations (10.2) and (10.3).

Now let ℓ be a line, not on A, B, C, $AA_1 \cap BB_1$, $AA_2 \cap BB_2$, $AA_3 \cap BB_2$; and let $L := BC \cap \ell$, $M := CA \cap \ell$, $N := AB \cap \ell$. By the projective Ceva theorem (Theorem 4.33),

$$R(B, C; A_4, L)R(C, A; B_4, M)R(A, B; C_4, N) = -1, \tag{10.4}$$

because AA_1, BB_1, CC_1 are concurrent. By the projective Menelaus theorem (Theorem 4.34),

$$R(B, C; A_6, L)R(C, A; B_6, M)R(A, B; C_6, N) = 1. \qquad (10.5)$$

We now apply the cocycle identity (Theorem 4.20) twice:

$R(B, C; A_3, L)R(C, A; B_3, M)R(A, B; C_3, N)$

$= R(B, C; A_3, A_4)R(B, C; A_4, L) \cdot R(C, A; B_3, B_4)R(C, A; B_4, M) \cdot$

$R(A, B; C_3, C_4)R(A, B; C_4, N)$ (cocycle id.)

$= -R(B, C; A_3, A_4)R(C, A; B_3, B_4)R(A, B; C_3, C_4)$ (by (10.4))

$= -R(B, C; A_5, A_6)R(C, A; B_5, B_6)R(A, B; C_5, C_6)$ (by (10.1)–(10.3))

$= -R(B, C; A_5, L)R(B, C; A_6, L) \cdot R(C, A; B_5, M)R(C, A; M, B_6) \cdot$

$R(A, B; C_5, N)R(A, B; N, C_6)$ (cocycle id.)

$= -R(B, C; A_5, L)R(C, A; B_5, M)R(A, B; C_5, N)$ (by (10.5)).

Therefore,

$$R(B, C; A_3, L)R(C, A; B_3, M)R(A, B; C_3, N)$$
$$= -R(B, C; A_5, L)R(C, A; B_5, M)R(A, B; C_5, N). \qquad (10.6)$$

By the projective Ceva theorem (Theorem 4.33), AA_2, BB_2, CC_2 are concurrent if and only if the right-hand side of Equation (10.6) is equal to 1 (because $AA_2 = AA_5$, etc.). By the projective Menelaus theorem (Theorem 4.34), A_3, B_3, and C_3 are collinear if and only if the left-hand side of Equation (10.6) is equal to -1. Thus, the result follows.   □

Jeřábek's proof of the '$\implies$' direction of Theorem 10.34
Using the notation of the first proof $(A, B, C, A_1, B_1, C_1, A_2, B_2, C_2)$, let P be $AA_1 \cap BB_1$ and Q be $AA_2 \cap BB_2$. For the direction we will prove, we assume that P is on CC_1 and Q is on CC_2, and show that $A_1A_2 \cap BC$, $B_1B_2 \cap CA$, $C_1C_2 \cap AB$ are collinear.

Let g be the central collineation fixing A, B, and C and taking P to Q. Let t be the tangent to the conic at A_1, u be the tangent to the conic at B_1, v be the tangent to the conic at C_1. The proof involves two steps:

(1) $t \cap CB$, $u \cap AC$, $v \cap BA$ are collinear;
(2) $t^g = A_1A_2$, $u^g = B_1B_2$, $v^g = C_1C_2$.

Step 1. Consider the hexagon $A_1A_1C_1CBB_1$. By the five-point Pascal theorem (Theorem 6.66), $t \cap CB$, $B' = A_1C_1 \cap BB_1$, $C' = C_1C \cap B_1A_1$ are collinear. Similarly, $u \cap AC$, $C' = C_1C \cap B_1A_1$, $A' = A_1A \cap C_1B_1$ are collinear, as are

$v \cap BA$, A', and B'. Hence, t, BC, $B'C'$ are concurrent, as are u, CA, $C'A'$ and v, AB, $A'B'$. But ABC and $A'B'C'$ are in perspective from P, so $t \cap CB$, $u \cap AC$, $v \cap BA$ are collinear (by Desargues' theorem (Theorem 2.20)).

Step 2. Define a projectivity from the pencil on A_1 to the pencil on A_2, by

$$m \mapsto A_2(m \cap BC)^g.$$

This projectivity defines a conic on A_1, A_2, A, B, and C; that is, the given conic. Under this projectivity, t corresponds to A_1A_2, because the tangent at the first point always corresponds to the join of the two points in a Steiner conic (while the join of the two points maps under the projectivity to the tangent at the second point). Thus, $(t \cap BC)^g = A_1A_2 \cap BC$. Similarly, $(t \cap AB)^g = A_1A_2 \cap AB$, $(t \cap CA)^g = A_1A_2 \cap CA$ (there was nothing special about BC). So $t^g = A_1A_2$.

Similarly, $u^g = B_1B_2$, $v^g = C_1C_2$.

Conclusion: $A_1A_2 \cap BC$, $B_1B_2 \cap CA$, $C_1C_2 \cap AB$ are collinear. ◻

VÁCLAV JEŘÁBEK (1845–1931) was a Czech mathematician specialising in constructive geometry and a professor of the Czech Realschule in Brno, 1881–1907.

Another proof of Theorem 10.27 Let $A_0 := DA_1 \cap BC$, $B_0 := DB_1 \cap CA$, and $C_0 := DC_1 \cap AB$. Suppose first that D is on the given conic or the polar line of P. The Frégier involution t_P with Frégier point P and the Frégier involution t_D with Frégier point D commute, and their product $t_P t_D$ maps A to A_2, B to B_2, C to C_2, where $DA_0 \cap S = \{A_1, A_2\}$, $DB_0 \cap S = \{B_1, B_2\}$, $DC_0 \cap S = \{C_1, C_2\}$. Note that $(A_1, B_1, C_1) = (A_2, B_2, C_2)$ in the case that D lies on the conic. Hence, AA_2, BB_2, CC_2 are concurrent. (The point of concurrency is the pole of DP in the case that D does not lie on the conic, and D itself otherwise.) By Theorem 10.34, $A_1A_2 \cap BC$, $B_1B_2 \cap CA$, and $C_1C_2 \cap AB$ are collinear. Since $A_1A_2 = DA_1$, $B_1B_2 = DB_1$, and $C_1C_2 = DC_1$, we have the forward direction of the result that A_0, B_0, and C_0 are collinear.

Conversely, suppose that A_0, B_0, and C_0 are collinear. By Theorem 10.34, AA_2, BB_2, CC_2 are concurrent in a point Q, where $A_2 := A_1^D$, $B_2 := B_1^D$, and $C_2 := C_1^D$. Suppose D does not lie on the conic. Now PD maps A to A_2, B to B_2, and C to C_2, but we also know that Q does too. So PDQ fixes A, B, and C, and hence $PDQ = 1$ (as the group of the conic is sharply 3-transitive). Therefore, PD is an involution and hence P and D commute. So D lies on the polar line of P. ◻

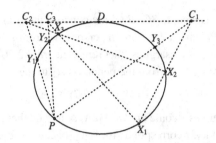

Figure 10.14 Details for the proof of Theorem 10.35.

The following is an important and beautiful consequence of Theorem 10.34. It provides us with an interesting property on the involutions of the projective general linear group $PGL(2, \mathbb{R})$.

Theorem 10.35

If $g_1: (D)(X_1 X_2)(P Y_3)\dots, g_2: (D)(X_2 X_3)(P Y_1)\dots, g_3: (D)(X_3 X_1)(P Y_2)\dots$ are three involutions on the projective line, so is $(X_1 Y_1)(X_2 Y_2)(X_3 Y_3)\dots$.

Proof We first identify the projective line $PG(1, \mathbb{R})$ and its cross-ratio preserving group of symmetries $PGL(2, \mathbb{R})$ with a conic $\mathcal{K}$ of $PG(2, \mathbb{R})$ (and the same group of symmetries; see Chapter 5).

As Frégier involutions of $\mathcal{K}$, g_1, g_2, g_3 all fix D, so their centres $C_1 := X_1 X_2 \cap P Y_3$, $C_2 := X_2 X_3 \cap P Y_1$, $C_3 := X_3 X_1 \cap P Y_2$ are collinear (in the tangent at D) (see Figure 10.14). With $X_1 =: A, X_2 =: B, X_3 =: C, A_1 = B_1 = C_1 = P$, $Y_1 = A_2, Y_2 = B_2, Y_3 = C_2$ in the statement of Theorem 10.34, we have AA_1, BB_1, CC_1 concurrent (in P) and

$$A_1 A_2 \cap BC = PY_1 \cap X_2 X_3 = C_2,$$
$$B_1 B_2 \cap CA = PY_2 \cap X_3 X_1 = C_3,$$
$$C_1 C_2 \cap AB = PY_3 \cap X_1 X_2 = C_1$$

collinear in the tangent at D. So, by Theorem 10.34, AA_2, BB_2, CC_2 are concurrent. That is, $(A A_2)(B B_2)(C C_2)\dots$ is an involution; which after relabelling means that $(X_1 Y_1)(X_2 Y_2)(X_3 Y_3)\dots$ is an involution. □

From Theorem 10.34, we can prove a 1927 result of Papelier.

Theorem 10.36 (Papelier, 1927, p. 42, art. 53)
A transversal ℓ meets the sides BC, CA, AB of a triangle ABC in the points A', B', C', respectively. Let O be another point of ℓ. If A'', B'', C'' are the images

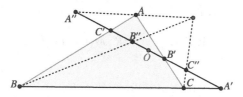

Figure 10.15 Papelier's theorem (Theorem 10.36).

of A', B', C', respectively, under a half-turn g about O, then $A'A''$, $B'B''$, $C'C''$ are concurrent (see Figure 10.15).

Proof Note that g is a harmonic homology, so the conic $\mathcal{K}$ on A, B, C with centre O is stabilised by g, and so contains $A_1 = A''$, $B_1 = B''$, $C_1 = C''$. Let A_2 be the other point of $\mathcal{K}$ on A_1A', B_2 be the other point of $\mathcal{K}$ on B_1B', C_2 be the other point of $\mathcal{K}$ on C_1C'. Then, by Theorem 10.34, since $A_1A_2 \cap BC = A'$, $B_1B_2 \cap CA = B'$, and $C_1C_2 \cap AB = C'$ are collinear, A_1A_2, B_1B_2, C_1C_2 are concurrent; that is, $A'A''$, $B'B''$, $C'C''$ are concurrent.   □

* * *

We begin with an elegant and not-very-well-known result of Norman Macleod Ferrers that involves involutions on the three sides of a triangle. It leads to the Lehmer–Daus theorem (Theorem 10.38) and its many consequences.

Theorem 10.37 (Ferrers, 1857a)
Let ABC be a triangle, and let P, P' be points on AB; Q, Q' be points on BC; R, R' be points on CA. Suppose D is a point on no side of ABC, $A' := AD \cap BC$, $B' := BD \cap CA$, $C' := CD \cap AB$ and that

- *$(P P')(A B)(C') \ldots$ is an involution t_1,*
- *$(Q Q')(B C)(A') \ldots$ is an involution t_2, and*
- *$(R R')(C A)(B') \ldots$ is an involution t_3.*

Then there is a (possibly degenerate) conic on P, P', Q, Q', R, R'. Hence, P, Q, R are collinear if and only if P', Q', R' are collinear.

Proof Let ℓ be the trilinear polar of D with respect to ABC; $L := BC \cap \ell$, $M := CA \cap \ell$, $N := AB \cap \ell$. Then

$$\text{R}(Q, L; B, C)\text{R}(Q', L; B, C) = 1$$

since t_2 fixes L. Similarly,

$$\mathrm{R}(R, M; C, A)\mathrm{R}(R', M; C, A) = 1$$

and

$$\mathrm{R}(P, N; A, B)\mathrm{R}(P', N; A, B) = 1.$$

Thus,

$$\mathrm{R}(P, N; A, B)\mathrm{R}(P', N; A, B)\mathrm{R}(Q, L; B, C)\cdot$$
$$\mathrm{R}(Q', L; B, C)\mathrm{R}(R, M; C, A)\mathrm{R}(R', M; C, A) = 1.$$

So, by the projective Carnot theorem (Theorem 7.16), there is a (possibly degenerate) conic on P, P', Q, Q', R, R'. □

NORMAN MACLEOD FERRERS (1829–1903) is mostly known for the diagrams that represent partitions of integers, so-called *Ferrers diagrams*. He was also known during his time as an outstanding lecturer, particularly in his home institution, Caius College (Cambridge University).

Notice what the dual of Theorem 10.37 states:

Let ABC be a triangle, let a, a' be lines on A, let b, b' be lines on B, and let c, c' be lines on C. Suppose D is a point on no side of ABC. Suppose
- $(a\,a')(AB\,AC)(AD)\ldots$ is an involution,
- $(b\,b')(BC\,BA)(BD)\ldots$ is an involution, and
- $(c\,c')(CA\,CB)(CD)\ldots$ is an involution.

Then a, a', b, b', c, c' lie on a (possibly degenerate) line conic. Hence, a, b, c are concurrent if and only if a', b', c' are concurrent.

The final sentence of the above is the following fundamental theorem.

Theorem 10.38 (Lehmer–Daus theorem (Lehmer, 1911; Daus, 1936))
Let ABC be a triangle, t_A, t_B, t_C be involutions on the pencils of lines on A, B, C, respectively, such that there is a (real or imaginary) point D with AD fixed by t_A, BD fixed by t_B, and CD fixed by t_C (see Figure 10.16). Suppose $(AB)^{t_A} = AC$, and $(BC)^{t_B} = BA$, and $(CA)^{t_C} = (CB)$. If P is a point on no side of ABC, then $(PA)^{t_A}$, $(PB)^{t_B}$, $(PC)^{t_C}$ are concurrent.

The following proof uses Theorem 10.35 and the converse of Pappus' involution theorem (Theorem 6.13).

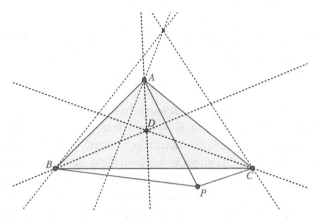

Figure 10.16 Theorem 10.38 for the case where t_A, t_B, t_C are reflections about angle bisectors.

Proof Consider the line $\ell := DP$. Let

$$X_1 := AB \cap \ell, \qquad\qquad Y_1 := (PA)^{t_A} \cap \ell,$$
$$X_2 := AC \cap \ell, \qquad\qquad Y_2 := (PB)^{t_B} \cap \ell,$$
$$X_3 := BC \cap \ell, \qquad\qquad Y_3 := (PC)^{t_C} \cap \ell.$$

Now t_A is an involution on the pencil of lines on A and it fixes D. The map $\bar{t}_A : X \mapsto (XA)^{t_A} \cap \ell$ on the points of ℓ does the following:

$$D^{\bar{t}_A} = D,$$
$$X_1^{\bar{t}_A} = (AB)^{t_A} \cap \ell = AC \cap \ell^{t_A} = X_2,$$
$$Y_1^{\bar{t}_A} = PA \cap \ell = P.$$

Hence, $\bar{t}_A$ is the involution $(D)(X_1 X_2)(P Y_3)\ldots$ on ℓ. Likewise, t_B and t_C induce the involutions $(D)(X_2 X_3)(P Y_1)\ldots$ and $(D)(X_3 X_1)(P Y_2)$ on ℓ (respectively). By Theorem 10.35, $(X_1 Y_1)(X_2 Y_2)(X_3 Y_3)\ldots$ is an involution on ℓ, and so, by the converse of the Pappus involution theorem (Theorem 6.13), $(PA)^{t_A}$, $(PB)^{t_B}$, $(PC)^{t_C}$ are concurrent. $\qquad\square$

Theorem 10.38 was rediscovered in (van Lamoen, 2001, p. 158, Theorem 4).

PAUL HAROLD DAUS (1894–1973) was a student of Derrick Norman Lehmer, and his PhD (Berkeley 1921) was on continued fractions.

Let A and B be two distinct points and take two concurrent lines a and b on A and B, respectively, and let D be their point of intersection. Consider the map χ on points not equal to A or B, given by

$$\chi: P \mapsto AP^{\sigma_a} \cap BP^{\sigma_b}.$$

Notice that χ maps any point on AB (apart from A and B) to the same point C. In fact, C is $AB^{\sigma_a} \cap BA^{\sigma_b}$. Moreover, the points of AC (other than C) are mapped to B and the points of BC (other than C) are mapped to A. The point D is fixed by χ. The point C is also missing from the domain of χ since C^{σ_a} and C^{σ_b} both lie on AB. Now if ℓ is a line incident with A, then (it can be shown that) χ maps ℓ (i.e., the points of ℓ apart from A) to another line, ℓ^{σ_a} in fact. Similarly, if ℓ is incident with B, then χ maps it to ℓ^{σ_b}. Finally, if ℓ is incident with C, χ maps ℓ to $\ell^{\sigma_{CD}}$.

These mappings of lines to lines are special cases. It turns out (see Theorem 10.39) that χ is a **quadratic transformation**[2]; it maps a line to a non-degenerate conic when the line is not incident with A, B, or C.

Theorem 10.39 (Lehmer, 1911)
Let A and B be two distinct points and let t_A, t_B be involutions on the pencils of lines on A, respectively, such that there is a (real or imaginary) point D with AD fixed by t_A and BD fixed by t_B. Let $C := (AB)^{t_A} \cap (BA)^{t_B}$. For each point P of the plane, except A, B, and C, let P' be the point $(AP)^{t_A} \cap (BP)^{t_B}$. Then

(i) the image of a line not through a vertex of $\triangle ABC$, under the map $P \mapsto P'$, is a non-degenerate conic on A, B, and C;
(ii) the image of a line through a vertex of $\triangle ABC$, under the map $P \mapsto P'$, is a line through that vertex.

Proof Let ℓ be a line not on A, B, or C, and let $E := \ell \cap AC$. Let ϕ_ℓ be the perspectivity with centre ℓ from pencil(A) to pencil(B). The map

$$g: m \mapsto m^{t_A \phi_\ell t_B}$$

yields a projective correspondence from pencil(A) to pencil(B). By Theorem 6.59, the points of intersection $a \cap a^\rho$, where $a \in$ pencil(A), form a (possibly degenerate) conic $\mathcal{K}$. Moreover, if ρ is not a perspectivity, then $\mathcal{K}$ is non-degenerate; otherwise, $\mathcal{K}$ is the line AB.

Now t_A maps AB to AC; ϕ_ℓ maps AC to BE; and t_B does not map BE to AB, since $BE \neq BC$, so g does not fix AB. Also, t_A maps AC to AB; ϕ_ℓ maps

---

[2] See also Theorem 5.2 for another example of a quadratic transformation.

AB to AB; and t_B maps BA to BC, so g maps AC to BC. Therefore, g is not a perspectivity, since AB is not fixed. □

Another proof of Theorem 10.38 For each pair of vertices of the triangle, we have the following 'quadratic maps':

$$\rho_{AB} : X \mapsto (AX)^{t_A} \cap (BX)^{t_B},$$
$$\rho_{BC} : X \mapsto (BX)^{t_B} \cap (CX)^{t_C},$$
$$\rho_{CA} : X \mapsto (CX)^{t_C} \cap (AX)^{t_A}.$$

Let P' be the point $P^{\rho_{AB}}$. By Theorem 10.39, the image of CP under ρ_{AB} is a line, and this line is CP'. To be more precise,

$$
\begin{aligned}
(AB \cap CP)^{\rho_{AB}} &= (A(AB \cap CP))^{t_A} \cap (B(AB \cap CP))^{t_B} \\
&= (AB)^{t_A} \cap (AB)^{t_B} \\
&= CA \cap BC \\
&= C,
\end{aligned}
$$

and P maps to P'. By Theorem 10.39, the image of BP under ρ_{AB} is a line, and this line is BP'. To see this, note that $CA \cap BP$ lies on the line CA, and $(CA)^{t_A} = BC$. Therefore,

$$(CA \cap BP)^{\rho_{AB}} = BC \cap (B(CA \cap BP))^{t_B} = B.$$

Likewise, the image of AP under ρ_{AB} is the line AP'. This argument shows that $(AP)^{t_A} \cap (BP)^{t_B}$ lies on $(AP)^{\rho_{AB}}$, $(BP)^{\rho_{AB}}$, and $(CP)^{\rho_{AB}}$.

Now if we use symmetry in the argument above, we have

$$P'C = (CP)^{\rho_{AB}} = (BP)^{\rho_{BC}} = (AP)^{\rho_{CA}}.$$

We have $P^{\rho_{CA}} = (AP)^{\tau_A} \cap (CP)^{\tau_C}$ and hence $P^{\rho_{CA}}$ is incident with $(CP)^{\tau_C}$. Similarly, $P^{\rho_{BC}} = (BP)^{\tau_B} \cap (CP)^{\tau_C}$ and hence $P^{\rho_{BC}}$ is incident with $(CP)^{\tau_C}$. Hence, $P^{\rho_{CA}}$ and $P^{\rho_{BC}}$ are both incident with $(CP)^{\tau_C}$. Now $P^{\rho_{CA}}$ is incident with $(AP)^{\rho_{CA}}$, which is equal to $P'C$. Likewise, $P^{\rho_{BC}}$ is incident with $(BP)^{\rho_{BC}} = P'C$. So if $P^{\rho_{CA}}$ and $P^{\rho_{BC}}$ are different, then $(CP)^{\tau_C} = P'C$ and $(AP)^{\tau_A}, (BP)^{\tau_B}, (CP)^{\tau_C}$ are concurrent in the point P'. Otherwise, if $P^{\rho_{CA}} = P^{\rho_{BC}}$, then $(BP)^{t_B} \cap (CP)^{t_C} = (CP)^{t_C} \cap (AP)^{t_A}$, and, again, $(AP)^{\tau_A}, (BP)^{\tau_B}, (CP)^{\tau_C}$ are concurrent. □

In 1858, Todhunter shows that if ellipses are inscribed in a triangle each with one focus lying on a fixed straight line, the locus of the other focus is a conic passing through the vertices of the triangle (Todhunter, 1855, p. 281). This is essentially the isogonal conjugate special case of the first theorem of Lehmer (1911). A more general result is due to Cayley (1861). If I, J are any two points on the line at infinity, and conics be inscribed in a triangle each with the intersection of a tangent on I and a tangent on J on a fixed straight line, the locus of the intersection of the other tangent on I and the other tangent on J is a conic passing through the vertices of the triangle.

The following is an *affine* theorem, and we attribute it to De Longchamps (1866, p. 124).

Theorem 10.40 (Isotomic conjugate theorem)
*Let ABC be a triangle and P be a point on no side of ABC. Let $D := AP \cap BC$, $E := BP \cap CA$, $F := CP \cap AD$, and let D', E', F' be the reflections of D, E, F in the midpoints of $\overline{BC}$, $\overline{CA}$, $\overline{AB}$, respectively. Then AD', BE', CF' are concurrent (in the **isotomic conjugate** of P with respect to ABC).*

Proof of Theorem 10.38 using isotomic conjugates There is a collineation ϕ of $PG(2, \mathbb{R})$ fixing A, B, and C and mapping D to the centroid G of ABC (for some choice of the line at infinity). Let $A' := AP^\phi \cap BC$, $B' := BP^\phi \cap CA$, $C' := CP^\phi \cap AB$, and let A'', B'', C'' be the reflections of A', B', C' in the midpoints of $\overline{BC}$, $\overline{CA}$, $\overline{AB}$, respectively. By the isotomic conjugate theorem (Theorem 10.40), AA'', BB'', CC'' are concurrent in the isotomic conjugate O of P^ϕ (with respect to ABC).

Now $\phi^{-1} t_A \phi$ interchanges AB and AC, and fixes the median AG. Let $M := AG \cap BC$ (the midpoint of $\overline{BC}$). If X is a point on BC, then

$$\mathrm{R}(AC, AB; (AX)^{\phi^{-1} t_A \phi}, AM) = \mathrm{R}((AB)^{\phi^{-1} t_A \phi}, (AC)^{\phi^{-1} t_A \phi}; (AX)^{\phi^{-1} t_A \phi}, (AM)^{\phi^{-1} t_A \phi})$$
$$= \mathrm{R}(AB, AC; AX, AM).$$

Let X' be the image of X under the half-turn about M. Now the half-turn about M is induced by a harmonic homology μ and hence

$$\mathrm{R}(AB, AC; AX, AM) = \mathrm{R}((AB)^\mu, (AC)^\mu; (AX)^\mu, (AM)^\mu)$$
$$= \mathrm{R}(AC, AB; AX', AM).$$

It follows that $(AX)^{\phi^{-1} t_A \phi} = AX'$. In particular, $(AA')^{\phi^{-1} t_A \phi} = AA''$. Therefore,

$$(AA'')^{\phi^{-1}} = (AA')^{\phi^{-1} t_A} = (A(AP \cap BC))^{t_A} = (AP)^{t_A}.$$

Similarly, $(BB'')^{\phi^{-1}} = (BP)^{t_B}$ and $(CC'')^{\phi^{-1}} = (CP)^{t_C}$, and, hence, $O^{\phi^{-1}}$ lines on each line $(AP)^{t_A}$, $(BP)^{t_B}$, $(CP)^{t_C}$. □

Theorem 10.41

Let ABC be a triangle and ℓ be a line on no vertex of ABC. Then the isogonal conjugate of ℓ is a conic $\mathcal{K}$. Let g be the isogonal conjugate map from ℓ to $\mathcal{K}$. Let P_1, P_2, P_3, P_4 be distinct points of ℓ on no side of ABC. Then

$$R(P_1, P_2; P_3, P_4) = R_{\mathcal{K}}(P_1^g, P_2^g; P_3^g, P_4^g).$$

Proof Take $A(1, 0, 0)$, $B(0, 1, 0)$, $C(0, 0, 1)$ so that g is the restriction to ℓ of the map $(x, y, z) \mapsto (yz, xz, xy)$. Let ℓ be $[a, b, c]$. Then $c \neq 0$ and ℓ is not on C. So the points $P(t)$ of ℓ on no side of ABC have coordinates $(1, t, -(a/c) - (b/c)t)$ for $t \neq 0$. Now $P(t)^g = (1, 1/t, 1/(-(a/c) - (b/c)t))$ and so

$$P(1/t)^g = (1, t, t/((-a/c)t - (b/c))),$$

which is a fractional linear map, and thus preserves cross-ratio. $\square$

Remark: We can do this without coordinates and in more generality by using the proof of Lehmer's theorem (Theorem 10.39). This proof shows that the map from ℓ to $\mathcal{K}$ (in more generality using two involutions) is a map preserving cross-ratio; namely, there is a projectivity ϕ from ℓ to pencil(B) taking AX to BX^g and hence

$$R(P_1, P_2; P_3, P_4) = R(AP_1, AP_2; AP_3, AP_4)$$
$$= R(BP_1^g, BP_2^g; BP_3^g, BP_4^g)$$
$$= R_C(P_1^g, P_2^g; P_3^g, P_4^g).$$

Note that it follows that g extends uniquely to a map defined on all of ℓ which preserves cross-ratio.

The quadratic transformation $g: m \mapsto m^{t_A \phi_{\ell} t_B}$ from ℓ to $\mathcal{K}$ in the proof of Lehmer's theorem (Theorem 10.39), preserves cross-ratio.

Worked Example 10.42

Problem (Eves, 1983) Let ABC be a triangle and m a line not passing through A, B, or C.

(i) Prove that the isogonal conjugate of m is an ellipse, parabola, or hyperbola accordingly as m meets the circumcircle of ABC in zero, one, or two points.

(ii) Prove that the isotomic conjugate of m is an ellipse, parabola, or hyperbola accordingly as m meets E in zero, one, or two points, where E is the ellipse through A, B, C having the centroid of triangle ABC as centre. $\square$

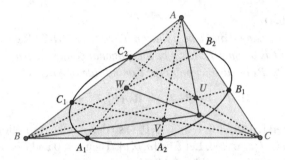

Figure 10.17 Theorem 10.43.

Proof

(i) By Theorem 10.41, the isogonal conjugate of m is a conic. Recall that the isogonal conjugate of the circumcircle of ABC is the line at infinity. Hence, the isogonal conjugate of m is an ellipse, parabola, or hyperbola according as m meets the circumcircle of ABC in zero, one, or two points.

(ii) Under an affine transformation sending ABC to an equilateral triangle $A'B'C'$, m is mapped to a line m' and E is mapped to the circumcircle of $A'B'C'$. But, since $A'B'C'$ is equilateral, the isotomic conjugate of m' with respect to $A'B'C'$ is also its isogonal conjugate. Since isogonal conjugacy is an affine invariant, applying (i) gives (ii). □

Theorem 10.43

Let ABC be a triangle; A_1, A_2 be points on BC; B_1, B_2 be points on CA; C_1, C_2 be points of AB; $U := BB_1 \cap CC_2$, $V := CC_1 \cap AA_2$, $W := AA_1 \cap BB_2$ (see Figure 10.17). Then the following are equivalent:

(a) A_1, A_2, B_1, B_2, C_1, C_2 *lie on a conic;*
(b) AA_1, AA_2, BB_1, BB_2, CC_1, CC_2 *lie on a dual conic;*
(c) AU, BV, CW *are concurrent.*

Proof The equivalence of (b) and (c) follows from the dual of Pascal's theorem (Theorem 6.28) and the dual of the Braikenridge–Maclaurin theorem (Theorem 6.43). Statements (a) and (b) are equivalent by duality in the Lehmer–Daus theorem (Theorem 10.38) and Ferrers' theorem (Theorem 10.37). □

We have already seen the equivalence of (a) and (c) in Theorem 10.43 as Hopkins' theorem (Theorem 6.31). We remark that we can obtain most of the results of Fox and Goggins (2003) from (c) in Theorem 10.43.

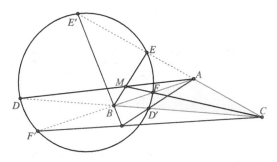

Figure 10.18 Ferriot–Terquem theorem (Theorem 10.44).

The equivalence of (a) and (b) in Theorem 10.43 has an amazing history. First, it is Theorem 6.51, which we attribute to Möbius. It also appears as Russell (1893, p. 96, ex. 1). In 2011, it was conjectured by Bradley.[3] In 2012, it was proved analytically by Szilasi (2012). In 2013, it was proved synthetically by Baralić.[4] In 2015, Baralić published it (Baralić, 2015). Before then, Szilasi posted *A purely projective proof of Bradley's conjecture* at http://math.unideb.hu/media/szilasi-zoltan/bradley.pdf. Yet it appears in a well-known text as Exercise 7 on page 117 of Veblen and Young (1965).

Theorem 10.44 (Ferriot–Terquem theorem (Ferriot, 1838, pp. 25–8, para. XVII, Terquem, 1842, p. 403))
Let ABC be a triangle, M be a point on no side of ABC, MA ∩ BC = D, MB ∩ CA = E, MC ∩ AB = F (see Figure 10.18). Let K be the circumcircle of DEF. Let K ∩ AB = {D, D′}, K ∩ BC = {E, E′}, K ∩ CA = {F, F′}. Then AD′, BE′, CF′ are concurrent.

Proof By Theorem 10.43 above, since D, D', E, E', F, F' lie on a conic,

$$CD, CD', AE, AE', BF, BF'$$

touch a conic. Since *CD, AE, BF* are concurrent, this dual conic is degenerate, so *AD′, AE′, AF′* are concurrent. □

[3] Bradley, 2011.
[4] Baralić, 2013.

OLRY TERQUEM (1782–1862) was a French Jewish mathematician and co-founder of *Bulletin de Bibliographie, d'Histoire et de Biographie de Mathématiques* (with Gerono of *Nouvelles annales de mathématiques*).

The above argument also proves the following generalisation:

Theorem 10.45
Let ABC be a triangle, M be a point on no side of ABC, $MA \cap BC = D$, $MB \cap CA = E$, $MC \cap AB = F$. Let K be a conic on D, E, F (and none of A, B, C). Let $K \cap AB = \{D, D'\}$, $K \cap BC = \{E, E'\}$, $K \cap CA = \{F, F'\}$. Then AD', BE', CF' are concurrent.

The ' $\Longrightarrow$ ' direction of the following result is due to Möbius (1827, p. 423, para. 278).

Theorem 10.46 (Cazamian, 1895, p. 35)
Let A, B, C, D, E, F be six points, no three collinear. Let $U := BC \cap DE$, $V := CA \cap EF$, $W := AB \cap FD$, $X := AB \cap EF$, $Y := BC \cap FD$, $Z := CA \cap DE$. Then UX, VY, WZ are concurrent if and only if A, B, C, D, E, F lie on a conic.

Proof Recall from Theorem 6.73 that the vertices of two triangles lie on a conic if and only if their sides lie on a dual conic. We apply this theorem to the triangles ABC and DEF noting that $AB = XW$, $BC = UY$, $CA = VZ$, $DE = UZ$, $EF = VX$, $FD = WY$, so $(XW)(WY)(YU)(UZ)(VZ)(VX)$ is a Brianchon hexagon with vertices $XWYUZV$. □

Worked Example 10.47
Our basic configuration of three involutions on pencils at the vertices of a triangle ABC with a point D on a (possibly imaginary) fixed line of all the involutions can be extended by adding the other (possibly imaginary) fixed line at each of the three vertices. These three new lines are the sides of a triangle and, adding in D, we obtain a quadrangle whose diagonal triangle is ABC. Figure 10.19 demonstrates this in the case that D is the incentre of ABC and the involutions are reflections about internal angle bisectors. The three new lines are the external angle bisectors.

Now the statement of Theorem 10.38 is:

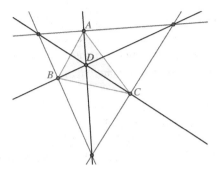

Figure 10.19 The given triangle as a diagonal triangle of a quadrangle.

Given a point P on no side of ABC, the harmonic conjugates of AP with respect to the sides of the complete quadrangle on A, BP with respect to the sides of the complete quadrangle on B, and CP with respect to the sides of the complete quadrangle on C are concurrent.

Each of these lines is the cross-axis of a perspectivity with centre P from one side of the complete quadrangle to the opposite side, so this is the 'cross-axes are concurrent' theorem 6.39. □

The following result is the *Lehmer–Daus version* of Oai Thanh Dao's generalisation of the Newton–Gauss theorem.

Theorem 10.48
Let ABC be a triangle, and let the lines ℓ on A, m on B, n on C be concurrent. Let t_A be the unique projectivity on the pencil of lines on A interchanging AB and AC and fixing ℓ; t_B be the unique projectivity on the pencil of lines on B interchanging AB and BC and fixing m; t_C be the unique projectivity on the pencil of lines on C interchanging BC and AC and fixing n. Let ℓ' also be fixed by t_A, m' also be fixed by t_B, and n' also be fixed by t_C. Let $A' := \ell' \cap BC$, $B' := m' \cap CA$, $C' := n' \cap AB$. Then for all points Q, $(QA)^{t_A} \cap QA'$, $(QB)^{t_B} \cap QB'$, $(QC)^{t_C} \cap QC'$ are collinear.

Proof We can use a collineation of $\mathrm{PG}(2, \mathbb{R})$ so that the point of concurrency of ℓ, m, and n is the centroid of ABC; the involutions t_A, t_B, t_C are then uniquely determined and if $a := BC$, $b := CA$, $c := AB$ and d is the trilinear polar of Q with respect to ABC, the result follows from the Newton–Gauss line theorem for the complete quadrilateral arising from the quadrangle $abcd$. □

Theorem 10.49 (Dual of Corollary 8.11 (Steiner, 1828a, p. 39))
The reflections in the angle bisector at that vertex of the joins of a point on no side to the vertices of a triangle are concurrent or parallel. They are parallel if and only if the point lies on the circumcircle.

Proof Let a, b, c be the angle bisectors at A, B, C, respectively. Let P be a point on no side of the triangle ABC. Let t_A, t_B, t_C be the reflections in the angle bisectors at A, B, C, respectively. Now $(AB)^{t_A} = AC$ and $(AB)^{t_B} = BC$. Therefore, $C = (AB)^{t_A} \cap (AB)^{t_B}$ and hence, by Theorem 10.38, $(AP)^{t_A}$, $(BP)^{t_B}$, $(CP)^{t_C}$ are concurrent in a point P'. Now the image of the line at infinity under the map $P \mapsto P'$ of Theorem 10.39 is a conic $\mathcal{K}$ on A, B, C. Furthermore, the circular points I and J are interchanged by this map and so the conic $\mathcal{K}$ passes through them. Therefore, $\mathcal{K}$ is a circle; the circumcircle of ABC. Therefore, the lines are $(AP)^{t_A}$, $(BP)^{t_B}$, $(CP)^{t_C}$ parallel in $\mathbb{R}^2$ only when P' lies on the circumcircle of $\triangle ABC$. $\square$

The Langr monthly problem E879 (1949) below is another corollary of Lehmer–Daus since the affine reflections in the medians that have direction that of the side they bisect induce the required involutions. Moreover, Ferrers' theorem (Theorem 10.37) shows that the conic exists.

Corollary 10.50 (Lee et al., 1949, problem E879 by Joseph Langr)
Let S_1, S_2, S_3 be the midpoints of three concurrent Cevians of a triangle $\triangle ABC$. Let $S_2 S_3$, $S_3 S_1$, $S_1 S_2$ meet the sides BC, CA, AB in A_1, B_1, C_1; A_2, B_2, C_2; A_3, B_3, C_3, respectively. Then A_1, B_2, C_3 are collinear and A_2, A_3, B_3, B_1, C_1, C_2 lie on a conic.

The following generalises both isogonal conjugates, when a, b, c are angle bisectors of ABC, and isotomic conjugates, when a, b, c are the medians.

Theorem 10.51
Let ABC be a triangle, t_A, t_B, t_C be involutions on the pencils of lines on A, B, C, respectively, such that there is a (real or imaginary) point D with AD fixed by t_A, BD fixed by t_B, and CD fixed by t_C. Let a be a line on A, b be a line on B, c be a line on C, $a' = a^{t_A}$, $b' = b^{t_B}$, $c' = c^{t_C}$. Let $C' := a' \cap b$, $B' = c' \cap a$, and $A' = b' \cap c$. Then:

(i) *AA', BB', CC' are concurrent. (Gale–Gardner theorem (Gale, 1998))*
(ii) *$(b \cap c')A'$, $(c \cap a')B'$, $(a \cap b')C'$ are concurrent. (Robson, 1923; Strange, 1974)*

Proof

(i) Let ℓ' be the trilinear polar of D with respect to ABC; and let $L := BC \cap \ell'$, $M := CA \cap \ell'$, $N = AB \cap \ell'$. From t_A,

$$R(AB, AC; AD, AL) = -R(AB, AC; AD', AL),$$

since t_A fixes the harmonic conjugate AL of AD with respect to $\{AB, AC\}$, and so $R(B, C; D, L) = R(C, B; D', L)$. Similarly,

$$R(C, A; E, M) = R(A, C; E', M),$$
$$R(A, B; F, N) = R(B, S; F', N).$$

Now AM', BE, CF' are concurrent in A', so, by the projective Ceva theorem (Theorem 4.33),

$$R(B, C; M', L)R(C, A; E, M)R(A, B; F', N) = 1.$$

Now AD', BN', CF are concurrent in B', so, by the projective Ceva theorem (Theorem 4.33),

$$R(B, C; D', L)R(C, A; N', M)R(A, B; F, N) = 1.$$

Now AD, BE', CO' are concurrent in C', so, by the projective Ceva theorem (Theorem 4.33),

$$R(B, C; D, L)R(C, A; E, M)R(A, B; O', N) = 1.$$

So

$$R(B, C; M', L)R(C, A; E, M)R(A, B; F', N) \cdot$$
$$R(B, C; D', L)R(C, A; N', M)R(A, B; F, N) \cdot$$
$$R(B, C; D, L)R(C, A; E, M)R(A, B; O', N) = 1.$$

But

$$R(B, C; D, L)R(B, C; D', L) = 1,$$
$$R(C, A; E, M)R(C, A; E', M) = 1,$$
$$R(A, B; F, N)R(A, B; F', N) = 1,$$

and so $R(B, C; M', L)R(C, A; N', M)R(A, B; O', N) = 1$. So by the projective Ceva theorem (Theorem 4.33), AM', BN', and CO' are concurrent, that is, AA', BB', CC' are concurrent.

(ii) This proof is adapted from Robson (1923). Let $C' := a' \cap b$, $B' := c' \cap a$, and $A' := b' \cap c$. Let $R := a \cap b'$, $Q := c \cap a'$, $P := b \cap c'$, $U := b' \cap a'$, $V := a' \cap c'$, $O := QB' \cap RC'$. Then BP, BA' are paired by t_B, and CP, CA' are paired by t_C. Therefore, by the Lehmer–Daus theorem (Theorem 10.38), AP, AA' are paired by t_A. Thus,

$$\begin{aligned}
R(AB, AR; AA', AU) &= R(AB, a; AA', a')\\
&= R((AB)^{t_A}, a^{t_A}; (AA')^{t_A}, (a')^{t_A})\\
&= R(AC, a'; AP, a)\\
&= R(AC, AV; AP, AB')
\end{aligned}$$

and so $R(C'B, C'R; C'A', C'U) = R(QC, QV; QP, QB') = R(QP, QB'; QC, QV)$. Now $QV = a' = C'U$ and

$$\begin{aligned}
C'B \cap QP &= b \cap QP = P,\\
C'R \cap QB' &= OR \cap OQ = O,\\
C'A' \cap QC &= C'A' \cap QA' = A'.
\end{aligned}$$

So $R(C'P, C'O; C'A', a') = R(QP, QO; QA', a')$ and therefore it follows from Lemma 4.24 that P, O, and A' are collinear. So the result follows as PA', QB', and RC' are concurrent in the point O. □

The Gale–Gardner theorem was first published in Gale (1996) then redis-covered by van Lamoen (2001). Both these results appear in Baker (1943), note 2, entitled 'A construction, given by Professor F. Morley, for an equilat-eral triangle, from any given triangle'. This builds on Chapter XIV, Sections 3 and 4, where it is shown that six of the relevant lines touch a conic. Baker also shows that the points of concurrence of AA', AB', AC', of AA'', AB'', AC'', and of $(AA')^{t_A}$, $(BB')^{t_B}$, $(C')^{t_C}$ are collinear.

Second proof of the Robson–Strange theorem, part (ii) of Theorem 10.51
This proof is an adaptation of Robson's proof of Morley's theorem (Rob-son, 1923). First we show that $(AA')^{t_A} = AA''$. Now $BA' = b'$ and so $(BA')^{t_B} = b$. Similarly, $CA' = c$, so $(CA')^{t_C} = c'$. We have $b \cap c' = A''$ and so, by the Lehmer–Daus theorem (Theorem 10.38), $(AA')^{t_A}$, $(BA')^{t_B}$, $(CA')^{t_C}$ are concurrent. So $(AA')^{t_A} = AA''$ as required.

We will show that $A''A'$, $B''B'$, $C''C'$ are concurrent. Let $U := a \cap b$, $V := a \cap c$, $O := B''B' \cap C''C'$. Now $R(AB, AC'; AA'', AU) = R(AC, AV; AA', AB'')$ by virtue of t_A, so $R(B, C'; A'', U) = R(C, V; A', B'') = R(A', B''; C, V)$. Hence,

$$R(C''B, C''C'; C''A'', C''U) = R(B'A', B'B''; B'C, B'V).$$

Now $C''U = a = B'V$, so $C''B \cap B'A' = A'$, $C''C' \cap B'B'' = O$, and $C''A'' \cap B'C = A''$ are collinear. Hence, $A''A'$, $B''B'$, $C''C'$ are concurrent. □

The dual of Ferrers' theorem (Theorem 10.37) is:

Let ABC be a triangle, a, a' be lines on A, b, b' be lines on B, c, c' be lines on C. Suppose D is a point on no side of ABC and suppose

$$(a\,a')(AB\,AC)(AD)\ldots \text{ is an involution,}$$
$$(b\,b')(BA\,BC)(BD)\ldots \text{ is an involution,}$$
$$(c\,c')(CA\,CB)(CD)\ldots \text{ is an involution.}$$

Then a, a', b, b', c, c' lie on a (possibly degenerate) line conic. Hence, a, b, c are concurrent if and only if a', b', c' are concurrent.

Notice how remarkably similar the theorem appears now in comparison to the Lehmer–Daus theorem (Theorem 10.38). The involutions listed above are precisely t_A, t_B, t_C in the statement of Theorem 10.38. Indeed, the final sentence of this is the Lehmer–Daus theorem (Theorem 10.38), that a', b', c' are concurrent. Now if we also apply Brianchon's theorem (Theorem 6.44) to the preceding sentence (that a, a', b, b', c, c' lie on a line conic), then we have the Robson–Strange theorem (Theorem 10.51(ii)). Therefore,

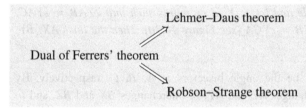

Another proof of Gale–Gardner, part (i) of Theorem 10.51

Let $P := c \cap AB$, $P' := c' \cap AB$, $Q := a \cap BC$, $Q' := a' \cap BC$, $R := b \cap CA$, $R' := b' \cap CA$. Let ℓ be the trilinear polar of D with respect to ABC; $L := BC \cap \ell$, $M := CA \cap \ell$, $N := AB \cap \ell$. Then,

$$(P, N; A, B)(P', N; A, B)(Q, L; B, C)(Q', L; B, C)(R, M; C, A)(R', M; C, A) = 1.$$

Let $P'' := C(a \cap b') \cap AB$, $Q'' := A(b \cap c') \cap BC$, $R'' := B(c \cap a') \cap CA$. Then $a, b', C(a \cap b')$ are concurrent, so (by Theorem 4.33)

$$(P'', N; A, B)(Q, L; B, C)(R', M; C, A) = -1.$$

Also $b, c', A(b \cap c')$ are concurrent, so

$$(P', N; A, B)(Q'', L; B, C)(R, M; C, A) = -1.$$

Additionally, $c, a', B(c \cap a')$ are concurrent, so

$$(P, N; A, B)(Q', L; B, C)(R'', M; C, A) = -1.$$

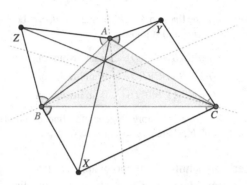

Figure 10.20 Jacobi's theorem (Theorem 10.52).

Hence, $(P'', N; A, B)(Q'', L; B, C)(R'', M; C, A) = -1$. Thus, AQ'', BR'', CP'' are concurrent, that is, $A(b \cap c')$, $B(c \cap a')$, $B(c \cap a')$ are concurrent.   □

Theorem 10.52 (Jacobi's theorem (Jacobi, 1825))
Let ABC be a triangle and let X, Y, Z be points such that $\angle ZAB = \angle YAC$, $\angle ZBA = \angle XBC$, $\angle XCB = \angle YCA$ (see Figure 10.20). Then the lines AX, BY, CZ are concurrent.

Proof Let t_A, t_B, t_C be the angle bisectors at A, B, C, respectively. By assumption, t_A interchanges AZ and AY, t_B interchanges BX and BZ, and t_C interchanges CY and CX. Let D be the incentre. Then D is fixed by each one of t_A, t_B, t_C. Let $a = AY$, $b := BZ$, and $c := CX$. Now $\angle XCB = \angle YCA$ and $\angle YAC = \angle ZAB$, and so t_A interchanges AY and AZ. In other words, $a' = a^{t_A} = AZ$. Likewise, $b' = BX$ and $c' = CY$. Therefore, $A' = b' \cap C = X$, $B' = c' \cap a = Y$, $C' = a' \cap b = Z$. By Theorem 10.51, AX, BY, CZ are concurrent.   □

We have the following special case of Theorem 10.52.

Corollary 10.53 (Fermat–Torricelli theorem)
*Let ABC be a triangle in $\mathbb{R}^2$, and erect equilateral triangles ABC', BCA', CAB' externally on the sides of ABC. Then AA', BB', CC' are concurrent (in the **Fermat–Torricelli point**).*

Fermat posed the problem of minimising the sum of the distances to the vertices of a triangle to Torricelli in a letter; Torricelli solved the problem

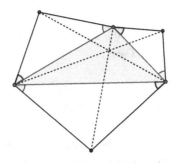

Figure 10.21 Kiepert's theorem (Theorem 10.54).

around 1640, and it was published in his collected works in 1659. The point that minimises the distance is the Fermat–Torricelli point if the triangle has no angle greater than $2\pi/3$; otherwise it is the vertex at which the greatest angle occurs.

The following generalisation of the Fermat–Torricelli theorem (Theorem 10.53) includes concurrence of the medians and of the altitudes as special cases, as well as Morley's theorem (Theorem 8.18).

Corollary 10.54 (Kiepert, 1869)
Erect similar isosceles triangles on each edge of a triangle ABC (see Figure 10.21). Then the lines joining each summit of these new triangles to the opposite vertex of the original triangle are concurrent.

Proof We have the hypotheses of Theorem 10.52 plus the extra conditions that $\angle ZAB = \angle YAC = \angle ZBA = \angle XBC = \angle XCB = \angle YCA$. So the the lines joining each summit of these isosceles triangles to the opposite vertex of the original triangle are concurrent (by Theorem 10.52). $\square$

Corollary 10.55 (Napoleon's theorem)
If equilateral triangles are constructed externally on the sides of any arbitrary triangle, the centroids of those equilateral triangles also form an equilateral triangle.

Proof The points X, Y, Z in Jacobi's theorem (Theorem 10.52) are the centroids of the equilateral triangles erected on the sides of $\triangle ABC$. It is also apparent that the triangles $\triangle ZAB$, $\triangle YAC$, and $\triangle XBC$ are isosceles triangles. So the result follows from Kiepert's theorem (Theorem 10.54). $\square$

Theorem 10.56

Let $\mathcal{K}$ be the Kiepert hyperbola of a triangle ABC. Let G be the centroid and H be the orthocentre of ABC. Then:

(a) The Kiepert hyperbola $\mathcal{K}$ contains the Fermat–Torricelli points E, F.

(b) On $\mathcal{K}$, $\{E, F\}$ harmonically separates $\{G, H\}$.

Proof The angle $\theta = \pm\pi/3$ gives the Fermat–Torricelli points by Napoleon's theorem (Theorem 10.55), as then the triangles erected on the sides are equilateral. The reflection with axis BC induces a Frégier involution of $\mathcal{K}$, in effect mapping the point parametrised by θ to that parametrised by $-\theta$, thus, fixing G and H and interchanging E and F. So, $\{E, F\}$ harmonically separates $\{G, H\}$ on $\mathcal{K}$. □

Similar applications of Theorem 10.52 yield a 1928 result of Lemoine and a 1941 result of Coşniţă.

Corollary 10.57 (Émile Lemoine, 1928; de Villiers, 1996)
The lines respectively joining the vertices A, B, and C of a given triangle ABC with the incentres D, E, F of the triangles ABI, BCI, and CAI (where I is the incentre of $\triangle ABC$) are concurrent.

Proof Now I is on the bisector of angle $\angle BAC$ and D is on the bisector of angle $\angle BAI$, so $\angle BAD = \angle BAC/4$. Similarly, $\angle CAF = \angle BAC/4$, $\angle DBA = \angle ABC/4 = \angle CBE$, and $\angle ECB = \angle ACB/4 = \angle ACF$. So Jacobi's theorem (Theorem 10.52) applies, and thus AE, BF, and CD are concurrent. □

Corollary 10.58 (Coşniţă, 1941)
The lines joining the vertices A, B, and C of a given triangle ABC with the circumcentres of the triangles ABO, BCO, and CAO (where O is the circumcentre of $\triangle ABC$), respectively, are concurrent.

Proof This time we apply Jacobi's theorem (Theorem 10.52) with DEF as the inner triangle and DAF, FCE, EBD as the triangles erected on the edges. To do this, we must show that $\angle BDE = \angle ADF$ et cetera. Now $d(O, A) = d(O, B)$ and $d(D, A) = d(D, B)$, so $\angle BDO = \angle ADO$. Similarly, $d(O, E) = d(O, B)$ and $d(D, E) = d(B, E)$, and $d(D, A) = d(D, O)$ and $d(F, O) = d(F, A)$, so $\angle BDE = \angle BDO/2$ and $\angle ADF = \angle ADO/2$, giving $\angle BDE = \angle ADF$. Similarly, $\angle BED = \angle CEF$ and $\angle CFE = \angle AFD$. By Jacobi's theorem (Theorem 10.52), AE, BF, and CD are concurrent. □

Theorem 10.59 (Kariya's theorem)
Let I be the incentre of a triangle ABC, and let X, Y, Z be the points where the incircle of △ABC touches the sides BC, CA, AB, respectively. If A', B', C' are three points on the rays $\overrightarrow{IX}$, $\overrightarrow{IY}$, $\overrightarrow{IZ}$, respectively, such that $d(I, A') = d(I, B') = d(I, C')$, then the lines AA', BB', and CC' are concurrent.

Proof Reflection g in the angle bisector of $\angle BAC$ fixes I, interchanges AB and AC, and so maps the perpendicular IZ to AB on I to the perpendicular IY to AC on I. Since $d(I, B') = d(I, C')$, the point B' is on the ray $\overrightarrow{IY}$, and C' is on the ray $\overrightarrow{IY}$, it follows that $(B')^g = C'$. Hence, angles $\angle CAB'$ and $\angle BAC'$ are congruent. Similarly for the other vertices. So Jacobi's theorem (Theorem 10.52) applies: AA', BB', and CC' are concurrent. □

Theorem 10.59 has a long history. It was discovered independently by Auguste Boutin (1890b,a), and by Virginio Retali (1896). The result only became well known with Kariya's paper (Kariya, 1904).

Corollary 10.60
Let I be the incentre of a triangle ABC, and let A', B', C' be the reflections of I through sides BC, CA, AB, respectively. Then the lines AA', BB', and CC' are concurrent.

Actually, a generalisation of this result was obtained before Kariya by Émile Lemoine in Section 4 of Lemoine (1889, p. 202). Note that this is earlier than all those who found the weaker Kariya's theorem, which corresponds to M being the incentre (below).

Theorem 10.61 (Lemoine, 1889)
Let ABC be a triangle, M a point in its plane on no side of ABC, and X, Y, Z the orthogonal projections of M on BC, CA, AB, respectively. If A', B', C' are points on the rays $\overrightarrow{MX}$, $\overrightarrow{MY}$, and $\overrightarrow{MZ}$, respectively, such that $d(M, X)d(M, A') = d(M, Y)d(M, B') = d(M, Z)d(M, C')$, then AA', BB', CC' are concurrent.

Proof Exercise 10.3. □

André Auric gave another generalisation of Kariya's theorem.

Theorem 10.62 (Auric, 1915)
Let ABC be a triangle, M a point in its plane on no side of ABC, and X, Y, Z the orthogonal projections of M on BC, CA, AB, respectively. If A', B', C' are points on the rays $\overrightarrow{MX}$, $\overrightarrow{MY}$, and $\overrightarrow{MZ}$, respectively, such that

$$d(M, X)/d(M, A') = d(M, Y)/d(M, B') = d(M, Z)/d(M, C'),$$

then AA', BB', CC' are concurrent.

Proof Exercise 10.4. □

We restate the first part of the dual of Theorem 10.37.

Corollary 10.63
Let $ABCD$ be a quadrangle, t_A be the involution on the pencil of lines on A fixing AD and interchanging AB and AC, t_B be the involution on the pencil of lines on B fixing BD and interchanging AB and BC, t_C be the involution on the pencil of lines on C fixing CD and interchanging BC and AC. Let ℓ, ℓ' be a pair of the involution t_A, let m, m' be a pair of the involution t_B, let n, n' be a pair of the involution t_C. Then ℓ, ℓ', m, m', n, n' touch a conic.

Another proof of Morley's theorem (Theorem 8.18) First, ABC is the given triangle in Morley's theorem, and D is its incentre. Let t_A, t_B, t_C be isogonal involutions (i.e., reflections about angle bisectors); D is fixed by each one. Moreover, t_A interchanges AB and AC, t_B interchanges AB and BC, and t_C interchanges BC and AC. Now set ℓ, ℓ' to be the angle trisectors at A, m, m' to be the angle trisectors at B, and n, n' to be the angle trisectors at C. Then Corollary 10.63 implies that ℓ, ℓ', m, m', n, n' touch a conic $\mathcal{K}$. So $A'B''C'A''B'C''$ is a Brianchon hexagon for $\mathcal{K}$. The angles of the diagonals of the Brianchon hexagon at the point of concurrency are (easily) calculated to be $2\pi/3$, and it is then observed that they are perpendiculars of the sides of the Morley triangle. □

Exercises

10.1 Recall Baker's definition of a *conjugate P'* of a point P with respect to a quadrangle $ABCD$ (see Theorem 6.39 and ensuing material). Show that we can also use the Lehmer–Daus theorem (Theorem 10.38) to define P'.

10.2 Let $A' = AD \cap BC$, $B' = BD \cap CA$, $C' = CD \cap AB$, so that $A'B'C'D'$ is a quadrilateral with diagonal points A, B, C. Let
   - a be the harmonic conjugate of AD with respect to $\{AB, AC\}$,
   - b be the harmonic conjugate of BD with respect to $\{AB, BC\}$,
   - t_A be the map on pencil(A) fixing AD and a and taking m to the harmonic conjugate of m with respect to $\{a, AD\}$,

- t_B be the map on pencil(B) fixing BD and b and taking m to the harmonic conjugate of m with respect to $\{b, BD\}$.

Show that t_A, t_B are involutions. Moreover, show that Theorem 6.39 implies the Lehmer–Daus theorem (Theorem 10.38).

10.3 Prove Lemoine's theorem (Theorem 10.61) using the Gale–Gardner theorem (Theorem 10.51).

10.4 Prove Auric's theorem (Theorem 10.62) using the Gale–Gardner theorem (Theorem 10.51).

10.5 Use Theorem 10.38 to prove the Ferriot–Terquem theorem (Theorem 10.44). (Hint: The involutions are those interchanging a pair of sides and fixing the line on the Gergonne point.)

10.6 Use Theorem 10.43 to prove the following converse to the Ferriot–Terquem theorem:

Theorem 10.64 (Steiner, 1828a, p. 40)
Let M and M' be isogonal points, with respect to a triangle ABC, and draw perpendiculars from M to each side of ABC, and similarly for M'. Then the six feet of these perpendiculars are concyclic (on a circle with centre the midpoint of M and M').

10.7 (a) Use Theorem 10.43 to prove the following result. (Hint: First show there is a conic $\mathcal{K}$ lying on D, D^*, E, E^*, F, F^*.)

Theorem 10.65
Let ABC be a triangle, t_A, t_B, t_C be involutions on the pencils of lines on A, B, C, respectively, such that there is a (real or imaginary) point X with AX fixed by t_A, BX fixed by t_B, and CX fixed by t_C. Let a be a line on A, b be a line on B, and c be a line on C. Let $D := a \cap BC$, $D^* := a^{t_A} \cap BC$, $E := b \cap CA$, $E^* := a^{t_B} \cap CA$, $F := c \cap AB$, $F^* := c^{t_C} \cap AB$, $A_0 = DF^* \cap D^*E$, $B_0 = D^*E \cap E^*F$, $C_0 = E^*F \cap DF^*$. Then AA_0, BB_0, CC_0 are concurrent.

(b) Deduce the following result from (a).

Corollary 10.66 (Kouřilová and Röschel, 2013, Theorem 2.1)
Let ABC be a triangle, and let U_0 be a point on BC, let V_0 be a point on CA, and let W_0 be a point on AB. Let U_1 be the image of U_0 under the half-turn about the midpoint of $\overline{BC}$, and likewise for V_1 and W_1. Let $A_0 := U_0W_1 \cap U_1V_0$, $B_0 := U_1V_0 \cap V_1W_0$, and $C_0 := V_1W_0 \cap U_0W_1$. Then AA_0, BB_0, CC_0 are concurrent.

11

From Perspective to Projective

Perspective is nothing else than the seeing of an object behind a sheet of glass, smooth and quite transparent, on the surface of which all the things may be marked that are behind this glass. All things transmit their images to the eye by pyramidal lines, and these pyramids are cut by the said glass. The nearer to the eye these are intersected, the smaller the image of their cause will appear.

Leonardo da Vinci (1480) in da Vinci et al. (1970, p. 343)

11.1 Prologue

SOMETHING HAPPENED in Florence six hundred years ago, something so unique and miraculous that it changed our world forever. We call it the Renaissance, a rebirth of ancient art and learning. Yet it was more than a rebirth or rediscovery of ancient secrets; it was a first birth, the beginning of a modern consciousness, a modern way of seeing and representing the world around us.

Paul Robert Walker (2009, p. 4)

The greatest technological achievement of the Renaissance was printing. The greatest *mathematical* achievement of the Renaissance was perspective. This step was first taken by artists and architects. The perhaps surprising idea of an artist as a scientist in historical research goes back at least to the neo-Kantian philosophers of the late nineteenth century (e.g., Cohen, 1889) and was continued in the work of the Italian historian of geometry Gino Loria (1908). Already in the early Renaissance, art foreshadowed mathematics, and therefore its research and achievements in relation to the construction of perspective formed the paradigm which later, especially in the seventeenth century, was unified by mathematicians and practitioners of optics, turned into

288

an axiom, and canonised as a technique. The essential contribution of early modern art was the systematisation of its principles that allowed the development of representational geometry as a self-contained mathematical discipline, independent of art.

Surrounded as we are by photography, film, and television, it is not obvious to us that interpretation of a perspective image is a learned skill. Here the encounter of Jesuit missionaries with the Chinese in the seventeenth and eighteenth centuries is instructive. Wu Li (1632–1781), who was already one of the acknowledged *Six Masters of the Early Qing Dynasty* when he became a convert to Christianity (and ultimately a Jesuit priest), never embraced the European way of painting. Wu Li was critical of European painting and observed: 'Our painting does not seek physical likeness and does not depend on fixed patterns ... we call it divine and untrammeled. Their is concentrates entirely on the problems of light and dark, front and back, and the fixed patterns of physical likeness' (Arnold, 2003, p. 2).

Despite the many decades of modern art, it is even less obvious to us that a painter might have to be taught that the aim of naturalistic painting is to reproduce a scene as it might be viewed, at one instant, out of a window; and yet, that is what Leon Battista Alberti explained in his book *Della Pittura* in 1435 (Alberti et al., 2005). (This led to three recent jocular transpositions of his metaphor into the computer age: Samuel Edgerton referred to Alberti's book as *Windows 1435* in a serious academic paper in 2000 (Edgerton, 2000); it was called the original version of *what you see is what you get* by Rehkämper (2003); and the title of Friedberg (2006) is *The Virtual Window: From Alberti to Microsoft*.)

11.2 Naturalism in Art

It was a turning point in cultural history when artists linked the horizon to the human gaze and ceased relating it to a transcendental gaze that looked down at the world from above, as had been the case in medieval times. The 'discovery of the horizon' ... coincides chronologically with centering paintings on an observing subject.

Hans Belting (2007, pp. 138–9)

The desire to naturally represent a scene as its appears from a particular viewpoint is a necessary precondition for the invention of linear perspective. This precondition was absent in the medieval period, when art was more abstract, more symbolic and allegorical, and when rarely not of this form, concerned with *things as they are rather than things as they appear*. It had been satisfied in the ancient classical period (although Plato condemned mimetic art as an imitation of mere appearance, thrice removed from reality), which raises the question of whether or not linear perspective was known to the ancients?

There is a distinction between the illusion of perspective and the use of the theory of linear perspective. This point is nicely made in the study by Field et al. (1989) of the (see Figure 11.1) Trinity fresco of Masaccio in Santa Maria Novella in Florence, painted circa 1426. (Incidentally, the architect of the church Santa Maria Novella was Leon Battista Alberti.) It is universally agreed by critics that this work is a successful representation of a three-dimensional object on a plane, especially with regard to the barrel-vaulted ceiling it depicts. (An account of this is also given in Field (1997, chapter 3).) Aware that measurements from reproductions introduce too much error in trying to determine whether or not Masaccio used linear perspective and vanishing points, Field made a visit to the actual fresco. A colleague had asked her some question, to which she had responded that 'we should need scaffolding for that', unaware that the question had originated from a friend of the colleague whose job it was to look after the fabric of the Santa Maria Novella church in Florence. So, with

Figure 11.1: Masaccio (1401–28), *Holy Trinity*[1] (c. 1425–6), Santa Maria Novella (Florence).

a team, she actually got to examine the fresco on scaffolding, close up, to see the brushwork, and the mass of indentation and scribed, snapped and sinopia lines that he had used to mark his design in the fresh plaster. They could take measurements, with an accuracy comparable to that available to Masaccio. Their conclusions were that the fresco is visually right and mathematically wrong, in that it doesn't use linear perspective in any way that isn't very approximate.

---

[1] Available at https://commons.wikimedia.org/wiki/File:Masaccio,_trinit%C3%A0 .jpg.

Many authors claim that in classical times, linear perspective was used **because** the illusion of perspective was successfully achieved, especially in stage design in Greek plays. But there is a blog post by Carla Schodde (2013) where a similar conclusion to Field's about Masaccio's Trinity fresco is reached about a fresco from the Villa of Publius Fannius Synistor, preserved by ash in AD 79. (That fresco is now in the Metropolitan Museum of Art, New York.) Thus, there is a huge (and false) assumption in deciding that because the illusion of perspective exists in a painting, the painter used the theory of linear perspective. Once this distinction is understood, it is possible to claim that, on the balance of the evidence, the theory of linear perspective was invented in Renaissance Florence.

When did the change in viewing the world between the medieval and the Renaissance view first occur? The evidence makes it clear that it was among Italian Renaissance humanists. We therefore must say a few words about humanism – the cultural and intellectual movement of the Renaissance that emphasised human potential to attain excellence, promoted the direct study of the literature of classical Greece and Rome, and concerned itself with the question: What is the true nature of humanity? Italian scholars held that the historical methods of the humanists implied nothing less than a new view of the world. According to Voigt (1859), the Italian poet Petrarch (1304–74) was the first Renaissance humanist. Paul Johnson (2000, p. 34) states that Petrarch was 'the first to put into words the notion that the centuries between the fall of Rome and the present had been an age of darkness'. Petrarch saw himself in surprisingly modern terms, as a human being with free choice and almost limitless possibilities.

On 26 April 1336, Petrarch climbed Mount Ventoux, near Avignon, just to take in the view. According to (Grau, 2003, p. 35), his description of this experience marked a turning point in how the world was viewed.[2] At the summit, he mused over where perception gives out. On the horizon, he could see the snow-covered Alps. Now, a horizon implies a beholder and, with that, a sense of self-awareness. While contemplating the horizon, looking outwards, his attention moved inwards, towards the source of vision. Then Petrarch opened his pocket-sized copy of Augustine's *Confessions* to a random page: 'Where I first fixed my eyes it was written: "And men go about to wonder at the heights of the mountains, and the mighty waves of the sea, and the wide sweep of rivers, and the circuit of the ocean, and the revolution of the stars, but themselves they consider not"' (Maleuvre, 2011, p. 144).

---

[2] Petrarch was anticipated in his consideration of the horizon by Roger Bacon, seven decades earlier, in 1267: 'Every point of the earth is the centre of its own horizon' (Bacon and Burke, 1928, volume II, p. 272).

Once the desire for naturalism in art had re-arisen in the fourteenth century, a period began of intuitive development of techniques, some of which in hindsight seem significant in our tale. The medieval adoption of God's point of view was put aside in the works of Cimabue, Giotto, Duccio, and the Lorenzetti brothers, Pietro and Ambrogio, and thus the adoption of a particular viewpoint was made. (This could be called *the invention of the observer* or it could be summarised by the matter-of-fact observation that the world appears differently from different points of view.) Reflection on the nature of perspective will teach us that whatever presents itself to the eye, to perception, is no more than subjective appearance. A God-centred art slowly gave way to a human-centred art.

Figure 11.2a Ambrogio Lorenzetti (c. 1290–1348), *Presentation of Christ at the Temple,*[a] Uffizi Gallery (Florence), 1342.

Figure 11.2b Ambrogio Lorenzetti (c. 1290–1348), *The Annunciation,*[b] Pinacoteca Nazionale (Siena), 1344.

[a] Available at https://commons.wikimedia.org/wiki/File:Ambrogio_Lorenzetti_-_ The_Presentation_in_the_Temple_-_WGA13480.jpg.
[b] The painting is in the public domain.

Two paintings by Ambrogio Lorenzetti make innovations that will matter to us: the *Presentation of Christ at the Temple* (1342) uses a checkerboard floor pattern which becomes a style called *pavimento* (see Figure 11.2a) and in *The Annunciation* (1344) the lines on the ground perpendicular to the picture plane are represented as concurrent lines (see Figure 11.2b). Later painters also represent lines *not on the ground* perpendicular to the picture plane as passing through the same point; the anonymous miniature of Petrarch (Figure 11.3) is representative. (A detail of this work is presented as Andersen (2007, figure I.8), where an analysis of this phenomenon follows.)

---

[3] The painting is in the public domain, available at www.alamy.com/francesco-petrarca-nello-studio-2-image343963024.html.

11.3 Brunelleschi

> The world having for so long been without artists of lofty soul or inspired talent, heaven ordained that it should receive from the hand of Filippo the greatest, the tallest, and the finest edifice of ancient and modern times, demonstrating that Tuscan genius, although moribund, was not yet dead.

Giorgio Vasari, *The Lives of the Artists*, 1550 (Vasari et al., 2008, pp. 110–111)

The next important transition is to a monocular view which it seems came with Filippo Brunelleschi, for which the motivation probably came from optics and from the camera obscura. It is a standard reductionist tendency to first understand one eye and then move on to study stereo (or binocular) vision, especially in the light of experience of people with only one functioning eye having serviceable vision.

Figure 11.3: Unknown artist, *Petrarch in His Study* (mid-fourteenth century),[3] Biblioteca Trivulziana (Milan), Codex 905.

The date of the birth of the theory of linear perspective is sometime between 1401 and 1413, and it arises when the architect, sculptor, and artisan-engineer Filippo Brunelleschi (1377–1446) produced a work of art that no longer survives; it was last heard of in 1494, when it was listed among the effects of the Medici ruler of Florence, Lorenzo the Magnificent, on his death. Our only real description of it was written in the 1480s, by Brunelleschi's biographer the architect and mathematician Antonio Manetti, whose account of it is puzzling and unsatisfactory (see Manetti, 1970). No record exists from Brunelleschi's experiments written by Brunelleschi himself. There have been numerous attempts to reconstruct exactly what Brunelleschi did, and why, for, by the accounts of contemporaries, this object represented the birth of perspective painting. All attempts at reconstruction run into difficulties. The object was a painting on a wooden panel perhaps about 12 inches square. It showed the Baptistry in Florence, an octagonal building, and something of the buildings on either side. The top part of the painting, where the sky would have been, was covered in burnished silver. In this painting, in the centre near the bottom, Brunelleschi made a hole, and the viewer was invited to look through the back of the painting. According to Manetti (2007, p. 44), '[Brunelleschi] had made a hole in the panel ... which was as small as a lentil on the painting side ... and on the back it opened pyramidally, like a woman's straw hat, to the size of a ducat or a little more.'

Standing in exactly the right place, on the west side of the then unfinished cathedral, looking towards the Baptistry, the viewer was intended to see through the hole a mirror that reflected the painting itself so that the front of the painting was seen in the mirror overlapping with the real Baptistry; by raising and lowering the mirror the viewer could convince themselves that the painting looked just like the real thing. Because they were looking at both the painting and the real world through one eye, the painting would look nearly three-dimensional, and the real world more two-dimensional, so the two would become more like each other. In the burnished silver of the painting the sky was reflected, so that clouds (if there were any) could be seen; reversed on the silver, the clouds would have been reversed again in the mirror so that they, too, would have corresponded to reality.

The evidence that Brunelleschi did this work by 1413 is based on a letter in the Florentine archives from the poet Domenico da Prato which described Brunelleschi as an 'ingenious perspective man'.

That in this sterile place devoid of everything good and fertile, where asps and basilisks trouble me, on the contrary I am grieved, will not surprise you who, accompanied by the magnificent will of your Flower, enjoy occasional meetings with that perspective expert and ingenious man, Filippo di ser Brunellesco, remarkable for virtues and fame. ... By such brilliant lights you are able to see your steps, and not by those insane and rough betrayers, whose stupid shouts have already made me deaf.

Domenico da Prato, letter to Alessandro Rondinelli, 6 August 1413.

Why would Brunelleschi have left no written account? We can learn something of this from Prager (1968). This concerns a notebook of the artist and engineer Mariano di Jacopo (1382–c. 1453), called *Taccola*. In it there is a quote from Brunelleschi from a conversation with Taccola in Siena, circa 1427–30, in a chapter written sometime between 1433 and 1450 (Hyman, 1974, p. 31):

Do not share your inventions with many, share them only with few who understand and love the sciences. To disclose too much of one's inventions and achievements is one and the same thing as to give up the fruit of one's ingenuity. Many are ready, when listening to the inventor, to belittle and deny his achievements, so that he will no longer be heard in honourable places, but after some months or a year they use the inventor's words, in speech or writing or design. They boldly call themselves the inventors of the things that they first condemned, and attribute the glory of another to themselves.

Brunelleschi probably passed the method verbally to Masaccio, Masolino, and Donatello, who used it in their works. According to Manetti, Brunelleschi, accompanied by his pupil Donatello, had made a study of the surviving classical buildings in Rome in 1403–4, a study that involved taking measurements and drawing plans and elevations. Brunelleschi's ingenious construction serves a double function: it demonstrates that art can successfully imitate nature, so that the two are almost indistinguishable; and it demonstrates that even when art is at its most objective, we make it and we find ourselves in it.

11.4 Optics

[N]othing can be seen except by light.

Lorenzo Ghiberti, cited in Argan and Robb (1946, p. 101)

If we get our awareness of parallelism through touch, as by running our fingers along a simple molding, there is no question of the sensuous return that parallel lines do not meet. If, however, we get our awareness of parallelism through sight, as when we look down a long colonnade, there is no doubt of the sensuous return that parallels do converge and will meet if they are far enough extended.

William M. Ivins Jr. (1938, p. 8)

It has been argued that both Brunelleschi and Alberti were applying to painting the theories of medieval optics, which ultimately derived from an eleventh-century Arabic author, ibn al-Haytham (c. 965–c. 1040), known in the West as *Alhazen*, whose works were available in both Latin and Italian translations. These works on optics were about 'perspective', a term that meant 'the science of sight'.

Graziella Federici Vescovini, in a number of papers (Federici Vescovini, 1965, 1980, 1990, 1998), has convincingly argued that an anonymous translation in Italian from the middle of the fourteenth century of the entire works of ibn al-Haytham in the Vatican Library (MS Vat. no. 4595) was consulted by Lorenzo Ghiberti. Philological analysis has shown that MS. London, British Library, Royal 12.G.VII served as the Latin matrix for the Italian version (Goldstein and Smith, 1993). This text was discovered by Enrico Narducci (1871). Federici Vescovini (1998, p. 68) claims that this Italian manuscript may have been central to the development of perspective in Italian Renaissance painting. (See also Smith (2001).)

The theory of optics was written about by the Ancient Greeks, with al-Kindi and al-Haytham making substantial advances in the medieval Arab period. Ibn al-Haytham refuted Aristotle's theory of the rays emanating from the eyes.[5] Ibn al-Haytham invented all the tenets of the scientific method except peer-reviewed publications; he was without peers. He devised a scientific solution to ancient controversies over the nature of vision, light, and colour, which were disputed between the classical mathematicians (exponents of Euclid and Ptolemy) and the Aristotelian physicists. Ibn al-Haytham's research in optics (including his studies in catoptrics and dioptrics, respectively on the principles and instruments of the reflection and refraction of light) also benefited from the investigations of his predecessors in the Archimedean-Apollonian tradition of ninth-century Arab polymaths, like Banu Musa and Thabit ibn Qurra, and of tenth-century mathematicians, like al-Quhi, al-Sijzi, and ign Sahl. This led to his work appearing in versions in Latin as *Perspectiva*. Later this became *Perspectiva naturalis* to distinguish it from the perspective of the architects and painters.

ABŪ 'ALĪ AL-ḤASAN IBN AL-ḤASAN IBN AL-HAYTHAM (c. 965–c. 1040) invented all the tenets of the scientific method (except peer review) in the context of his work on optics. Some of his mathematical works are: The circle, its transformations and its properties; Analysis and synthesis: the founding of analytical art; A new mathematical discipline: the Knowns; The geometrisation of place; Analysis and synthesis: examples of the geometry of triangles; Completion of the conics; On the construction of the regular heptagon; Treatise on the determination of the lemma of the heptagon; The division of the straight line used by Archimedes; On a solid numerical problem; Treatise on lunes; Treatise on the quadrature of the circle; Exhaustive treatise on the figures of lunes; On the measurement of the paraboloid; On the measurement of the sphere; On the sphere which is the largest of all the solid figures having equal perimeters and on the circle which is the largest of all the plane figures having equal perimeters. His most famous other works are the Optics and his works on astronomy. He also contributed to number theory.

In inventing his device to show people his plans for the Baptistry in Florence, Brunelleschi incorporated the point of view into representative painting for the first time. Perspectivist pictures have a point from which they should be viewed; it is that from which the painting will be most realistic. Moreover, in using a mirror in that device, he brought out the idea of projection on a plane. And he made people aware of the frame, which Alberti referred to as

---

[5] A schoolmate of the older author, when asked how to answer a person who claimed that the rays of vision emanated from the eyes, answered: *Take him to a dark room. Ask him what he 'sees'*.

the window through which the painting represented the view. These Renaissance Florentines knew of the Arabic work on optics via the work of Biagio da Parma (also known as *Pelacani*), which was studied in Florence. The most important idea taken from optics by the Renaissance painters is that of the rectilinear propagation of light. Propositions 4 and 5 of Euclid's *Optics* establish the relationship between size perception and distance perception on the basis of the visual angle subtended by the object. This idea was mentioned in Plato's *Republic*:

SOCRATES: *The body which is large when seen near, appears small when seen at a distance?*

Proposition VI claims that *parallel lines, when seen from a distance, appear not to be equidistant from one another*. This will influence later work on perspective. The visual pyramid/cone and the use of a stationary viewpoint are also important legacies of optics that were taken on board.

Roger Bacon (1219/20–92) became a key figure in advocating experimental science in the West;[6] he also wrote a text on optics, which he called *Perspectiva*. Roger Bacon introduced the term as a title of part V of his *Opus Maius*, thus originating the tradition of perspectiva in the West. Bacon had a strong influence on his fellow Franciscan John Peckham, later Archbishop of Canterbury, who wrote a most popular treatise, *Perspectiva Communis* (1279), and on the Silesian scholar Witelo and his long and equally influential treatise *Perspectiva* (1273). A printed edition of ibn al-Haytham's Latin version of the Optics was established by Friedrich Risner in 1572 in Basel, under the title *Opticae Thesaurus*, which was eventually consulted by seventeenth-century scientists and philosophers such as Kepler, Descartes, Huygens, and possibly even Newton. Witelo influenced Kepler in his seminal 1604 text on optics *Ad Vitelonem Paralipomena quibus Astronomiae Pars Optica Traditur* (Kepler, 1604).

The perspectivist tradition persisted through to the fourteenth century. The most influential treatise of that period was by Biagio da Parma (Pelican): the *Questiones Perspectivae* (1389). Pelacani's treatise influenced Brunelleschi, Alberti, Piero della Francesca, and other Renaissance artists and architects in their creation of perspective drawing. No one before Alberti had had the courage to treat the primary issues of perspective as a purely mathematical problem. Alberti's contribution to perspective was developed entirely around the principles of proportion. In the work of these Renaissance Florentine artists, the

---

[6] In 1948, in a famous paper, Alexandre Koyré argued that the scientific method found its earlier roots in the legacy of ibn al-Haytham (Alhazen) in optics, which resulted in the flourishing of the perspectivism of Franciscan scholars in the European Middle Ages, in addition to the application of their experimental methods.

vanishing point and the vanishing line first appear. Samuel Y. Edgerton (2009) argues that the perspective window of Alberti, the mirror, and the telescope led to a new way of seeing the world, which in turn led to science.

Beginning in the decades around the turn of the fourteenth century, Western Europeans evolved a new way, more purely visual and quantitative than the old, of perceiving time, space, and the material environment.

Plato and Aristotle remained influential throughout the Middle Ages and Renaissance, one more than another at a given time, never one alone. Aristotelian thought was dominant before the Renaissance. The foundation of the Platonic Academy of Florence under the sponsorship of Cosimo de Medici and the leadership of Marsilio Ficino in c. 1438–9 and its survival to the death of Lorenzo de Medici in 1492 gave new impetus to the study of the ideas of Plato in the original Greek. That gave a setting for a new faith in the Platonic ideas, as expressed in *The Republic*, that numbers 'have the power of leading us toward reality' (Plato, 1941, p. 236) that 'geometry is knowledge of the eternally existent' (Plato, 1941, p. 238).

Thus, during the Renaissance, Platonic thought resurged, but for the first time with an admixture of influences from two other sources. The first was ibn al-Haytham and his use of experiments. The second was Archimedes, with the desire for the first time not just to admire his treatises but also to emulate him. This last is particularly true of Galileo, but it is also true of all his Italian predecessors in mechanics in the second half of the sixteenth century: Cardano, Tartaglia, Benedetti, Commandino, Guidobaldo, and Baldi.

11.5 Mirrors

Within the rich history of how the mirror can function to provoke transformations in perception and consciousness, Brunelleschi's use of a mirror in his early fifteenth-century demonstration of the effect of single point perspective looms large.

Charles H. Carman (2016, p. 111)

If *we* have any knowledge of [the world], we derive it from the symbolism and the mirror of [our] mathematical knowledge.

Cusanus, *De Possest*, I.43 (Hopkins, 2007, p. 113, emphasis in original)

From the Florentine artist Antonio di Pietro Averlino, known as *Filarete*, who wrote a work on architecture (see Spencer, n.d.) completed around 1461

– our earliest source – we hear that Brunelleschi had come up with his new method of perspective representation as a result of studying mirrors. The mirror makes it easy to answer the question of how big the object being painted looks from here. For a painter, a mirror is a good tool for measuring the size of an image; rather than measuring angles and distance, measuring an image on a mirror is much simpler. The easiest way to decide the size of an image of an object on a canvas at a certain distance is to use a mirror.

The use of mirrors picks out the picture plane. Now we have the elements of linear perspective: the centre of projection (or eye), the rectilinear nature of light from optics, and the picture plane to receive the image. These are also all the elements of the pinhole camera, also known as the camera obscura, which may have played a role in the work of the first person to write a treatise on linear perspective.

11.6 Camera Obscura

The earliest European references to optics did not occur until about 1270, with essays by Erazmus Ciolek Witelo and Roger Bacon – but by the Renaissance, the camera obscura was clearly one of the common optical toys of the time. Here is how it works: A light source, 'outside', is directly cast through a small hole and then is produced as an 'image' on some flat surface 'inside' the darkened room. Carefully now note the effects and the ways in which the camera transforms an 'object' into an 'image'. Our 'Alberti' now can reproduce in a drawing this 'automatically-reduced-to-Renaissance-perspective' image, emerge from the dark room and invert the drawing, and get the appreciative gasp of 'verisimilitude' his audience noted. One can see from this simple example how thoroughly so-called Renaissance perspective is 'technologically' produced. Its mathematization in one sense follows the optical effect.

Don Ihde (2008, p. 385)

We began this section with a reference to photography. Is it an accident that linear perspective, lying at the origin of projective geometry, is so useful in photography? Here the rediscovery of the application of projective geometry to photography by those working on computer vision in the 1970s and 1980s (first discovered by the researchers in photogrammetry in the second half of the nineteenth century) is a salutary lesson (see Buchanan, 1988 and Sturm, 2011). It is probable that some artists used pinhole cameras in studies for some of their works, even in the fifteenth century. If this is the case, the usefulness of projective geometry alluded to above would not be accidental.

First, then, what is the camera obscura? In simple terms, it is a very large pinhole camera; the literal meaning of the term is 'dark room'. The earliest extant written record of the camera obscura is to be found in the writings of Mozi (470 BC–390 BC), a Chinese philosopher and the founder of Mohism. Mozi correctly asserted that the image in a camera obscura is flipped upside down because light travels in straight lines from its source. His disciples developed this into a minor theory of optics (Needham, 1962). Euclid's *Optics* (c. 300 BC) mentioned the camera obscura as a demonstration that light travels in straight lines. In the fourth century (AD), Theon of Alexandria, in his recension of the pseudo-Euclidean *Catoptrics*, observed that 'candlelight passing through a pinhole will create an illuminated spot on a screen that is directly in line with the aperture and the centre of the candle' (Steffens, 2007, p. 71).

In the AD sixth century, the Byzantine-Greek mathematician and architect Anthemius of Tralles (most famous for designing the Hagia Sophia) used a type of camera obscura in his experiments (Huxley, 1959). There were later experiments by the Arabs al-Kindi and ibn al-Haytham. The camera obscura was first thoroughly described in ibn al-Haytham's Optics in 1037, which was the first major systematic treatise on light and optics. He probably used the camera obscura to observe a solar eclipse. This may have been the first anticipation of an early modern scientific use of the camera (Ihde, 2008). Yet the *Encyclopaedia Britannica* (1929, vol. 4, p. 658) has this: 'The first practical step towards the development of the camera obscura seems to have been made by ...Leon Battista Alberti, in 1437 ...[I]n a fragment of... biography... published in Muratori's *Rerum Italicarum Scriptores* (xxv. 296)... it is stated that he produced wonderfully painted pictures ...[via a] small closed box through a very small aperture.'

That advances in realism and accuracy in the history of Western art since the Renaissance were primarily the result of optical instruments such as the camera obscura, the camera lucida, and curved mirrors, rather than solely due to the development of artistic technique and skill, is now known as the Hockney–Falco thesis, after the artist David Hockney and the physicist Charles M. Falco (see Hockney, 2001 and Falco, 2007). But, as the quote from the 1929 *Encyclopaedia Britannica* makes clear, Hockney and Falco were merely the latest to rediscover that 'the Renaissance in both art and science was embodied through technologies, with the camera obscura being one favourite optical toy' (Ihde, 2008, p. 384).

11.7 Leon Battista Alberti

Alberti... produced wonderfully painted pictures, which [he] exhibited... with great verisimilitude. These demonstrations were of two kinds, one nocturnal, showing the moon and bright stars, the other diurnal, for day scenes.

Vita di Leon Battista Alberti (part of an anonymous fragment In Muratori's 1751 *Rerum Italicarum Scriptores*, vol. XXV, p. 296; translation taken from *Encyclopaedia Britannica*, 1929, vol. 4, p. 658)

Alberti transferred the old metaphor of the eye as a 'window of the soul' (Heraclitus...) to the painting, describing it as a window.

Hans Belting (2007, p. 142)

The Italian humanist, architect, poet, priest, linguist, philosopher, and cryptographer Leon Battista Alberti (1404–72) wrote the first book on perspective in 1435, *'Della pittura'* (see Alberti et al., 2005). In it, he presents linear perspective as a point-to-point mapping, under which:

- lines not through the centre of projection map to lines;
- lines parallel to the picture plane, in a different plane through the centre of perspective, map to parallel lines.

He picks out a horizontal ground plane, taking the picture plane as vertical, and picks out the ground line of intersection of the ground and picture planes, and then notes that under linear perspective

- lines parallel to the ground line map to lines parallel to the ground line;
- orthogonal lines to the picture plane in the ground plane map to lines converging at a point.

He picks out the horizon line as a horizontal line on the convergence point. He uses diagonals to check accuracy, probably employing the camera obscura. Alberti also gives the first known definition of what we would call a perspective projection, recognising that the closest point in the picture plane to the eye is the convergence point for orthogonals.

According to Morris Kline (1972, p. 232), describing Alberti's *Della pittura*:

Alberti conceived the principle that became the basis for a mathematical system of perspective adopted and perfected by his artist successors. He proposed to paint what one eye sees, though he was well aware that in normal vision both eyes see the same scene from slightly different positions. ... His basic principle was explained in the following terms. Between the eye and the scene he interposed a glass screen standing upright. He then imagined lines of light running from the eye or station point to each point in the scene itself. ... Where the rays pierced the glass screen, the picture plane, he imagined points marked out; the collection of points he called a section. The significant fact about it is that it creates the same impression on the eye as the scene itself, because the same lines of light come from the section as from the original scene. Hence the problem of painting realistically is to get a true section on the glass screen or, in practice, on a canvas. Of course that section depends on the position of the eye and the position of the screen. This means merely that different paintings of the same scene can be made ...

Kline continues (p. 233):

Beyond introducing the concepts of projection and section, Alberti raised a very significant question. If two glass screens are interposed between the eye and the scene itself, the sections on them will be different. Further, if the eye looks at the same scene from two different positions and in each case a glass screen is interposed between the eye and the scene, again the sections will be different. Yet all these sections convey the original figure. Hence they must have some properties in common. The question is: What is the mathematical relation between any two of these sections, or what mathematical properties do they have in common? *This question became the starting point of the development of projective geometry.* (our emphasis)

11.8 The Mathematics of Perspective

A man must look at the world on his own terms, and the way he looks at it depends on where he is standing. This is the true meaning of perspective, and Filippo Brunelleschi may have been the first to fully understand the implications of this idea.

Paul Robert Walker (2009, p. 37)

Five things pertain to this 'perspective': The first is the eye, that sees. Next is the thing that is seen. Third is the space between them. Fourth [is that] everything is seen by straight lines, which are the shortest lines. Fifth is the parts of the thing that is seen.

Albrecht Dürer, expansion of a list from Piero della Francesca, *Book of Measurements*, 1525 (cited in Elkins, 2018, p. 50)

Slowly attention turned from practical workshop methods of producing pictures in linear perspective to the reasons these methods worked and their connection with geometry.

Piero della Francesca (c. 1415–92) wrote the first treatise *De prospettiva pingendi* (della Francesca, [c. 1474] 2021) on the mathematics of linear perspective. In it, he notes that it takes lines at 45 degrees (diagonals) to lines through another convergence point on the horizon line (indeed two; one for SW/NE and one for NW/SE). He traces a curve by constructing points on it (an ellipse). He also gives a proof that orthogonals not in the ground plane converge to the same point (Book I, Propositions 8 and 14).

According to Wittkower (1953, pp. 285–7), in the early 1490s, Leonardo da Vinci made an important contribution by formulating the connection between *proportion in perspective* and musical consonances, by determining the ratio of each projection to the objective height of each object, obtaining a harmonic progression, for the case where the distances between the eye, the intersection, and the objects in space are equal. See Richter and Da Vinci (1970, p. 157, para. 102), translated as: 'I give the degrees of the objects seen by the eye as the musician does the notes heard by the ear.'

Jean Pèlerin (pseudonym Viator) included in *De artificiali perspectiva* a two-point perspective in a geometric diagram (Pèlerin, 1504). Federico Commandino (1509–75) published a commentary on Ptolemy's Planisphaerium in 1558 (*In planisphaerium Ptolemaei commentarius*, Commandino, 1558) that contains an entirely geometrical deduction showing the correctness of a perspective construction (Andersen, 1990, p. 25). He also recognises that there is a connection between Ptolemy's stereographic projection and the perspective construction of the artists and architects.

Among all the many constructions, practitioners preferred one type especially, namely, that by distance points. Jean Cousin (1500–before 1593) in his *Livre de perspective* (Cousin, 1560) gives an implicit proof of a distance point construction (Andersen, 2007, pp. 175, 182–3). Cousin also employs convergence points for general horizontal lines, not just diagonals and orthogonals, and they occur on the horizon (Andersen, 2007, p. 179).

The architect and painter Giacomo Barozzi da Vignola (1507–73) took for granted that the convergence points for the diagonals lie on the line known as the horizon, which is the horizontal line through the principal vanishing point (see da Vignola, c.1530–73 and Andersen, 2013). Vignola's application of the convergence rule for the diagonals seems to have inspired him to conclude that lines other than orthogonals and diagonals can have convergence points (in his diagrams). Vignola also had some work on three point perspective, where there are three convergence points that are not collinear. (This doesn't always happen in perspective renderings of rectangular buildings, because vertical lines are usually depicted as parallels.) His commentator, the mathematician Egnazio

Danti (1536–86), started to wonder about converging points and included in a long introduction to Vignola's text the following interesting set of definitions:

Definition V: Perspective parallel lines are those that will meet at the horizontal point [a point on the horizon].

Definition VII: A distance point is that at which all the diagonals arrive.

Definition X: Principal parallel lines are those which will all converge at the principal point of perspective (principal vanishing point).

Definition XI: Secondary parallel lines are those which will be united at the horizontal line at their particular points apart from the principal point.

But Danti's diagrams contradict his words.

Giovanni Battista Benedetti (1530–90) described his thoughts about perspective in the treatise *De rationibus operationum perspectivae* (On the reasons for the operations in perspective) published in 1585 (Benedetti, 1585). In this, he proves the correctness of a couple of perspective constructions, one of which was later also presented by Guidobaldo. Moreover, he shows some understanding of general convergence points, but, similarly to what Danti did in his diagrams, Benedetti draws one pair of sides in the image of a horizontal rectangle having no sides parallel to the picture plane as lines converging on the horizon, whereas the second pair does not have a convergence point on the horizon. Benedetti also anticipates the converse of Desargues' theorem on perspective triangles (for non-coplanar triangles in three dimensions; he proves it while working out the perspective image of a parallelogram). (In chapter XI, in the course of proving something else, he proves the converse of Desargues' theorem, without stating it. (See Field, 1985 and Field, 1997, chapter 7, especially pp. 167–71.). On p. 124, Benedetti points out the sufficiency of pointwise constructions.

11.9 Guidobaldo Del Monte

A vanishing point turns up in the following important result, which can, following Andersen, be called the *vanishing point theorem*.

The perspective image of a line ℓ that is not parallel to the picture plane passes through the vanishing point of ℓ, and so do the perspective images of all lines parallel to ℓ.

Guidobaldo (Monte, 1600) gives several proofs of this theorem, the last of which implicitly uses Proposition 7 in Book XI of Euclid's *Elements*. During this proof, he also proves what Andersen calls the *main theorem of perspective*:

The image of any line ℓ that is not parallel to the picture plane is determined by the vanishing point of ℓ and the intersection of ℓ with the picture plane.

He applies this result in Book 2, Propositions 2 and 11, determining the image of a line and the image of a line segment.

Guidobaldo's text implicitly contains the general concept of a vanishing line, but explicitly only one occurs – the horizon. In Book 1, Proposition 33, Corollary 2, he shows that the vanishing points of sets of parallel lines in the ground plane lie on a line parallel to the ground line and that all horizontal lines have their vanishing points on this line.

Guidobaldo's implicit consideration of a general vanishing line is in Book 1, Proposition 35 where he shows that the vanishing points of lines situated in a non-horizontal plane that is not parallel to the picture plane lie on the intersection of the picture plane and the plane through the eye point parallel to the plane they lie in.

11.10 The Infinitude of Space

> It is the conception of space as infinite . . . that underlies Leon Battista Alberti's perspective construction. Addressed primarily to painters and those interested in understanding the craft of painting, his theory of perspective teaches us to create convincing representations of what we see, as it appears. What paintings represent then are not the objects themselves but their inevitably subjective appearances. Implicit in all such appearances is a particular point of view. All appearance is relative to the subject seeing.
>
> Karsten Harries (2001, p. 66)

The Age of the Medici, originally released in Italy as *L'età di Cosimo de Medici (The Age of Cosimo de Medici)*, is a 1973 three-part TV series about the Renaissance in Florence, directed by Roberto Rossellini. The series was shot in English in the hope of securing a North American release, which it failed to achieve, and was later dubbed into Italian and shown on state television. The films are: *Cosimo de Medici, The Power of Cosimo*, and *Leon Battista Alberti: Humanism.*

That last film pairs the humanist writer Leon Battista Alberti and the theologian/philosopher Nicholas Cusanus, invoking the storied but undocumented belief that they knew each other. Rossellini's film offers a stunning evocation of the tenacity of the story of this relationship.

Included is a meeting between Alberti and Cusanus, supposed to have taken place in Florence at the time of the council that fleetingly reunited the Eastern and the Western Churches. If such a meeting ever took place, it was in the mind of Johannes Kepler. He takes the *principle of continuity* from ideas of Cusanus, and points at infinity from ideas that grew out of Renaissance painters' work on perspective, first explicated by Leon Battista Albert in 1436, and puts them together to explain that a parabola has two foci, one of which is at infinity. Five decades later in 1668, his great inheritor in the application of conics to physics, Newton, invents the reflecting telescope based on the same principle, namely, that light rays emanating from the focus at infinity would be reflected to the other focus.

A key aspect of the Aristotelian cosmos is that it was finite in size. This was adopted in the medieval Christian perspective.

11.11 Does the Infinity of Space Follow from the Use of Perspective Methods?

As Harries (2001, p. 19) remarks, Koyre (1957) concludes with the suggestion that all these changes can be subsumed under just one or perhaps two closely related developments: they can be understood as a result of the destruction of the finite world of the medievals and of the geometrisation of space characteristic of modern science.

One thing that developed simultaneously with linear perspective and the evolution of projective geometry in the fifteenth and sixteenth centuries was the belief in the infinitude of space. Some authors have argued that the first was the source of the second. Certainly, Kepler's use of points at infinity gives some evidence for this. Space can perhaps be traced back to the use of 'spatium' by Lucretius (98–55 BC) in his didactic poem *'De rerum natura'*. Lucretius is from the philosophical school of Epicurus. His work was lost in medieval times, and rediscovered by Poggio Bracciolini in 1417 in the library of a monastery in southern Germany, possibly Fulda. Stephen Greenblatt in *The Swerve* (Greenblatt, 2011) argues that this recovery was an important factor amongst the causes of the Renaissance.

In medieval times, Thomas Bradwardine (c. 1290–1349) provided all the points required by geometry by lodging them in God's imagination, and Hasdai Crescas (1340–1410), a Catalonian rabbi, asserted the existence of an infinite vacuum consisting of 'three abstract dimensions, divested of body' in his book *'Or Adonai'*. The appearance of several 'celestial novelties' in the 1570s and 1580s – the nova of Cassiopeia in 1572, and the comets of 1577, 1580, 1582, and 1585 – helped to pave the way for new ideas. Bernardino Telesio (1509–88) in *De rerum natura iuxta propria principia (On the Nature of things according to their own principles)* (1565, 1570, and 1586; see also Telesio, 1587) gives a non-Aristotelian conception of space and time that contributed to the establishment of the modern concept of infinite, homogeneous, and absolute space and time – crucial developments for the affirmation of modern physics, but he does not assert the infinitude of space. Francesco Patrizi (1529–97) in *Nova de universis philosophia* asserted the infinitude of space (Patrizi, 1591). Giordano Bruno (1548–1600), a Dominican monk, is the only sixteenth-century philosopher who adhered to Copernicus' cosmology. Bruno's infinite universe (Bruno and Gosnell, 2014) has the same composition throughout and obeys the same laws throughout the entirety of its infinite spatial and temporal extension. He was burnt at the stake for heresy. A. M. Paterson (1970, p. 198) says that while we no longer have a copy of the official papal condemnation of Bruno, his heresies included 'the doctrine of the infinite universe and the innumerable worlds'. Bruno asserted that each star is like our sun, and this seems to be the origin of his ideas about an infinite universe.

Nicholas Cusanus argued against the existence of a centre of the universe and therefore against absolute motion; he considered that only relative motion was possible. Alexandre Koyre in (1957, p. 18) represents it as 'the astonishing transference to the universe of the pseudo-Hermetic characterization of God: "a sphere of which the center is everywhere, and the circumference nowhere"'. The origin of this metaphor goes back to the pre-Socratics; to Xenophanes, Parmenides, and Empedocles. Much later, the twelfth-century theologian Alan of Lille is said to have discovered the formula 'God is an intelligible sphere, whose center is everywhere and whose circumference nowhere' in a fragment attributed to Hermes Trismegistus, the purported author of the *Hermetic Corpus*, a series of sacred texts that are the basis of Hermeticism (cited in Harries, 2001, p. 30):

Every figure, without deviation from the rules of its construction, can be made by continuous modifications to come nearer and nearer to coincidence with figures of which the rules of construction are different. For example, the larger you make the circumference of a circle, the nearer an arc of it is to a straight line; the arc, therefore, of a circle than which there can be no greater will be actually a straight line. If supposed infinite, then, the curve and the straight line coincide.

... a polygon inscribed in a circle becomes more and more similar to it as it has more angles, though it never becomes equal even by multiplication of angles to infinity, unless it is resolved into identity with the circle ...

Cusanus, one of the first in Western Europe to study Greek after its revival, knew the ancient neo-Platonic thought to some extent directly. The knowledge he chiefly shows (Cusani, 1913 [1440]) is of the positions transmitted by 'Dionysius the Areopagite' and by commentators like Chalcidius, the translator of Plato's *Timaeus*. Cusanus puts forward the doctrine of the coincidence of opposites, and says that it occurs when contemplating the infinite. Kepler says that the point at infinity on a line is on both sides of that line. The themes of perspective and infinity are linked in the thought of Cusanus (Whittaker, 1925). The traditional hierarchical conception of the cosmos depends on the notion of a centre. As this idea is called into question, so is the idea of a hierarchically ordered cosmos. The thesis of cosmic homogeneity that Cusanus presupposes remains very much with us. The challenge that it mounted the hierarchical conception of the cosmos that had ruled medieval thought also became evident evident: no longer is there any reason to divide the cosmos into a sublunar sphere that knows death and decay and a superlunar realm that knows only the perfection of untiring circular motion.

Cusanus calls for a mathematisation of the science of nature in a shift from the heterogeneity of the immediately experienced world to the homogeneity of a world subjected to the measure of number. With Cusanus, this privileging of mathematics does not have its foundation in the nature of things; instead, as he points out in *De Beryllo* (para. 55 of (Hopkins, 1998)), it relates to the nature of human understanding: 'Plato had considered that [claim], assuredly he would have found that our mind, which constructs mathematical entities, has these mathematical entities, which are in its power, more truly present with itself than as they exist outside the mind.' Blumenberg (1985) presents Cusanus as a modern on the basis that the knowledge of one's ignorance is a central element in the modern idea of science.

11.12 Points at Infinity

> Passively receiving illuminations like any instrument, the eye, in Kepler's understanding, is not merely comparable to camera obscura; it is one. The pupil has taken the place of Alberti's perspectival window; the cornea is now nothing but a lens and the retina nothing but a screen. Kepler was the first to declare that genuine vision occurs when the pupil of the eye is exposed most closely to the arriving ray of light. In this understanding of vision the human observer has disappeared from the treatment of optics. The naturalisation of the eye, in which the natural and artificial became one, separated experience from its objects. Turned into an optical instrument the eye no longer furnished the observer with genuine representation of visible objects. It became a mere screen on which an anonymous image is projected. In the case of the eye, the screen was identified as retina and the projected image as picture.

> Dalibor Vesely (2014, p. 67)

Kepler begins his work on conics in *Ad Vitellionem paralipomena quibus astronomiae aprs optica traditur* (see *De coni sectionibus* [On the sections of a cone] in Kepler, 1604, Chapter IV, Section 4, pp. 92–6) the focal property of the parabola, whose proof goes back to Diocles' On burning mirrors (c. 200 BC) (who in turn credits its first statement to Dositheus of Colonus (c. 250 BC), to whom Archimedes' Quadrature of the Parabola, On the Sphere and Cylinder, On conoids and spheroids, and On spirals were addressed). The focal property of the parabola is that at any point of a parabola the tangent makes equal angles with a parallel to the axis and the line from the point to the focus. (The theorem also appears in works of Anthemius of Tralles (AD sixth century), al-Kindi, *Kitab fi al-Shu'a'at* (AD ninth century), ibn Sahl (984), ibn al-Haytham (c. 1000), Roger Bacon (1268), and Witelo (1270), and all but Bacon include a proof.) It is in this work that Kepler first names the foci of the conic sections. This is applied to burning mirrors, as the rays of the sun are nearly parallel when they reach the Earth, and it seems that this gives Kepler the idea that parallel lines meet at infinity. Kepler says that, by the principle of analogy (which we call the principle of continuity), the parabola is intermediate between the hyperbola and the ellipse. It seems as though, starting from a cone with base a circle, he takes a tangent line to that circle parallel to the base, and considers the sections of the cone by the planes on that line and not on the vertex. Thus, he is the first to consider a pencil of conics. By rotating the plane around the tangent line, there is a continuous variation of the conics of the pencil, which would correspond to the variation of the parameter of the pencil of conics in a modern analytic treatment. When the plane is parallel to a generator of the cone, it meets the cone in a parabola, whereas another plane will give a hyperbola or ellipse, depending on which side of our 'parallel plane' it lies. Moreover, if in an animation of the focal property one were to rotate the plane in this way, one of the foci would move off to infinity as the plane tended to the one parallel to the generator. According to Knobloch (2000, p. 300), Kepler also says: 'There is an affinity between all sections in the sense of an analogy: in such a way, one comes from the straight line via the hyperbola, parabola, and ellipse to the circle. The flattest hyperbola is a line; the most acute hyperbola is the parabola. The most acute (infinite) ellipse is the parabola; the flattest ellipse is the circle.'

The focus at infinity of a parabola is on the axis in either direction, and thus the two opposite extremities of every infinite straight line are taken as points that coincide. Kepler was deeply influenced in the principle of analogy by Nicholas Cusanus, one of whose doctrines is the (somewhat paradoxical) coincidence of opposites, a doctrine that had considerable influence on Carl Jung. Thus, the infinite straight line, together with its point at infinity, is a closed figure like a circle; and, here again, Cusanus anticipates this, saying that a circle of infinite radius is a straight line (cited in Rosenfeld, 2005, p. 744):

A circle has one focus A which is its centre. An ellipse has two foci B, C, equidistant from the center of the figure, and the more acute the ellipse the greater this distance. A parabola has one center D inside the figure and the other must be imagined to be lying on the axis of the section inside or outside the parabola, at an infinite distance from the first, so that the lines HG and IG, drawn from that blind focus to an arbitrary point G of the section, are parallel to the axis DK. In the case of a hyperbola, the outer focus F is the closer to the inner focus E the more obtuse the hyperbola, and the outer focus for one of two opposite sections is the inner focus for the other and conversely.

The parabola is partly of the infinite sections and partly of the finite, being the intermediate among them, and its arms do not spread out, like that of the hyperbola, but brought near parallelism, whilst the hyperbola tends more and more to its asymptotes. To perform the continuous change from one conic to another in his system, Kepler imagined keeping one focus fixed and moving the other focus along the common axis to form all conics in the pencil. When the moving focus approaches the fixed focus, the ellipse tends to a circle. When the moving focus goes off to infinity, the ellipse tends to a parabola, then, pushing it further, the moving focus reappears on the other side of the axis, and the parabola becomes a hyperbola, and finally, when the two foci of the hyperbola approach each other the hyperbola degenerates into a pair of (coincident) lines.

In 1604, via experiments, Galileo had discovered that the paths of balls rolling down an inclined plane are parabolas, and then that projectile motion is via a parabola. He used this discovery to produce accurate ballistics tables. Some of these experiments dated back to 1593 and were joint with his patron and mentor Guidobaldo del Monte. Simon Gindikin (2007, p. 38) writes:

Galileo's discoveries must have startled his contemporaries. The conic sections – ellipses, parabolas, and hyperbolas, the acme of Greek geometry – seemed to be the fruit of mathematical fantasy, unrelated to reality. And here Galileo showed that parabolas inevitably arise in a perfectly "worldly" situation. (Even in the 19th century, Laplace presented an application of the conic sections as a most unexpected use of pure mathematics.) It is remarkable that literally at the same time, conic sections arose in a completely different problem and in a no less surprising way. In 1604-1605 Johannes Kepler (1571-1630) discovered that Mars moves along an ellipse, with the sun at one focus (within ten years Kepler extended this to all the planets). This is an important coincidence and for us the two discoveries go hand in hand, but it is likely that before Newton no one seriously put these results together. Moreover, Galileo did not accept Kepler's law and did not communicate his own discovery to Kepler, in spite of their regular correspondence (only published after Kepler's death).

Remembering that Desargues' work was lost for a long time, it is very clear that Kepler had a huge influence on Poncelet, since Poncelet takes up *both* the principle of continuity *and* points at infinity.

11.13 Line at Infinity

Until Guidobaldo, none of the mathematical treatments of linear perspective had concentrated on the restriction of the domain of linear perspective to a plane. Until this is done, the vanishing points will not truly be vanishing, and will only be convergence points. There are three main processes leading to this restriction. The first, appearing in many projective geometry texts, is that of landscape painting, where the visible parts of the Earth's surface are to first approximation a plane, and the horizon line is literally painting the horizon of vision. Here, if two sets of straight railway lines are painted, they will be represented by two pairs of lines meeting on the horizon. The second is pavimento, mentioned in Section 11.2. And the third is *anamorphosis.*

Once perspective painters painted portraits of wealthy individuals who had perspective paintings on their walls, it was inevitable that they would discover that a perspective image of a perspective image need not be a perspective image. This led to anamorphosis – the disguise of objects via this method, such as the skull in the front of Holbein's *The Two Ambassadors* from 1533 (see Figure 11.4). The earliest known use of anamorphosis is in Piero della Francesca's *De prospectiva pingendi*, (della Francesca, [c. 1474] 2021). Renaissance obsession with symbolism led to a fascination with anamorphosis. Whereas the mathematics of central perspective attempted to represent the real on a flat surface in a way that appeared spatial, anamorphosis in contrast allowed representations that thwarted any attempt at a spatial reading.

The Paris monastery of the Minims, with which the famous mathematician Mersenne was associated, became a leading centre of speculation concerning optics and perspective, with a striking emphasis on problems of anamorphic composition. A number of large anamorphic frescoes were painted at the time. Niceron, who like Mersenne was a Minim, painted two such frescoes in the cloister of the monastery of the Minims in Paris: one representing St John the Evangelist, a repetition of a work he had done for the Minims in Rome two years before, the other a St Magdalen, begun in 1645. There was a secret element to this culture, which was only exposed in a 1638 work of Niceron, *La perspective curieuse* (Nicéron, 1638). This book contains nothing less than an attempt to create a complete classification of all perspective-based forms of representation on the basis of the conic section theory alongside other forms such as conical or catoptrical anamorphoses on conical, spherical, and cylindrical surfaces.

---

[8] The painting is in the public domain, available at https://en.wikipedia.org/wiki/File:
Hans_Holbein_the_Younger_-_The_Ambassadors_-_Google_Art_Project.jpg.

Figure 11.4 Hans Holbein the Younger (c. 1497–1543), *The Ambassadors* (1533), National Gallery (London).[8] If the observer places their head at the bottom-left or top-right corner of the painting, then the obscured skull on the floor appears accurate.

- Anamorphoses reveal an unsuspected deeper meaning in a seemingly superficial appearance.
- Anamorphosis leads to the idea of replacing an object by its picture.

As Kvasz (2008, pp. 120–1) put it, in his work from 1636 to 1648 Desargues replaces the object by its image:

Gérard Desargues, the founder of projective geometry came up with an excellent idea. He replaced the object with its picture. So while the painters formulated the problem of perspective as a relation between the picture and reality, Desargues formulated it as a problem of the relation between two pictures. The center of projection represents the point of view from which the two pictures make the same impression. Desargues found a way to give to the term infinity a clear, unambiguous and verifiable meaning. Desargues' replacement of reality by its picture makes it possible to study transformations of the plane on which the objects are placed independently of the objects themselves. It is important to realize one basic difference between the horizon in a perspectivist painting and in a picture of projective geometry. In projective geometry the horizon is a straight line, which means it belongs to the language. So instead of the Euclidean looking from nowhere onto a homogeneous world, or the perspectivist watching from outside, for Desargues the point of view is explicitly incorporated into language.

Desargues makes at least two breakthroughs. The first is the invention of the line at infinity, vital to projective geometry. The second is *invariance*. As Massey (2016, p. 111) explains, 'Desargues radically reconceived geometric space. ... Focusing on those elements of perspective that maintained the property of constancy under any number of modifying operations, he searched not for what was "changed by perspective" but for those mathematical relationships that remained the same, for example, the potential "projective properties" of extended space.'

Desargues discovers the idea of studying the properties of figures left invariant under central projection. This can be seen as taking up Alberti's question, in the light of the restriction of the domain of linear perspective to a plane. Like Kepler, Desargues seeks generality in his theories. And he achieves it, bringing together many of the results of Apollonius' Conics under one umbrella.

11.14 Geometry as the Study of Space

Geometry, in ancient times, was the study of geometric figures: triangles, circles, parallelograms, and the like, but by the eighteenth century it had become the study of space.

Judith Grabiner (2009, p. 6)

One useful way to frame the divide between ancient and modern geometry is to use the emergence of the consideration of space itself as an object of geometrical investigation. Greek mathematics understood geometry as a study of figures conceived against an amorphous background space whose definition was outside the limits of the theory. This was superseded by a conception of space itself endowed with geometrical properties, with the object of characterising the structures and features of geometrical space via axioms.

According to Hans Blumenberg (1989), there are two inventions whereby the modern age cleanly breaks with the past: Florentine linear perspective and absolute space. They are the same thing in many ways since one is unthinkable without the other and vice versa.

That mathematicians in the Renaissance found their means to live not in the universities but in the patronage of wealthy clients had the inevitable consequence that applied mathematics was more important in the Renaissance than it was in Medieval times, and so we see mathematics developed in stone-cutting, in ballistics, in fortification, in the study of sundials, in painting. (In particular, the development of descriptive geometry was influenced by stonemasons, architecture, and stone-cutting.) For geometry, this emphasised three dimensions over two. This was very important for the later scientific revolution, but it's worth pointing out that even plane projective geometry has its natural setting in three dimensions, as in the study of the invariants of central projection. Thus, putting more emphasis on solid geometry can be seen as a necessary precondition to evolving a projective perspective. Plato's laments that '[not enough is known about solid geometry] and for two reasons: in the first place, no government places value on it; this leads to a lack of energy in the pursuit of it, and it is difficult. In the second place, students cannot learn it unless they have a teacher. But then a teacher can hardly be found' (put in Socrates' mouth) and that '[the] ludicrous state of solid geometry made me pass over this branch' turn out to be prescient.

Before the seventeenth century, geometry was the study of figures. After it, it was the study of space. Kepler's use of actual infinity in the study of the focus of the parabola was important in this expansion of vision. But so was the work of the perspective painters.

Newton needed absolute space as a reference frame in order to argue that there is a difference between real and apparent accelerations. The truth of Newton's first and second laws implies the truth of the parallel postulate for absolute space, as pointed out in unpublished work of Lagrange from 1806 (see Grabiner, 2009).

Monge was the greatest teacher of his era. Seeing geometry as the study of space and not of figures had certain consequences one being that space was infinite and another that transformations were allowed into geometry (contrary to Aristotle's view that there is no motion in geometry). Orthogonal projections and central projections took centre stage, and, in the hands of Monge's illustrious students, projective geometry flowered. One striking aspect of this revival of synthetic geometry is the attempt of its most prominent representatives – Gaspard Monge, Lazare Carnot, Michel Chasles, Jean-Victor Poncelet, Charles Dupin, and Charles Brianchon – to broaden the scope of traditional geometric methods through the introduction of explicitly physicalist techniques and concepts such as transformation, projection, and continuity (Daston, 1986).

11.15 Epilogue

In the Renaissance and the period of modern science, technology drove discovery: the mirror, the telescope, the camera obscura, the mechanical clock, the microscope, and the cannon, amongst others (even the balance scale, the yardstick, and the hourglass). Later, in a slow process, geometrical justifications were constructed for advances first driven by technology. This had the effect of applying physical considerations to geometry. The early modern advances in physics turned geometry into the study of space. These physicalist considerations led to the introduction of both symmetry and more general transformations into geometry, in the hands of Legendre and Monge at the turn of the nineteenth century. Desargues had prefigured this somewhat nearly two centuries earlier. In order to make this concord between the world and geometry, Aristotle's about the finiteness of space had to be set aside.

The use of mirrors and pinhole cameras drove the invention of linear perspective. But it had to be preceded by a desire to represent the world naturally, from one point of view.

The introduction of the use of conics into mechanics by Galileo and Kepler and its flowering in the hands of Newton thus play a role in this evolution of geometry, as does Kepler's transformation of optics from an ancient theory of sight into a theory of light.

The move from two-dimensional representations of a three-dimensional world in art to projections between planes was driven by at least three sources: landscape painting, the use of pavimento techniques, and anamorphosis. It led to the first appearance of the concept of invariance in geometry, which had been somewhat prefigured by a question of Alberti about the relationship between different perspective representations.

With invariance, symmetry, and transformations in hand, during the nineteenth century the evolution of the concept of transformation geometry could take place, with Möbius prefiguring Klein in this.

Thus, a complex process, building on ancient and medieval insights about geometry, optics, and pinhole cameras, evolved in such a way as to transform art, physics, and philosophy, and at the same time produced the seeds of both projective geometry and transformation geometry – and even of group theory.

Moreover, this is all part of the processes of the birth of modern science and of the mathematisation of the world. The adoption of the observer and adoption of measurement are important parts of all of these developments. Measurement in book-keeping was also important (and transformed economics). Cartography also mattered.

To a modern eye, there is a discord between using geometry to do science and applying scientific observations to modify geometry. Since geometry had been viewed as proceeding from axioms for nearly two millennia when the action considered in this section begins, revisiting the axioms via experiment at the same time as using consequences of the axioms to determine results looks like a conflict. But in the period in question, both were seen as being about truth. And since geometry gives certain truth and experiment reveals truth, they can't be in conflict. The process, seen to be about clarifying truth, is possibly best viewed as a complex dialogue between the disciplines of geometry and physics, rather than the simple influence of one upon the other.

12

Remarks on the History of Projective Geometry

In the decades leading up to the period of relativity theory the architecture of space was revolutionized. Until then the mathematical imagination, and with it all of scientific thinking, had been dominated by a single book. No text, other than the Bible, had done more to shape the thinking of the West than Euclid's *Elements*. Since the invention of printing alone, more than one thousand editions have appeared. Yet the mathematical framework the Elements espoused grants an unfounded privilege to one view, excluding the very idea of non-Euclidean geometries. The roots of a more flexible attitude to geometry reach back to the Renaissance creators of linear perspective, but the development of their first insights into the modern discipline of 'projective geometry' had to await the work of great mathematicians such as Poncelet (1788–1867), Cayley (1821–95) and Klein (1849–1925). By the time of Einstein, non-Euclidean geometries and the even more comprehensive theory of projective geometry had broken the grip of Euclid on mathematical and spatial thinking, and a new imagination of space could be born.

Arthur Zajonc (1995, pp. 272–3).[1]

Unlike many areas of mathematics, projective geometry has a clear birth date: the publication of the book *Traité des propriétés projectives des figures* by Jean-Victor Poncelet (1788–1867) in 1822 (Poncelet, 1822). The preface details the book's impossibly romantic origins in a prisoner of war camp in Saratov between March 1813 and June 1814 after Poncelet's capture in the Battle of Krasnoi on 19 November during Napoleon's retreat from Moscow in the 1812 campaign. (Tolstoy's *War and Peace* and Tchaikovsky's *1812 Overture* are also about this campaign.) His notebook, written in prison, was published in two volumes in 1862 and 1864, as part of the book *Applications d'analyse et*

---

[1] Zajonc, Arthur. 1995. *Catching the Light: The Entwined History of Light and Mind*. Oxford University Press paperback. Oxford University Press. Reprinted with permission.

de géométrie (Poncelet, 1862–4). The 'Traité' marked the emergence of projective geometry as a theory: a way of doing Euclidean geometry. Later, after
the emergence of non-Euclidean geometry in the late 1820s, and its eventual
acceptance by the 1870s, projective geometry was viewed as a geometry in its
own right.

Before projective geometry emerged as a theory in its own right, it had a
pre-history in results that in hindsight belong to the subject, such as major
results from Apollonius' *Conics* and invariance of cross-ratio in Pappus' *Collection*, and then a major breakthrough in the seventeenth century with the
point at infinity in Kepler's work, and the line at infinity and important theorems in Desargues' work. And after emerging as a theory, it came to full
flower after the acceptance of non-Euclidean geometry in the 1870s (for which
see Voelke, 2005 and Volkert, 2013) when geometry acquired a plural. The
nineteenth century was the golden age of projective geometry.

It is often claimed that there were no results in projective geometry
between the work of Desargues, Pascal, and de la Hire, and 1800 – but
this is false. Isaac Newton (1643–1727) used projective methods both
in his *Principia* (1687) (where he shows results about conic sections in
Book I and introduces general projective transformations, see Newton,
1999) and in his work on cubics published as an appendix to his *Opticks*
in 1704. Luigi Ruggero Ventimiglia published a book, *Enodationes duo-
decimo problematic a Geometra post tabular latente propositorum*, in
1690 and another, *Geometram Quaero*, in 1692 which showed that he
had been following a programme of projective study of analytic geometry interrupted by his premature death at the age of 28 in 1698. Saccheri
published a book, *Quaesita Geometrica*, in 1693 in which he rediscovered
many important theorems of Desargues, de la Hire, and others. Vincenzo
Viviani published a book on solid loci in 1701, which also had some projective character in parts. Colin Maclaurin (1698–1746) in his *Geometria
Organica* in 1720 (Maclaurin, 1720) followed up on Newton's work on
cubics, as did James Stirling (1717) in 1717, Jean Paul de Gua de Malves
in 1740, and Patrick Murdoch in 1749, and Maclaurin again (posthumously) in 1748 (*De Linearum Geometricarum Proprietatius Generalibus
Tractatus*; Maclaurin, 1748). In 1723, Simson decoded a passage in Pappus' *Collection* that implies the dual of Desargues' theorem. In the 1730s,
Maclaurin and Braikenridge had a dispute about priority for proving the
converse of Pascal's theorem. In 1735, Simson published the first textbook

on geometry containing the statements of Desargues' and Pascal's theorems. In 1764 (in a three-part paper, the third part published in 1767) and in 1779 (in a book), Étienne Bézout published his theorem that we now interpret as being about intersections of plane algebraic curves; he had been anticipated by Newton (*Principia*, Book I, proof of Lemma XXVIII) and Maclaurin (1720) and the result was simultaneously discovered by Euler (1766).

Part II

Real Projective 3-Space

Projective geometry has opened up for us with the greatest facility new territories in our science, and has rightly been called the royal road to its own particular field of knowledge.

Felix Klein, as paraphrased in Bell (1986, p. 206)

13

Fundamental Aspects of Real Projective Space

Historically, points at infinity, lines at infinity, and the plane at infinity were first thought of by Kepler, Desargues, and Poncelet, respectively.

H. M. S. Coxeter (1974, p. 109)

Extended Euclidean space is the incidence structure with points the points of $\mathbb{R}^3$, together with the **points at infinity**, each of which is a parallel class of lines of $\mathbb{R}^3$, and lines the lines of $\mathbb{R}^3$, together with the **lines at infinity**, each of which is a parallel class of planes of $\mathbb{R}^3$ and incidence inherited from $\mathbb{R}^3$ (see Table 13.1).

A point at infinity is incident with every line of the corresponding parallel class and is incident with a line at infinity if and only if that parallel class of lines consists of lines in planes of the parallel class of planes. To this, we could add the planes: the planes of $\mathbb{R}^3$ together with the plane at infinity and all points at infinity (and no other point) and all lines at infinity (and no other lines) are incident with the plane at infinity. However, the planes can be reconstructed from the points and the lines as subspaces, so we don't *need* to add them.

Theorem 13.1
The incidence structure of points at infinity and lines at infinity of extended Euclidean space is isomorphic to $\mathsf{PG}(2, \mathbb{R})$.

Proof The map that takes each parallel class of lines of $\mathbb{R}^3$ to its member on the origin and each parallel class of planes of $\mathbb{R}^3$ to its member on the origin is an isomorphism. $\qquad \square$

Table 13.1 *Extra elements of extended Euclidean space.*

Points	parallel classes of lines of $\mathbb{R}^3$
Lines	parallel classes of planes of $\mathbb{R}^3$
Planes	one plane: the set of parallel classes of lines and planes

This is quite a revelation. Initially, the real projective plane seemed quite exotic, and now we've discovered that it is isomorphic to the parallel classes of lines of $\mathbb{R}^3$ and the parallel classes of planes of $\mathbb{R}^3$, which is much more familiar. Moreover, duality in $PG(2, \mathbb{R})$ is natural in this setting; it is induced by 'perpendicularity'. See Chapter 1 where this was explored at length.

Real projective space $PG(3, \mathbb{R})$ is the incidence structure with points the one-dimensional subspaces of $\mathbb{R}^4$ and lines the two-dimensional subspaces of $\mathbb{R}^4$ and incidence is the inclusion relation. As for the plane case:

Theorem 13.2
Extended Euclidean space and $PG(3, \mathbb{R})$ *are isomorphic.*

Proof Exercise 13.1.   □

The set of points of $PG(3, \mathbb{R})$ contained in a three-dimensional subspace of $\mathbb{R}^4$ is called a **plane** of $PG(3, \mathbb{R})$.

Theorem 13.3
In $PG(3, \mathbb{R})$,
 (i) any two points are incident with a unique line;
 (ii) any two planes are incident with a unique line;
(iii) a plane and a non-incident line meet in a point.

Proof For (i), we observe that any two 1-dimensional subspaces of $\mathbb{R}^4$ span a unique 2-dimensional subspace. Likewise, for (ii), any two 3-dimensional subspaces of $\mathbb{R}^3$ meet in a unique 2-dimensional subspace. Finally, (iii) follows for noting that any three-dimensional subspace of $\mathbb{R}^3$ and a two-dimensional subspace it doesn't contain meet in a one-dimensional subspace.   □

The projective space $PG(3, \mathbb{R})$ can also be coordinatised in a similar way to $PG(2, \mathbb{R})$. This time, points and planes are readily in correspondence via this coordinatisation (i.e., by swapping round and square parentheses).

★ We denote a point by homogeneous coordinates (x, y, z, w), $x, y, z, w \in \mathbb{R}$, not all zero.
★ We denote a plane by homogeneous coordinates $[a, b, c, d]$, $a, b, c, d \in \mathbb{R}$, not all zero. This is the plane whose points (x, y, z, w) satisfy $ax+by+cz+dw = 0$.

A **collineation** of $PG(3, \mathbb{R})$ is a bijection of the points taking lines to lines. The set of all collineations of $PG(3, \mathbb{R})$ is a group under composition, which we denote by $\text{Aut}(PG(3, \mathbb{R}))$.

Theorem 13.4

(i) The join of two fixed points is a fixed line.

(ii) The intersection of a fixed plane and a fixed line is a fixed point.

(iii) The intersection of two fixed planes is a fixed line.

Proof Exercise 13.2. □

Worked Example 13.5

We will show that the point-wise stabiliser of a plane and a non-incident line is trivial. Suppose a collineation g of $PG(3, \mathbb{R})$ fixes all of the points of a plane π, and all of the points of a line ℓ, with ℓ not incident with π. Let X be a point of $PG(3, \mathbb{R})$ not incident with ℓ or π. If A and B are two points of ℓ, not incident with π, then we can project A and B to π via X: define $A' := AX \cap \pi$ and $B' := BX \cap \pi$. Then X is precisely the point of intersection of AA' and BB'. Since all points A, A', B, B' are fixed, by Theorem 13.4, g fixes AA' and BB'. Therefore,

$$X^g = (AA' \cap BB')^g = (AA')^g \cap (BB')^g = AA' \cap BB' = X$$

and so every point of $PG(3, \mathbb{R})$ is fixed by g. Therefore, g is the identity. □

Given a 4×4 invertible matrix A, the map $(x, y, z, w) \mapsto (x, y, z, w)A$ is a collineation. Two such matrices A, B define the same collineation if and only if $A = cB$, for $c \in \mathbb{R}$ with $c \neq 0$, because the coordinates are homogeneous. The projective general linear group $PGL(4, \mathbb{R})$ is the group induced by the action of the group $GL(4, \mathbb{R})$ of all 4×4 invertible matrices on the points of $PG(3, \mathbb{R})$. We show now that all planes of real projective space have the same status.

Theorem 13.6

$PGL(4, \mathbb{R})$ *acts transitively on the planes of* $PG(3, \mathbb{R})$.

Proof Given two planes π, π', take a basis $\{v_1, v_2, v_3\}$ for π and extend it to a basis $\{v_1, v_2, v_3, v_4\}$ for $\mathbb{R}^4$, and also take a basis $\{w_1, w_2, w_3\}$ for m and extend it to a basis $\{w_1, w_2, w_3, w_4\}$ for $\mathbb{R}^4$. There is an invertible matrix A with $v_i A = w_i$, for $i = 1, 2, 3, 4$, and this induces a collineation of $PG(3, \mathbb{R})$ taking π to π'. □

Theorem 13.7

Let π *be a plane of* $PG(3, \mathbb{R})$. *Then the incidence structure of points in* π *and lines contained in* π, *with the restricted incidence, is isomorphic to* $PG(2, \mathbb{R})$.

Proof This is an immediate consequence of Theorems 13.2 and 13.6. □

Example 13.8

Let $c \in \mathbb{R}^*$. *Then* $g : (x, y, z, w) \mapsto (x + cw, y, z, w)$ *defines a collineation of* $PG(3, \mathbb{R})$. *Indeed, we could write it in matrix form:*

$$(x, y, z, w) \begin{bmatrix} 1 & 0 & 0 & 0 \\ 0 & 1 & 0 & 0 \\ 0 & 0 & 1 & 0 \\ c & 0 & 0 & 1 \end{bmatrix}.$$

Now g fixes the point C: $(1, 0, 0, 0)$, *and also fixes every plane incident with* C. *To see this, note that every plane incident with* C *is of the form* $[0, s, t, u]$ *for some* $s, t, u \in \mathbb{R}$ *(not all zero). Then*

$$\begin{bmatrix} 1 & 0 & 0 & 0 \\ 0 & 1 & 0 & 0 \\ 0 & 0 & 1 & 0 \\ c & 0 & 0 & 1 \end{bmatrix} \begin{bmatrix} 0 \\ s \\ t \\ u \end{bmatrix} = \begin{bmatrix} 0 \\ s \\ t \\ u \end{bmatrix}.$$

Moreover, every point of the plane A: $[0, 0, 0, 1]$ *is fixed. Such a collineation is a* **central collineation**, *and more particularly an* **elation**.

An **axis** of a collineation is a plane fixed pointwise. Dually, a **centre** of a collineation is a point such that every plane on it is fixed.

Theorem 13.9

(i) *A collineation has a centre if and only if it has an axis.*

(ii) *A non-identity collineation has at most one centre and at most one axis.*

Proof Analogous to the proof of Theorem 2.12.   □

A collineation is **central** if it has a centre. A **homology** is a central collineation with centre not on its axis, whereas an **elation** has its centre on its axis.

Example 13.10

Let $c \in \mathbb{R}^*$ *with* $c \neq 1$. *Then* $(x, y, z, w) \mapsto (cx, y, z, w)$ *is a homology with centre* $(1, 0, 0, 0)$ *and axis* $[1, 0, 0, 0]$.

The central collineations with centre $(a_1, a_2, \ldots, a_n)$ and axis $[b_1, b_2, \ldots, b_n]$ have (up to a non-zero scalar multiple) matrices of the form

$$I + t(b_1, b_2, \ldots, b_n)^\top (a_1, a_2, \ldots, a_n)$$

for some non-zero scalar t. To see why, note that such a map fixes $[b_1, b_2, \ldots, b_n]$ pointwise and adds a multiple of $(a_1, a_2, \ldots, a_n)$ to every vector, so has centre $(a_1, a_2, \ldots, a_n)$.

We can recover real affine space from real projective space by deleting a plane and all points and lines incident with it, reversing the procedure we first used to construct real projective space. The group $\mathsf{AGL}(3, \mathbb{R})$ is the set of all affine transformations of real affine 3-space.

Theorem 13.11

Let π be a plane of $\mathsf{PG}(3, \mathbb{R})$. Then the stabiliser of π in $\mathsf{PGL}(4, \mathbb{R})$ acting on the points not on π is permutationally isomorphic to $\mathsf{AGL}(3, \mathbb{R})$, acting on the points of $\mathsf{AG}(3, \mathbb{R})$.

The reader will be able to prove Theorem 13.11 once seeing the connection to $\mathsf{AG}(3, \mathbb{R})$ in Worked Example 13.12. We can obtain $\mathsf{AG}(3, \mathbb{R})$ from $\mathsf{PG}(3, \mathbb{R})$ by 'removing' a plane π from $\mathsf{PG}(3, \mathbb{R})$.

Worked Example 13.12

Let π be the plane $[0, 0, 0, 1]$. Then each point of $\mathsf{PG}(3, \mathbb{R})$ not incident with π has homogeneous coordinates $(x, y, z, 1)$. The 'affine points' incident with a line of $\mathsf{PG}(3, \mathbb{R})$, that is, incident with just a point of π, are of the form

$$(x, y, z, 1) + \lambda(x', y', z', 0) = (x + \lambda x', y + \lambda y', z + \lambda z', 1), \quad \lambda \in \mathbb{R}.$$

The planes σ not equal to π are of the form $[a, b, c, d]$, where $(a, b, c) \neq (0, 0, 0)$. By simply removing the last coordinate, we obtain a copy of $\mathsf{AG}(3, \mathbb{R})$:

point	$(x, y, z, 1)$	$\longrightarrow$	(x, y, z)
line	$(x + \lambda x', y + \lambda y', z + \lambda z', 1), \lambda \in \mathbb{R}$	$\longrightarrow$	$(x + \lambda x', y + \lambda y', z + \lambda z'),$ $\lambda \in \mathbb{R}$
plane	$[a, b, c, d], (a, b, c) \neq (0, 0, 0)$	$\longrightarrow$	$aX_1 + bX_2 + cX_3 + d = 0.$

$\square$

$$* * *$$

A **frame** of $\mathsf{PG}(3, \mathbb{R})$ is a set of five points, no four coplanar. The **fundamental frame** is $\{(1, 0, 0, 0), (0, 1, 0, 0), (0, 0, 1, 0), (0, 0, 0, 1), (1, 1, 1, 1)\}$.

Theorem 13.13

$\mathsf{PGL}(4, \mathbb{R})$ *acts regularly on ordered frames of* $\mathsf{PG}(3, \mathbb{R})$.

Proof Analogous to the proof of Theorem 2.5. $\square$

Worked Example 13.14

We will show that $\mathsf{PGL}(4, \mathbb{R})$ acts transitively on pairs of skew lines of $\mathsf{PG}(3, \mathbb{R})$. Let ℓ and m be skew lines of $\mathsf{PGL}(4, \mathbb{R})$, and suppose u and v are vectors of $\mathbb{R}^4$ giving homogeneous coordinates for two points incident with ℓ; likewise, let x and y be representatives for two points incident with m. Then the matrix M with rows u, v, x, y has full rank and maps the points $(1, 0, 0, 0)$,

$(0, 1, 0, 0)$, $(0, 0, 1, 0)$, $(0, 0, 0, 1)$ to u, v, x, y, accordingly. So the collineation g induced by M maps the pair of lines $X_3 = X_4 = 0$ and $X_1 = X_2 = 0$ to ℓ and m.   □

Theorem 13.15

$PGL(4, \mathbb{R})$ *is transitive on triples of pairwise skew lines.*

Proof First, $PGL(4, \mathbb{R})$ acts transitively on pairs of skew lines of $PG(3, \mathbb{R})$ by Worked Example 13.14. Consider three skew lines ℓ, m, n of $PG(3, \mathbb{R})$. We can assume that ℓ and m have a nice representation. That is, we can assume that ℓ is the span of the points with coordinates $(1, 0, 0, 0)$ and $(0, 1, 0, 0)$, and we can assume that m is the span of the points with coordinates $(0, 0, 1, 0)$ and $(0, 0, 0, 1)$. Now write n as the row-space of a 2×4 matrix, but in block form in the following way: $\begin{bmatrix} S & T \end{bmatrix}$ where S and T are 2×2 matrices. Since n is skew to both ℓ and m, it follows that S and T both have full rank. (Note that ℓ is the row-space of $\begin{bmatrix} I & O \end{bmatrix}$ and m is the row-space of $\begin{bmatrix} O & I \end{bmatrix}$, where O and I denote the 2×2 zero and identity matrices, respectively.) So we can define a collineation g of $PGL(4, \mathbb{R})$ induced by the matrix

$$\begin{bmatrix} S^{-1} & O \\ O & T^{-1} \end{bmatrix}.$$

Notice that g fixes ℓ and m. Moreover, g maps n to the line that is the row-space of $\begin{bmatrix} I & I \end{bmatrix}$. So we have shown that any three skew lines of $PG(3, \mathbb{R})$ can be mapped via a collineation to ℓ, m, and the line spanning $(1, 0, 1, 0)$ and $(0, 1, 0, 1)$. We have also shown that $PGL(4, \mathbb{R})$ acts transitively on triples of skew lines of $PG(3, \mathbb{R})$.   □

Theorem 13.16 (Fundamental theorem of projective geometry)

$Aut(PG(3, \mathbb{R})) = PGL(4, \mathbb{R})$.

Proof We will 'bootstrap' the argument for the planar case and use Theorem 2.8. Let $G := Aut(PG(3, \mathbb{R}))$ and let $H := PGL(4, \mathbb{R})$. Fix a plane π of $PG(3, \mathbb{R})$ and a point P not incident with π. Then the stabiliser $H_{\pi, P}$ of π and P is isomorphic to $GL(3, \mathbb{R})$. To see why, we can choose coordinates so that π: $[0, 0, 0, 1]$ and P: $(0, 0, 0, 1)$. Then $H_{\pi, P}$ consists of matrices of the form

$$\begin{bmatrix} a & b & c & 0 \\ d & e & f & 0 \\ g & h & i & 0 \\ 0 & 0 & 0 & 1 \end{bmatrix}.$$

Moreover, the permutation group $H_{\pi, P}^{\pi}$ induced on the points of π is isomorphic to $PGL(3, \mathbb{R})$. Now $G_{\pi, P}$ acts on the points and lines of π and induces collineations of $\mathsf{PG}(2, \mathbb{R})$ through its identification with π. By Theorem 2.9, the permutation group $G^\pi_{\pi,P}$ induced by $G_{\pi,P}$ on the points of π is contained in $\mathsf{PGL}(3, \mathbb{R})$. Since $H^\pi_{\pi,P} \leqslant G^\pi_{\pi,P}$, it follows that $H^\pi_{\pi,P} = G^\pi_{\pi,P}$. The next step is to show that $H_{\pi,P} = G_{\pi,P}$, and, therefore, $G = H$ (by Theorem 2.8). (We leave it to the reader to verify that H acts transitively on 'antiflags' – pairs of non-incident points and planes.) So it suffices to show that the kernels of the respective actions are equal. The kernel of the action of $H_{\pi,P}$ on the points of π is precisely the scalar matrices of $\mathsf{GL}(3, \mathbb{R})$ – in the argument above, the matrices of the kernel of the action have the form

$$
\tau_\mu := \begin{bmatrix} \mu & 0 & 0 & 0 \\ 0 & \mu & 0 & 0 \\ 0 & 0 & \mu & 0 \\ 0 & 0 & 0 & 1 \end{bmatrix}
$$

where μ is non-zero. Let g be an element of the kernel of the action of $G_{\pi,P}$ on the points of π. Then g fixes Q: $(1, 0, 0, 0)$ and the line PQ. So there exists non-zero λ such that $(1, 0, 0, 1)^g = (\lambda, 0, 0, 1)$. Then $g\tau_\lambda^{-1}$ fixes three collinear points, namely P, Q, and $(1, 0, 0, 1)$, and hence $g\tau_\lambda^{-1}$ fixes every point of PQ. By Worked Example 13.5, $g\tau_\lambda^{-1}$ is the identity collineation, and hence $g = \tau_\lambda$. Therefore, the kernels of the actions of $H_{\pi,P}$ and $G_{\pi,P}$ on the points of π are equal. □

A **plane-pencil** of lines, also called a **flat-pencil** or simply a pencil of lines (depending on the context), of $\mathsf{PG}(3, \mathbb{R})$ is a set of lines in a plane π incident with a point P.

Theorem 13.17
$\mathsf{PGL}(4, \mathbb{R})$ *acts transitively on the plane-pencils of* $\mathsf{PG}(3, \mathbb{R})$.

Proof We already have, by Theorem 13.6, that $\mathsf{PGL}(4, \mathbb{R})$ acts transitively on the planes of $\mathsf{PG}(3, \mathbb{R})$, so fix a plane π. The stabiliser of π in $\mathsf{PGL}(4, \mathbb{R})$ induces $\mathsf{PGL}(3, \mathbb{R})$ on the points and lines of π, and so the result follows by noting that $\mathsf{PGL}(3, \mathbb{R})$ acts transitively on the points of π. □

The **dual** of a statement about 3-space is the statement that results after interchanging *point* and *plane*, *intersection* and *join*, and making the necessary linguistic adjustments. In 3-space, lines are self-dual.

The Principle of Duality[1] for $\mathsf{PG}(3, \mathbb{R})$:
The dual of every theorem about $\mathsf{PG}(3, \mathbb{R})$ is also a theorem.

---

[1] Again, this is due to Poncelet (1822) and Gergonne (1825/6).

We prove the principle of duality by noting that the map taking a point (a, b, c, d) to a plane $[a, b, c, d]$ preserves incidence. As a simple application, notice that the principle of duality and Theorem 13.6 imply that $\mathsf{PGL}(4, \mathbb{R})$ acts transitively on the points of $\mathsf{PG}(3, \mathbb{R})$.

An incidence-preserving map from the points to the planes of $\mathsf{PG}(3, \mathbb{R})$ is called a **duality**. A duality of order two is called a **polarity**. The map inter-changing point and plane coordinates is a polarity, called the **standard duality**. Given a polarity, a point, line, or plane is **absolute** if it is incident with its image under the polarity. In $\mathsf{PG}(2, \mathbb{R})$, there are two polarities up to equivalence (Theorem 3.2); in 3-space, we have four.

Theorem 13.18

There are four conjugacy classes of polarities of $\mathsf{PG}(3, \mathbb{R})$, *with representatives as Table 13.2 shows:*

Table 13.2 *Representative polarities of* $\mathsf{PG}(3, \mathbb{R})$.

Description	Map	# absolute points	# absolute lines
standard	$(a, b, c, d) \mapsto [a, b, c, d]$	0	0
orthogonal	$(a, b, c, d) \mapsto [a, b, c, -d]$	∞	0
orthogonal	$(a, b, c, d) \mapsto [a, b, -c, -d]$	∞	∞
null	$(a, b, c, d) \mapsto [-d, -c, b, a]$	all	∞

Proof Suppose that Δ is a polarity of $\mathsf{PG}(3, \mathbb{R})$. Then the product $g = \delta\Delta$ of Δ with the standard duality δ is a collineation, so, by Theorem 13.16, $g \in \mathsf{PGL}(4, \mathbb{R})$: there exists $A \in \mathsf{GL}(4, \mathbb{R})$ with g mapping the point (x, y, z, w) to the point $(x, y, z, w)A$. Now g has the action $[a, b, c, d] \mapsto [a, b, c, d]A^{-\top}$ on the planes, and so $\Delta = \delta g$ maps the point (e, f, g, h) to the plane $[e, f, g, h]A$ and the plane $[a, b, c, d]$ to the point $(a, b, c, d)A^{-\top}$. Since Δ has order two, it follows that $AA^{-\top}$ is a scalar multiple of the identity. By replacing A by a scalar multiple (which doesn't change g) we may assume that $AA^{-\top} = \pm I$, that is, $A = \pm A^{\top}$. Therefore, A is symmetric or skew-symmetric. If A is skew-symmetric, then Δ is conjugate to $(a, b, c, d) \mapsto [-d, -c, b, a]$. Now a symmetric real matrix can be orthogonally diagonalised: there is a basis of eigenvectors of A. So, up to conjugacy in the correlation group, we may assume that A is diagonal. Moreover, by scaling each eigenvector we may assume that the diagonal entries are ± 1 or 0; since A is invertible, 0 does not occur. Now by permuting the

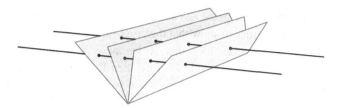

Figure 13.1 The cross-ratio of four planes.

columns and replacing A by $-A$ if necessary, we may assume that $A = I$ or that $A = \mathrm{diag}(1, 1, 1, -1)$ or that $A = \mathrm{diag}(1, 1, -1, -1)$. □

The **cross-ratio** of four collinear points of $\mathrm{PG}(3, \mathbb{R})$ is defined in a similar way as in the plane. Let ℓ be a line of $\mathrm{PG}(3, \mathbb{R})$, and let $O(x, y, z, w)$ and $I(x', y', z', w')$ be the **reference points** of ℓ. The **cross-ratio** $\mathrm{R}(P, Q; R, S)$ of four points P, Q, R, S of ℓ, where

$$P(x + px', y + py', z + pz', w + pw'), \quad Q(x + qx', y + qy', z + qz', w + qw'),$$
$$R(x + rx', y + ry', z + rz', w + rw'), \quad S(x + sx', y + sy', z + sz', w + sw')$$

is defined to be

$$\frac{(r - p)(s - q)}{(r - q)(s - p)}$$

with suitable adjustments made for when some of these four points coincide with the reference points O and I. Again, just as in the plane case, cross-ratio is preserved by **projectivities** (Exercise 13.6), where the **perspectivities** here are defined as the mapping between points of two planes π and π', via a **centre** point $O: X \mapsto XO \cap \pi'$.

There are two ways to define a cross-ratio for four collinear planes. The first is to apply a duality, and compute the cross-ratio of points to the images. The second (equivalent) way is to consider a line intersecting all four planes not on their common line, take the four points of intersection, and compute the cross-ratio of the four points (see Figure 13.1). For both methods, we need to check that the definition is well-defined. We will adopt the second method. So suppose we have four planes $\pi_1, \pi_2, \pi_3, \pi_4$ all incident with a line ℓ. Suppose we have two distinct lines m and m' that intersect all four planes, but not on ℓ. Let $M_i := \pi_i \cap m$ and $M_i' := \pi_i \cap m'$, for each $i \in \{1, 2, 3, 4\}$. The M_i are four collinear points, and likewise for the M_i'. If the two lines m and m' span a plane σ, then Corollary 4.10 applied to the plane σ (thought of as a copy of $\mathrm{PG}(2, \mathbb{R})$) implies that $\mathrm{R}(M_1, M_2; M_3, M_4) = \mathrm{R}(M_1', M_2'; M_3', M_4')$. If, however, m and m' are skew, then we take a transversal line n, and apply the above argument twice: once to the plane mn, and then again to $m'n$. So essentially we are using two projectivities to show that the cross-ratios are equal.

We will see in Exercise 13.7 that if we have four collinear planes $\pi_1, \pi_2, \pi_3, \pi_4$ and a fifth transversal plane π_5, then $R(\pi_1, \pi_2; \pi_3, \pi_4)$ coincides with the cross-ratio of line intersections when we think of π_5 as a copy of $PG(2, \mathbb{R})$:

$$R(\pi_1, \pi_2; \pi_3, \pi_4) = R(\pi_5 \cap \pi_1, \pi_5 \cap \pi_2; \pi_5 \cap \pi_3, \pi_5 \cap \pi_4).$$

By definition of the cross-ratio of planes, we have the following observation:

Theorem 13.19
Let A, B, C, D be collinear points, and let ℓ be a line not incident with these points. Then $R(\ell A, \ell B; \ell C, \ell D) = R(A, B; C, D)$.

13.1 The Complex Projective Plane

We touched on the complex projective plane briefly in Section 6.2, and we explore it more here, since we will need the notion of the complex extension of the plane at infinity in 3-space. The incidence structure with points the set of all one-dimensional subspaces of $\mathbb{C}^3$, lines the set of all two-dimensional subspaces of $\mathbb{C}^3$, and incidence inclusion is the **complex projective plane**, denoted by $PG(2, \mathbb{C})$. This projective plane shares many of the properties that $PG(2, \mathbb{R})$ has, and we remind the reader of these below. In $PG(2, \mathbb{C})$:

- Any two points are incident with a unique line.
- Any two lines meet in a unique point.
- A point has homogeneous coordinates (x, y, z), $x, y, z \in \mathbb{C}$, not all zero.
- A line has homogeneous line coordinates $[a, b, c]$, $a, b, c \in \mathbb{C}$, not all zero.
- A point (x, y, z) is incident with the line $[a, b, c]$ if and only if $ax+by+cz = 0$.
- The **dual** of a statement about the plane is the statement that results after interchanging *point* and *line*, *collinear* and *concurrent*, *intersection* and *join*, and making the necessary linguistic adjustments. The dual of every theorem about $PG(2, \mathbb{C})$ is also a theorem.
- A **conic** is the set of zeros in $PG(2, \mathbb{C})$ of a complex homogeneous quadratic equation $Q(x, y, z) = 0$ in three variables. It is **non-degenerate** if $Q(x, y, z)$ doesn't factor.

The last of these properties is particularly interesting. The fundamental theorem of algebra implies that any line meets a conic $\mathcal{K}$ in one or two points (i.e., never zero points). In particular, the pole–polar property simplifies: for a point P on $\mathcal{K}$, there is a unique tangent on P, but if P does not lie on $\mathcal{K}$, we always have two tangents on P. So there is no such thing as *internal* points to $\mathcal{K}$; all points off $\mathcal{K}$ look alike.

Recall from Section 6.2 that a **collineation** of PG(2, $\mathbb{C}$) is a bijection of the points taking lines to lines. The set of all collineations of PG(2, $\mathbb{C}$) is a group under composition, which we denote by Aut(PG(2, $\mathbb{C}$)). The projective general linear group PGL(3, $\mathbb{C}$) is the group induced by the action of the group of 3×3 invertible complex matrices GL(3, $\mathbb{C}$) on the points of PG(2, $\mathbb{C}$), but it is not equal to Aut(PG(2, $\mathbb{C}$)). The full collineation group of PG(2, $\mathbb{C}$) is twice as big as PGL(3, $\mathbb{C}$), since entrywise complex conjugation induces a collineation (Exercise 13.8).

Theorem 13.20
PGL(3, $\mathbb{C}$) *acts transitively on non-empty, non-degenerate conics of* PG(2, $\mathbb{C}$).

Proof Let $Q(x, y, z)$ be a homogenous quadratic polynomial over $\mathbb{C}$ that does not factor. Then $f((x, y, z), (x', y', z')) = Q(x + x', y + y', z + z') - Q(x, y, z) - Q(x', y', z')$ is a symmetric bilinear form, and its Gram matrix A (the symmetric matrix with $f((x, y, z), (x', y', z')) = (x, y, z)A(x, y, z)^{\top})$ is non-singular. Thus, applying the Gram–Schmidt process to a basis for $\mathbb{C}^3$ with respect to f produces an orthonormal basis for f. Now the transition matrix P between the standard basis and the orthonormal basis has $PAP^{\top} = I$. Hence, every non-degenerate conic of PG(2, $\mathbb{C}$) is projectively equivalent to $x^2 + y^2 + z^2 = 0$. □

Identify $\mathbb{R}^2$ with $\mathbb{C}$. Compute the cross-ratio of four points A, B, C, D, disregarding whether they are collinear or not, by taking the following as a definition of *cross-ratio* in this representation:

$$R(A, B; C, D) := \frac{(C - A)/(C - B)}{(D - A)/(D - B)} \in \mathbb{C}.$$

Draw the circle passing through B, C, D and the circle passing through A, C, D, or the line for three collinear points. The **argument** of the cross-ratio

$$\arg R(A, B; C, D)$$

is the angle between the two circles where they meet at C.[2] The reader may note a connection here with the Casey angle.

---

[2] There's some ambiguity both in which angle to take, since there are two choices, and in what counts as positive or negative. This can be resolved by checking for each circle the order the three marked points come in, and whether the fourth point is inside or outside the circle.

13.2 The Absolute Conic

The analogue of the circular points for the plane is the absolute conic for space. Recall that in extended Euclidean space, the points and lines of the plane at infinity are the parallel classes of lines and the parallel classes of planes, respectively. Given two parallel classes of lines of $\mathbb{R}^3$, there is a representative of each class that passes through the origin, say ℓ and ℓ'. So we can write these two lines as one-dimensional subspaces of $\mathbb{R}^3$:

$$\ell = \{\lambda(x, y, z) : \lambda \in \mathbb{R}\},$$
$$\ell' = \{\lambda(x', y', z') : \lambda \in \mathbb{R}\}.$$

The points $[\ell]$ and $[\ell']$ of the plane at infinity project to the points (x, y, z) and (x', y', z') of $\mathsf{PG}(2, \mathbb{R})$.

Now ℓ and ℓ' are perpendicular if and only if

$$xx' + yy' + zz' = 0.$$

Hence, from the viewpoint of our points (x, y, z) and (x', y', z') of $\mathsf{PG}(2, \mathbb{R})$, the lines ℓ and ℓ' are perpendicular if and only if (x, y, z) and (x', y', z') are conjugate with respect to the conic defined by this degree 2 homogeneous equation. However, if we then suppose $(x', y', z') = (x, y, z)$, we obtain the absolute points of a conic with equation $x^2 + y^2 + z^2 = 0$; an empty conic! So, as for the plane case, we can extend to $\mathsf{PG}(2, \mathbb{C})$ if it suits us to have non-trivial solutions to this equation; just as it was convenient to have the *circular points* in order to define circles of the Euclidean plane.

Consider $\mathsf{PG}(3, \mathbb{R})$. For the rest of this section, we stipulate a plane π and an empty conic C of π. Now π is isomorphic to $\mathsf{PG}(2, \mathbb{R})$, which we can embed in $\mathsf{PG}(2, \mathbb{C})$, and C defines a conic $\overline{C}$ of the complex projective plane, which we will call the **absolute conic**. We will canonically use $w = 0$ for π and $x^2 + y^2 + z^2 = 0$ for C. An **ellipsoid** of $\mathbb{R}^3$ is an elliptic quadric disjoint from the plane at infinity.

Theorem 13.21
An ellipsoid of $\mathbb{R}^3$ is a sphere if and only if it contains the absolute conic at infinity.

Proof Translations of $\mathbb{R}^3$ act trivially on the plane at infinity (both the real and the complex plane), and rotations about the origin are given by orthogonal matrices, and so stabilise the absolute conic $x^2 + y^2 + z^2 = 0$. Thus, we may assume, by translating and rotating, that the ellipsoid $\mathcal{E}$ is centred at the origin and aligned with the axes, and therefore has an equation of the form

$$\frac{X^2}{a^2} + \frac{Y^2}{b^2} + \frac{Z^2}{c^2} = 1.$$

Hence, a, b, c are the lengths of the semi-principal axes of $\mathcal{E}$. This has homogeneous form

$$\frac{x^2}{a^2} + \frac{y^2}{b^2} + \frac{z^2}{c^2} = w^2.$$

Thus, it meets the plane $w = 0$ at infinity in

$$\frac{x^2}{a^2} + \frac{y^2}{b^2} + \frac{z^2}{c^2} = 0,$$

which is the absolute conic at infinity in the complex projective plane if and only if $a^2 = b^2 = c^2$. This is equivalent to $\mathcal{E}$ having equation $X^2 + Y^2 + Z^2 = a^2$. This is precisely when $\mathcal{E}$ is the sphere centred at the origin with radius a. □

Recall that $\overline{C}$ denotes the absolute conic arising from an empty conic C. The group $\mathrm{Sim}(\mathbb{R}^3)$ consists of all similarities of $\mathbb{R}^3$.

Theorem 13.22

Let π be a plane of $\mathrm{PG}(3, \mathbb{R})$ *and C be an empty conic of π. Then* $(\mathrm{PG}(3, \mathbb{R}) \setminus \pi, \mathrm{PGL}(4, \mathbb{R})_{\overline{C}})$ *is permutationally isomorphic to* $(\mathbb{R}^3, \mathrm{Sim}(\mathbb{R}^3))$.

Proof This follows from Theorem 13.21 and a little known fact about similarities. Carathéodory's theorem (see Carathéodory, 1937) states that a bijection on $\mathbb{R}^2$ takes circles to circles if and only if it is a similarity. The three-dimensional analogue follows from this result: that a bijection $\mathbb{R}^3$ takes spheres to spheres if and only if it is a similarity. □

Thus, we can reconstruct real affine space from real projective space by distinguishing a plane, and we can reconstruct Euclidean space from real affine space by distinguishing an empty conic in that plane. Two concurrent lines of $\mathrm{PG}(3, \mathbb{R})$ not lying in π are **perpendicular** if and only if they meet π in points that are conjugate with respect to C.

In elementary vector calculus, one of the central geometric ideas is the *normal* to a plane. We will not use vector equations to define the normal to a plane; rather, we will use beautiful synthetic geometry and the absolute conic at infinity. Consider a plane π' that is not equal to π; we think of π' as an arbitrary plane of Euclidean space. Now π' meets π in a line ℓ, and, with respect to C, we let P be the pole of ℓ. Let m be a line through P. Then m is perpendicular to every line of π'. This allows us to describe the parallel class of perpendicular lines to π' as those lines having direction P, the pole of $\pi \cap \pi'$. Thus, we have:

Theorem 13.23

Two planes of $\mathrm{PG}(3, \mathbb{R})$, *not equal to π, are perpendicular if their lines at infinity are conjugate with respect to C.*

We now define an operation $\wedge$ on affine lines ℓ and m. Let ℓ_∞ and m_∞ be the intersections of ℓ and m with π, and let $\perp$ be the polarity of the absolute conic in π. Then we define

$$\ell \wedge m := \langle \ell_\infty^\perp \cap m_\infty^\perp, \ell \rangle \cap \langle \ell_\infty^\perp \cap m_\infty^\perp, m \rangle.$$

If ℓ and m are parallel, then $\ell_\infty \cap m_\infty = \ell_\infty$ and $\ell \wedge m$ is equal to the full projective space. Otherwise, if ℓ and m are non-parallel, then $\ell \wedge m$ is a line. Moreover:

Lemma 13.24

If ℓ and m are non-parallel lines, then $\ell \wedge m$ is the unique perpendicular transversal to ℓ and m.

Proof An affine line n that is perpendicular to ℓ has its point n_∞ at infinity conjugate to ℓ_∞, with respect to the absolute conic C. Likewise, n_∞ is conjugate to m_∞. This means that n_∞ lies on the lines ℓ_∞ and m_∞, and hence is equal to the intersection $\ell_\infty^\perp \cap m_\infty^\perp$. Thus, the plane $\langle n, \ell \rangle$ is equal to $\langle n_\infty, \ell \rangle$, and, similarly, $\langle n, m \rangle$ is equal to $\langle n_\infty, m \rangle$. So $\ell \wedge m = \langle n_\infty, \ell \rangle \cap \langle n_\infty, m \rangle$ and, therefore, $n = \ell \wedge m$. $\qquad\square$

Moreover, once we visit line geometry and Plücker coordinates in Chapter 16, we will be able to prove the following result (Corollary 16.6):

$$\ell \wedge (m \wedge n) \text{ is perpendicular to } (m \wedge (\ell \wedge n)) \wedge (n \wedge (m \wedge \ell)).$$

A highlight of the realisation of $\mathbb{R}^3$ as $\mathsf{PG}(3, \mathbb{R})$ equipped with an absolute conic is a proof of the following result about *skew right-angled hexagons* in $\mathbb{R}^3$.

Theorem 13.25 (Petersen–Morley theorem (compare with Petersen, 1898; Baker, 1936b))

Given three lines p, q, r, let p' be the unique normal to q and r, q' that to r and p, and r' that to p and q. Then the lines a, b, c, normal respectively to the pairs of lines p and p', q and q', r and r', have a common normal.

Proof (Todd, 1936) By hypothesis, $p' = q \wedge r$, $q' = r \wedge p$, $r' = p \wedge q$. Since a is perpendicular to p and p', we have $a = p \wedge p'$, and likewise for the lines b and c. So

$$a = p \wedge (q \wedge r), \quad b = q \wedge (r \wedge p), \quad c = r \wedge (p \wedge q).$$

Therefore, by Corollary 16.6 (which is paraphrased above), a is perpendicular to $b \wedge c$. Hence, either b is parallel to c, or a, b, c have a common perpendicular transversal. $\qquad\square$

JULIUS PETERSEN (1839–1910) was a professor at the University of Copenhagen. He was a pioneer in graph theory, coming to that topic via work in invariant theory, but also worked in geometry, analysis, differential equations, mathematical physics, and mathematical economics.

Exercises

13.1 Prove Theorem 13.2.

13.2 Prove Theorem 13.4.

13.3 Show that $PGL(4, \mathbb{R})$ has precisely two orbits on pairs of distinct lines: pairs of coplanar lines and pairs of skew lines.

13.4 Show that four skew lines of $PG(3, \mathbb{R})$ have two transversal lines.

13.5 Four points P, Q, R, X of $PG(3, \mathbb{R})$ lie in a plane π. Suppose Y and Z are two points not on π, collinear with X. The planes QRZ, RPZ, PQZ meet YP, YQ, YR, respectively, in P', Q', R' and the plane $P'Q'R'$ meets π in the line ℓ. Prove that the position of ℓ depends only on P, Q, R, and X and is independent of the positions of Y and Z.

13.6 Using an analogous argument to that of the proof of Pappus' theorem (Theorem 4.7), show that projectivities of $PG(3, \mathbb{R})$ preserve cross-ratio of collinear points of $PG(3, \mathbb{R})$.

13.7 Suppose we have four collinear planes $\pi_1, \pi_2, \pi_3, \pi_4$ and a fifth transversal plane π_5. Show that

$$R(\pi_1, \pi_2; \pi_3, \pi_4) = R(\pi_5 \cap \pi_1, \pi_5 \cap \pi_2; \pi_5 \cap \pi_3, \pi_5 \cap \pi_4),$$

where the second cross-ratio is regarded as a cross-ratio of lines of $PG(2, \mathbb{R})$, embedded into π_5 in $PG(3, \mathbb{R})$.

13.8 Show that $\mathrm{Aut}(PG(2, \mathbb{C}))$ is the semi-direct product $PGL(3, \mathbb{C}) \rtimes \langle \iota \rangle$ where ι is the complex conjugation map (on homogeneous coordinates). (Hint: Use the FToPG (Theorem 13.16) and Theorem 2.8.)

13.9 Using $PG(3, \mathbb{R})$ and an absolute conic C, prove that two spheres of $\mathbb{R}^3$ meet in a circle.

14

Triangles and Tetrahedra

Think geometrically, prove algebraically.

Silverman and Tate (2015, p. 277).[1]

14.1 Triangles

The 3D-version of Desargues' theorem yields a short proof of the 2D-version (Theorem 2.20).

Theorem 14.1 (Desargues' theorem (1648), (Desargues, 1951))
Given triangles ABC and A'B'C' of $\mathsf{PG}(3, \mathbb{R})$ *in perspective from a point O, the lines AB and A'B' meet in a point P, the lines AC and A'C' meet in a point Q, the lines BC and B'C' meet in a point R, and P, Q, and R are collinear.*

Proof We break this proof into two cases depending on whether the planes A, B, C and A', B', C' are equal or not.

1 Suppose the planes spanned by A, B, C and A', B', C' are distinct. Then they meet in a line and this line is incident with P, Q, and R.

2 Suppose the planes spanned by A, B, C and A', B', C' are equal (to π). Then the lines AC and $A'C'$ meet in a point Q, the lines BC and $B'C'$ meet in a point R. Choose a point O' not on π, and a point D, not equal to O, not in π on $O'C$. Now consider the plane π' on O, O', and C. This contains C', and so the line $O'C'$ and also the line OD. Thus, these two lines meet in a unique point D'. Now triangles ABD and $A'B'D'$ are in perspective from O and span distinct planes, so by (1), the lines AD and AD' meet in a point Q', the lines BD and $B'D'$ meet in a point R', and P, Q', and R' are collinear in a line ℓ. But now P, Q, and R all lie on the line $\langle O, \ell \rangle \cap \pi$. $\quad\square$

[1] Reprinted by permission from Springer International Publishing: *Rational Points on Elliptic Curves*. Appendix A.2. Silverman, Joseph H., and Tate, John T. © 2015.

Corollary 14.2
Desargues' theorem (Theorem 2.20) holds in PG(2, ℝ).

Proof Let ABC and $A'B'C'$ be triangles of PG(2, ℝ) in perspective from a point O. We can embed the points and lines of PG(2, ℝ) via the map $(x, y, z) \mapsto (x, y, z, 0)$, into the plane $X_4 = 0$ (i.e., the points having fourth coordinate zero). This embedding preserves incidence of points and lines, and so, by Theorem 14.1, the points $AB \cap A'B'$, $AC \cap A'C'$, $BC \cap B'C'$ are collinear. □

Theorem 14.3 (Converse of Desargues' theorem in space)
Let ABC and $A'B'C'$ be triangles of PG(3, ℝ). *If $AB \cap A'B'$, $AC \cap A'C'$, $BC \cap B'C'$ are collinear, let $V := ABA' \cap ACA' \cap BCB'$. Then V is on $ABA' \cap ACA' = AA'$, $ABA' \cap BCB' = BB'$, and $ACA' \cap BCB' = CC'$. So, if ABC and $A'B'C'$ are in perspective from a line, then they are in perspective from a point.*

Proof This (beautiful) proof appears in Hodge and Pedoe (1994, p. 196). Let BC and $B'C'$ meet in P, CA and $C'A'$ meet in Q, and AB and $A'B'$ meet in R. Suppose P, Q, R lie on a line ℓ – that is, ABC and $A'B'C'$ are in perspective from ℓ. First, we may suppose without loss of generality that no side of ABC is equal to a side of $A'B'C'$, otherwise the result is trivially true. So neither (B, B', R) nor (C, C', Q) can be a collinear triple of points because if (say) B, B', R were incident with a line m, then A is incident with $BR = m$ and A' is incident with $B'R = m$, thus implying that $AB = A'B'$.

Second, the two triangles $BB'R$ and $CC'Q$ are in perspective from P (because P, Q, R are collinear). Now, $B'R$ and $C'Q$ meet only in A', and RB and QC meet only in A. Hence, by Desargues' theorem in the plane (Theorem 2.20), BB' and CC' meet on AA'. That is, ABC, $A'B'C'$ are in perspective. □

14.2 Tetrahedra

A **tetrahedron** is the natural three-dimensional analogue of a triangle (see Figure 14.1). It can be defined as four points, not lying in a plane, and no three collinear. This then implies that there are four triangular faces, yielding six *edges* in total, and four vertices – when we think of the closed geometric object bound by these points as a polyhedron.

In projective space, we can have one face of a tetrahedron at infinity. So, for example,

$$(1, 0, 0, 0), (0, 1, 0, 0), (0, 0, 1, 0), (0, 0, 0, 1)$$

Figure 14.1: A tetrahedron.

is a tetrahedron where the first three vertices yield a triangle at infinity, and the fourth vertex of this tetrahedron corresponds to the origin in $\mathbb{R}^3$. So, to visualise this tetrahedron in Euclidean space, think of the x, y, z axes of $\mathbb{R}^3$ as three edges of the tetrahedron emanating from $(0, 0, 0)$, and they meet the plane at infinity in the vertices of a triangle at infinity.

Theorem 14.4 (Klein's theorem (Klein, 1870))
If two tetrahedra are in perspective from a point, then they are in perspective from a plane. Thus, the six pairs of corresponding edges of the two tetrahedra meet in six coplanar points and the four pairs of corresponding faces meet in four coplanar lines.

Proof Let the two tetrahedra be $P_1 P_2 P_3 P_4$ and $P_1' P_2' P_3' P_4'$, and let the lines $P_1 P_1', P_2 P_2', P_3 P_3', P_4 P_4'$ meet in the centre of perspectivity O. Two corresponding edges $P_i P_j$ and $P_i' P_j'$ then clearly intersect; call the point of intersection P_{ij}. The points P_{12}, P_{13}, P_{23} all lie on the same line since the triangles $P_1 P_2 P_3$ and $P_1' P_2' P_3'$ are in perspective from O, by Desargues' theorem (Theorem 14.1). By similar reasoning applied to the other pairs of perspective triangles, we find that the following triples of points are collinear:

$$(P_{12}, P_{13}, P_{23}); (P_{12}, P_{14}, P_{24}); (P_{13}, P_{14}, P_{34}); (P_{23}, P_{24}, P_{34}).$$

The first two triples have the point P_{12} in common, and hence determine a plane; each of the other two triples has a point in common with each of the first two. Hence, all the points P_{ij} lie in the same plane π. The lines of the four triples just given are the lines of intersection of the pairs of corresponding faces of the tetrahedra. Thus, the two tetrahedra are in perspective from π. □

As for Desargues' theorem, the dual of Theorem 14.4 is its own converse.

Corollary 14.5
If two tetrahedra are in perspective from a point O, then they are in perspective from a plane π, and there is a homology with centre O and axis π taking the first tetrahedron to the second.

Corollary 14.6
If two tetrahedra are in perspective from a point, then they are in perspective from a plane which meets the edges of the tetrahedra in a complete quadrilateral.

A homology h of $\mathrm{PG}(3, \mathbb{R})$ is **harmonic** if, for every point P not on the axis π and not equal to the centre O of h, we have $(O, OP \cap \pi, P, P^h)$ harmonic.

Theorem 14.7 (Stephanos, 1879)

Given a tetrahedron in $\mathsf{PG}(3, \mathbb{R})$ *and a point M on no face of the tetrahedron, consider the tetrahedron with vertices the intersections of the lines joining M with a vertex with the opposite side. Then there is a harmonic homology with centre M interchanging the two tetrahedra.*

Proof Let $ABCD$ be a tetrahedron in $\mathsf{PG}(3, \mathbb{R})$ and M be a point on no face of $ABCD$. Let $A' = AM \cap BCD$, $B' = BM \cap ACD$, $C' = CM \cap ABD$, $D' = DM \cap ABC$. Let the transversal to DA and BC on M meet DA in U and BC in X, the transversal to DB and CA on M meet DB in V and CA in Y, the transversal to DC and AB on M meet DC in W and AB in Z; let U' be the harmonic conjugate of U with respect to A, D; V' be the harmonic conjugate of V with respect to B, D; W' be the harmonic conjugate of W with respect to C, D; X' be the harmonic conjugate of X with respect to B, C; Y' be the harmonic conjugate of Y with respect to A, C; and Z' be the harmonic conjugate of Z with respect to A, B. We must show that U', V', W', X', Y', Z' are coplanar in π, $A'B'C'D'$ is a tetrahedron, and $ABCD$ and $A'B'C'D'$ are in perspective from the point M and from the plane π. Choose coordinates so that $A(1, 0, 0, 0)$, $B(0, 1, 0, 0)$, $C(0, 0, 1, 0)$, $D(0, 0, 0, 1)$, $M(1, 1, 1, 1)$. Then

$A'(0, 1, 1, 1), B'(1, 0, 1, 1), C'(1, 1, 0, 1), D'(1, 1, 1, 0), U(1, 0, 0, 1), X(0, 1, 1, 0),$
$V(0, 1, 0, 1), Y(1, 0, 1, 0), W(0, 0, 1, 1), Z(1, 1, 0, 0), U'(-1, 0, 0, 1), V'(0, -1, 0, 1),$
$W'(0, 0, -1, 1), X'(0, -1, 1, 0), Y'(-1, 0, 1, 0), Z'(-1, 1, 0, 0),$

and the last six points lie on the plane π: $[1, 1, 1, 1]$. Moreover, AB meets $A'B'$ in $Z'(-1, 1, 0, 0)$, which is on π, and by permuting the coordinates, we see that $ABCD$ and $A'B'C'D'$ are in perspective from π. □

CYPARISSOS STEPHANOS (1857–1917) was a professor at the National and Kapodistrian University of Athens, the National Technical University of Athens, and the Hellenic Naval Academy. Stephanos was an invited speaker at each of the first five international congresses of mathematicians.

Theorem 14.8 (von Staudt, 1856, p. 21, para. 35)

Let ABCD be a tetrahedron in $\mathsf{PG}(3, \mathbb{R})$ *and ℓ be a line on no vertex and contained in no face of ABCD. Then*

$$\mathrm{R}(\ell \cap BCD, \ell \cap ACD; \ell \cap ABD, \ell \cap ABC) = \mathrm{R}(\ell A, \ell B; \ell C, \ell D).$$

Proof Let $P = \ell B \cap AD$, $Q = \ell C \cap AD$. Then

$$R(\ell A, \ell B; \ell C, \ell D) = R(A, P; Q, D) = R(BCA, BCP; BCQ, BCD)$$
$$= R(BCA \cap \ell, BCP \cap \ell; BCQ \cap \ell, BCD \cap \ell).$$

Now B, P, $BCP \cap \ell$ are all in $\ell B \cap ABD$, so are collinear; thus, $BCP \cap \ell = ABD \cap \ell$; similarly, $BCQ \cap \ell = ACD \cap \ell$. Thus,

$$R(\ell A, \ell B; \ell C, \ell D) = R(BCA \cap \ell, ABD \cap \ell; ACD \cap \ell, BCD \cap \ell)$$
$$= R(\ell \cap BCD, \ell \cap ACD; \ell \cap ABD, \ell \cap ABC). \qquad \square$$

Theorem 14.9 (Möbius, 1828)
Given two tetrahedra of $\mathsf{PG}(3, \mathbb{R})$ *such that every face but one contains a vertex of the other tetrahedron, and such that no vertex of the first tetrahedron lies on an edge of the second tetrahedron, then the remaining face also contains a vertex of the other tetrahedron.*

Proof Let $ABCD$ be the first tetrahedron; $A'B'C'D'$ be the second, with A' on the plane BCD, B' on the plane ACD, C' on the plane ABD, D' on the plane ABC, A on the plane $B'C'D'$, B on the plane $A'C'D'$, and C on the plane $A'B'D'$. Consider the line ℓ of intersection of the two planes ABC and $A'B'C'$. Since A is on the plane $B'C'D'$, $B'C'$ and AD' meet in a point P', and since D' is on the plane ABC, P' is on ℓ. Similarly, the point Q' of intersection of $C'A'$ and BD' is on ℓ and the point R' of intersection of $A'B'$ and CD' is on ℓ. Let $P = BC \cap \ell$, $Q = CA \cap \ell$, $R = AB \cap \ell$. Let $p = BC$, $p' = AD'$, $r = AB$, $r' = CD'$. By Pappus' involution theorem (Theorem 6.12) applied to the plane ABC, $(PP')(QQ')(RR')$ are in involution. By the converse of Pappus' involution theorem (Theorem 6.13) applied to the plane $A'B'C'$, the lines PA', QB', RC' are concurrent. The point where they meet is on the planes $A'BC$, $B'CA$, $C'AB$ and therefore is D. Thus, D is on the plane $A'B'C'$. $\qquad \square$

Two tetrahedra that inscribe each other, such as in Theorem 14.9, is called a **Möbius configuration**. The point–plane configuration of two Möbius tetrahedra is abstractly isomorphic to the Möbius–Kantor 8_3 configuration. By embedding $\mathsf{PG}(2, \mathbb{R})$ as a plane in $\mathsf{PG}(3, \mathbb{R})$, Theorem 14.9 leads to a result in the plane.

Theorem 14.10 (Möbius, 1828)
Let ABC, FGH be triangles, ℓ a line of $\mathsf{PG}(2, \mathbb{R})$, *$A' = \ell \cap BC$, $B' = \ell \cap CA$, $C' = \ell \cap AB$, $F' = \ell \cap GH$, $G' = \ell \cap HF$, $H' = \ell \cap FG$. Then, if $A'F$, $B'G$, $C'H$ are concurrent, so are AF', BG', and CH'.*

Proof Embed $PG(2, \mathbb{R})$ in $PG(3, \mathbb{R})$ and choose a point D such that $ABCD$ is a tetrahedron. Let π be the plane on D on ℓ. In π choose F, G, H. Let $F = \pi \cap A'D$, $G = \pi \cap B'D$, and $H = \pi \cap C'D$. Let ℓ be the intersection of AGH, BHF, and CFG. Then seven of the eight incidences of the *Möbius configuration* in Theorem 14.9 hold. Hence, ℓ lies in ABC. Now $F' = GH \cap A'B'C'$. Then $AF' = AGH \cap ABC$, so ℓ is in AF'. Similarly, ℓ is in BG' and CH'. Thus, AF', BG', CH' are concurrent. □

Example 14.11 (Reye's configuration (Reye, 1882))
Let $\mathcal{P}$ be the following set of 12 points of $PG(3, \mathbb{R})$:

$(1, 0, 0, 0)$	$(0, 1, 0, 0)$	$(0, 0, 1, 0)$	$(0, 0, 0, 1)$
$(1, 1, 1, 1)$	$(1, 1, -1, -1)$	$(1, -1, 1, -1)$	$(1, -1, -1, 1)$
$(1, -1, -1, -1)$	$(1, -1, 1, 1)$	$(1, 1, -1, 1)$	$(1, 1, 1, -1)$.

Let Π be the following set of 12 planes of $PG(3, \mathbb{R})$:

$[1, 1, 0, 0]$	$[1, 0, 1, 0]$	$[1, 0, 0, 1]$	$[0, 1, 1, 0]$
$[0, 0, 1, 1]$	$[0, 1, 0, 1]$	$[1, -1, 0, 0]$	$[1, 0, -1, 0]$
$[1, 0, 0, -1]$	$[0, 1, -1, 0]$	$[0, 0, 1, -1]$	$[0, 1, 0, -1]$.

*Then $(\mathcal{P}, \Pi)$ is such that every point is on six of the planes and every plane is on six of the points, which is called **Reye's configuration**.*

Theorem 14.12
If a set $\mathcal{P}'$ of 12 points and a set Π' of 12 planes of $PG(3, \mathbb{R})$ are given together with a bijection $\phi: \mathcal{P} \cup \Pi \to \mathcal{P}' \cup \Pi'$, where $(\mathcal{P}, \Pi)$ is Reye's configuration with $\mathcal{P}^\phi = \mathcal{P}'$ and $\Pi^\phi = \Pi'$ and if there exist $P \in \mathcal{P}$ and $\pi \in \Pi$ with P and π incident and such that ϕ is incidence-preserving except possibly at (P, π), then ϕ is incidence-preserving. In other words, if all but one of the incidences hold, then the remaining incidence must hold.

Proof Each pair of planes of Π is mutually incident with three points which must map under ϕ and each such set of three points is on three planes of Π. So each such set of three points (a *line* of $(\mathcal{P}, \Pi)$) must map under ϕ to three points of a line of $PG(3, \mathbb{R})$; as two planes determine a line. If ℓ is a line of $(\mathcal{P}, \Pi)$ on P and π, and π_1, π_2 are the planes of Π on ℓ other than π, and P_1, P_2 are the points of $\mathcal{P}$ on ℓ other than P, then P^ϕ is incident with π_1^ϕ and π_2^ϕ and so with

Figure 14.2 A desmic configuration. The third tetrahedron has faces the xy, yz, zx planes of $\mathbb{R}^3$ and the plane at infinity.

their intersection ℓ^ϕ, and π^ϕ is incident with P_1^ϕ and P_2^ϕ and so with their join ℓ^ϕ; hence, P^ϕ is incident with π^ϕ. □

A collection $\{S_1, S_2, S_3\}$ of three non-empty sets of points in a projective space is a **desmic configuration** if S_i and S_j are in perspective from every element of S_k, for every permutation $\begin{pmatrix} 1 & 2 & 3 \\ i & j & k \end{pmatrix}$ (see Figure 14.2). Stephanos constructed a triad of desmic tetrahedra in 1879 in real projective 3-space. Note the connection with Reye's configuration!

Theorem 14.13 (Triads of desmic tetrahedra (Stephanos, 1879))
The following set is a triad of desmic tetrahedra in $\mathsf{PG}(3, \mathbb{R})$:

$$\{\{(1, 0, 0, 0), (0, 1, 0, 0), (0, 0, 1, 0), (0, 0, 0, 1)\},$$
$$\{(1, 1, 1, 1), (1, 1, -1, -1), (1, -1, 1, -1), (1, -1, -1, 1)\},$$
$$\{(1, -1, -1, -1), (1, -1, 1, 1), (1, 1, -1, 1), (1, 1, 1, -1)\}\}.$$

Proof The setwise stabiliser G in $\mathsf{PGL}(4, \mathbb{R})$ of this configuration has structure $((A_4 \times A_4) : C_2) : C_2$. Indeed, it is generated by the following elements (given by representative matrices):

$$\begin{bmatrix} -1 & 0 & 0 & 0 \\ 0 & 1 & 0 & 0 \\ 0 & 0 & 1 & 0 \\ 0 & 0 & 0 & 1 \end{bmatrix}, \begin{bmatrix} 0 & 0 & 0 & 1 \\ 1 & 0 & 0 & 0 \\ 0 & 1 & 0 & 0 \\ 0 & 0 & 1 & 0 \end{bmatrix}, \begin{bmatrix} 0 & 1 & 0 & 0 \\ 1 & 0 & 0 & 0 \\ 0 & 0 & 1 & 0 \\ 0 & 0 & 0 & 1 \end{bmatrix}, \begin{bmatrix} 1 & 1 & 1 & 1 \\ -1 & -1 & 1 & 1 \\ -1 & 1 & -1 & 1 \\ -1 & 1 & 1 & -1 \end{bmatrix}.$$

In particular, G acts transitively on the set of three sets. The stabiliser in G of one of the three tetrahedra is generated by the first three matrices above, and acts transitively on the vertices of the fixed tetrahedron. So it suffices to show that S_2 and S_3 are in perspective from P: $(1,0,0,0)$. This is clear since the following triples of points are collinear:

$$\{P, (1,1,1,1), (1,-1,-1,-1)\} \qquad \{P, (1,1,-1,-1), (1,-1,1,1)\}$$
$$\{P, (1,-1,1,-1), (1,1,-1,1)\} \qquad \{P, (1,-1,-1,1), (1,1,1,-1)\}. \qquad \square$$

The following theorem is for the reader that has some knowledge of a real projective space in arbitrary dimension. These projective spaces will be explored in detail in Part III. In the proof of the following, we will need the projective analogue of the so-called *Sylvester–Gallai theorem* (see Aigner and Ziegler, 2014, ch. 11, p. 74). We state the projective analogue:

If a finite set K of points in $\mathsf{PG}(2,\mathbb{R})$ is not a subset of a line, then there exists a line of $\mathsf{PG}(2,\mathbb{R})$ containing precisely two points of K.

Theorem 14.14 (Edelstein and Kelly, 1966)
If a desmic configuration exists in a real projective space, then the space has dimension at most 3.

Proof Suppose we have a desmic configuration $\{S_1, S_2, S_3\}$ in a real projective space $\mathsf{PG}(n,\mathbb{R})$ where $n \geqslant 4$. Notice that this configuration has the property that any line that meets two of the three sets must meet the third. Consider a finite pencil $\mathcal{F}$ of lines in $\mathsf{PG}(3,\mathbb{R})$, not all in the same plane. Next, consider intersecting each element of $\mathcal{F}$ by a plane π; thus, we have a finite set of points $\mathcal{F}_\pi$ in a copy of $\mathsf{PG}(2,\mathbb{R})$. By the Sylvester–Gallai theorem, there is a line of π containing precisely two points of $\mathcal{F}_\pi$. Therefore, $\mathcal{F}$ contains a pair of lines such that the plane they span contains none of the other lines.

We now take a pair of points P and Q of $\cup S_i$ so that P and Q are from different S_i. The points of $\cup S_i \setminus \{P, Q\}$ define a pencil of planes Π with axis PQ (for each point R, take PQR). A section of this pencil by a properly chosen 3-space κ defines a finite pencil of lines in κ not all in a plane and so (by the above observation) there are two lines of this pencil that span a plane σ not containing any other lines of the pencil. Now σ and PQ span a 3-space γ such that the points of $\cup S_i$ in γ are on precisely two planes of the pencil Π. Each of these two planes contains at least one point of $\cup S_i \setminus \{P, Q\}$. It is straightforward to see that if a collection of two or more finite non-empty disjoint sets in $\mathsf{PG}(3,\mathbb{R})$ lies on two planes and not on one, then there is a line intersecting precisely two of the sets; a contradiction. Therefore, $n \leqslant 3$. $\qquad \square$

Peter Borwein proved the conjecture of Edelstein and Kelly that:

Theorem 14.15 (Borwein, 1983)
Every desmic configuration of a three-dimensional real projective space is equivalent to the Stephanos configuration of a triad of desmic tetrahedra.

We do not give Borwein's proof since it requires establishing many pages of intricate results. The authors do not know of a short proof of this result. We now give a converse to Corollary 14.6 to Klein's theorem (Theorem 14.4).

Theorem 14.16
Given a tetrahedron in $\mathrm{PG}(3, \mathbb{R})$ and a plane on no vertex of the tetrahedron ABCD, the six points of intersection of that plane with the edges of the tetrahedron form a complete quadrilateral. Moreover, if UVW is the diagonal triangle of this quadrilateral, then

(i) *the triangular faces BCD, ACD, ABD, BCD of the tetrahedron are each in perspective with the triangle UVW from points A_1, B_1, C_1, D_1, respectively;*

(ii) *the tetrahedra ABCD, $A_1B_1C_1D_1$ are in perspective from U, V, W and from a fourth point O.*

Proof (Robinson, 1940) First of all, the six points of intersection of the plane with the edges of the tetrahedron form a complete quadrilateral, by Klein's theorem (Theorem 14.4), and in particular Corollary 14.6. We can order the labels of UVW (the diagonal triangle), so that we can write $N := BC \cap VW$, $Q := CA \cap UW$, and $P := AB \cap UV$. So P, Q, N are collinear as they lie on one side of the complete quadrilateral. Hence, AU, BV, CW intersect in a point, D_1 say. Similarly, from the triangles BCD, UVW it can be shown that BW, CV, DU meet in a point A_1; from the triangles ACD, UVW that AW, CU, DV meet in a point B_1; and from the triangles ABD, UVW that AV, BU, DW meet in a point C_1. Hence, the triangles BCD, ACD, ABD, BCD are in perspective with the triangle UVW from the points A_1, B_1, C_1, D_1, respectively. This proves (i).

From above, $U = AD_1 \cap A_1D$ and $U = BC_1 \cap B_1C$. Hence, the tetrahedra $ABCD, A_1B_1C_1D_1$ are in perspective from U, and, by a similar argument, also from V and W. Since AD_1 and A_1D intersect (at U), also, DD_1 and AA_1 intersect. Similarly, DD_1 and BB_1 intersect, and also DD_1 and CC_1, but BB_1 and CC_1 are not in the plane AA_1DD_1. Therefore, AA_1, BB_1, CC_1, DD_1 meet in the same point, O. This proves (ii).   □

Hence, $ABCD, A_1B_1C_1D_1, OUVW$ form a triad of desmic tetrahedra. The above theory also offers a proof of Theorem 2.21 (see Exercise 14.3).

Exercises

14.1 Two tetrahedra $XYZT$, $PQRS$ in PG(3, $\mathbb{R}$) are such that PQ and RS meet XY and ZT, QR and PS meet YZ and XT, and RP and QS meet ZX and YT. Show that any face of one cuts the other tetrahedron in a complete quadrilateral whose diagonals are the edges of that face, and hence that any edge of one is divided harmonically by the points in which it meets the two edges of the other. Show that PT, QZ, RY, SX are concurrent.

14.2 $ABCD$ is a tetrahedron in PG(3, $\mathbb{R}$) and S is a point, not lying on any of its faces. The transversals from S to the pairs of opposite edges BC, AD; CA, BD; AB, CD meet those edges, respectively, in P, P'; Q, Q'; R, R'. Prove that AP, BQ, CR, DS are concurrent.

14.3 Let ABC and $A'B'C'$ be two triangles of PG(2, $\mathbb{R}$) in perspective from a point P. Let $A'' := BC' \cap B'C$, $B'' := CA' \cap C'A$, and $C'' := AB' \cap A'B$. Embed PG(2, $\mathbb{R}$) as a plane π of PG(3, $\mathbb{R}$). Show that the three triangles ABC, $A'B'C'$, and $A''B''C''$, can be considered as the projections of desmic tetrahedra. Deduce Theorem 2.21.

15

Reguli and Quadrics

Geometry is the science that deals with the properties of space. It differs essentially from pure mathematical domains such as the theory of numbers, algebra, or the theory of functions. The results of the latter are obtained through pure thinking ... The situation is completely different in the case of geometry. I can never penetrate the properties of space by pure reflection, much as I can never recognize the basic laws of mechanics, the law of gravitation or any other physical law in this way. Space is not a product of my reflections. Rather, it is given to me through the senses. I thus need my senses in order to fathom its properties. I need intuition and experiment, just as I need them in order to figure out physical laws, where also matter is added as given through the senses.

David Hilbert (1891, p. 86).[1]

15.1 Reguli

In this section, we will explore the basic properties of collections of skew lines in $PG(3, \mathbb{R})$. We will see that a set of three skew lines generates a **regulus**, a particular infinite set of mutually skew lines, and that the set of transversals to this regulus is another regulus. We will see in the next section that a regulus also uniquely defines a **quadric** of $PG(3, \mathbb{R})$, and so reguli are important objects in three-dimensional geometry.

Theorem 15.1 (Dandelin–Gallucci theorem (Dandelin,1822, 1824/5, 1826; Dandelin and Gergonne, 1825/6; Gallucci, 1906))
If 2 sets of 4 mutually skew lines are given in $PG(3, \mathbb{R})$ and 15 of the 16 pairs of lines consisting of one from each set of 4 are not skew, then the 16th pair is also not skew.

[1] Reprinted by permission from Springer Nature: *Axiomatization in Hilbert's Early Career*. Corry, Leo. © 2004.

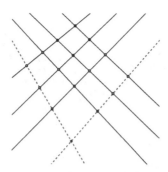

Figure 15.1 The Dandelin–Gallucci theorem.

Proof Let $\ell_1, \ell_2, \ell_3, \ell_4$ be mutually skew lines and m_1, m_2, m_3, m_4 be mutually skew lines, such that ℓ_i meets m_j in P_{ij} for $1 \leqslant i \leqslant 4, 1 \leqslant j \leqslant 4, (i, j) \neq (4, 4)$ (see Figure 15.1). Then $P_{11}, P_{21}, P_{32}, P_{42}, P_{13}, P_{23}, P_{34}$ and the intersection P of the plane $P_{11}P_{21}P_{34}$ with m_4 is a Möbius configuration if and only if P is incident with the plane $P_{13}P_{23}P_{34}$. Hence, by Theorem 14.9, P is incident with the plane $P_{13}P_{23}P_{34}$. Hence, P is incident with the intersection of $P_{11}P_{21}P_{34}$ and $P_{13}P_{23}P_{34}$, which is ℓ_4. □

A **Dandelin skew hexagon** is a hexagon $ABCDEF$ of $\mathsf{PG}(3, \mathbb{R})$ such that AB, CD, EF are skew and BC, DE, FA are skew and transversal to AB, CD, EF.

GERMINAL PIERRE DANDELIN (1794–1847) was a mililtary man and a professor of mining engineering in Liège. Dandelin's best work was in geometry. He was elected to the the Royal Belgium Academy of Science and received the Knight's Cross of the Order of Leopold.

GENEROSO GALLUCCI (1874–1941) was a high school teacher and a school inspector who also had teaching assignments at the University of Naples. He was a member of the Academy of Science of Naples and of the *Accademia Pontaniana*.

Another proof of Pappus' theorem (Theorem 2.18) Embed $\mathsf{PG}(2, \mathbb{R})$ as a plane π of $\mathsf{PG}(3, \mathbb{R})$. Choose lines a with $x \cap \pi = X$, y skew to x with $y \cap \pi = Y$. Let x' be the transversal to x and y on X', y' be the transversal to x and y on Y',

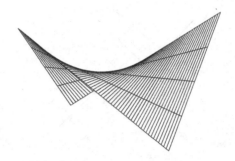

Figure 15.2 Some of the lines of a regulus of PG(3, $\mathbb{R}$).

z' be the transversal to x and y on Z'. Let z be the transversal to x' and y' on Z. Then, by the Dandelin–Gallucci theorem (Theorem 15.1) applied to the sets $\{\ell, X, Y, Z\}$ and $\{\ell', x', y', z'\}$, z and z' meet. Now consider the Dandelin skew hexagon with sides $xy'zx'yz'$. Intersecting pairs of planes that correspond to opposite vertices in this skew hexagon give the three lines that can be obtained by joining the three points $x' \cap z$, $y' \cap z$, $z' \cap y$, and so these lines lie in the plane π' spanned by these three points. (To see this, consider the planes $\langle x, x' \rangle$ and $\langle y', z \rangle$: their intersection contains $x' \cap z$ and $x \cap z'$; consider the planes $\langle x', y \rangle$ and $\langle z, z' \rangle$: their intersection contains $x' \cap z$ and $y \cap z'$; consider the planes $\langle y, y' \rangle$ and $\langle z', x \rangle$: their intersection contains $y \cap z'$ and $x \cap y'$.)

Each side of the hexagon $AA'BB'CC'$ lies in a plane corresponding to a vertex of the skew hexagon, so the intersections of the opposite sides lie in π'. Hence, the intersections of opposite sides of $XX'YY'ZZ'$ lie in $\pi \cap \pi'$; that is, they are collinear. $\square$

A **regulus** is the set of transversal lines to three skew lines of PG(3, $\mathbb{R}$) (see Figure 15.2).

Theorem 15.2
*The set of transversals to a regulus is a regulus, called the **opposite regulus**.*

Proof By Theorem 15.1, any line transversal to three lines of a regulus is transversal to all of them. $\square$

The **double-six** of Schläfli is a configuration of 30 points and 12 lines with every point on 2 of the lines, and every line on 5 of the points (see Figure 15.3). If we label the lines $\ell_1, \ldots \ell_6$ and $m_1, \ldots, m_6$, then ℓ_i and ℓ_j and m_i and m_j are skew for $i \neq j$ while ℓ_i and m_j are skew if and only if $i = j$. The proof of Theorem 15.3 (below) follows Yamashita's proof (Yamashita, 1954) and is a beautiful application of the Dandelin–Gallucci theorem (Theorem 15.1).

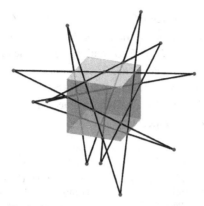

Figure 15.3 Schläfli's double-six configuration constructed from taking two lines on each face of a cube (in a regular pattern).

Theorem 15.3 (Double-six (Schläfli, 1858))

Suppose there are six pairwise skew lines $\ell_1, \ldots \ell_6$ in $\mathsf{PG}(3, \mathbb{R})$ such that no four of them lie in a regulus, and another five pairwise skew lines $m_1, \ldots m_5$ such that ℓ_i meets m_j for all distinct (i, j), and ℓ_i is skew to m_i for all i. Then, there exists another line m_6 such that ℓ_i meets m_6 if and only if $i \neq 6$. Moreover, if all but one of the incidences hold for a double-six embedded in $\mathsf{PG}(3, \mathbb{R})$, then the remaining incidence must hold.

Proof Since ℓ_1 meets m_2, m_3, m_4 but not m_1, m_1 is not in the regulus generated by m_2, m_3, m_4. Hence, ℓ_5 and ℓ_6 are the only transversals to m_1, m_2, m_3, m_4. Similarly, ℓ_4 and ℓ_6 are the only transversals to m_1, m_2, m_3, m_5, ℓ_3 and ℓ_6 are the only transversals to m_1, m_2, m_4, m_5, ℓ_2 and ℓ_6 are the only transversals to m_1, m_3, m_4, m_5, ℓ_1 and ℓ_6 are the only transversals to m_2, m_3, m_4, m_5. Similarly, $\ell_2, \ell_3, \ell_4, \ell_5$ cannot lie together in a regulus. Now m_1 is a transversal to $\ell_2, \ell_3, \ell_4, \ell_5$, so there is a unique other transversal m_6 to $\ell_2, \ell_3, \ell_4, \ell_5$. We must show that ℓ_1 meets m_6 and that ℓ_6 and m_6 are skew. Since $\ell_3, \ell_4, \ell_5, \ell_6$ do not lie together in a regulus, m_1 and m_2 are their only transversals, so ℓ_6 and m_6 are skew. We repeatedly apply Theorem 15.1 to show that ℓ_1 meets m_6.

Let S be the point S of ℓ_5 with the plane π spanned by the three points $m_3 \cap \ell_4 = P$, $m_4 \cap \ell_3 = Q$, and $m_5 \cap \ell_6 = R$. Let m_6 be the transversal to ℓ_4 and ℓ_1 on S. Let d_3 be the transversal to m_2 and m_3 on Q, and d_4 be the transversal to m_2 and m_4 on P. Let c_5 be the transversal to ℓ_1 and ℓ_5 on R, and c_6 be the transversal to ℓ_1 and ℓ_6 on S. By Theorem 15.1, d_3 meets c_5 in a point E. Similarly, c_5 and d_4 meet in F, c_6 and d_4 meet in G, and c_6 and d_3 meet in H. Let $E' = m_5 \cap \ell_3, F' = m_5 \cap \ell_4, G' = m_6 \cap \ell_4, O = m_2 \cap \ell_1, T = PR \cap SQ$.

The point O lies on the planes ERE' and EQE', so that O, E, E' are collinear. Similarly, O, G, G' are collinear. The points E, G, and T lie on the plane PRF and also on the plane SQH, so E, G, and T are collinear. Hence, the points E', G', and T lie on the plane EOG. These points E', G', and T lie also on the plane PRF, so they are collinear. Therefore, in the plane $E'TS$, the line $G'S = m_5$ meets the line $E'Q = \ell_3$. Thus, m_6 meets ℓ_1. Similarly, replacing m_2 by m_1 and ℓ_1 by ℓ_2, m_6 also meets ℓ_2. □

LUDWIG SCHLÄFLI (1814–95) was a professor at the University of Bern, and a winner of the Steiner prize of the Berlin Academy in 1870. He was an expert in the theory of special functions, especially the gamma function and the Bessel functions, and he made contributions of note to celestial mechanics and also anticipated the work of Dedekind on elliptic modular functions.

15.2 Quadrics and Polarities

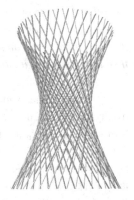

Figure 15.4: Two parallel classes of lines lying on a hyperbolic quadric.

We begin with a simple and instructive example. Let $Q(x, y, z, w) := x^2 + y^2 - z^2 - w^2$. Then the zero-set of Q gives the homogeneous coordinates for points lying on a *hyperbolic quadric* Q of $\mathsf{PG}(3, \mathbb{R})$ (see Figure 15.4). We also see that this zero-set contains all the points of some lines of $\mathsf{PG}(3, \mathbb{R})$: for example, the line $(x = z) \cap (y = w)$ is fully contained in Q since the points on this line are of the form (u, t, u, t) and

$$Q(u, t, u, t) = u^2 + t^2 - u^2 - t^2 = 0.$$

On the other hand, if we instead define Q by $Q(x, y, z, w) := x^2 + y^2 + z^2 - w^2$, it is now not possible for a line to be contained in its zero-set, for it is simply the projective completion of the unit sphere. So there are two types of (non-degenerate) quadric in $\mathsf{PG}(3, \mathbb{R})$, one having lines crowding its surface (*hyperbolic*), the other devoid of lines (*elliptic*).

A **quadric** is the set Q of zeros in $\mathsf{PG}(3, \mathbb{R})$ of a real homogeneous quadratic equation $Q(x, y, z, w) = 0$ in four variables. It is **non-degenerate** if $Q(x, y, z, w)$ does not factor over $\mathbb{C}$. It may be empty. We usually mean (by default) a non-empty, non-degenerate quadric when we say *quadric*. Define f by

$$f((x, y, z, w), (x', y', z', w')) := Q(x + x', y + y', z + z', w + w') - Q(x, y, z, w)$$
$$- Q(x', y', z', w').$$

Then f is bilinear, and can be written as

$$f((x,y,z,w),(x',y',z',w')) = (x,y,z,w)A(x',y',z',w')^\top,$$

for a 4×4 symmetric matrix A.

Theorem 15.4

Let A be a symmetric matrix, and let Q be the quadric defined by A. (That is, the points of Q have coordinates (x,y,z,w) satisfying $(x,y,z,w)A(x,y,z,w)^\top = 0$.) Then Q is non-degenerate if and only if $\det(A) \neq 0$.

Proof The proof is similar to the proof of Theorem 3.5. □

Let Q be a quadric of $\mathsf{PG}(3,\mathbb{R})$. The properties of points, lines, planes with respect to Q emulate the properties of points and lines $\mathsf{PG}(2,\mathbb{R})$ with respect to a non-degenerate conic. If we restrict the quadratic form Q to a subspace of $\mathbb{R}^3$, we will have induced a quadratic form on that subspace. Thus, our knowledge of the possible conics, degenerate or non-degenerate, in $\mathsf{PG}(2,\mathbb{R})$ gives us a set of possible ways a plane of $\mathbb{R}^3$ can meet a quadric.

Worked Example 15.5

Consider the hyperbolic quadric Q defined by $x^2 + y^2 - z^2 - w^2 = 0$. The plane $w = 0$ leaves us an equation $x^2 + y^2 - z^2 = 0$ for the points of intersection with Q. Therefore, this plane meets the quadric in the points of a non-degenerate conic. However, if we take the plane $y = w$, the points (x,y,z,w) of intersection satisfy $x^2 - z^2 = 0$, and hence we have two lines for the intersection. □

A plane π is:

TANGENT to Q at P if $\pi \cap Q$ is either $\{P\}$ or a pair of lines intersecting at P;
SECANT to Q if $\pi \cap Q$ is a non-degenerate, non-empty conic of π;
EXTERNAL to Q if $\pi \cap Q = \varnothing$.

You will notice that we have used the word 'the' when we reference a tangent plane to Q at a point P. To see that there is just one tangent plane at P, we need to show that there is just one *degenerate* plane incident with P. By a *degenerate* plane π, we mean that the restriction of Q to π is a degenerate form. Now π is degenerate if there is a point X of π such that for every point Y of π, we have $XAY^\top = 0$ (when X and Y are written in homogeneous coordinates). Indeed, this condition defines π because the set of vectors Z such that $XAZ^\top = 0$ forms a subspace of $\mathbb{R}^4$, which must be a proper subspace of $\mathbb{R}^4$ since Q is non-degenerate. Indeed, this point X is the point P, upon inspecting the degenerate conic intersection. Moreover, it is not possible for a plane to meet a quadric in a double-line, and we leave this as an exercise for the reader (see Exercise 15.13).

We also extend the language we used for points of the plane with respect to a conic, to points of PG(3, $\mathbb{R}$) with respect to a quadric:

- A point is **internal** to a quadric Q if it lies on no tangent plane to Q.
- A point is **external** to Q if it lies on more than one tangent plane to Q.

The map $\tau\colon (x, y, z, w) \mapsto [a, b, c, d]$, where $(a, b, c, d) = (x, y, z, w)A$ is a map from the points of PG(3, $\mathbb{R}$) to the planes of PG(3, $\mathbb{R}$). When Q is non-degenerate, a point–plane pair interchanged by τ are said to be **pole** and **polar**. A polar plane to a point P of Q is the unique tangent plane to Q at P. A pole of a tangent plane is the point at which it is tangent (its **point of contact**).

Theorem 15.6

*Let Q be a non-degenerate quadric of PG(3, $\mathbb{R}$). The pole–polar relationship with respect to Q defines a **polarity** τ of PG(3, $\mathbb{R}$) (a duality with square the identity). Moreover, τ interchanges*

- *points P of Q with the tangent plane to Q at P,*
- *external points E with the secant plane joining the points of contact of the tangent planes on E, and*
- *internal points with external planes.*

Proof Let Q be the quadratic form defining Q, and let f be the bilinear form associated to Q. Let A be the symmetric matrix of f. Let τ be the map $(x, y, z, w)^\tau := [a, b, c, d]$, where $(a, b, c, d) = (x, y, z, w)A$. Dually, we define $[a, b, c, d]^\tau := (x', y', z', w')$, where $(x', y', z', w') = (a, b, c, d)A^{-\top}$. Then it is straightforward to check that τ is incidence-preserving, and so is a duality on the points, lines, and planes of PG(3, $\mathbb{R}$). Moreover, if (x, y, z, w) represents a point of PG(3, $\mathbb{R}$), and $(a, b, c, d) = (x, y, z, w)A$, then $(x, y, z, w)^{\tau^2} = [a, b, c, d]^\tau = (x', y', z', w')$ where

$$(x', y', z', w') = (a, b, c, d)A^{-\top} = (x, y, z, w)AA^{-\top} = (x, y, z, w)AA^{-1} = (x, y, z, w)$$

since A is symmetric. So τ is a polarity.

Let P be a point of PG(3, $\mathbb{R}$). If P lies in Q, then P is zero under the form f. In particular, P is incident with its image under τ because

$$0 = (x, y, z, w)A(x, y, z, w)^\top = (x, y, z, w)^\tau \cdot (x, y, z, w).$$

So τ interchanges P with its tangent plane at P.

Suppose P is an internal point. Then P does not lie on any tangent plane to Q. Let $\pi := P^\tau$, and suppose by way of contradiction that π contains a point X of Q. Then P lies on X^τ, which is the tangent plane to Q at X; a contradiction. Therefore, π is an external plane. So τ interchanges internal points with external planes.

Finally, suppose P is an external point, that is, P lies on two tangent planes ρ and σ to Q. Let $R := \rho^\tau$ and $S := \sigma^\tau$. Then R and S are points of Q. This means that P^τ lies on the line RS because τ is incidence-preserving. Indeed, P^τ contains all of the points of contact of the tangent planes on P (with Q). So τ interchanges external points with secant planes. □

A point P of $\mathrm{PG}(3, \mathbb{R})$ is **absolute** with respect to a polarity τ if and only if P is incident with P^τ. The set of absolute points of the polarity arising from a quadric is precisely the set of points of that quadric. A non-degenerate, non-empty quadric of $\mathrm{PG}(3, \mathbb{R})$ is **hyperbolic** if it contains a line; otherwise **elliptic**.

Theorem 15.7 *(i) Any hyperbolic quadric is projectively equivalent to*

$$x^2 + y^2 - z^2 - w^2 = 0.$$

(ii) Any elliptic quadric is projectively equivalent to

$$x^2 + y^2 + z^2 - w^2 = 0.$$

Proof By Theorem 13.18, there are four conjugacy classes of polarities of $\mathrm{PG}(3, \mathbb{R})$, with respect to the conjugation action of $\mathrm{PGL}(4, \mathbb{R})$. The standard polarity has no absolute points, and the null polarity has every point absolute. So there are two types of quadric arising from the two types of orthogonal polarity. Just as in Worked Example 3.12, the action of $\mathrm{PGL}(4, \mathbb{R})$ on quadrics, as sets of points, is permutationally isomorphic to the action of $\mathrm{PGL}(4, \mathbb{R})$ by conjugation on the polarities arising from quadrics. Finally, note that $x^2 + y^2 - z^2 - w^2 = 0$ contains the line $(x = z) \cap (y = w)$, and that $x^2 + y^2 + z^2 - w^2 = 0$ is the projective completion of the unit sphere of $\mathbb{R}^3$, and therefore contains no line. □

Theorem 15.8
Let C be a non-degenerate, non-empty conic of a plane π of $\mathrm{PG}(3, \mathbb{R})$. Then:

(i) there is a hyperbolic quadric $\mathcal{H}$ of $\mathrm{PG}(3, \mathbb{R})$ with $\mathcal{H} \cap \pi = C$; and
(ii) there is an elliptic quadric $\mathcal{E}$ of $\mathrm{PG}(3, \mathbb{R})$ with $\mathcal{E} \cap \pi = C$.

Proof The plane $z = 0$ meets the hyperbolic quadric $\mathcal{H}_1 : x^2 + y^2 - z^2 - w^2 = 0$ and the elliptic quadric $\mathcal{E}_1 : x^2 + y^2 + z^2 - w^2 = 0$ in the non-degenerate, non-empty conic $C' : x^2 + y^2 - w^2 = 0$. Since all non-degenerate, non-empty conics of the plane are projectively equivalent (Theorem 3.13), there is a collineation g of PG(3, $\mathbb{R}$) taking C' to C. Now $\mathcal{H}_1^g$ and $\mathcal{E}_1^g$ are the desired quadrics. $\qquad \square$

Theorem 15.9
The set of points on the lines of a regulus is a hyperbolic quadric. Hence, every hyperbolic quadric contains exactly two reguli, which are opposite to one another.

Proof First, PGL(4, $\mathbb{R}$) is transitive on triples of pairwise skew lines (by Theorem 13.15), so we may consider the opposite regulus of $\ell: (x = 0) \cap (y = 0)$, $m: (z = 0) \cap (w = 0)$, and $n: (x = z) \cap (y = w)$. The unique transversal to ℓ and m on the point $(1, a, 1, a)$ of n is $[a, -1, 0, 0] \cap [0, 0, a, -1] = \{(s, as, t, at) : s, t \in \mathbb{R}, (s, t) \neq (0, 0)\}$ and the unique transversal to the point $(0, 1, 0, 1)$ of n is $[1, 0, 0, 0] \cap [0, 0, 1, 0] = \{(0, s, 0, t): s, t \in \mathbb{R}, (s, t) \neq (0, 0)\}$ and all these points are on the hyperbolic quadric $xw = yz$ (and, conversely, every point on the hyperbolic quadric $xw = yz$ has this form). $\qquad \square$

Not only are the points of a regulus a hyperbolic quadric (by Theorem 15.9) but there is a unique hyperbolic quadric on three lines of the regulus. Note also that the proof below represents lines differently this time, just to emphasise the two ways we are working with lines in PG(3, $\mathbb{R}$).

Theorem 15.10
Let S be a triple of three skew lines of PG(3, $\mathbb{R}$). Then there is a unique hyperbolic quadric $\mathcal{H}$ of PG(3, $\mathbb{R}$) whose line set contains S.

Proof By Theorem 13.15, we may suppose the set S is the following lines, represented as row spaces of 2×4 matrices:

$$\ell: \begin{bmatrix} I & O \end{bmatrix}, \quad m: \begin{bmatrix} O & I \end{bmatrix}, \quad n: \begin{bmatrix} I & I \end{bmatrix}$$

where I and O are the 2×2 identity and zero matrices, respectively. We already saw in the proof of Theorem 15.9 that the quadric defined by $xw = yz$ contains these three lines. We will show that the converse is true. Let M be a symmetric 4×4 matrix written in block form as

$$\begin{bmatrix} A & B \\ B^\mathsf{T} & C \end{bmatrix}$$

where A, B, C are 2×2 matrices, and A and C are symmetric. Suppose ℓ, m, n lie on the quadric defined by M. Then

$$O = \begin{bmatrix} I & 0 \end{bmatrix} M \begin{bmatrix} I \\ 0 \end{bmatrix} = A, \qquad\qquad O = \begin{bmatrix} 0 & I \end{bmatrix} M \begin{bmatrix} 0 \\ I \end{bmatrix} = C,$$

$$O = \begin{bmatrix} I & I \end{bmatrix} M \begin{bmatrix} I \\ I \end{bmatrix} = A + B^\top + B + C.$$

These equations imply that $A = O$, $C = O$, and $B^\top + B = O$, and hence B is (up to scalar) equal to the matrix $\begin{bmatrix} 0 & 1 \\ -1 & 0 \end{bmatrix}$. Therefore,

$$M = \begin{bmatrix} 0 & 0 & 0 & 1 \\ 0 & 0 & -1 & 0 \\ 0 & -1 & 0 & 0 \\ 1 & 0 & 0 & 0 \end{bmatrix},$$

or, in other words, M defines the quaric $xw - yz = 0$, as required. $\qquad\square$

We can use the theory of quadrics to prove Pascal's theorem in the plane.

Theorem 15.11 (Pascal's theorem revisited (Pascal, 1639))
If A, B, C, A', B', C' lie on a conic C of $\mathrm{PG}(2, \mathbb{R})$, then the intersections of the opposite sides of the hexagon $AB'CA'BC'$ are collinear.

Proof Embed $\mathrm{PG}(2, \mathbb{R})$ as a plane π of $\mathrm{PG}(3, \mathbb{R})$. Then (by Theorem 15.8) there is a hyperbolic quadric $\mathcal{H}$ of $\mathrm{PG}(3, \mathbb{R})$ with of $\mathcal{H} \cap \pi = C$. Let $\mathcal{R}, \mathcal{R}'$ be the reguli of lines contained in $\mathcal{H}$. Let a be the line of $\mathcal{R}$ on A, b be the line of $\mathcal{R}$ on B, c be the line of $\mathcal{R}$ on C, a' be the line of $\mathcal{R}'$ on A', b' be the line of $\mathcal{R}'$ on B', c' be the line of $\mathcal{R}'$ on C'. Now consider the Dandelin skew hexagon with sides $ab'ca'bc'$. Intersecting pairs of planes that correspond to opposite vertices in this skew hexagon give the three lines that can be obtained by joining the three points $a' \cap c$, $b' \cap a$, $c' \cap b$, and so these lines lie in the plane π' spanned by these three points. (To see this, consider the planes $\langle a, a' \rangle$ and $\langle b', c \rangle$: their intersection contains $a' \cap c$ and $a \cap c'$. Consider the planes $\langle a', b \rangle$ and $\langle c, c' \rangle$: their intersection contains $a' \cap c$ and $b \cap c'$. Consider the planes $\langle b, b' \rangle$ and $\langle c', a \rangle$: their intersection contains $b \cap c'$ and $a \cap b'$.)

Each side of the hexagon $AA'BB'CC'$ lies in a plane corresponding to a vertex of the skew hexagon, so the intersections of the opposite sides lie in π'. Hence, the intersections of opposite sides of $AA'BB'CC'$ lie in $\pi \cap \pi'$; that is, they are collinear. $\qquad\square$

Call two lines of PG(3, $\mathbb{R}$) **conjugate** with respect to an elliptic quadric if

- they lie in a plane and are conjugate with respect to the conic of intersection of the plane and the elliptic quadric, or
- they are skew and harmonically separate their polar lines (i.e., transversal lines exist and each meets the lines in two pairs separating each other harmonically).

In 1873, Felix Klein was getting very close to the notion of *isomorphism* via the *Erlanger Programme* and Hesse's *Übertragungsprinzip* (transfer principle). He decided to develop an analogue of Pascal's theorem (Theorem 6.28) for real elliptic quadrics. Using Hesse's *Übertragungsprinzip*, Klein constructed a tortuous path from a complex conic to a real elliptic quadric, and Pascal's theorem on the conic became the following:

Theorem 15.12 (Klein (1883) (see also Klein, 1883))
Given six points A, B, C, D, E, F on an elliptic quadric in PG(3, $\mathbb{R}$), there are unique lines conjugate to: AB and DE; AF and CD; BC and EF and these three lines have a line conjugate to all three of them.

In Euclidean geometry, Pascal's theorem for the conic yields a *three reflection theorem* for the half-turns determined by the three points of intersection of diagonals. If H_P is the half-turn with centre P, then if $ABCDEF$ is a hexagon inscribed in a conic, with $X := AB \cap DE$, $Y := BC \cap EF$, $Z := CD \cap FA$, then the product $H_X H_Y H_Z$ is an involution. Klein was very close to a three reflection theorem. The involutions

$$(AB)(DE)\ldots, \quad (AF)(CD)\ldots, \quad (BC)(EF)\ldots$$

fix every point of the line conjugate to both lines, fix every point of the polar line, and map a point on neither line to its harmonic conjugate with respect to the intersections with the lines of the unique transversal to the lines on the point. Thus, the involutions determine and are determined by the lines; and the product of the three involutions corresponds to the line conjugate to all three of them. (The corresponding lines are secant; their polar lines are external.) So Klein was very close to obtaining the following nice result:

Theorem 15.13
Given six points A, B, C, D, E, F on an elliptic quadric in PG(3, $\mathbb{R}$), there are unique lines conjugate to AB and DE; AF and CD; BC and EF and these three lines have a common point.

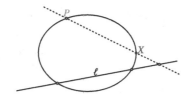

Figure 15.5 Projection of a conic onto a line.

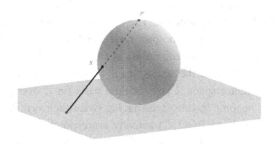

Figure 15.6 Stereographic projection.

We will leave the details of the proof as an exercise for the reader, but a hint is to first recognise that Pascal's theorem (Theorem 6.28) is true over the complex numbers, in PG(2, $\mathbb{C}$). Then, one *transfers* the properties of a conic of PG(2, $\mathbb{C}$) to an elliptic quadric of PG(3, $\mathbb{R}$). Worked Example 15.14 introduces the necessary lines of transfer between the two spaces, commonly known as *stereographic projection*.

Worked Example 15.14

Consider a conic $\mathcal{K}$ of PG(2, $\mathbb{C}$), and let P be a point of $\mathcal{K}$, and let ℓ be a line not incident with P (see Figure 15.5). So ℓ meets $\mathcal{K}$ in two points A and B, or one point with multiplicity 2. At any rate, we can map the points of $\mathcal{K} \setminus \{P\}$ to the points of ℓ via projection: $X \mapsto PX \cap \ell$. In other words, we map a point X of $\mathcal{K}$ to the point of PX on $\mathcal{K}$, different from P.

The points A and B are fixed, and this map is a bijection. Moreover, this map preserves cross-ratio. This is the complex analogue of the development of cross-ratio on a conic in Section 5.4.

Now think of ℓ as a real plane! The field $\mathbb{C}$ has degree 2 over $\mathbb{R}$ and so we can associate the (affine) points of ℓ as pairs of real numbers, the real and imaginary parts of a complex number. Embed the projective completion π of this plane into PG(3, $\mathbb{R}$), and consider an elliptic quadric Q tangent to π (see Figure 15.6). Let P' be a point of Q not incident with π. Then we can project the other points of Q to the plane π, and vice versa: $X \mapsto P'X \cap \pi$.

So we have a cross-ratio-preserving sequence of maps:

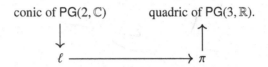

It turns out that Pascal's theorem for a conic of $PG(2, \mathbb{C})$ translates to Theorem 15.13 via this correspondence. The interested reader might also ponder another way to connect the two spaces via the Klein correspondence. Via 'field reduction', we can associate points of $PG(2, \mathbb{C})$ with (disjoint) lines of $PG(5, \mathbb{R})$: the subspaces of the vector space $\mathbb{C}^3$ can be naturally mapped to subspaces of the vector space $\mathbb{R}^6$ by taking a basis for $\mathbb{C}$ over $\mathbb{R}$, except that the $\mathbb{C}$-subspaces become $\mathbb{R}$-subspaces having twice their original dimension. The points of a conic of $PG(2, \mathbb{C})$ actually correspond to singular lines for an elliptic quadric of $PG(5, \mathbb{R})$. Under the Klein correspondence, these lines correspond to point–plane incident pairs of an elliptic quadric of $PG(3, \mathbb{R})$.

We refer the interested reader to Rowe (1989) on Klein's employment of Hesse's transfer principle.   □

We visited **skew projection** in Chapter 5, and here we explore it more deeply. Let ℓ, m be skew lines of $PG(3, \mathbb{R})$. Then every point P not on ℓ or m lies on a unique transversal t_P to ℓ and m: the plane $P\ell$ meets the plane Pm in the line t_P. Next, fix a plane π. We can then define a map ρ from points P (not on ℓ or m) to points of π by $P \mapsto t_P \cap \pi$. This is an example of a **quadratic transformation** and dates back to the work of Steiner (1832, pp. 251–70), and perhaps Poncelet (1822, Chapter 2, Section III), though we could not find an explicit statement of skew projection in Poncelet's *Traité*, just a rather in-depth analysis of the properties of quadrics.

Theorem 15.15
The skew projection map ρ defined above takes lines skew to the given two lines, to conics of π.

Proof Let n be a line skew to ℓ and m. Then n, ℓ, m determine a regulus and hyperbolic quadric $\mathcal{H}$ (by Theorem 15.10). The result follows by noticing that the plane π meets $\mathcal{H}$ in a conic (see Figure 15.7).   □

Recall that a line either lies in a given plane or intersects it in one point. So in skew projection, we know that none of the three lines ℓ, m, n can be disjoint from the fixed plane π, and no two of the lines are coplanar. Therefore, we have two cases:

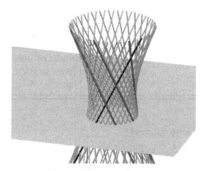

Figure 15.7 Skew projection from the view of the quadric generated by the three skew lines.

1. one line lies in the fixed plane and the other two intersect it in one point;
2. all three lines intersect the fixed plane in one point.

Let $\mathcal{K}$ be the conic of π cut out by the quadric defined by the three skew lines. In the first case, $\mathcal{K}$ is degenerate, and is simply two intersecting lines. In the second case, there is one more subtlety to consider. If the three skew lines intersect the plane in three collinear points, then $\mathcal{K}$ is degenerate and we again obtain two intersecting lines. In the case that the three points of intersection $L := \ell \cap \pi$, $M := m \cap \pi$, $N := n \cap \pi$ are non-collinear, we will see that $\mathcal{K}$ is non-degenerate. Suppose not, and so $\mathcal{K}$ is two intersecting lines. Without loss of generality, LM is one of these lines. However, the polar image of LM is again LM (for it is a line of the quadric) and so the polar image $\pi^{\perp}$ of π lies on LM. Therefore, the polar image $L^{\perp}$ of L is the plane $\langle LM, \ell \rangle$, and likewise the polar image $M^{\perp}$ of M is the plane $\langle LM, m \rangle$. Let $X := \ell \cap m$. Then $\langle \ell, m \rangle$ is the polar image $X^{\perp}$ of X (because X lies on the quadric), and so $\langle LM, \ell \rangle = X^{\perp}$ because $L, M, \ell \subset X^{\perp}$. Likewise, $\langle LM, m \rangle = X^{\perp}$ and so the lines ℓ and m lie in the same plane; which is a contradiction. Therefore, $\mathcal{K}$ is non-degenerate. So we have:

Theorem 15.16

If the three skew lines meet the plane π in three non-collinear points, then skew projection generates a non-degenerate conic. Otherwise, skew projection generates a degenerate conic in the form of two intersecting lines.[2]

A **bundle of lines** is the set of all lines through a point.

---

[2] The first author thanks his former student Dominique Douglas-Smith for working out the details of this result as part of her honours dissertation.

Table 15.1 *The quotient space* $\mathsf{PG}(3, \mathbb{R})/P$
isomorphic to $\mathsf{PG}(2, \mathbb{R})$.

Points	Bundle of lines on P
Lines	Planes incident with P
Incidence	Natural

Theorem 15.17 (Quotient spaces)
*Let P be a point of $\mathsf{PG}(3, \mathbb{R})$. Then the incidence structure with 'points' the bundle of lines on P, and 'lines' the planes on P, is isomorphic to $\mathsf{PG}(2, \mathbb{R})$, called the **quotient space** of $\mathsf{PG}(3, \mathbb{R})$ by P (and denoted by $\mathsf{PG}(3, \mathbb{R})/P$).*

Proof Since the collineation group of $\mathsf{PG}(3, \mathbb{R})$ is transitive on points, we may assume that $P: (0, 0, 0, 1)$. So a line incident with P can be written uniquely as PQ where Q has homogeneous coordinates $(a, b, c, 0)$. Similarly, a plane incident with P can be written uniquely in dual homogeneous coordinates as $[r, s, t, 0]$. The isomorphism we desire is

$$PQ \mapsto (a, b, c), \quad \text{where } Q = (a, b, c, 0)$$

$$[r, s, t, 0] \mapsto [r, s, t]$$

and we leave the rest to the reader to verify that this is an isomorphism of geometries. □

We can study dualities taking the quotient space of $\mathsf{PG}(3, \mathbb{R})$ by P to the quotient space of $\mathsf{PG}(3, \mathbb{R})$ by Q. These dualities will map every line on P to a plane on Q, and every plane on P to a line on Q.

Theorem 15.18 (Projective generation of quadrics (Seydewitz, 1847))
Let ϕ be a duality taking the quotient space of $\mathsf{PG}(3, \mathbb{R})$ by P to the quotient space of $\mathsf{PG}(3, \mathbb{R})$ by Q, $P \neq Q$. Then the union of $\ell \cap \ell^\phi$ over the lines ℓ on P is a (possibly degenerate) quadric on both P and Q, and any non-degenerate quadric on P and Q can be represented in this form.

Proof Let P be $(1, 0, 0, 0)$ and Q be $(0, 1, 0, 0)$. Any plane through Q will have coordinates of the form $[e, 0, f, g]$; let the point R, with coordinates (x, y, z, w), be on both the plane and the corresponding line through P. The line corresponds to the point (y, z, w) of the quotient space by P, and the duality is represented algebraically by the equation

$$(e, f, g) = (ay + bz + cw, a'y + b'z + c'w, a''y + b''z + c''w).$$

The condition that R is on both the plane and the corresponding line is

$$(ay + bz + cw)x + (a'y + b'z + c'w)z + (a''y + b''z + c''w)w = 0.$$

This the most general form of a homogeneous quadratic in which the terms in x^2 and y^2 are missing (subject only to the condition that the matrix of the duality is invertible). Hence, every non-degenerate quadric passing through P, Q can be represented in this way.[3] □

We remind the reader that if G is a permutation group on a set Ω, and if Δ is a subset of Ω, then G_Δ is the setwise stabiliser of Δ in G, and G_Δ^Δ is the permutation group induced on Δ, which is isomorphic to $G_\Delta / G_{(\Delta)}$ where $G_{(\Delta)}$ is the kernel of the action of G_Δ on Δ.

Theorem 15.19

The stabiliser in $\mathsf{PGL}(4, \mathbb{R})$ *of a hyperbolic quadric is isomorphic to*

$$(\mathsf{PGL}(2, \mathbb{R}) \times \mathsf{PGL}(2, \mathbb{R})) \rtimes C_2$$

where the cyclic group of order 2 in this semi-direct product permutes the two factors of the direct product. If $\mathcal{R}$ is a regulus, then $(\mathcal{R}, \mathsf{PGL}(4, \mathbb{R})_\mathcal{R}^\mathcal{R})$ is permutationally isomorphic to $(\mathsf{PG}(1, \mathbb{R}), \mathsf{PGL}(2, \mathbb{R}))$.

A **quadric cone** of $\mathsf{PG}(3, \mathbb{R})$ is a set of points on a line from a fixed point V to a point P of a fixed non-empty, non-degenerate conic of a plane π, not on V. The point V is called the **vertex** of the cone.

Theorem 15.20

$\mathsf{PGL}(4, \mathbb{R})$ *acts transitively on quadric cones of* $\mathsf{PG}(3, \mathbb{R})$.

Proof First, $\mathsf{PGL}(4, \mathbb{R})$ is transitive on planes of $\mathsf{PG}(3, \mathbb{R})$, by Theorem 13.6. The stabiliser of a plane π of $\mathsf{PG}(3, \mathbb{R})$ is isomorphic to $\mathsf{AGL}(3, \mathbb{R})$, by Theorem 13.11. Next, $\mathsf{AGL}(3, \mathbb{R})$ acts transitively on the points not incident with π, which follows from the transitivity of the affine group on the points of affine space, but the reader can also verify this easily as an exercise. Finally, the stabiliser of π is also transitive on conics of π by Theorem 3.13. □

Given distinct quadrics $F(x, y, z, w) = 0$ and $G(x, y, z, w) = 0$, the **pencil** of quadrics they generate is the set of quadrics of the form

$$\lambda \cdot F(x, y, z, w) + \mu \cdot G(x, y, z, w) = 0,$$

where $\lambda, \mu \in \mathbb{R}$, $(\lambda, \mu) \neq (0, 0)$. A pencil is uniquely determined by two quadrics belonging to the pencil. Some members of the pencil will be degenerate.

---

[3] The condition that the matrix of the duality is invertible can be seen as a condition of the form $db''c' + eb'' + fc + g \neq 0$ for some fixed d, e, f, g, and $b'' + c'$ is the coefficient h of wz in the equation of the quadric, giving a quadratic in c' that must fail to be satisfied for some c'.

Example 15.21

The hyperbolic quadrics defined by $xy = zw$ and $x(y + x) = z(z + w)$ share the line $x = z = 0$. The degenerate members of the pencil

$$\lambda \cdot (xy - zw) + \mu \cdot (x(y + x) - z(z + w)), \quad (\lambda, \mu) \in \mathbb{R}^2 \setminus \{(0, 0)\}$$

can be determined by computing the determinant of the Gram matrix

$$\begin{bmatrix} \mu & \frac{\lambda+\mu}{2} & 0 & 0 \\ \frac{\lambda+\mu}{2} & 0 & 0 & 0 \\ 0 & 0 & -\mu & \frac{-\lambda-\mu}{2} \\ 0 & 0 & \frac{-\lambda-\mu}{2} & 0 \end{bmatrix}$$

which is $\frac{(\lambda+\mu)^2 \lambda - \mu^2}{4}$, showing that $(xy - zw) + (x(y + x) - z(z + w))$ and $(xy - zw) - (x(y + x) - z(z + w))$ are the degenerate members of the pencil. Each is a cone; the first has vertex $(0, 0, 0, 1)$ and the second has vertex $(0, 1, 0, 0)$.

15.3 Eight Associated Points

We saw in Theorem 5.17 that five points of $\mathsf{PG}(2, \mathbb{R})$, no three collinear, determine a non-degenerate conic. So two non-degenerate conics $\mathcal{K}_1$ and $\mathcal{K}_2$ can meet in at most four points, and these four points will be in *general position*: no three will be collinear. These points determine a pencil of conics, and so any conic on these four points will be of the form $\lambda \mathcal{K}_1 + \mu \mathcal{K}_2 = 0$.

Recall that a quadric of $\mathsf{PG}(3, \mathbb{R})$ is the zero-set of a degree 2 polynomial of the form

$$Ax^2 + By^2 + Cz^2 + Dw^2 + Exy + Fyz + Gzw + Hwx + Ixz + Jyw.$$

There are 10 coefficients, and so a quadric is determined by 9 *general* points. Now three quadrics (Q_1, Q_2, Q_3) in *general position* will meet in eight points. We will see that any quadric lying on seven of these points will be of the form $\lambda Q_1 + \mu Q_2 + \nu Q_3 = 0$ and will pass through the eighth point.

Theorem 15.22 (Theorem of the eight associated points (Lamé, 1818))
Given eight distinct points which are the set of intersections of three quadrics of $\mathsf{PG}(3, \mathbb{R})$, all quadrics through any subset of seven of the points must pass through the eighth point.

Proof Let the points be $P_1, P_2, \ldots, P_8$, and the quadrics be $F = 0$, $F^* = 0$, and $F' = 0$. Let $G = 0$ be a quadric on the seven points $P_1, P_2, \ldots P_7$.

If $G = aF + bF^* + cF'$, then $G = 0$ contains the point P_8 also, since this lies on the three quadrics $F = 0, F^* = 0$, and $F' = 0$. To prove our theorem by contradiction, we therefore assume that G is not linearly dependent on F, F^*, and F'. The system of quadrics $aF + bF^* + cF' + dG = 0$ can now be made to satisfy three linear conditions, and can therefore be made to contain any three given points of $\mathrm{PG}(3, \mathbb{R})$. We show that this assumption leads to a contradiction.

We first note that of the eight points P_i no three can be collinear, since the three quadrics which defined the set of points would contain the line of collinear points, and therefore intersect more than eight points. Again, no four of the seven points $P_1, \ldots, P_7$ can be coplanar. For suppose that P_1, P_2, P_3 and P_4 are coplanar. We can find a quadric of the system above to contain two points P, Q in the plane of the four points, where P and Q do not lie on any conic through the four points P_1, P_2, P_3 and P_4, and also to contain an arbitrary point R of $\mathrm{PG}(3, \mathbb{R})$. This quadric must be reducible, one component being the plane through the four points P_1, P_2, P_3, P_4. If it were not reducible, it would cut this plane in a conic through the points P_1, P_2, P_3, P_4, P, Q.

The other component of the quadric must be a plane, and this plane has to contain the points P_5, P_6, P_7, and R. We have already ruled out the possibility that P_5, P_6, and P_7 are collinear. There is consequently a unique plane through these three points, and this cannot contain R, which is an arbitrarily chosen point of $\mathrm{PG}(3, \mathbb{R})$. This contradiction shows that with our hypothesis, the points P_1, P_2, P_3, and P_4 cannot be coplanar. We now have that no four of the seven points $P_1, P_2, \ldots, P_7$ coplanar. Choose three points P, Q, and R in the plane $P_1 P_2 P_3$ such that the six points do not lie on a conic. We then find a quadric of the system given above to contain P, Q, and R. This quadric will contain the plane $P_1 P_2 P_3$, and is therefore reducible into two planes. The other plane must contain the points P_4, P_5, P_6, and P_7. But these are not coplanar. We therefore have a contradiction. □

We call a set of eight points as in the theorem **associated**. This theorem was rediscovered by Hesse (1840a).

Another proof of Möbius' theorem (Theorem 14.9) Consider the quadrics:

Q_1:	the product of the planes BCD and $B'C'D'$;
Q_2:	the product of the planes CDA and $C'D'A'$;
Q_3:	the product of the planes DAB and $D'A'B'$.

Then $Q_1 \cap Q_2 \cap Q_3 = \{A, B, C, D, A', B', C', D'\}$. So we have a set of eight associated points, and thus the quadric Q_4: the product of the planes ABC and $A'B'C'$, which contains the seven points A', B', C', D', A, B, C, also contains the eighth point D, by Theorem 15.22. Thus, $D \in A'B'C'$. □

Exercises

15.1 Given a skew hexagon in $PG(3, \mathbb{R})$, show that the locus of the vertex of a quadric cone tangent to the sides of the hexagon is a non-degenerate ruled quadric containing the diagonals of the hexagon as generators.

15.2 An involution is set up on the lines of a regulus of $PG(3, \mathbb{R})$ with fixed lines ℓ and m. Also, A is a point on no line of the regulus and the transversal from A to ℓ and m is a line n whose polar line is n'. Prove that the transversals from A to pairs of lines of the regulus in involution all lie in the plane containing A and n'.

15.3 Four coplanar points are taken on a non-degenerate plane section of a non-degenerate hyperbolic quadric Q of $PG(3, \mathbb{R})$. Prove that the 16 intersections of the 8 generators through them lie 4 in each of 4 planes and that the tetrahedron with faces these 4 planes is self-polar with respect to Q.

15.4 Given a non-degenerate quadric of $PG(3, \mathbb{R})$, prove that, if the lines joining the vertices of one tetrahedron to the poles of the faces of another tetrahedron meet in a point, then the lines joining the vertices of the second tetrahedron to the poles of the faces of the first tetrahedron also meet in a point.

15.5 Given a non-degenerate quadric of $PG(3, \mathbb{R})$, prove that a line meets its polar line with respect to the quadric if and only if the line is a tangent to the quadric. (Here, lines contained in a quadric are considered as tangent lines to the quadric.)

15.6 A variable line ℓ through a given point is perpendicular to its polar line with respect to a given quadric. Prove that ℓ generates a quadric cone.

15.7 Prove that the perpendiculars from B, C, D to the opposite faces of the tetrahedron $ABCD$ meet the perpendicular in the plane of the triangle BCD through the orthocentre of this triangle. Hence, show that the perpendiculars from the vertices of a tetrahedron to the opposite faces lie on a quadric Q, and that the perpendiculars through the orthocentres also lie on Q. Show that each face of the tetrahedron meets Q in a rectangular hyperbola, and deduce that Q contains a set of three mutually perpendicular generators.

15.8 Prove that the plane sections of a quadric are projected from a given point O of the quadric into conics passing through two (possibly imaginary) given points of the plane π of projection. If these fixed points are interpreted as the circular points in π, show that

(i) the centre of a circle in π is the intersection with π of the line joining O to the pole of the plane of the corresponding section of the quadric;

(ii) orthogonal circles arise from conjugate planes.

15.9 If S, S_1, S_2 are three conics on a quadric such that S is tangent to S_1 and S_2 at P and Q, respectively, show that PQ passes through the vertex of one of the cones through S_1 and S_2. Hence, show that there are eight circles which are tangent to three given circles.

15.10 Prove that the locus of the vertex of a cone which touches the sides of a skew hexagon of $\mathrm{PG}(3, \mathbb{R})$ consists of the planes containing two consecutive sides of the hexagon and a quadric. In what circumstances is the quadric degenerate? Show that the locus is the whole space if and only if the hexagon lies on a quadric, or is a plane hexagon circumscribing a conic.

15.11 This exercise is based on a result of Serret (1862): *If a quadric is tangent to the four edges of a skew quadrilateral, then the points of tangency are coplanar.* Suppose every edge of a tetrahedron is tangent to a given quadric. Prove that the three line segments joining the points of tangency of the three pairs of opposite edges of the tetrahedron are concurrent.

15.12 A quadric Q of $\mathrm{PG}(3, \mathbb{R})$ circumscribes a tetrahedron T, so that each face F of T meets Q in a non-degenerate conic.

(a) Show that the vertices of the triangle formed by the tangent lines to $F \cap Q$ are in perspective with the vertices of F from a point O.

(b) Show that, if the tangent plane to Q at one vertex of T passes through the point O belonging to the opposite face of T, then the tangent planes at all four vertices have this property.

(c) Show that, if the four points O arising from the four faces of T are coplanar, then Q is a cone, and conversely.

15.13 Let Q be a non-degenerate quadric of $\mathrm{PG}(3, \mathbb{R})$ and let π be a plane. Without using any characterisation of the quadrics of $\mathrm{PG}(3, \mathbb{R})$, show that π cannot meet Q in a double-line.

15.14 Show that the stabiliser in $\mathrm{PGL}(4, \mathbb{R})$ of an elliptic quadric of $\mathrm{PG}(3, \mathbb{R})$ is isomorphic to $\mathrm{PGL}(2, \mathbb{C}) \rtimes C_2$.

16

Line Geometry

The work of the Italian School [of algebraic geometry], and in particular that of C. Segre and E. Bertini, rests on four correlated pillars: Plücker's so-called line-geometry, which was interpreted as a starting point towards a new idea of multi-dimensional space in which the 'points' were able to assume different meanings; Grassmann's point of view with regard to 'abstract' linear space as the natural framework for any particular geometrical interpretation; von Staudt's notion of the 'coordinatization' of synthetic projective space as a fundamental tool for the systematic translation of geometrical concepts into algebraic ones (and vice versa); and finally, algebraic ideas presented by Weierstraßand Frobenius, but suitably translated into geometrical terms.

Aldo Brigaglia (1996, pp. 155–6).[1]

Duality means that points and planes are on an equal footing in $PG(3, \mathbb{R})$. We can see three-dimensional real projective space either as built out of points or as built out of planes, as the fundamental underlying objects. It was Julius Plücker in 1868 in his *Neue Geometrie des Raumes, gegründet auf die Betrachtung der geraden Linie als Raumelement* (New geometry of space, based upon the consideration of the straight line as a space element) who first conceived of *lines* as the fundamental elements that three-dimensional real projective space can be built from. Since the totality of lines of $PG(3, \mathbb{R})$ forms a four-dimensional object, this gives a way to use low-dimensional geometric thinking to study higher dimensions.

16.1 Linear Independence of Lines and Plücker Coordinates

A line of $PG(3, \mathbb{R})$ is defined by two points, say (x_1, x_2, x_3, x_4) and (y_1, y_2, y_3, y_4). There are eight coordinates in total; however, not all eight items

[1] Reprinted by permission from Springer Nature: *The Influence of H. Grassmann on Italian Projective N-Dimensional Geometry*. Brigaglia, Aldo. © 1996.

of information are needed to determine the given line. For instance, if x_1 is non-zero, we can *row-reduce* our pair of points and replace the second one with $(0, y_2', y_3', y_4')$; we have reduced our eight items to seven. We know that not all of the y_i' are zero, so we may suppose for the sake of argument that y_2' is non-zero, and *row-reduce* again: we replace the first point with $(x_1', 0, x_3', x_4')$. So we have reduced the amount of information to six items. In general, this is all that's needed, but we do not need to employ a case analysis to determine when coordinates are zero or not. Plücker's innovation is that there is a simple and clever way to derive a sixtuple that uniquely defines a line of $\mathrm{PG}(3, \mathbb{R})$.

We denote a line ℓ joining the points (x_1, x_2, x_3, x_4) and (y_1, y_2, y_3, y_4) by **homogeneous line coordinates**

$$(x_1y_2 - x_2y_1, \; x_1y_3 - x_3y_1, \; x_1y_4 - x_4y_1, \; x_2y_3 - x_3y_2, \; x_4y_2 - x_2y_4, \; x_3y_4 - x_4y_3). \tag{16.1}$$

This is well-defined: if (x_1', x_2', x_3', x_4') and $(y_1', y_2', y_3', y_4')'$ are also distinct points on ℓ, then

$$(x_1y_2 - x_2y_1, x_1y_3 - x_3y_1, x_1y_4 - x_4y_1, x_2y_3 - x_3y_2, x_4y_2 - x_2y_4, x_3y_4 - x_4y_3) =$$
$$(x_1'y_2' - x_2'y_1', x_1'y_3' - x_3'y_1', x_1'y_4' - x_4'y_1', x_2'y_3' - x_3'y_2', x_4'y_2' - x_2'y_4', x_3'y_4' - x_4'y_3').$$

For, if $x_i' = ax_i + by_i$ and $y_i' = cx_i + dy_i$ for $i = 1, 2, 3, 4$ with $a, b, c, d \in \mathbb{R}$ with $ad - bc \neq 0$, then $x_i'y_j' - x_j'y_i' = (x_iy_j - x_jy_i)(ad - bc)$, for $1 \leqslant i < j \leqslant 4$. These are called the **Plücker coordinates** of the line ℓ.

JULIUS PLÜCKER (1801–68) was a professor in physics at Bonn, who worked in geometry before 1847 and after 1865, but in physics in the intermediate period. Plücker took a major step towards the discovery of the electron in 1858 by noticing the effect of a magnetic field on cathode rays, which he interpreted as a negative electric charge on the particles making up the cathode ray. He also greatly advanced analytic projective geometry, and the first clear explanation of duality in projective geometry, and pioneered an approach to geometry in which the line is a primitive element.

Theorem 16.1 (Klein, 1868)
Two lines ℓ and ℓ', with Plücker coordinates $(\ell_1, \ell_2, \ldots, \ell_6)$ and $(\ell_1', \ell_2', \ldots, \ell_6')$, respectively, are coplanar if and only if $\ell_1\ell_6' + \ell_6\ell_1' + \ell_2\ell_5' + \ell_5\ell_2' + \ell_3\ell_4' + \ell_4\ell_3' = 0$.

Proof We let ℓ be the span of the points (x_1, x_2, x_3, x_4) and (y_1, y_2, y_3, y_4), and we let ℓ' be the span of the points (x_1', x_2', x_3', x_4') and (y_1', y_2', y_3', y_4') so that the Plücker coordinates of these lines are (respectively)

$$(x_1y_2 - x_2y_1, x_1y_3 - x_3y_1, x_1y_4 - x_4y_1, x_2y_3 - x_3y_2, x_4y_2 - x_2y_4, x_3y_4 - x_4y_3),$$
$$(x_1'y_2' - x_2'y_1', x_1'y_3' - x_3'y_1', x_1'y_4' - x_4'y_1', x_2'y_3' - x_3'y_2', x_4'y_2' - x_2'y_4', x_3'y_4' - x_4'y_3').$$

Consider the determinant

$$\begin{vmatrix} x_1 & x_2 & x_3 & x_4 \\ y_1 & y_2 & y_3 & y_4 \\ x_1' & x_2' & x_3' & x_4' \\ y_1' & y_2' & y_3' & y_4' \end{vmatrix}.$$

If this determinant is expanded into 2×2 minors, we have

$$\begin{vmatrix} x_1 & x_2 & x_3 & x_4 \\ y_1 & y_2 & y_3 & y_4 \\ x_1' & x_2' & x_3' & x_4' \\ y_1' & y_2' & y_3' & y_4' \end{vmatrix} = + \begin{vmatrix} x_3 & x_4 \\ y_3 & y_4 \end{vmatrix}\begin{vmatrix} x_1' & x_2' \\ y_1' & y_2' \end{vmatrix} - \begin{vmatrix} x_2 & x_4 \\ y_2 & y_4 \end{vmatrix}\begin{vmatrix} x_1' & x_3' \\ y_1' & y_3' \end{vmatrix} + \begin{vmatrix} x_2 & x_3 \\ y_2 & y_3 \end{vmatrix}\begin{vmatrix} x_1' & x_4' \\ y_1' & y_4' \end{vmatrix}$$
$$+ \begin{vmatrix} x_1 & x_4 \\ y_1 & y_4 \end{vmatrix}\begin{vmatrix} x_2' & x_3' \\ y_2' & y_3' \end{vmatrix} - \begin{vmatrix} x_1 & x_3 \\ y_1 & y_3 \end{vmatrix}\begin{vmatrix} x_2' & x_4' \\ y_2' & y_4' \end{vmatrix} + \begin{vmatrix} x_1 & x_2 \\ y_1 & y_2 \end{vmatrix}\begin{vmatrix} x_3' & x_4' \\ y_3' & y_4' \end{vmatrix},$$

which is precisely $\ell_1\ell_6' + \ell_6\ell_1' + \ell_2\ell_5' + \ell_5\ell_2' + \ell_3\ell_4' + \ell_4\ell_3'$. So the result follows by noting that the (4×4)-determinant above is zero precisely when ℓ and ℓ' are coplanar. □

Corollary 16.2

The Plücker coordinates $(\ell_1, \ell_2, \dots, \ell_6)$ *of any line* ℓ *of* $\mathsf{PG}(3, \mathbb{R})$ *satisfy the equation* $\ell_1\ell_6 + \ell_2\ell_5 + \ell_3\ell_4 = 0$.

We will see in Section 20.1 that the Plücker coordinates of lines can be considered points of a quadric in five-dimensional projective space. Hence, some of the results in this chapter will have their proofs delayed until later.

Worked Example 16.3

Let P be a point incident with a plane π, of $\mathsf{PG}(3, \mathbb{R})$. A line ℓ that is incident with P and π has its Plücker coordinates of the form

$$\ell_i = \lambda a_i + \mu b_i$$

where a and b are two fixed lines of the plane-pencil determined by P and π.

To see why, let c be a line not incident with P, so that the lines of the plane-pencil on P and π are determined by the points of intersection with c. Fix two points C and C' of c. So the lines of the plane-pencil (on P and π) can each be written as $\langle P, \lambda C + \mu C' \rangle$ where $\lambda, \mu \in \mathbb{R}$ (not both zero). Give P homogeneous coordinates (p_1, p_2, p_3, p_4). Likewise, the homogeneous coordinates of C and

C' are (c_1, c_2, c_3, c_4) and (c'_1, c'_2, c'_3, c'_4), respectively. So the Plücker coordinates of the line $\langle P, \lambda C + \mu C' \rangle$ are

$$\left(p_1(\lambda c_2 + \mu c'_2) - p_2(\lambda c_1 + \mu c'_1), \ldots, p_3(\lambda c_4 + \mu c'_4) - p_4(\lambda c_3 + \mu c'_3) \right)$$

$$= \left(\lambda(p_1 c_2 - p_1 c_1) + \mu(p_1 c'_2 - p_1 c'_1), \ldots \lambda(p_3 c_4 - p_4 c_3) + \mu(p_3 c'_4 - p_4 c'_3) \right)$$

$$= \lambda(p_1 c_2 - p_1 c_1, \ldots, p_3 c_4 - p_4 c_3) + \mu(p_1 c'_2 - p_1 c'_1, \ldots, p_3 c'_4 - p_4 c'_3).$$

With $a := \langle P, C \rangle$ and $b := \langle P, C' \rangle$, the observation then follows. $\qquad \square$

Worked Example 16.4

Let ℓ, m, n be three mutually skew lines. Then it turns out that a line r of the regulus determined by ℓ, m, n has its Plücker coordinates of the form

$$r_i = \lambda \ell_i + \mu m_i + \nu n_i.$$

The reader can use a similar argument to that adopted in Worked Example 16.4, where we express each line as a span of two points, and write the Plücker coordinates in terms of the homogeneous coordinates of these points. In the end, we have the simple expression above for the lines of a regulus. $\qquad \square$

We will see that the set of lines, when written in their Plücker coordinates, satisfying one linear equation of the form

$$\alpha_1 \ell_1 + \alpha_2 \ell_2 + \alpha_3 \ell_3 + \alpha_4 \ell_4 + \alpha_5 \ell_5 + \alpha_6 \ell_6 = 0$$

form a **linear complex**. If we instead have two independent linear equations, we end up with a **linear congruence**, and if we have three independent linear equations, we obtain a **regulus**. A subtle point arises here: in all cases we must stipulate that the lines also satisfy $\ell_1 \ell_6 + \ell_2 \ell_5 + \ell_3 \ell_4 = 0$, as a global assumption.

Define addition of lines ℓ and m via addition of their Plücker coordinates. Recall that the operation $\wedge$ on affine lines ℓ and m was defined as follows. Let ℓ_∞ and m_∞ be the intersections of ℓ and m with the plane at infinity π, and let $\perp$ be the polarity of the absolute conic in π. Then

$$\ell \wedge m := \langle \ell_\infty^\perp \cap m_\infty^\perp, \ell \rangle \cap \langle \ell_\infty^\perp \cap m_\infty^\perp, m \rangle.$$

We have the following *Jacobi identity* for lines. When ℓ and m are non-parallel, $\ell \wedge m$ is the unique perpendicular transversal to ℓ and m.

Theorem 16.5

Let ℓ, m, and n be three affine lines that are mutually non-parallel. Then

$$\ell \wedge (m \wedge n) = m \wedge (\ell \wedge n) + n \wedge (m \wedge \ell).$$

Proof Write ℓ as the span of an affine point $(1, x, y, z)$ and $\ell_\infty = (0, a, b, c)$. Then the Plücker coordinates of ℓ are $(a, b, c, xb - ya, xc - za, zb - yc)$. So we can think of the Plücker coordinates of ℓ as consisting of ℓ_∞ for the first three coordinates (minus the 0 at the front) together with the vector cross product of the triples (x, y, z) and (a, b, c). So we will write ℓ as $(\ell_\infty, \ell_\infty \times \ell_{\text{aff}})$ where we identify ℓ_∞ with (a, b, c) and $\ell_{\text{aff}} = (x, y, z)$. Likewise, we can represent m and n in the same way for a choice of m_{aff} and n_{aff}.

A straightforward calculation shows that $\ell \wedge m$ has Plücker coordinates

$$(\ell_\infty \times m_\infty, \ell_\infty \times m_{\text{aff}} + \ell_{\text{aff}} \times m_\infty).$$

So, using this generic formula, we can work out the right-hand side:

$$m \wedge (\ell \wedge n) + n \wedge (m \wedge \ell)$$

$$= m \wedge (\ell_\infty \times n_\infty, \ell_\infty \times n_{\text{aff}} + \ell_{\text{aff}} \times n_\infty) + n \wedge (m_\infty \times \ell_\infty, m_\infty \times \ell_{\text{aff}} + m_{\text{aff}} \times \ell_\infty)$$

$$= (m_\infty \times (\ell_\infty \times n_\infty) + n_\infty \times (m_\infty \times \ell_\infty), m_\infty \times (\ell_\infty \times n_{\text{aff}} + \ell_{\text{aff}} \times n_\infty)$$

$$+ m_{\text{aff}} \times (\ell_\infty \times n_\infty) + n_\infty \times (m_\infty \times \ell_{\text{aff}} + m_{\text{aff}} \times \ell_\infty) + n_{\text{aff}} \times (m_\infty \times \ell_\infty)),$$

which by the Jacobi identity for the vector cross product is equal to

$$(\ell_\infty \times (m_\infty \times n_\infty), \ell_\infty \times (m_\infty \times n_{\text{aff}} + m_{\text{aff}} \times n_\infty) + \ell_{\text{aff}} \times (m_\infty \times n_\infty)) = \ell \wedge (m \wedge n). \quad \square$$

Corollary 16.6
$\ell \wedge (m \wedge n)$ *is perpendicular to* $(m \wedge (\ell \wedge n)) \wedge (n \wedge (m \wedge \ell))$.

Proof Let $a = \ell \wedge (m \wedge n)$, $b = m \wedge (\ell \wedge n)$, and $c = n \wedge (m \wedge \ell)$. Then in Plücker coordinates, we have

$$a \cdot (b \wedge c) = (b + c) \cdot (b \wedge c) = b \cdot (b \wedge c) + c \cdot (b \wedge c) = 0. \quad \square$$

We now define **linear dependence** and **linear independence** of lines (see Table 16.1). If two lines are coplanar, the lines of the pencil containing them both are said to be **linearly dependent** on them. If two lines are skew, the only lines **linearly dependent** on them are the two lines themselves. On three skew lines are **linearly dependent** the lines of the regulus on which they lie.

If $\ell_1, \ldots \ell_n$ are any number of lines and $m_1, \ldots m_k$ are lines such that m_1 is linearly dependent on two or three of $\ell_1, \ldots \ell_n$, m_2 is linearly dependent on two or three of $\ell_1, \ldots \ell_n, m_1$ and so on, m_k being linearly dependent on two or three

Table 16.1 *Lines linearly dependent on two or three lines.*

Case	Linearly dependent lines
Two coplanar lines ℓ, m	lines of the pencil (on $\ell \cap m$ and within ℓm)
Two skew lines ℓ, m	$\{\ell, m\}$
Three skew lines ℓ, m, n	regulus containing ℓ, m, n

of $\ell_1, \ldots \ell_n, m_1, \ldots, m_{k-1}$ then m_k is said to be **linearly dependent** on $\ell_1, \ldots \ell_n$. A set of n lines no one of which is linearly dependent on the $n-1$ others is said to be **linearly independent**. In other words, a line m_k is linearly dependent on $\{\ell_1, \ell_2, \ldots, \ell_n\}$ if there exist lines $m_1, m_2, m_3, \ldots, m_{k-1}$ such that m_1 is linearly dependent on two or three lines of $\{\ell_1, \ell_2, \ldots, \ell_n\}$ and each m_i is linearly dependent on two or three lines of the set $\{\ell_1, \ell_2, \ldots, \ell_n, m_1, m_2, \ldots, m_{i-1}\}$.

The following inheritance property for linear independence of lines follows from the definition.

Lemma 16.7

Any non-empty subset of a linearly independent set of lines is linearly independent.

This complicated definition, taken from Veblen and Young (1965, Section 106) is equivalent to a simpler algebraic one.

Theorem 16.8

A set of lines of $\mathsf{PG}(3, \mathbb{R})$ is linearly dependent or linearly independent accordingly as their Plücker coordinates are linearly dependent or linearly independent.

Proof Suppose m_k is linearly dependent on $\{\ell_1, \ell_2, \ldots, \ell_n\}$, so that there exist lines $m_1, m_2, m_3, \ldots, m_{k-1}$ such that m_1 is linearly dependent on two or three lines of $\{\ell_1, \ell_2, \ldots, \ell_n\}$ and each m_i is linearly dependent on two or three lines of the set $\{\ell_1, \ell_2, \ldots, \ell_n, m_1, m_2, \ldots, m_{i-1}\}$. The theorem will be proved if we understand how this recursion operates. In order to have linear dependence of a line on two or three lines of a set, at each step, we must form pencils of lines out of two concurrent lines and reguli out of three mutually skew lines, of the given set. In the first case, suppose ℓ and m are coplanar, and suppose n is a line of the pencil, on $P := \ell \cap m$ and within ℓm. By Worked Example 16.3, there exist λ, μ such that the Plücker coordinates of n satisfy $\ell_i = \lambda \ell_i + \mu m_i$ for each $i \in \{1, \ldots, 6\}$. So the Plücker coordinates of n, as a vector, is linearly dependent upon the Plücker coordinates of ℓ and m.

Now suppose we have three mutually skew lines ℓ, m, n, and suppose r is a line of the regulus determined by these three lines. By Worked Example 16.3, there exist λ, μ, ν such that the Plücker coordinates of r satisfy

$$r_i = \lambda \ell_i + \mu m_i + \nu n_i$$

for each $i \in \{1, \ldots, 6\}$. So the Plücker coordinates of r, as a vector, are linearly dependent upon the Plücker coordinates of ℓ, m, and n. □

Theorem 16.9

The set of all lines of $\mathsf{PG}(3, \mathbb{R})$ *linearly dependent on three linearly independent lines is:*

(i) *a regulus,*
(ii) *a bundle of lines,*
(iii) *the set of all lines of a plane, or*
(iv) *the union of two pencils having distinct points and planes but a common line.*

Proof Let $G = \mathsf{PGL}(4, \mathbb{R})$. Then G has two orbits on lines of $\mathsf{PG}(3, \mathbb{R})$: pairs of skew lines and pairs of concurrent lines (Exercise 13.3). Consider two skew lines ℓ and m. Then the setwise stabiliser $G_{\{\ell, m\}}$ has three orbits on lines, apart from $\{\ell, m\}$, namely, (i) the lines skew to ℓ and m, (ii) the transversal lines to ℓ and m, (iii) the lines concurrent to just one of ℓ or m. All three cases give rise to triples of linearly independent lines.

- If n is a line skew to both ℓ and m, then the lines linearly dependent on ℓ, m, n form a regulus (by definition).
- If n is a transversal line to ℓ and m, concurrent in L and M, respectively, then the set of lines dependent on ℓ, m, n is the union of two pencils, on L and M, in the planes ℓn and mn, respectively.
- If n is concurrent with ℓ in the point L, but skew to m, then m meets ℓn in a point Q, and the set of lines dependent on ℓ, m, n consists of the pencil of lines on Q in the plane Lm, and the pencil of lines on L in the plane Qn.

Consider two lines ℓ and m meeting in a point P. Then the setwise stabiliser $G_{\{\ell, m\}}$ has five orbits on lines, apart from $\{\ell, m\}$, namely, (i) the lines on P not in ℓm, (ii) the lines skew to ℓ and m, (iii) the transversal lines to ℓ and m not incident with P, (iv) the lines concurrent to just one of ℓ or m, (v) the lines on P in the same plane as ℓm.

The fifth case does not give rise to a triple of linearly independent lines. If any of the pairs of lines are skew, then we have addressed it in the argument above. So we are left with cases (i) and (iii).

- If n is a line on P not in ℓm, then the lines linearly dependent on ℓ, m, n are all the lines incident with P; a bundle of lines.
- If n is a transversal line to ℓ and m not incident with P, then the lines linearly dependent on ℓ, m, n are the lines lying in ℓm; the lines in a plane. □

16.2 Linear Congruences

Let us first recall a few facts about the possible relationships between a small number of lines. First, $PGL(4, \mathbb{R})$ acts transitively on lines, and transitively on pairs of skew lines (Worked Example 13.14). More generally, $PGL(4, \mathbb{R})$ has two orbits on pairs of distinct lines: pairs of coplanar lines and pairs of skew lines (Exercise 13.3). In the previous section, we implicitly determined the possible orbits of $PGL(4, \mathbb{R})$ on triples of distinct lines:

  (i) three concurrent lines of a plane-pencil ($*$)
 (ii) a trilateral
(iii) three concurrent lines, not spanning a plane
 (iv) two skew lines and a transversal
  (v) two coplanar lines and a line skew to both lines
 (vi) three skew lines.

The asterisk above indicates that the triple of lines is linearly dependent. Knowing the possible orbits of $PGL(4, \mathbb{R})$ on triples of linearly independent lines leads to Theorem 16.9, where we should note that cases (iv) and (v) above yield the same type of set of lines that are linearly dependent on the triple (i.e., a union of two pencils). We now consider four and five linearly independent lines, where new and interesting structures arise; surprisingly, it is a much smaller range of possibilities than orbits on quadruples or quintuples of lines would suggest.

The set of all lines of $PG(3, \mathbb{R})$ linearly dependent on **four** linearly independent lines is a **linear congruence**. Given a regulus, a line of the regulus is called a **generator** of the regulus and a line of the opposite regulus is called a **directrix** of the regulus.

Theorem 16.10

A linear congruence is one of the following (see Table 16.2):

Table 16.2 *Linear congruences.*

Name	Description
elliptic	lines linearly dependent on four linearly independent skew lines, such that no one of them meets the regulus containing the other three
hyperbolic	all lines meeting two skew lines
parabolic	all generators and tangent lines of a given regulus which meet a fixed directrix of the regulus
degenerate	lines consisting of a bundle of lines and a plane of lines, the centre of the bundle being on the plane

A line that has points in common with all lines of a congruence is called a **directrix** of the congruence.

We delay the proof of Theorem 16.10 and forthcoming results on linear congruences until we visit the Klein correspondence in Section 20.1.

Theorem 16.11
A parabolic congruence consists of all lines on corresponding points and planes in a projectivity between the points and planes on a line. The directrix is a line of the congruence.

A **spread** of $PG(3, \mathbb{R})$ is a partition of the points into lines. A spread is **regular** if, for every line not in the spread, the set of lines of the spread meeting that line is a regulus.

Theorem 16.12
An elliptic linear congruence of $PG(3, \mathbb{R})$ is a regular spread of $PG(3, \mathbb{R})$, and, conversely, a regular spread of $PG(3, \mathbb{R})$ is an elliptic linear congruence of $PG(3, \mathbb{R})$.

16.3 Linear Complexes

The set of all lines of $PG(3, \mathbb{R})$ linearly dependent on **five** linearly independent lines is a **linear complex**. A linear complex is **special** if it contains a line that meets all of its elements (called its **directrix**); otherwise **general**. It will turn out that there are only two linear complexes up to symmetry. Again, we delay the proofs of the following results until we visit the Klein correspondence (Section 20.1).

Theorem 16.13 (Classification of special linear complexes)
A special linear complex consists of a line and all lines meeting it; any two special linear complexes are projectively equivalent.

Theorem 16.14 (Classification of general linear complexes)
(a) Any set S of lines of $PG(3, \mathbb{R})$ such that, for every point P, the lines of S on P are a pencil is a general linear complex, and conversely.

(b) The lines of a general linear complex are the lines of a null polarity, and conversely. Any two general linear complexes are projectively equivalent.

All the lines of a linear complex which pass through a point P lie in a plane π, and all the lines which lie in a plane π pass through a point P except for the points and planes on the directrix for a special complex. The point P is called the **null point** of the plane π and π is called the **null plane** of P with regard to the complex. The correspondence between the points and the planes of space thus established is called a **null system**. If the complex is general, the null system is a **null polarity**.

Worked Example 16.15
Recall that the map ρ given by

$$(a, b, c, d) \mapsto [-d, -c, b, a]$$

defines a null polarity of $\mathrm{PG}(3, \mathbb{R})$. Consider a line ℓ and points (x_1, x_2, x_3, x_4) and (y_1, y_2, y_3, y_4) incident with it. For ℓ to be an absolute line with respect to the above null polarity ρ, we require that it be fixed by the polarity. Consider a point P of the form $(x_1 + \lambda y_1, x_2 + \lambda y_2, x_3 + \lambda y_3, x_4 + \lambda y_4)$ on ℓ, where $\lambda \in \mathbb{R}$. Then the image of P under ρ is the plane with dual coordinates

$$[-x_4 - \lambda y_4, -x_3 - \lambda y_3, x_2 + \lambda y_2, x_1 + \lambda y_1].$$

This plane P^ρ is incident with ℓ precisely when (x_1, x_2, x_3, x_4) and (y_1, y_2, y_3, y_4) lie in P^ρ, which is when

$$x_1 y_4 - x_4 y_1 + x_2 y_3 - x_3 y_2 = 0. \tag{16.2}$$

Notice that this equation is independent of λ. Now the Plücker coordinates of ℓ are

$$(p_1, \ldots, p_6) := (x_1 y_2 - x_2 y_1, x_1 y_3 - x_3 y_1, x_1 y_4 - x_4 y_1,$$
$$x_2 y_3 - x_3 y_2, x_4 y_2 - x_2 y_4, x_3 y_4 - x_3 y_4).$$

Equation 16.2 is precisely that $(p_1, \ldots, p_6)$ satisfies the homogeneous linear equation $p_3 + p_4 = 0$. So we see that the linear complex defined by ρ corresponds to lines that have Plücker coordinates that satisfy $p_3 + p_4 = 0$. □

Theorem 16.16
(i) A linear complex is the set of all lines of $\mathrm{PG}(3, \mathbb{R})$ whose Plücker coordinates satisfy a homogeneous linear equation, and conversely.

(ii) A linear congruence is the set of all lines of $PG(3, \mathbb{R})$ *whose Plücker coordinates satisfy two linearly independent homogeneous linear equations, and conversely.*

More generally, we can define an **algebraic complex** of lines of $PG(3, \mathbb{R})$ to be the set of all lines of $PG(3, \mathbb{R})$ whose Plücker coordinates satisfy a homogeneous polynomial equation $F(x_1, x_2, x_3, x_4, x_5, x_6) = 0$, provided that this set is proper. The **degree** of an algebraic complex is the maximum number of lines (over $\mathbb{C}$) of the complex in a pencil, taken over the set of pencils for which this number is finite. (It will equal the degree of F.) The set of Plücker coordinates of lines of $PG(3, \mathbb{C})$ satisfying $F(x_1, x_2, x_3, x_4, x_5, x_6) = 0$ is then an algebraic manifold of dimension 3. So an algebraic complex of lines of $PG(3, \mathbb{R})$ is the set of *real* lines of $PG(3, \mathbb{C})$ whose Plücker coordinates lie on an algebraic manifold of dimension 3. A linear complex is an algebraic complex of degree 1, and conversely. The set of all lines tangent to a quadric is an algebraic complex of degree 2.

An **algebraic congruence** of lines of $PG(3, \mathbb{R})$ is the set of real lines of $PG(3, \mathbb{C})$ whose Plücker coordinates lie on an algebraic surface (an algebraic manifold of dimension 2). For example, the intersection of two complexes of lines, neither of which contains the other, is an algebraic congruence, called a **complete intersection**. The **order** of a congruence is the maximum number of lines (over $\mathbb{C}$) of the congruence on a point, taken over the set of points for which this number is finite. The **class** of a congruence is the maximum number of lines (over $\mathbb{C}$) of the congruence in a plane, taken over the set of planes for which this number is finite. Thus, a linear congruence has order one and class one. A linear congruence is a complete intersection of two linear complexes. A bundle of lines is an algebraic congruence of order one and class zero, and is not a complete intersection. A plane of lines is an algebraic congruence of order zero and class one, and is not a complete intersection.

16.4 Twisted Cubics

Twisted cubics are interesting curves of 3-space, and are in one sense the three-dimensional analogue of conics. They are the curves of $PG(3, \mathbb{R})$ of minimal degree that do not lie in a (hyper)plane, and they are determined by six points, no four coplanar. Other incarnations of twisted cubics are as sets of singular points of congruences defined by a projectivity between two *bundles* of planes: in analogy with conics being the points of intersection of projectively corresponding lines of two pencils.

A **bundle of planes** is the set of all planes bundle(P) through a point.

Theorem 16.17

Let ϕ be a projectivity between two bundles bundle$(P) \to$ bundle(Q) *of planes of* PG$(3, \mathbb{R})$ *with $P \neq Q$ which has no fixed lines and no fixed planes. The set*

$$C = \{\pi^{\phi} \cap \pi : \pi \in \text{bundle}(P)\}$$

of all lines of intersection between elements of the domain and their images is a congruence of lines of the first order and the third class. Moreover:

(i) *Every point is either* **regular** *(contains exactly one line of C) or* **singular** *(the set of lines of C on that point forms a quadric cone).*

(ii) *Every plane that is not on P or Q contains between one and three singular points; and if three singular points are contained, they form a triangle.*

(iii) *Every plane that is not on P or Q contains between one and three lines of C; and if three lines are contained, they form a triangle.*

Proof Let P and Q be the vertices of the cones, π be a plane on neither P nor Q. Then ϕ induces a map $\pi \cap \pi' \mapsto \pi \cap (\pi')^{\phi}$ of π which is induced by a collineation ψ. Note that the cones are contained in C, so that P and Q are singular points of C, and that any point on PQ other than P and Q lies on the unique line PQ of C. Take a point R of C, not on PQ. A line on C on R corresponds to a plane π' on PR with $(\pi')^{\phi}$ on QR. Choose a plane on R and the preimage of QR under ϕ; this gives such a line. So every point lies on at least one line of C. If $(PR)^{\phi} = QR$, then all the planes on PR give such lines. Intersecting these lines with π gives the set of points $\ell \cap \ell^{\phi}$ for ℓ on $PR \cap \pi$, which is a conic by the Chasles–Steiner theorem (Theorem 6.59), since ϕ having no fixed lines and no fixed planes implies that this is not a perspectivity. Hence, the set of lines of C on R forms a quadric cone in this case. Singular points in π correspond to fixed points of ϕ and singular lines in π correspond to fixed lines of ϕ. Moreover, ϕ is not a central collineation, so has between one and three fixed points, and, if three, they form a triangle, and, dually, ϕ has between one and three fixed lines, and, if three, they form the sides of a triangle. $\qquad\square$

Worked Example 16.18

Consider the bundle of planes $\mathcal{A}$, $\mathcal{B}$ on A: $(1, 0, 0, 0)$, B: $(0, 0, 0, 1)$, respectively. Define the following projectivity ϕ from $\mathcal{A}$ to $\mathcal{B}$:

$$[0, u, -s, t]^{\phi} := [u, -s, t, 0].$$

We now calculate the lines of $\{\pi \cap \pi^{\phi} : \pi \in \mathcal{A}\}$. The intersection of $[0, u, -s, t]$ and $[u, -s, t, 0]$ is the line

$$\langle (-st, -tu, 0, u^2), (s^2 - tu, su, u^2, 0) \rangle.$$

The set of all of these lines (running over $s, t, u \in \mathbb{R}$, not all zero) defines a congruence C. Indeed, the Plücker coordinates of the lines of the congruence are of the form

$$\left(t^2, st, s^2 - tu, tu, -su, u^2\right)$$

which is isomorphic to the **Veronese surface** (in projective 5-space). The singular points of the congruence C are those points of $\mathsf{PG}(3, \mathbb{R})$ that lie on more than one line of C. One of these points is A; another is B. The lines of C on A are of the form $\langle A, (0, t^2, st, s^2) \rangle$, which defines a quadric cone. Likewise, the lines of C on B, namely those of the form $\langle B, (s^2, su, u^2, 0) \rangle$, define a quadric cone. The other singular points lie on a line of each of these two quadric cones, so we are looking for points that are written in two ways:

$$(\lambda, t^2, st, s^2) \equiv (\sigma^2, \sigma\mu, \mu^2, \kappa)$$

for some $\lambda, \kappa, \sigma, \tau, s, t$. Now $t^2 = \sigma\mu$ and so $st\sigma^2 = t^4$, and, hence, $s\sigma^2 = t^3$ when t is non-zero. Therefore, every singular point can be written in the form $(t^3, t^2 s, ts^2, s^3)$ for some $s, t \in \mathbb{R}$. If we simplify this expression so that it only has one free variable, we obtain the canonical example of a twisted cubic $\{(1, s, s^2, s^3) : s \in \mathbb{R}\} \cup \{(0, 0, 0, 1)\}$. □

The set of singular points of a congruence defined by a projectivity between two bundles that has no fixed lines or planes is called a **twisted cubic**.

Theorem 16.19

Two quadric cones with distinct vertices that share a generator and with distinct tangent planes at that generator meet in the union of a twisted cubic on the vertices and the line joining their vertices. Conversely, given a twisted cubic and distinct points on it, the twisted cubic can be represented in this form for cones with these points as vertices.

Proof Let P and Q be the vertices of the cones, π be a plane on neither P nor Q, the intersection of the cone with vertex P with π be C, and the intersection of the cone with vertex P with π be C'. Let $PQ \cap \pi = R$. Let ψ be a collineation of π fixing R and taking C to π be C', and let ϕ be the projectivity induced by ψ taking bundle$(P) \to$ bundle(Q). Then, by Theorem 16.17, the set

$$C = \{\pi^\phi \cap \pi : \pi \in \text{bundle}(P)\}$$

of all lines of intersection between elements of the domain and their images is a congruence of lines of the first order and the third class containing both cones. A point on both cones and not on the common generator is singular; the points P, Q are the only points on PQ that are singular. Therefore, the intersection of the cones is contained in the union of PQ and the twisted cubic of singular points of C. Conversely, if a point R of C is singular, then both PR and QR are in C, so R is in the intersection of the cones. □

Corollary 16.20

PGL$(4, \mathbb{R})$ *acts transitively on twisted cubics of* PG$(3, \mathbb{R})$.

Proof We simply observe that PGL$(4, \mathbb{R})$ acts transitively on pairs of points, and hence pairs of quadric cones (by Theorem 15.20). □

Theorem 16.21

Two hyperbolic quadrics that share a generator meet in the union of a twisted cubic and that generator.

Proof The pencil of quadrics containing the two hyperbolic quadrics contains two quadric cones. (The condition for degeneracy of a quadric of the pencil is a quadratic in the parameter of the pencil.) The intersection of hyperbolic quadrics equals that of the cones (for they both generate the same pencil), which have different tangent planes at the common generator, so Theorem 16.19 is sufficient to complete the proof. □

We also have a degenerate version of Theorems 16.19 and 16.21, for example, a cylinder and a cone, sharing a generator, meet in the union of a twisted cubic and that generator. See Exercise 16.9.

We say that a set S of points of PG$(3, \mathbb{R})$ are in **general position** if every quadruple of elements of S spans the entire space.

Theorem 16.22

In PG$(3, \mathbb{R})$, *given a line ℓ and the pencil of quadrics through eight points in general position, the locus of the points conjugate to ℓ with respect to the quadrics of the pencil is a quadric, ruled by the polar lines of ℓ.*

Proof Each of the quadrics in the pencil is a linear combination of two fixed quadrics in the pencil, so each of the polar lines of ℓ with respect to a quadric in the pencil is a linear combination of the polar lines of ℓ with respect to this linear combination. Thus, the polar lines of ℓ with respect to the quadrics in the pencil form a regulus. □

Theorem 16.23

In PG$(3, \mathbb{R})$, *given a plane π and the pencil of quadrics through eight points in general position, the locus of the poles of π with respect to the quadrics of the pencil is a twisted cubic.*

Proof By Theorem 16.22, the lines of π determine a pencil of quadrics, whose base consists of the poles of π with respect to the quadrics in the pencil. Since this is an infinite base, it is a twisted cubic.   □

Theorem 16.24

Given six points of $\mathsf{PG}(3, \mathbb{R})$, *no four coplanar, there is a unique twisted cubic containing them.*

Proof In the quotient space $\mathsf{PG}(3, \mathbb{R})/P$, the lines PQ, PR, PS, PT, PU form a set of five points, no three collinear, and so lie on a unique conic. Therefore, there is a unique cone with vertex P containing Q, R, S, T, U. Similarly, there is a unique cone with vertex Q containing P, R, S, T, U. The intersection of the cone on P and the cone on Q, with the points on PQ other than P and Q removed, is a twisted cubic on P, Q, R, S, T, U, by Theorem 16.19. Also, by Theorem 16.19, any twisted cubic on P, Q, R, S, T, U is contained in the intersection of these two cones, so equals this one.   □

Theorem 16.25

Let C be a twisted cubic of $\mathsf{PG}(3, \mathbb{R})$. *Then*

$$C = \{(P_1(s,t), P_2(s,t), P_3(s,t), P_4(s,t)): s,t \in \mathbb{R}, (s,t) \neq (0,0)\},$$

where P_1, P_2, P_3, P_4 are linearly independent homogeneous cubic polynomials in the variables x and y over $\mathbb{R}$, and, conversely, every such set is a twisted cubic. Hence, any twisted cubic of $\mathsf{PG}(3, \mathbb{R})$ *is projectively equivalent to*

$$C' = \{(1, t, t^2, t^3): t \in \mathbb{R}\} \cup \{(0, 0, 0, 1)\}.$$

Proof That C' is a twisted cubic follows from Theorem 16.19, by consideration of the cones $y^2 = xz$ and $z^2 = yw$. Then the image of C' under an element A of $\mathsf{GL}(3, \mathbb{R})$ has the form $\{(P_1(s,t), P_2(s,t), P_3(s,t), P_4(s,t)): s,t \in \mathbb{R}, (s,t) \neq (0,0)\}$ with $P_i(s,t) = a_{i1}s^3 + a_{i2}s^2t + a_{i3}st^2 + a_{i4}t^3$, $1 \leqslant i \leqslant 4$. So it only remains to show that every twisted cubic has this form. It is sufficient to show that, given six points P, Q, R, S, T, U, no four coplanar, there is a twisted cubic of this form on those six points. If $G(x, y)$ is a homogeneous polynomial of degree 4 with distinct zeros (μ_i, ν_i), $1 \leqslant i \leqslant 4$, in $\mathsf{PG}(1, \mathbb{R})$, then the polynomials $H_i(x, y) = G(x, y)/(\mu_i x - \nu_i y)$, $1 \leqslant i \leqslant 4$, form a basis for the vector space of homogeneous polynomials of degree 3 in x and y (if there were a linear dependence $\sum a_i H_i(x, y) = 0$ then substituting (μ_i, ν_i) shows that $a_i = 0$). Thus, $\{(H_1(s,t), H_2(s,t), \ldots, H_4(s,t)): s,t \in \mathbb{R}, (s,t) \neq (0,0)\}$ is a twisted cubic passing through all the coordinate points $(1, 0, 0, 0), (0, 1, 0, 0), (0, 0, 1, 0), (0, 0, 0, 1)$ of $\mathsf{PG}(3, \mathbb{R})$. In addition, if μ_i and ν_i

are non-zero, then the points corresponding to $(s, t) = (1, 0)$ and $(s, t) = (0, 1)$ are $(\mu_1^{-1}, \ldots, \mu_4^{-1})$ and $(\nu_1^{-1}, \ldots, \nu_4^{-1})$, and any two points forming a frame with the coordinate points can be obtained uniquely in this way. Conversely, any twisted cubic through the coordinate points can be written in this way. □

Given a twisted cubic C and a point P of C, there is a unique cone K with vertex P containing C, and there is a unique generator ℓ of K intersecting C exactly in $\{P\}$. Moreover, ℓ is the **tangent line** to C at P, and the tangent plane to K at ℓ is **the osculating plane** to C at P.

Example 16.26
The tangent line to the twisted cubic

$$C = \{(1, t, t^2, t^3) : t \in \mathbb{R}\} \cup \{(0, 0, 0, 1)\}$$

at $(0, 0, 0, 1)$ is $[1, 0, 0, 0] \cap [0, 1, 0, 0]$; the osculating plane to C at $(0, 0, 0, 1)$ is $[1, 0, 0, 0]$. Interchanging the first and last coordinates fixes C and so the osculating plane to C at $(1, 0, 0, 0)$ is $[0, 0, 0, 1]$. The map

$$\phi_c : (x, y, z, w) \mapsto (x, cx + y, c^2x + 2cy + z, c^3x + 3c^2y + 3cz + w)$$

fixes C as it fixes $(0, 0, 0, 1)$ and takes $(1, t, t^2, t^3)$ to $(1, t + c, (t + c)^2, (t + c)^3)$, so the osculating plane to C at $(1, t, t^2, t^3)$ is $[0, 0, 0, 1]^{\phi_t} = [t^3, -3t^2, 3t, -1]$.

Theorem 16.27
Given a twisted cubic, there is a unique polarity of $\mathsf{PG}(3, \mathbb{R})$ interchanging each point of the twisted cubic with the osculating plane at that point; it is a null polarity.

Proof Since a twisted cubic contains a frame, if the polarity exists, it is unique. The null polarity $(a, b, c, d) \mapsto [d, -3c, 3b, -a]$ maps $(1, t, t^2, t^3)$ to $[t^3, -3t^2, 3t, -1]$ and $(0, 0, 0, 1)$ to $[1, 0, 0, 0]$. □

Theorem 16.28
Let $\mathcal{K}$ be a twisted cubic of $\mathsf{PG}(3, \mathbb{R})$. Then $(\mathcal{K}, \mathsf{PGL}(4, \mathbb{R})_{\mathcal{K}})$ is permutationally isomorphic to $(\mathsf{PG}(1, \mathbb{R}), \mathsf{PGL}(2, \mathbb{R}))$.

Proof Consider the projective line $\mathsf{PG}(1, \mathbb{R})$ as $\mathbb{R} \cup \{\infty\}$ and $\mathsf{PGL}(2, \mathbb{R})$ the fractional linear transformations of it. Then the map

$$t \mapsto (1, t, t^2, t^3)$$

yields the necessary intertwining map for an action of $\mathsf{PGL}(2, \mathbb{R})$ on the points of $\mathcal{K}$. (Here, ∞ maps to $(0, 0, 0, 1)$.) Indeed, this map induces a four-dimensional representation of $\mathsf{GL}(2, \mathbb{R})$:

$$\begin{bmatrix} a & b \\ c & d \end{bmatrix} \mapsto \begin{bmatrix} a^3 & 3a^2b & 3axb^2 & b^3 \\ a^2c & a^2d + 2abc & b^2c + 2abd & b^2d \\ ac^2 & bc^2 + 2acd & ad^2 + 2bcd & bd^2 \\ c^3 & 3c^2d & 3cd^2 & d^3 \end{bmatrix}.$$

Moreover, the image of this one-to-one homomorphism yields (after factoring out scalars) the full stabiliser of $\mathcal{K}$ in $\mathrm{PGL}(4, \mathbb{R})$ (see Exercise 16.10). □

For a point E of a twisted cubic $\mathcal{K}$ of $\mathrm{PG}(3, \mathbb{R})$, we use the shorthand $\mathcal{K}/E$ for the images of the bundle of lines XE, with $X \in \mathcal{K} \backslash E$, under the quotient map to $\mathrm{PG}(3, \mathbb{R})/E$.

Corollary 16.29
Let $\mathcal{K}$ be a twisted cubic of $\mathrm{PG}(3, \mathbb{R})$, and let E be a point of $\mathcal{K}$. Then $\mathcal{K}/E$ is a conic.

Proof Since $\mathrm{PGL}(4, \mathbb{R})$ acts transitively on twisted cubics (Corollary 16.20), we may suppose that $\mathcal{K}$ is the canonical twisted cubic. By Theorem 16.28 and the fact that $\mathrm{PGL}(2, \mathbb{R})$ acts transitively on the projective line (see Chapter 4, footnote 3), the action of $\mathrm{PGL}(2, \mathbb{R})$ on $\mathcal{K}$ is transitive, and so we may suppose that $E = (0, 0, 0, 1)$. Now (the preimage of) $\mathcal{K}/E$ consists of all lines ℓ_t of the form $\langle (0, 0, 0, 1), (1, t, t^2, t^3) \rangle$. We can rewrite a line of this form with respect to a different basis: $\ell_t = \langle E, (1, t, t^2, 0) \rangle$. The map $\ell_t \to \mathrm{PG}(2, \mathbb{R})$ given by $\ell_t \mapsto (1, t, t^2)$, yields the quotient map from $\mathcal{K}/E$ to the points of the conic $y^2 = xz$ of $\mathrm{PG}(2, \mathbb{R})$. □

Theorem 16.30
Given four points A, B, C, D on a twisted cubic $\mathcal{K}$, for any four points E, F, E', F' on $\mathcal{K} \backslash \{A, B, C, D\}$,

$$R(AEF, BEF; CEF, DEF) = R(AE'F', BE'F'; CE'F', DE'F').$$

*(We call this the **cross-ratio** of A, B, C, D on $\mathcal{K}$.)*

Proof It is sufficient to prove that

$$R(AEF, BEF; CEF, DEF) = R(AEF', BEF'; CEF', DEF'),$$

since then $R(AEF', BEF'; CEF', DEF') = R(AE'F', BE'F'; CE'F', DE'F')$. This allows us to work in the quotient space over E, where the result follows from Theorem 5.18 applied to the conic $\mathcal{K}/E$. □

The tangent line t_P at a point P to a twisted cubic C is the generator of the cone at P not of the form PQ, for Q in C. The osculating plane π_P at P is the tangent plane to the cone K_P at P at the t_P.

Theorem 16.31 (P. W. Wood (1913) (see Wood, 1960, Section 14, p. 16))
Let P, Q, R, S be points on a twisted cubic C, Then P, Q harmonically divides R, S if and only if the intersection of the tangent line at P and the osculating plane at Q lies on QRS.

Proof Let t be an involution fixing C, P, and Q. Then t induces an involutory homology of the quotient space of P, fixing the conic K_P/P. Thus, there is a line on P fixed plane-wise by t, and it is not the tangent line at P. Now $X = t_P \cap \pi_Q$ is fixed by t. Similarly, $Y = t_Q \cap \pi_P$ is fixed by t. Thus, XQ and YP are fixed plane-wise by t. Moreover, every plane fixed by t is on one of these lines. If $R, S \in C$ are interchanged by t, then QRS is fixed by t, so contains QX (as P, Q, R, S are not coplanar, so this can't be YPQ). Hence, $t_P \cap \pi_Q$ is on QRS.

Conversely, if $t_P \cap \pi_Q$ is on QRS, then the involution fixing XQ and YP plane-wise fixes QRS, but not R or S, and contains three points of C, fixing Q. Thus, R and S are interchanged. □

Here we have three-dimensional analogues of Theorems 6.73 and 6.74. We do not give proofs here, as it requires more theory.

Theorem 16.32 (Hesse, 1840b)
If the vertices of two tetrahedra lie on a twisted cubic or are associated, then their sides touch a twisted cubic or are associated, and conversely.

Corollary 16.33 (Hesse, 1840b)
If the eight vertices of two tetrahedra, self-polar with respect to the same quadric, have no four points coplanar, then they lie on a twisted cubic or are associated; conversely, if the vertices of two tetrahedra lie on a twisted cubic or are associated, then there exists a quadric with respect to which they are both self-polar.

We also have the following result, since if the vertices of two tetrahedra lie on a twisted cubic, then there is a *net of quadrics* containing these vertices, and the net of corresponding dual quadrics contains the eight faces, so they lie on a twisted cubic.

Corollary 16.34 (von Staudt, 1856, pp. 377–8)
If the vertices of two tetrahedra lie on a twisted cubic, then their sides touch a twisted cubic.

Exercises

16.1 Let $\mathcal{L}$ be a general linear complex of lines of PG$(3, \mathbb{R})$, and let ℓ be a line not in $\mathcal{L}$.
   (a) Show that there is a line ℓ' so that every line of $\mathcal{L}$ meeting ℓ meets ℓ'.
   (b) Show that as ℓ varies in a regulus, so does ℓ'.

16.2 Prove the following:

Theorem 16.35
If two planes meet in a line of a linear congruence and neither contains a directrix, the other lines of the congruence meet the planes in corresponding points of a projectivity. Conversely, if two planes correspond via a projectivity in such a way that their line of intersection corresponds to itself, the lines joining corresponding points are in the same linear congruence.

16.3 Let Q be a non-degenerate ruled quadric of PG$(3, \mathbb{R})$ and $\mathcal{L}$ be a general linear complex of lines.
   (a) Show that the polar lines of the elements of $\mathcal{L}$ with respect to Q form a general linear complex $\mathcal{L}'$.
   (b) Show that $\mathcal{L}' = \mathcal{L}$ if and only if $\mathcal{L}$ contains a regulus on Q.

16.4 A projectivity is established between the points of two skew lines ℓ, m of PG$(3, \mathbb{R})$, P_1, P_2 are any two points of ℓ, and Q_1, Q_2 are the corresponding points of m.
   (a) Prove that the lines which meet $P_1 Q_2$, $P_2 Q_1$ belong to a linear complex $\mathcal{L}$, which is independent of the choice of P_1, P_2, and that all lines of $\mathcal{L}$ are thus obtained, as P_1, P_2 varies.
   (b) Show further that any general linear complex can be so defined in terms of two of its lines ℓ, m.

16.5 The tangent lines to a twisted cubic Γ of PG$(3, \mathbb{R})$ at four fixed points A, B, C, D are a, b, c, d, respectively. Prove that the four transversals drawn from any point P of Γ (not equal to A, B, C, or D) to the pairs of lines (a, b), (c, d), (AC, BD), (AD, BC) lie in one plane π.

16.6 Show that any five lines of PG$(3, \mathbb{R})$ belong to a linear complex $\mathcal{L}$, and that if the five lines are all tangent to a twisted cubic Γ, then every tangent line to Γ belongs to $\mathcal{L}$. In this case, prove that $\mathcal{L}$ is uniquely determined by Γ and that $\mathcal{L}$ contains no secant lines to Γ.

16.7 Given a twisted cubic $\mathcal{K}$ of PG$(3, \mathbb{R})$, distinct points P, Q, R, S of $\mathcal{K}$ and two chords ℓ, m to $\mathcal{K}$, show that $R(P\ell, Q\ell, R\ell, S\ell) = R(Pm, Qm, Rm, Sm)$.

16.8 Let $\ell_1, \ldots, \ell_5$ be linearly independent, pairwise skew lines of $PG(3, \mathbb{R})$. The transversals from a variable point P_1 on ℓ_1 to ℓ_2, ℓ_3 and ℓ_4, ℓ_5 meet those lines in P_2, P_3 and P_4, P_5. Show that the locus of $P_2 P_4 \cap P_3 P_5$, as P_1 varies on ℓ_1, is a line.

16.9 Let C be a cone of $PG(3, \mathbb{R})$ and let S be a cylinder sharing a generator with C. Show that $C \cap S$ is the union of the shared generator and a twisted cubic.

16.10 Show that the full stabiliser of the canonical twisted cubic $\{(1, t, t^2, t^3) : t \in \mathbb{R}\} \cup \{(0, 0, 0, 1)\}$ is the group of matrices of the form

$$\begin{bmatrix} a^3 & 3a^2 b & 3axb^2 & b^3 \\ a^2 c & a^2 d + 2abc & b^2 c + 2abd & b^2 d \\ ac^2 & bc^2 + 2acd & ad^2 + 2bcd & bd^2 \\ c^3 & 3c^2 d & 3cd^2 & d^3 \end{bmatrix}$$

where $ad - bc \neq 0$. (Hint: Use Theorem 2.8 in the action on ordered triples. Use the face on which $PGL(2, \mathbb{R})$ acts 3-transitively, and show that the stabiliser of the three points $(1, 0, 0, 0)$, $(1, 1, 1, 1)$, $(0, 0, 0, 1)$ within the stabiliser of the twisted cubic is trivial.)

17

Projections

> Programmers who use homogeneous coordinates for geometric computations are implicitly – and often unknowingly – working in the so-called projective space, a strange and wonderful world which only superficially resembles the Euclidean space we all know and love.

Jorge Stolfi (1991, p.1).[1]

17.1 Central Projections

The geometry underpinning the use of perspective in art is projective geometry, and in particular the theory of *central projections*. A **central projection** (or just **projection**) via a point P of a plane π to a plane π' (where P is not on π, π') is the map

$$Q \mapsto PQ \cap \pi'.$$

The point P is called **centre** of the projection.

Theorem 17.1

The projection ϕ via P of π to π' is an isomorphism of the projective plane π with the projective plane π'.

Proof We break the proof into three steps. First, note that ϕ is one-to-one and onto: For $Q \in \pi$, the line PQ meets π' in a unique point since it is not contained in π' and P is not in π'. Next, ϕ takes lines of π to lines of π': Let ℓ be a line of π. Then $\langle P, \ell \rangle$ is a plane which meets π' in a line, which is the image of

---

[1] Reprinted from *Oriented projective geometry. A framework for geometric computations*. Stolfi, Jorge. 1991, with permission from Elsevier.

ℓ under ϕ. Finally, ϕ preserves incidence: If Q is incident with ℓ, which is in turn incident with π, then PQ lies on $\langle P, \ell \rangle$, and so $PQ \cap \pi'$ is incident with $\langle P, \ell \rangle \cap \pi'$. □

Corollary 17.2
A composition of projections beginning and ending at π is a collineation of π.

Corollary 17.3
Central projections preserve cross-ratio.

The group induced on π by compositions of projections is called the **projective group** of π and denoted $\mathsf{Proj}(\pi)$.

Theorem 17.4
$(\pi, \mathsf{Proj}(\pi))$ *is permutationally isomorphic to* $(\mathsf{PG}(2, \mathbb{R}), \mathsf{PGL}(3, \mathbb{R}))$.

Proof By the FToPG (Theorem 13.16) and Corollary 17.2, $\mathsf{Proj}(\pi)$ is a subgroup of $\mathrm{Aut}(\pi)$, which in its action on the points of π is permutationally isomorphic with $(\mathsf{PG}(2, \mathbb{R}), \mathsf{PGL}(3, \mathbb{R}))$. So we just need to display enough compositions of projections to generate $\mathsf{PGL}(3, \mathbb{R})$. Projecting via P of π to π' and back again from a point Q gives a collineation of π with axis $\pi \cap \pi'$ and centre $PQ \cap \pi = C$, which is a homology. The different choices of Q on PC give all the homologies with centre C and axis $\pi \cap \pi'$. Thus, by varying π' and Q, we see that $\mathsf{Proj}(\pi)$ contains all homologies of π, and these generate $\mathsf{PGL}(3, \mathbb{R})$, by Exercise 2.7. □

So now the traditional definition of projective geometry as the study of properties invariant under projection as propagated by Poncelet (1822) is merely Klein's point of view on projective geometry from the 1872 Erlangen Programme, and now *it makes much more sense*. Central projections preserve incidence and take lines to lines, and this is all that needs to be exploited for a basic understanding of perspective in art. Consider an object plane π and an image plane π', and a centre of projection P. Distinguish a plane at infinity. Consider two lines ℓ and m of π which meet at the line at infinity of π; two parallel lines in the plane. What is their image in π'?

Recall from perspective in art, such as a drawing of train tracks leading to the horizon, that 'seemingly' parallel lines, when drawn, converge at a point on the horizon. This is the *vanishing point* for this parallel class of lines, as we shall see. Two distinct planes of $\mathsf{PG}(3, \mathbb{R})$ meet in a line. So a simple way to construct the image of lines is to realise that all we need to do is take the planes $\langle P, \ell \rangle$ and $\langle P, m \rangle$, and take the intersection of each plane with π' (see

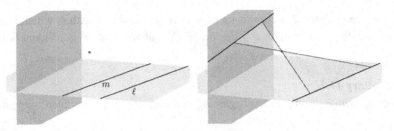

Figure 17.1 (i) Two parallel lines of the object plane; (ii) Projection of a line.

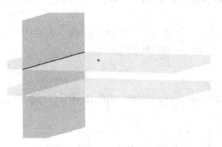

Figure 17.2 The horizon line.

Figure 17.1). These two lines of π' are the images of ℓ and m under central projection via P. Next, consider the plane σ through P that is parallel to the object plane, and let h be the intersection of σ with the image plane. This line is the **horizon line** or *vanishing line* (see Figure 17.2), since $\langle P, h \rangle = \sigma$ intersects the object plane at a line at infinity.

Consider a line ℓ of the object plane, and let X be the point at infinity on ℓ. Then the projection of X (via P) lies on h and is the **vanishing point** of ℓ.

Theorem 17.5 (Vanishing point theorem (Monte, 1600))
The perspective image of a line ℓ not parallel to the image plane passes through the vanishing point of ℓ and so do the images of all lines parallel to ℓ.

Proof Let ℓ be a line of the object plane. Since central projection preserves incidence, the image of ℓ under central projection passes through the vanishing point of ℓ. (Note that we have required that ℓ be not parallel to the image plane, otherwise its image will intersect the horizon line at infinity, which is not a problem in a projective extension of perspective art. The requirement is only in place for its application to the *affine* part of the image plane.) Any line m parallel to ℓ has the same point at infinity and so the image of m under central projection also passes through the vanishing point of ℓ.   □

Finally, we have the celebrated result of Guidobaldo del Monte.

Theorem 17.6 (Main theorem of perspective (Monte, 1600))
The image of any line not parallel to the image plane is determined by the vanishing point of ℓ and the intersection of ℓ with the image plane.

Proof Let ℓ be a line of the object plane and let ℓ' be its image under central projection via a point P. By Theorem 17.5, ℓ passes through the vanishing point of ℓ. Next, let X be the point of intersection of ℓ with the image plane. Then XP meets the image plane at X, and so X is fixed under central projection. Since central projection preserves incidence, ℓ' passes through X. Now the horizon line is parallel to the line o of intersection of the object and image planes, and since P does not lie in the object plane, the horizon line is not equal to o. So X is not equal to the vanishing point of ℓ, and two points determine a line, and, therefore, ℓ' is determined by X and the vanishing point of ℓ. □

17.2 Singular Projections and Singular Polarities

In the last section, we explored central projection that yields a one-to-one map between the sets of points of two planes. Now we remove the plane from the domain of this map and open it up to the whole space, apart from the central point P. The map is defined in the same way, except that it is no longer one-to-one. Given a point Q, its image is $PQ \cap \pi$, but the same is true for any point Q' on the line PQ (apart from P). The line PQ is collapsed to a single point of π.

Let us now be more formal. Given a point P and a plane π of $PG(3, \mathbb{R})$ with P not incident with π, the **singular projection** $PG(3, \mathbb{R}) \setminus \{P\} \to \pi$ with singular variety P onto π is the map $Q \mapsto PQ \cap \pi$. For any plane π' not on p, the restriction of the singular projection with singular variety P onto π to π' is the *central* projection from π' to π from P. Similarly, given a line ℓ and a line m of $PG(3, \mathbb{R})$ with ℓ and m skew, the **singular projection** $PG(3, \mathbb{R}) \setminus \{\ell\} \to m$ with **singular variety** ℓ onto m is the map

$$Q \mapsto \langle Q, \ell \rangle \cap m.$$

For any other line m' skew to ℓ, the restriction of the singular projection to m' with singular variety ℓ onto m is the projection from m' to m from ℓ, and yields a one-to-one correspondence between the points of m' and the points of m.

Worked Example 17.7
Let us consider a concrete example in homogeneous coordinates. Write points as '(w, x, y, z)' and let π be the plane $z = 0$. Let $P := (0, 0, 0, 1)$ and let $Q := (w, x, y, z)$ be a variable point, not equal to P. In order to find the image Q' of

Table 17.1 *Theorem 15.17:* $PG(3, \mathbb{R})/P$ *isomorphic to* $PG(2, \mathbb{R})$.

Points	Bundle of lines on P
Lines	Planes incident with P
Incidence	Natural

Table 17.2 *How ϕ acts on lines and planes incident with P.*

Line ℓ on P	$\rightarrow$	Point $\ell \cap \pi$ of π
Plane ρ on P	$\rightarrow$	Line $\rho \cap \pi$ of π

Q under singular projection to π (via P), we need only consider the following matrix:

$$\begin{bmatrix} a & b & c & 0 \\ w & x & y & z \\ 0 & 0 & 0 & 1 \end{bmatrix}$$

and when it has rank 2, we will have a point $(a, b, c, 0)$ of π collinear with both P and Q. This can only happen if the vector (a, b, c) is a scalar multiple of (w, x, y). So, in homogeneous coordinates, we see that $Q' = (w, x, y, 0)$. We can write the singular projection as a **singular** linear transformation:

$$(w, x, y, z) \mapsto (w, x, y, 0)$$

if we think of the points by their underlying vectors (i.e., as one-dimensional subspaces of $\mathbb{R}^4$). □

Recall from Theorem 15.17 that the lines and planes incident with P form an incidence geometry isomorphic with $PG(2, \mathbb{R})$ (see Table 17.1):

Let ϕ be the singular projection to a plane π, with singular variety P. All the points of a line ℓ passing through P will be mapped to the point $\ell \cap \pi$ (besides P itself). Similarly, if m is a line not incident with P, then all the points of $\langle m, P \rangle$ are mapped to the line $\langle m, P \rangle \cap \pi$. Hence, we have the following correspondence (Table 17.2):

So ϕ realises the isomorphism of $PG(3, \mathbb{R})/P$ with $PG(2, \mathbb{R})$, when we identify the points and lines of π with $PG(2, \mathbb{R})$. Given a prior identification of π with $PG(2, \mathbb{R})$, it turns out that singular projection induces a projectivity of $PG(2, \mathbb{R})$ (i.e., preserves cross-ratio). Indeed, if $\ell_1, \ell_2, \ell_3, \ell_4$ are four coplanar lines incident with P, then $R(\ell_1, \ell_2; \ell_3, \ell_4) = R(\ell_1 \cap \pi, \ell_2 \cap \pi; \ell_3 \cap \pi, \ell_4 \cap \pi)$.

A quadric cone with vertex V defines a map $PG(3, \mathbb{R}) \setminus \{V\}$ to the bundle bundle(V) of planes on V as follows: if $Q(x, y, z, w) = 0$ is the equation of the cone, and

$$f((x, y, z, w), (x', y', z', w')) = Q(x + x', y + y', z + z' + w + w') - Q(x, y, z, w)$$
$$- Q(x', y', z', w'),$$

then, for $P(x, y, z, w) \neq Q$, the map

$$P \rightarrow \{(x', y', z', w'): f((x, y, z, w), (x', y', z', w')) = 0\}$$

is called a **singular polarity**. The singular polarity given by a cone with vertex V can be seen as the composition of the singular projection with singular variety P onto π and a duality between π and $PG(3, \mathbb{R})/P$, for any plane π not on V. Just as a singular projection induces a projectivity of $PG(2, \mathbb{R}) \cong PG(3, \mathbb{R})/P$, a singular polarity induces a polarity on $PG(2, \mathbb{R}) \cong PG(3, \mathbb{R})/P$.

A special linear complex with directrix ℓ defines a map $PG(3, \mathbb{R}) \setminus \{\ell\}$ to the pencil of planes on ℓ via the null system, which is also called a **singular polarity**. Points on the same plane through ℓ have the same image, so the null system induces a map from the pencil of planes on ℓ to itself, and this induced map preserves cross-ratio.

Worked Example 17.8
Suppose we have a Euclidean space V and subspaces U and W with $U^\perp = W$. That is, U and W are orthogonal with respect to the standard scalar product. Now complete V to a projective space $\mathbb{P}V$ by adding a hyperplane H at infinity. Then the orthogonal complement map $\perp$ induces a polarity ρ on $\mathbb{P}V$, and, moreover, $(\mathbb{P}U \cap H)^\rho = \mathbb{P}W \cap H$. In the dual projective space $\mathbb{P}V^*$, there is a singular polarity ρ^* with radical H^* such that $((\mathbb{P}U \cap H)^*)^{\rho^*} = (\mathbb{P}W \cap H)^*$. The learned reader might notice that this establishes a connection between the foundations of Euclidean geometry and singular polarities. □

17.3 Computer Vision

The long-standing observation that, in the absence of such effects as reflection and refraction, and ignoring relativistic effects of gravity, light travels in straight lines,[2] coupled with the realisation that the study of straight lines is in

---

[2] Indeed, looking at the history of the link, we see that geometric optics can be traced back to Euclid and Ptolemy and then were further advanced by ibn al-Haytham, and the coupling of al-Haytham's ideas purloined by Witelo with the advances in pictorial representation of the Italian Renaissance by Kepler led to points at infinity in geometry, and then to projective geometry in the work of Desargues, surely influenced by anamorphic perspective. Rather than being surprised by the link, we should be surprised that it has been forgotten: twice forgotten (once before the nineteenth-century invention of photogrammetry and again before the mid-twentieth-century work on computer vision).

the main domain of projective geometry leads to the inexorable conclusion that projective geometry and vision are related. Computer vision is largely concerned with computations concerning conclusions about information extracted from the output of visual sensors, as well as the extraction and representation of this information. If we put aside the statistical questions concerning the accuracy of this information and the related problems concerning uncertain data, it becomes clear that the underlying model of a camera as a mapping between projective spaces of different dimensions is fundamental.

This is given a hint of verisimilitude by the widespread use of homogeneous coordinates in computer graphics. Often this use is presented solely as a method of simplifying calculations, without any consideration of the geometric consequences of the adoption of homogeneous coordinates, but they are essentially an analytic model of projective geometry.

Since computer vision is mostly concerned with the assembly of images that are made by photography or other forms of imaging, we will consider a simple camera and a geometric model for it. The *pinhole camera* has a simple aperture on its front face through which light enters, diffuses, and projects onto the back face, the *retinal plane* of the camera. This is very much like the *camera obscura*, linking geometric optics and the rise of perspective in the Renaissance.

Here we will give a simple model of the pinhole camera. Suppose we have a pinhole camera with the following attributes:

- it has a focal length f (i.e., the camera is a cube with side-lengths f);
- the focal plane (front face of the camera) lies in the plane $z = 0$ in $\mathrm{PG}(3, \mathbb{R})$;
- the retinal plane (back of the camera) lies in the plane $z = -fw$ in $\mathrm{PG}(3, \mathbb{R})$;
- the hole of the camera is at the origin $C = (0, 0, 0, 1)$.

In this coordinate system, the camera is modelled by the singular projection from C onto $\pi_R = [0, 0, 1, f]$. Let us now find the matrix representing this projection. Let π_S be the plane for which the scene lies (it must be different from π_R). Now if $P = (x, y, z, w)$ is a point on π_S, the line through C and P is

$$\ell = \{s(x, y, z, w) + t(0, 0, 0, 1) : (s, t) \neq (0, 0)\}$$

written as its set of points. The intersection of ℓ and π_R is the point (x', y', z', w') which satisfies $z' = -fw'$ and $(x', y', z', w') = s(x, y, z, w) + t(0, 0, 0, 1)$ for some $(s, t) \neq (0, 0)$. Solving this, we find that

$$(x', y', z', w') = (sx, sy, sz, -sz/f) \equiv (-fx, -fy, -fz, z).$$

So in homogeneous coordinates, T can be represented by the transformation

$$(x, y, z, w) \mapsto (x, y, z, w) \begin{bmatrix} -f & 0 & 0 & 0 \\ 0 & -f & 0 & 0 \\ 0 & 0 & -f & 1 \\ 0 & 0 & 0 & 0 \end{bmatrix} = (-fx, -fy, -fz, z).$$

The matrix above is called the **perspective projection matrix**. What we have done, in using homogeneous coordinates and a model in $PG(3, \mathbb{R})$, is obtained equations for the pinhole camera that are linear.

Worked Example 17.9

Suppose you have a pinhole camera that is a 12 cm cube with pinhole in the centre of one of the faces, and you would like to take a photo of a scene that is 5 m high by 2 m wide. Given that the camera has its base placed at the same height as the foot of the scene, and is pointed directly at the scene on the perpendicular bisector of the base of the scene, how far apart should the camera be from the scene such that the photograph includes the whole shot?

Solution: So $f = 12$ and we consider the extremal points of the scene (which lie on the plane $z = d$, for an unknown distance d):

$$A = (-200, -6, d, 1), \qquad B = (-200, 494, d, 1),$$
$$C = (200, -6, d, 1), \qquad D = (200, 494, d, 1).$$

The perspective projection matrix for our camera is

$$M = \begin{bmatrix} -12 & 0 & 0 & 0 \\ 0 & -12 & 0 & 0 \\ 0 & 0 & -12 & 1 \\ 0 & 0 & 0 & 0 \end{bmatrix}.$$

The images of our extremal points under M are:

$$A' = (2400, 72, -12d, d), \qquad B' = (2400, -5928, -12d, d),$$
$$C' = (-2400, -72, -12d, d), \qquad D' = (-2400, -5928, -12d, d),$$

which when translated back to affine coordinates[3] in $AG(3, \mathbb{R})$ are

$$\left(\tfrac{-2400}{d}, \tfrac{-72}{d}, 12\right), \left(\tfrac{-2400}{d}, \tfrac{5928}{d}, 12\right), \left(\tfrac{2400}{d}, \tfrac{72}{d}, 12\right), \left(\tfrac{2400}{d}, \tfrac{5928}{d}, 12\right).$$

In order for our photograph to include the whole shot, we must make sure that these points lie within the square with vertices

$$(-6, -6, 12), (-6, 6, 12), (6, -6, 12), (6, 6, 12).$$

Clearly, the smallest d we can take is $5928/6 = 988$cm. $\qquad\qquad\square$

---

[3] That is, $w = 0$ is the plane at infinity.

Note that we can project the retinal plane onto $\mathsf{PG}(2, \mathbb{R})$ via the map

$$(-fx, -fy, -fz, z) \mapsto (-fx, -fy, z).$$

So sometimes we just omit the third column of the perspective projection matrix to incorporate this projection:

$$(x, y, z, w) \mapsto (x, y, z, w) \begin{bmatrix} -f & 0 & 0 \\ 0 & -f & 0 \\ 0 & 0 & 1 \\ 0 & 0 & 0 \end{bmatrix}.$$

We have seen above how we can model the pinhole camera, having particular focal and retinal planes, and a particular position of the aperture (or *optical centre*). To change these attributes we simply apply a collineation ρ that moves these three objects to the new positions, and then conjugate our perspective projection matrix by ρ to obtain the new perspective projection matrix.

Worked Example 17.10
What is the optical centre of a pinhole camera modelled by the perspective projection matrix M?

Solution: The optical centre $C = (x, y, z)$ for M is the point that satisfies $(x, y, z, 1)M = (0, 0, 0)$. □

Exercises

17.1 Give a 4×4 matrix for the projection T of $\mathsf{PG}(3, \mathbb{R})$ from the centre with homogeneous coordinates $(4, 3, 7, 1)$ onto the plane with homogeneous coordinates $[1, 1, -1, 1]$.

17.2 Given a camera that is a 12 cm cube with the pinhole in the centre of one of the faces, what distance must the front of the camera be from a scene that is 4 m wide by 8.56 m high, in order that the photograph include all of the scene in the shot, given that the camera has its base placed at the same height as the foot of the scene, and is pointed directly at the scene on the perpendicular bisector of the base of the scene?

18

A Glance at Inversive Geometry

> We place a spherical cage in the desert, enter it, and lock it. We perform an inversion with respect to the cage. The lion is then in the interior of the cage, and we are outside.

Petard (1938, p. 446).[1]

18.1 The Real Inversive Plane

In this chapter we develop a projective geometric viewpoint of the real inversive plane and its properties. Inversion in a circle is a transformation in the plane on the set of lines and circles. This unification of the lines and circles can reveal interesting properties about the Euclidean plane. Let K be a circle with centre O, and let A be a point of $\mathbb{R}^2$ not equal to O. Then all the circles passing through A and orthogonal to K pass through a second point A' (see Figure 18.1). This point A' is the inversion of A in K. Indeed, we have the following fundamental property of the Euclidean plane that follows from the so-called power of a point theorem: 'Every circle which passes through a pair of mutually inverse points with respect to another circle is orthogonal to that circle and, conversely, every circle orthogonal to a circle meets every diameter of that circle in a pair of mutually inverse points.'[2] If A lies on K, then A' is defined to be A. Note also that A' lies on the polar line of A (with respect to K). Now A' is not defined if we permit A to be the centre O of K; and so, to

[1] Petard, H. A Contribution to the Mathematical Theory of Big Game Hunting. *Amer. Math. Monthly*, 45(7), 446–447, 1938. Reprinted by permission of Taylor & Francis Ltd, www.tandf online.com

[2] More on this can be found in Coxeter (1971).

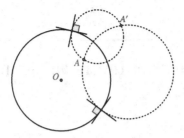

Figure 18.1 Inversion in a circle.

complete the picture, we add a single additional point ∞ to the plane and let it be the inversion of the centre of every circle, in the given circle.

Theorem 18.1
Every circle which passes through a pair of mutually inverse points with respect to another circle is orthogonal to that circle and, conversely, every circle orthogonal to a circle meets every diameter of that circle in a pair of mutually inverse points.

The *real inversive plane* is what results when we ensure that inversion in a circle induces a map on the points of the plane. We can reconstruct the real inversive plane I by taking the incidence structure with points the points of $\mathbb{R}^2$, together with a new symbol ∞, and with *circles* of two kinds: the circles of $\mathbb{R}^2$, and the union of each line of $\mathbb{R}^2$ with $\{\infty\}$. The incidence relation is just set membership. Now recall that any two points of $\mathbb{R}^2$ define a unique line, and three noncollinear points of $\mathbb{R}^2$ define a unique circle. Somewhat deliberately, this new incidence structure I combines these two properties into one:

Every three points of I are incident with a unique circle of I.

So just as the line at infinity of the projective plane ensures that pairs of lines are always concurrent, the point at infinity of the real inversive plane ensures that when we combine lines and circles, we have a uniform property with respect to triples of points. Likewise, we can also highlight another interesting property of I:

If P and Q are points, and C is a circle of I incident with P but not with Q, then there is a unique circle C' incident with Q such that $C \cap C' = \{P\}$.

Table 18.1 *Another model of the real inversive plane.*

Points:	points of $PG(1, \mathbb{C})$
Lines:	images of the *subline* $\{(x, 1): x \in \mathbb{R}\} \cup \{(1, 0)\}$ under $PGL(2, \mathbb{C})$

We now explore another model of the real inversive plane $\mathcal{I}$ as *real sublines* of the complex projective line. To be more specific, consider the following incidence structure (Table 18.1):

Worked Example 18.2

If we think of $\mathbb{C}$ as a plane, the **Gauss plane**, it turns out that the lines of $\mathcal{I}$ correspond to lines and circles of the Gauss plane, but with the addition of ∞ to the lines. To see this, note that an affine equation of a line in the Gauss plane is of the form $\alpha \text{Re}(z) + \beta \text{Im}(z) + \gamma = 0$. Now $\text{Re}(z) = \frac{1}{2}(z + \bar{z})$ and $\text{Im}(z) = \frac{1}{2i}(z - \bar{z})$, so we can rearrange the formula so that it is equivalent to

$$\delta z + \bar{\delta}\bar{z} + \gamma = 0$$

where $\delta = \frac{1}{2}(\alpha - i\beta)$. Likewise, a circle would have the form $(\text{Re}(z) - x_0)^2 + (\text{Im}(z) - y_0)^2 = r^2$ which, when rearranged, can be written in the form

$$z\bar{z} + \delta z + \bar{\delta}\bar{z} + \epsilon = 0$$

where $\delta = -x_0 + iy_0$ and $\epsilon = x_0^2 + y_0^2 - r^2$. So we have a unification of lines and circles of the Gauss plane; they can be written in the form

$$\alpha z\bar{z} + \beta z + \bar{\beta}\bar{z} + \gamma = 0.$$

Indeed, in this representation, it is straightforward to see that the image of the real line $z - \bar{z} = 0$ under the general Möbius group is the set of all lines and circles. □

Through this model, we will see that the automorphism group of the real inversive plane is isomorphic to the semi-direct product $\mathbb{M}$ of $PGL(2, \mathbb{C})$ with the group $\langle \bar{} \rangle$ generated by complex conjugation, called **the general Möbius group**, a result that we attribute to Möbius (1855) and Darboux (1880).

Theorem 18.3

The incidence structure above, consisting of the points of $PG(1, \mathbb{C})$ (as points) and the real sublines (as lines), is isomorphic to the real inversive plane $\mathcal{I}$.

Proof The necessary bijection from $PG(1, \mathbb{C})$ to $\mathcal{I}$ sends a point with homogeneous coordinates $(x + i \cdot y, 1)$ to the point (x, y) of $\mathcal{I}$, and maps $(1, 0)$ to ∞. Notice that the canonical subline $\ell_0: \{(x, 1): x \in \mathbb{R}\} \cup \{(1, 0)\}$ maps to the line $\ell: y = 0$, extended by ∞. We leave it to the reader to verify that real sublines map onto the lines and circles of $\mathbb{R}^2$. □

Now, since $\mathsf{PGL}(2,\mathbb{C}) \rtimes \langle \; ^- \; \rangle$ acts transitively on the real sublines, we see that $\mathbb{M}$ acts transitively on the circles of $\mathcal{I}$: so extended lines and circles of $\mathbb{R}^2$ are indistinguishable. A corollary of this is that the projectivity group of a line and the projectivity group of a conic are permutationally isomorphic, once we move to the projective setting. This is the key to Möbius' *Ubertragungsprinzip*, as a forerunner to the famous principle of Hesse.

Corollary 18.4
The stabiliser of a circle in the Möbius group $\mathbb{M}$ is permutationally isomorphic to the projectivity group of the conic.

The transitivity of the general Möbius group alongside the identification of $\mathsf{PG}(1,\mathbb{C})$ with $\mathcal{I}$ has another elegant consequence. Note that the image of four points under the general Möbius group has a real cross-ratio if and only if the original points have a real cross-ratio. The points of the subline $\mathsf{PG}(1,\mathbb{R})$ have a real cross-ratio, and the set of all circles of $\mathcal{I}$ lies within the orbit of this subline.

Corollary 18.5
Four points z_1, z_2, z_3, z_4 of $\mathsf{PG}(1,\mathbb{C})$ lie on a circle of the real inversive plane if and only if $\mathrm{R}(z_1, z_2; z_3, z_4)$ is a real number.

So from Corollary 18.5 and Theorem 6.11, we have:

Theorem 18.6 (Miquel's six-circle theorem (Miquel, 1844, Result 8))
Consider four points A, B, C, D on a circle of $\mathbb{R}^2$. Draw four more circles C_1, C_2, C_3, C_4 that pass through the pairs of points (A, B), (B, C), (C, D), and (D, A), respectively. Now consider the other intersections of C_i and C_{i+1} for $i = 1, \ldots, 4$ (indices modulo 4). These four intersections are again concircular.

We can also model the real inversive plane via secant plane sections of an elliptic quadric Q of $\mathsf{PG}(3,\mathbb{R})$ (Theorem 18.7). This is essentially the projective reverse of **stereographic projection** of the sphere in $\mathbb{R}^3$. Since $\mathsf{PGL}(4,\mathbb{R})$ acts transitively on elliptic quadrics (Theorem 15.7), we can consider a particular one, namely, that defined by the quadratic form $Q(x, y, z, w) = x^2 + y^2 + z^2 - w^2$. The elements of $\mathsf{PGL}(4,\mathbb{R})$ that *preserve* this form are precisely the elements that stabilise the associated elliptic quadric. So g preserves Q if $Q(P^g) = Q(P)$ for every point P. The isometry group of an elliptic quadric of $\mathsf{PG}(3,\mathbb{R})$ is the group of all $g \in \mathsf{PGL}(4,\mathbb{R})$ that preserve Q, and we denote it by $\mathsf{PGO}^-(4,\mathbb{R})$.

Theorem 18.7

Consider the following incidence geometry $\mathcal{I}_Q$ arising from an elliptic quadric Q of $\mathrm{PG}(3,\mathbb{R})$:

Points:	*the points of Q*
Lines:	*the secant plane sections of Q*

Then $\mathcal{I}_Q$ is isomorphic to the real inversive plane $\mathcal{I}$.

Proof Without loss of generality (Theorem 15.7), we can take Q to be the unit sphere, defined by the quadratic form $Q(x,y,z,w) = x^2 + y^2 + z^2 - w^2$. We can parametrise the points of Q as the **Riemann sphere** representation, as $\mathrm{PG}(1,\mathbb{C})$:

$$Q \to \mathrm{PG}(1,\mathbb{C}): \qquad (x,y,z,w) \mapsto \frac{x+yi}{w-z}, \qquad (0,0,1,1) \mapsto \infty.$$

This is essentially stereographic projection from the point $(0,0,1,1)$ onto the plane $\pi: z = 0$. The inverse map is

$$a+bi \mapsto \left(\frac{2a}{a^2+b^2+1}, \frac{2b}{a^2+b^2+1}, \frac{a^2+b^2-1}{a^2+b^2+1}, 1 \right), \qquad \infty \mapsto (0,0,1,1).$$

Now consider a secant plane given by plane coordinates $[r,s,t,u]$. In order for it to be a secant plane, we require that $(r,s,t,-u) \notin Q$. Now $a+bi$ corresponds to a point in this plane section if

$$r\frac{2a}{a^2+b^2+1} + s\frac{2b}{a^2+b^2+1} + t\frac{a^2+b^2-1}{a^2+b^2+1} + u = 0,$$

which, after rearranging, yields

$$2ar + 2bs + (a^2+b^2)(t+u) = t - u. \qquad (18.1)$$

If the secant plane contains the point $(0,0,1,1)$, then we have $t + u = 0$ and hence Equation (18.1) gives an equation of a line in the complex plane: $ra + sb = t$. If, however, the secant plane is not incident with $(0,0,1,1)$, then we can divide through by $t + u$ and we have

$$\left(a + \frac{r}{t+u} \right)^2 + \left(b + \frac{s}{t+u} \right)^2 = \frac{r^2 + s^2 + t^2 - u^2}{(t+u)^2},$$

a circle in the complex plane. $\qquad\square$

Since $\mathrm{PGL}(4,\mathbb{R})$ acts transitively on elliptic quadrics (Theorem 15.7), we can consider a particular one, namely, that defined by the quadratic form

$$Q(x, y, z, w) = x^2 + y^2 + z^2 - w^2.$$

The elements of $PGL(4, \mathbb{R})$ that *preserve* this form are precisely the elements that stabilise the associated elliptic quadric. So g preserves Q if $Q(P^g) = Q(P)$ for every point P. The isometry group of an elliptic quadric of $PG(3, \mathbb{R})$ is the group of all $g \in PGL(4, \mathbb{R})$ that preserve Q, and we denote it by PGO^- $(4, \mathbb{R})$.

Corollary 18.8
There is a permutational isomorphism between $(Q, PGO^-(4, \mathbb{R}))$ and $(\mathcal{I}, \mathbb{M})$.

Corollary 18.9
Inversion in a circle is represented by a harmonic homology with axis the corresponding secant plane of the circle and centre the pole of that plane with respect to the elliptic quadric.

18.2 Lester's Theorem

Let ABC be a triangle of $\mathbb{R}^2$, and let G be its centroid, H its orthocentre, and O is circumcentre. By Theorem 8.19, G, H, O lie together on the Euler line of ABC. We now recap an interesting property of the nine-point circle is explored further in Section 1.7 of Coxeter and Greitzer (1967). Let N be the centre of the nine-point circle (see Theorem 8.66) of a triangle ABC. Then N also lies on the Euler line, and N is the midpoint of $\overline{OH}$. We summarise the ratios of distances between points via Figure 18.2 (e.g., $d(N, G) : d(G, O) = 1 : 2$):

Theorem 18.10
Consider the following four points for a triangle ABC: the nine-point centre N, the circumcentre O, the centroid G, and the orthocentre H. Then $\{N, O\}$ harmonically separates $\{G, H\}$.

The Lemoine point of a triangle is the intersection of the *symmedians* of the given triangle or, equivalently, the isogonal conjugate of the centroid of the given triangle. The **orthocentroidal circle** is the circle with the line segment joining the orthocentre and the centroid as its diameter. So Theorem 18.10

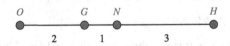

Figure 18.2   Ratios of distances of points on the Euler line.

(together with Apollonius' theorem (Theorem 4.38) shows that the nine-point centre and the circumcentre are mutually inverse in the orthocentroidal circle. We will use the results above to prove Lester's theorem.

Theorem 18.11 (Lester's theorem)
The nine-point centre N, the circumcentre O, and the Fermat–Torricelli points E, F lie on a circle.

Proof Let Q be the orthocentroidal circle of ABC, and let $\mathcal{L}$ be the circum-circle of NOE. Now, $\mathcal{L}$ and GH intersect in $\{N, O\}$, which are mutually inverse in Q (by Theorem 18.10). By Theorem 18.1, $\mathcal{L}$ is orthogonal to $\mathcal{K}$.

Let X, Y be the points at infinity on the asymptotes of $\mathcal{K}$, and let $L := YG \cap EF$, $M := YH \cap EF$. By Theorem 10.56,

$$-1 = \mathrm{R}_{\mathcal{K}}(E, F, G, H)$$
$$= \mathrm{R}(YE, YF, YG, YH) \quad \text{(by the Chasles–Steiner theorem (Theorem 6.61))}$$
$$= \mathrm{R}(E, F, L, M) \quad \text{(intersection with } EF\text{)}.$$

So $\{E, F\}$ harmonically separates $\{L, M\}$.

Now, by the Chasles–Steiner theorem (Theorem 6.61) and Theorem 10.56, we have

$$\mathrm{R}_{\mathcal{K}}(FG, FH; FE, t_F) = \mathrm{R}(G, H; E, F) = -1,$$

where t_F is the tangent at F to $\mathcal{K}$. Let $U := GH \cap EF$ and let $V := GH \cap \ell_\infty$. The tangent t_F at F is parallel to the Euler line GH (by Exercise 18.3), and hence $FV = t_F$. Then, via projection from F, we have

$$\mathrm{R}(G, H; U, V) = \mathrm{R}(FG, FH; FU, FV) = \mathrm{R}(FG, FH; FE, t_F) = -1,$$

and hence U is the midpoint of $\overline{GH}$. Therefore, LM (which is also EF) bisects $\overline{GH}$. We know that GL and HM are parallel because $YZ \cap GL = Y$, and $YZ \cap HM = Y$, where Z is the centre of $\mathcal{K}$. So $HMGL$ is a parallelogram. By Exercise 8.9, $HMGL$ is a rectangle because $\{E, F\}$ harmonically separates $\{L, M\}$. Therefore, L and M lie on Q, and, hence, E and F are mutually inverse in Q, and F lies on $\mathcal{L}$ (by Theorem 18.1). □

Exercises

18.1 Using inversive geometric tools, give an alternative proof of Theorem 5.22, that if A, B, C, D are four points on a conic $\mathcal{K}$, then $\{A, B\}$ *separates* $\{C, D\}$ if and only if $\mathrm{R}(A, B; C, D)_{\mathcal{K}} < 0$.

18.2 There is another model of the real inversive plane, given by reguli of an
      elliptic linear congruence $\mathcal{E}$. Prove the following result:

Theorem 18.12
*Consider the following incidence geometry $\mathcal{I}_{\mathcal{E}}$ arising from an elliptic
linear congruence $\mathcal{E}$ of $\mathrm{PG}(3, \mathbb{R})$:*

Points:	*the lines of $\mathrm{PG}(3, \mathbb{R})$*
Lines:	*the reguli of $\mathcal{E}$*

Then $\mathcal{I}_{\mathcal{E}}$ is isomorphic to the real inversive plane $\mathcal{I}$.

(This exercise is more advanced, and can be attempted once the reader
has understood the Klein correspondence. See Section 20.1.)

18.3 Show that the tangent at a Fermat–Torricelli point F (to the Kiepert
      hyperbola) is parallel to the Euler line.

Part III

Higher Dimensions

Many of my favorite examples come from my early training in classical projective geometry: the twisted cubic, the quadric surface, or the Klein representation of lines in 3-space.

Sir Michael Atiyah (2010, p. 370)

19

Generalising to Higher Dimensions

> Poncelet was the first mathematician to appreciate fully that projective geometry was a new branch of mathematics with methods and goals of its own.
>
> Morris Kline (1955, p. 842)

19.1 The Basic Properties of Higher-Dimensional Space

It is not too difficult to extrapolate the basic properties of $PG(3, \mathbb{R})$ to higher dimensional projective spaces. So we briefly describe what are the necessary generalisations for the reader to know, before moving on to the more interesting aspects of higher dimensional spaces. First of all, the d-**dimensional real projective space** $PG(d, \mathbb{R})$ is the incidence structure with points the one-dimensional subspaces of $\mathbb{R}^{d+1}$, lines the two-dimensional subspaces of $\mathbb{R}^{d+1}$, and incidence is the natural inclusion of subspaces. Just as in $PG(3, \mathbb{R})$, any two points P, Q of $PG(d, \mathbb{R})$ are incident with a unique line PQ, simply because any two 1-dimensional subspaces of $\mathbb{R}^{d+1}$ span a unique 2-dimensional subspace. A **subspace** of $PG(d, \mathbb{R})$ is a set S of points such that, whenever P, Q in S, $PQ \subseteq S$. The **projective dimension** of a subspace is the length of the longest chain of subspaces it contains. Points have projective dimension 0, lines have projective dimension 1, $PG(d, \mathbb{R})$ has projective dimension d. A **hyperplane** of $PG(d, \mathbb{R})$ is a subspace of projective dimension $d - 1$; a **secundum** of $PG(d, \mathbb{R})$ is a subspace of projective dimension $d-2$; a **plane** of $PG(d, \mathbb{R})$ is a subspace of projective dimension 2. Note that a subspace of $PG(d, \mathbb{R})$ of projective dimension e corresponds to a subspace of $\mathbb{R}^{d+1}$ of vector space dimension $e + 1$. Grassmann's identity for vector subspaces also holds for projective subspaces.

407

Theorem 19.1 (Grassmann's identity (Grassmann, 1862))

In $\mathrm{PG}(d, \mathbb{R})$, *given two subspaces* S *and* S', *their intersection* $S \cap S'$ *is a subspace, the intersection* $\langle S, S' \rangle$ *of all subspaces containing* S *and* S' *is a subspace, and* $\dim\langle S, S' \rangle + \dim S \cap S' = \dim S + \dim S'$.

Corollary 19.2

(i) Any two hyperplanes of $\mathrm{PG}(d, \mathbb{R})$ *are incident with a unique secundum.*

(ii) A hyperplane of $\mathrm{PG}(d, \mathbb{R})$ *and a non-incident line meet in a point.*

One can think of $\mathrm{PG}(d, \mathbb{R})$ simply as the subspace lattice of $\mathbb{R}^{d+1}$, and an interval in the lattice is again a subspace lattice of another $\mathbb{R}$-vector space. So we have the following result:

Theorem 19.3

Let S *be an* e-*dimensional subspace of* $\mathrm{PG}(d, \mathbb{R})$, $e > 1$. *Then the incidence structure of points in* S *and lines contained in* S, *with the restricted incidence, is isomorphic to* $\mathrm{PG}(e, \mathbb{R})$.

We denote a point by **homogeneous coordinates** $(x_1, x_2, \ldots, x_{d+1})$, with $x_1, x_2, \ldots, x_{d+1} \in \mathbb{R}$, not all zero. We denote a hyperplane by **homogeneous hyperplane coordinates** $[a_1, a_2, \ldots, a_{d+1}]$, $a_1, a_2, \ldots, a_{d+1} \in \mathbb{R}$, not all zero, and it is essentially shorthand for the equation $a_1 x_1 + a_2 x_2 + \cdots + a_{d+1} x_{d+1} = 0$ for the points contained in the hyperplane. That is, a point $(x_1, x_2, \ldots, x_{d+1})$ is incident with the hyperplane $[a_1, a_2, \ldots, a_{d+1}]$ if and only if $a_1 x_1 + \cdots + a_{d+1} x_{d+1} = 0$.

Just as in $\mathrm{PG}(2, \mathbb{R})$ and $\mathrm{PG}(3, \mathbb{R})$, a **collineation** of $\mathrm{PG}(d, \mathbb{R})$ is a bijection of the points taking lines to lines. The collineations group $\mathrm{PG}(d, \mathbb{R})$ is denoted $\mathrm{Aut}(\mathrm{PG}(d, \mathbb{R}))$. Given a $(d + 1)$-by-$(d + 1)$ invertible matrix A, the map $(x_1, x_2, \ldots, x_{d+1}) \mapsto (x_1, x_2, \ldots, x_{d+1})A$ is a collineation. The permutation group induced by the action of $\mathrm{GL}(d + 1, \mathbb{R})$ on the points of $\mathrm{PG}(2, \mathbb{R})$ is $\mathrm{PGL}(d + 1, \mathbb{R})$, which is isomorphic to the quotient group of $\mathrm{GL}(d, \mathbb{R})$ by the non-zero scalar matrices. Importantly, collineations of $\mathrm{PG}(d, \mathbb{R})$ map subspaces to subspaces and preserve projective dimension.

We can recover real affine space from real projective space by deleting a hyperplane and all points and lines incident with it. As for the plane and space cases, we have:

Theorem 19.4

Let $\mathcal{H}$ *be a hyperplane of* $\mathrm{PG}(d, \mathbb{R})$. *Then the stabiliser of* $\mathcal{H}$ *in* $\mathrm{PGL}(d + 1, \mathbb{R})$ *acting on the points not on* $\mathcal{H}$ *is permutationally isomorphic to* $\mathrm{AGL}(d, \mathbb{R})$, *acting on* $\mathbb{R}^d$.

A **frame** of $PG(d, \mathbb{R})$ is a set of $d + 2$ points, no $d + 1$ on a hyperplane. The **fundamental frame** is

$$\{(1, 0, 0, \ldots, 0), (0, 1, 0, \ldots, 0), \ldots, (0, \ldots, 0, 1, 0), (0, 0, \ldots, 0, 1), (1, 1, \ldots, 1)\}.$$

The proofs of the following results are very similar to those for the plane case.

Theorem 19.5
$PGL(d + 1, \mathbb{R})$ *acts regularly on ordered frames of* $PG(d, \mathbb{R})$.

Theorem 19.6 (Fundamental theorem of projective geometry)
$Aut(PG(d, \mathbb{R})) = PGL(d + 1, \mathbb{R})$.

Desargues' theorem in 3-space readily extends to higher dimensions:

Theorem 19.7 (Desargues (1648) (see Desargues, 1951))
Given triangles ABC and A′B′C′ of $PG(d, \mathbb{R})$, $d \geqslant 3$, *in perspective from a point O, the lines AB and A′B′ meet in a point P, the lines AC and A′C′ meet in a point Q, the lines BC and B′C′ meet in a point R, and P, Q, and R are collinear.*

Proof The configuration is contained in a subspace of dimension 3, so we may assume that the dimension is 3. The result now follows from Theorem 14.1.   □

Projective spaces, or subspaces therein, satisfying the conclusion of Desargues' theorem are called **Desarguesian**.

Corollary 19.8
In $PG(d, \mathbb{R})$, $d \geqslant 3$, *every subspace is Desarguesian.*

The **dual** of a statement about the space is the statement that results after interchanging *point* and *hyperplane*, *intersection* and *join*, and making the necessary linguistic adjustments. As before, the principle of duality holds, so that the dual of every theorem about $PG(d, \mathbb{R})$ is also a theorem. An incidence-preserving map from the points to the hyperplanes of $PG(d, \mathbb{R})$ is called a **duality**. A duality of order two is called a **polarity**. The map interchanging point and hyperplane coordinates is a polarity, called the **standard duality**. Given a polarity, a point is **absolute** if it is incident with its image under the polarity.

19.2 Quadrics and Polarities

A **quadric** of $\mathsf{PG}(d, \mathbb{R})$ is the set of zeros of a homogeneous quadratic polynomial in $d + 1$ variables. Two quadrics are **equivalent** if there is a collineation of $\mathsf{PG}(d, \mathbb{R})$ mapping one to the other. Recall that in the plane, all non-empty, non-degenerate quadrics are equivalent (Theorem 3.13). In 3-space, there are two non-empty, non-degenerate quadrics up to equivalence, and they fall into two classes: the hyperbolic quadrics and the elliptic quadrics. If we also include the degenerate examples, the list is longer. As we shall see, by Sylvester's theorem of inertia (Theorem 19.9), we can understand the possible equivalence classes of quadrics of 3-space by a representative matrix for a quadratic form. The values on the diagonals of these matrices, from $\{-1, 0, 1\}$, are what distinguish the types of quadrics.

Before we present J. J. Sylvester's famous theorem, we need one item of notation. The **singular radical** of a quadratic form $Q \colon \mathbb{R}^{d+1} \to \mathbb{R}$ is the set of vectors $v \in \mathbb{R}^{d+1}$ such that $Q(v) = 0$ and $f_Q(v, w) = 0$, for all $w \in \mathbb{R}^{d+1}$, where f_Q is the bilinear form associated to Q. So, for instance, the quadric cone in Table 19.1 has singular radical $\langle (0, 0, 0, 1) \rangle$, and the elliptic and hyperbolic quadrics have trivial singular radical. If the singular radical of Q is trivial, then Q is **non-degenerate** (see Exercise 19.1).

Theorem 19.9 (Sylvester's theorem of inertia (compare with Artin, 1991, Chapter 7, Section 2))
Let Q be a quadric of $\mathsf{PG}(d, \mathbb{R})$. Then there exist a collineation g of $\mathsf{PG}(d, \mathbb{R})$ and $\ell, m \in \{1, \ldots d + 1\}$ with $\ell + m \leqslant d + 1$ and $\ell \leqslant m$ and such that

Table 19.1 *Gram matrices for quadratic forms, for one representative of each class of non-empty quadrics in 3-space.*

Hyperbolic	Elliptic	Quadric cone
$\begin{bmatrix} 1 & 0 & 0 & 0 \\ 0 & 1 & 0 & 0 \\ 0 & 0 & -1 & 0 \\ 0 & 0 & 0 & -1 \end{bmatrix}$	$\begin{bmatrix} 1 & 0 & 0 & 0 \\ 0 & 1 & 0 & 0 \\ 0 & 0 & 1 & 0 \\ 0 & 0 & 0 & -1 \end{bmatrix}$	$\begin{bmatrix} 1 & 0 & 0 & 0 \\ 0 & 1 & 0 & 0 \\ 0 & 0 & -1 & 0 \\ 0 & 0 & 0 & 0 \end{bmatrix}$
$x^2 + y^2 - z^2 - w^2 = 0$	$x^2 + y^2 + z^2 - w^2 = 0$	$x^2 + y^2 - z^2 = 0$

Two collinear planes	Double plane
$\begin{bmatrix} 1 & 0 & 0 & 0 \\ 0 & -1 & 0 & 0 \\ 0 & 0 & 0 & 0 \\ 0 & 0 & 0 & 0 \end{bmatrix}$	$\begin{bmatrix} 1 & 0 & 0 & 0 \\ 0 & 0 & 0 & 0 \\ 0 & 0 & 0 & 0 \\ 0 & 0 & 0 & 0 \end{bmatrix}$
$x = y$ or $x = -y$	$x^2 = 0$

$$Q^g = Q_{\ell,m} = \left\{(x_1,\ldots,x_{d+1}): x_1^2 + \cdots + x_m^2 - (x_{m+1}^2 + \ldots x_{m+\ell}^2) = 0\right\}.$$

Moreover, if $Q_{\ell,m}$ and $Q_{\ell',m'}$ are equivalent then $\ell = \ell'$ and $m = m'$.

Proof There is an orthogonal basis $\{v_1,\ldots,v_{d+1}\}$ of $\mathbb{R}^{d+1}$ with respect to the associated bilinear form f of the underlying quadratic form Q of Q such that there exist $\ell, m \in \{1,\ldots d+1\}$ with $\ell+m \leqslant d+1$ and $f(v_i, v_i)$ is 1 if $i \in \{1,\ldots,m\}$, -1 if $i \in \{m+1,\ldots,m+\ell\}$, and 0 if $i \in \{m+\ell+1,\ldots,d+\ell\}$. Scaling the form by $\frac{1}{2}$ gives the required form for Q. If $m > \ell$, replacing the form by its negative (and permuting the variables) gives $\ell \leqslant m$. Now $n + 1 - \ell - m$ is the dimension of the singular radical of Q. If $Q_{\ell,m}$ and $Q_{\ell',m'}$ are equivalent then there is a basis $\{w_1,\ldots,w_{d+1}\}$ of $\mathbb{R}^{d+1}$ with respect to the associated bilinear form f of the underlying quadratic form Q of $Q_{\ell,m}$ such that $f(w_i, w_i)$ is 1 if $i \in \{1,\ldots,m'\}$, -1 if $i \in \{m'+1,\ldots,m'+\ell'\}$, and 0 if $i \in \{m'+\ell'+1,\ldots,d+\ell\}$. Now $\{v_1,\ldots,v_m,w_{m'+1},\ldots w_{d+1}\}$ is linearly independent, giving $m \leqslant m'$. Interchanging the roles of $Q_{\ell,m}$ and $Q_{\ell',m'}$ gives $m' \leqslant m$. Thus, m and $\ell + m$ (and hence ℓ) are determined by Q. Note that the vector space dimension of the largest subspace contained in the quadric is $d + 1 - \ell$. $\quad\square$

We say that $\ell+m$ is the **rank** and m is the **signature** of the quadratic form Q. Theorem 19.9 was first published in Sylvester (1852), but also (posthumously) by Carl Gustav Jacob Jacobi (who died in February, 1851) in Jacobi (1857).[1]

An **isometry** of a quadratic form Q is a collineation g of the ambient projective space such that $Q(P^g) = Q(P)$ for all points P. The isometries form a group, the **isometry group** Isom(Q) of Q.

Theorem 19.10
There are d conjugacy classes of polarities of $\mathrm{PG}(d,\mathbb{R})$, for d even, and $d + 1$ conjugacy classes of polarities of $\mathrm{PG}(d,\mathbb{R})$, for d odd. The representatives are

(i) the standard duality (with no absolute points)
$$(a_1,\ldots,a_{d+1}) \mapsto [a_1,\ldots,a_{d+1}],$$

(ii) the orthogonal polarities (with absolute points a non-degenerate quadric)
$$(a_1,\ldots,a_{d+1}) \mapsto [a_1,\ldots,a_m,-a_{m+1},\ldots,-a_{d+1}], \text{ for } \tfrac{d+1}{2} \leqslant m \leqslant d,$$

(iii) the null polarity (with every point absolute)
$$(a_1,\ldots,a_{d+1}) \mapsto [-a_{d+1},-a_d,\ldots,-a_{\frac{d+3}{2}},a_{\frac{d+1}{2}},\ldots,a_1], \text{ for } d \text{ odd}.$$

---

[1] According to Koenigsberger (1904, p. 412), it dates to summer 1847.

Proof Suppose that Δ is a polarity of $PG(d, \mathbb{R})$. Then the product $g = \delta\Delta$ of Δ with the standard duality δ is a collineation, so by Theorem 19.6, $g \in PGL(d + 1, \mathbb{R})$. That is, there exists $A \in GL(d + 1, \mathbb{R})$ with g mapping the point $(x_1, \ldots, x_{d+1})$ to the point $(x_1, \ldots, x_{d+1})A$. Now on hyperplanes, g has the action

$$[a_1, a_2, \ldots, a_{d+1}] \mapsto [a_1, a_2, \ldots, a_{d+1}]A^{-\top},$$

and so $\Delta = \delta g$ maps the point $(e_1, \ldots, e_{d+1})$ to the hyperplane $[e_1, \ldots, e_{d+1}]A$ and the hyperplane $[a_1, a_2, \ldots, a_{d+1}]$ to the point $(a_1, a_2, \ldots, a_{d+1})A^{-\top}$. Since Δ has order two, it follows that $AA^{-\top}$ is a scalar multiple of the identity. By replacing A by a scalar multiple (which doesn't change g) we may assume that $AA^{-\top} = \pm I$, that is, $A = \pm A^\top$. So A is symmetric or skew-symmetric. If A is skew-symmetric, then d is odd and Δ is conjugate to $(a_1, a_2, \ldots, a_{d+1}) \mapsto [-a_{d+1}, -a_d, \ldots, -a_{\frac{d+3}{2}}, a_{\frac{d+1}{2}}, \ldots, a_2, a_1]$. The rest of the proof follows that of Sylvester's theorem (Theorem 19.9).   □

JAMES JOSEPH SYLVESTER (1814–97) was denied a university appointment in his native England as he was Jewish. He thus first took a position at the Royal Military Academy in Woolwich, before becoming chair and professor in mathematics at Johns Hopkins University, where he founded the *American Journal of Mathematics* and was influential in establishing mathematical research in America. Together with Cayley and Salmon, he was one of the driving forces of the theory of invariants (called a *world-triumvirate* by Forsyth) and finally became Savilian Professor of Geometry at Oxford at the age of 68. He was also the second president of the London Mathematical Society.

19.3 Associated Secunda and Lines

Recall that a **secundum** of $PG(n, \mathbb{R})$ is a subspace of projective dimension $n - 2$. For example, a secundum of $PG(4, \mathbb{R})$ is a plane. The following concept was first introduced by Ludwig Schläfli in 1866: $n + 1$ secunda of $PG(n, \mathbb{R})$ are **associated** if every line that meets n of them meets all of them.

A **simplex** of $PG(n, \mathbb{R})$ is a set of $n + 1$ points, no n on a hyperplane. So for $n = 2$, a simplex is a triangle (three noncollinear points), and for $n = 3$, a simplex is a tetrahedron. We present, without proof, Schläfli's generalisation of Pascal's theorem (Theorem 6.28). The proof is perhaps too involved to explore here because the idea is to use **Plücker coordinates**, and it is a generalisation of what is presented in this book. We will be exploring the base case application of Plücker coordinates in Section 20.1, by studying lines of $PG(3, \mathbb{R})$ as points of $PG(5, \mathbb{R})$. Higher dimensional use of Plücker coordinates is an alternative method of analysing the relationships between faces, lines, and other subspaces. See Rupp (1929) for an excellent exposition of Schläfli's ideas.

Theorem 19.11 (Schläfli, 1866)
Given a non-degenerate quadric in $\mathsf{PG}(n, \mathbb{R})$ *and two simplices, which are self-polar with respect to a quadric, the secunda of intersection of corresponding faces are associated.*

By Hesse's theorem (Theorem 6.74), two triangles are self-polar with respect to a conic if and only if they lie together on a conic. Thus, the special case $n = 2$ corresponds to Pascal's theorem (Theorem 6.28). The special case $n = 3$ was stated by Michel Chasles (1837, p. 400) and first proved by Norman Macleod Ferrers (1857b) and independently by George Salmon the same year (Salmon, 1857; see also Salmon, 1882, pp. 141–2). A special feature occurs in four dimensions, as pointed out in 1888 by Corrado Segre: *given four lines in* $\mathsf{PG}(4, \mathbb{R})$ *such that every three have a unique transversal line and no line is transversal to all four, there is a unique fifth line such that any plane that meets the four lines also meets the fifth line.* Such a set of five lines are said to be **associated** in $\mathsf{PG}(4, \mathbb{R})$.

Cyparissos Stephanos showed that five associated lines in $\mathsf{PG}(4, \mathbb{R})$ have their *Grassmann–Plücker* coordinates linearly related (Stephanos, 1882). That is, that the fifth line associated with four given lines in 4-space, any three of which span, is the unique other line whose Grassmann–Plücker coordinates lie in the subspace spanned by the Grassmann–Plücker coordinates of the four lines.

We say that three lines of $\mathsf{PG}(4, \mathbb{R})$ are in **general position** if they are mutually skew, and no hyperplane contains all of them. Then three lines in general position have a unique transversal line (Exercise 19.2).

Theorem 19.12 (Segre, 1887, 1888a, 1888b)
Let a, b, c, d be four lines in $\mathsf{PG}(4, \mathbb{R})$ *such that any three are in general position. Let a′ be the unique transversal to b, c, d. Let b′ be the unique transversal to a, c, d. Let c′ be the unique transversal to a, b, d. Let d′ be the unique transversal to a, c, d. Then:* $\langle a' \cap b, b' \cap a \rangle$ *and* $\langle d' \cap c, c' \cap d \rangle$ *meet in a unique point N,* $\langle a' \cap c, c' \cap a \rangle$ *and* $\langle d' \cap b, b' \cap d \rangle$ *meet in a unique point M, and* $\langle a' \cap d, d' \cap a \rangle$ *and* $\langle b' \cap c, c' \cap b \rangle$ *meet in a unique point L. Moreover, L, M, and N are collinear in a line e.*

Proof Now $b' \cap c$ is in $\langle b, c \rangle$ and in b', which is in $\langle a, d \rangle$; while $c' \cap b$ is in $\langle b, c \rangle$ and in c', which is in $\langle a, d \rangle$, so $\langle b' \cap c, c' \cap b \rangle$ lies in the plane $\langle a, d \rangle \cap \langle b, c \rangle$. Similarly, $a' \cap d$ lies in $\langle a, d \rangle$ and in a', which lies in $\langle b, c \rangle$, and $d' \cap a$ lies in $\langle a, d \rangle$ and in d', which lies in $\langle b, c \rangle$. So $\langle a' \cap d, d' \cap a \rangle$ lies in the plane $\langle a, d \rangle \cap \langle b, c \rangle$. Being distinct lines of the same plane, $\langle a' \cap d, d' \cap a \rangle$ and $\langle b' \cap c, c' \cap b \rangle$ meet in a unique point L. Similarly, $\langle a' \cap c, c' \cap a \rangle$ and $\langle d' \cap b, b' \cap d \rangle$ meet in a unique point M and $\langle a' \cap b, b' \cap a \rangle$ and $\langle d' \cap c, c' \cap d \rangle$ meet in a unique point N. Let e be $\langle a, a' \rangle \cap \langle b, b' \rangle \cap \langle c, c' \rangle$. Then L lies on a transversal line to a and a', so lies in $\langle a, a' \rangle$, and on a transversal line to b and b', so lies in $\langle b, b' \rangle$, and on a transversal line to c and c', so lies in $\langle c, c' \rangle$. Thus, L lies on e. Similarly, M and N lie on e. We also note that e lies in $\langle d, d' \rangle$. Finally, e is a line, for otherwise e would be a plane meeting a, b, c, and d. $\qquad\square$

CORRADO SEGRE (1863–1924) was a professor in higher geometry at Turin, and leader of the Italian school of algebraic geometry at the turn of the twentieth century. He introduced the concept of semi-linear transformations, pioneered birational geometry, and is best remembered today for the algebraic varieties named after him, which he introduced.

Worked Example 19.13

Here is another argument that uses the Dandelin–Galluci theorem (Theorem 15.1), and the isomorphism between $\mathsf{PG}(3, \mathbb{R})$ and the quotient space $\mathsf{PG}(4, \mathbb{R})/P$ where P is a point.

Consider the quotient space of $\mathsf{PG}(4, \mathbb{R})$ by a point P on e. Since e is in $\langle a, a' \rangle$, the planes $\langle a, P \rangle$ and $\langle a', P \rangle$ map to concurrent lines in $\mathsf{PG}(4, \mathbb{R})/P$. Similarly, $\langle b, P \rangle$ and $\langle b', P \rangle$ meet in $\mathsf{PG}(4, \mathbb{R})/P$, $\langle c, P \rangle$ and $\langle c', P \rangle$ meet in $\mathsf{PG}(4, \mathbb{R})/P$, and $\langle d, P \rangle$ and $\langle d', P \rangle$ meet in $\mathsf{PG}(4, \mathbb{R})/P$. So we have the situation of the Dandelin–Galluci theorem (Theorem 15.1) in $\mathsf{PG}(4, \mathbb{R})/P$, and any plane on P meeting three of $\langle a, P \rangle$, $\langle b, P \rangle$, $\langle c, P \rangle$, $\langle d, P \rangle$ must meet the fourth. Thus, any plane of $\mathsf{PG}(4, \mathbb{R})$ meeting three of a, b, c, d and meeting e must meet the fourth. The visible symmetry among a, b, c, d, e implies the result. $\qquad\square$

Notice that Segre's theorem (Theorem 19.12) expresses when five lines of projective 4-space are in general position, beyond what can easily be said using linear algebra: not only must any three of them span the whole space, but also some plane must meet exactly four of them (that is, the fifth line must not be the special line such that all planes meeting the first four meet the fifth).

Recall Schläfli's double-six theorem (Theorem 15.3): given six pairwise skew lines $\ell_1, \ldots \ell_6$ in $\mathsf{PG}(3, \mathbb{R})$, and another five pairwise skew lines $m_1, \ldots m_5$ such that ℓ_i meets m_j for all distinct (i, j), and ℓ_i is skew to m_i for all i, there exists another line m_6 meeting each $\ell_1, \ldots \ell_5$. Here is another proof that uses Theorem 19.12, and is due to Baker (1920).

Here is an instructive quote from Baker's paper (Baker, 1920, p. 133): 'The naturalness of this figure lies in the fact that three lines in four dimensions have just one transversal.'

Another proof of Theorem 15.3. First we restate Schläfli's theorem (Theorem 15.3) so that the initial ingredient is a set of four skew lines. Let a, b, c, d be four skew lines of $\mathsf{PG}(3, \mathbb{R})$, and let a', b', c', d' be another set of four skew lines, each meeting three of the former set of four lines such that like labelled lines are not skew. For example, a' meets b, c, d. Recall (see Exercise 13.4) that four skew lines of $\mathsf{PG}(3, \mathbb{R})$ have two transversal lines, so let e and f be two transversals of a', b', c', d', and let e', f' be two transversals of a, b, c, d. Then Schläfli's theorem (Theorem 15.3) restated is that if e and f' meet, then e' and f meet.

Let $A := b' \cap c$, $B := c' \cap a$, $C := a' \cap b$, $A' := b \cap c'$, $B' := c \cap a'$, $C' := a \cap b'$. Let $P := a \cap d'$, $Q := b \cap d'$, $R := c \cap d'$, $P' := a' \cap d$, $Q' := b' \cap d$, $R' := c' \cap d$. Notice that these 12 points determine the initial configuration, and we can easily recover the given lines (e.g., a lies on C', B, and P). Now embed $\mathsf{PG}(3, \mathbb{R})$ as a hyperplane Σ in $\mathsf{PG}(4, \mathbb{R})$, and let O be a point not incident with Σ. We will show that we can slide along the cone subtended by this configuration and O, to replace these points by a similar set of points that instead determine a pair of sets of skew lines in general position. Since A, B, C, A', B', C' are six points of $\mathsf{PG}(3, \mathbb{R})$, no three collinear, we can scale the homogeneous coordinates of each point so that the following two relations hold (with abuse of notation; the point and the homogeneous coordinates for it being synonymous):

$$A + B + C + A' + B' + C' = 0$$

$$\alpha A + \beta B + \gamma C + \alpha' A' + \beta' B' + \gamma' C' = 0$$

for some $\alpha, \beta, \gamma, \alpha', \beta', \gamma' \in \mathbb{R}$. This just means that we can write B' and C' in terms of the first four points, which form a frame of $\mathsf{PG}(3, \mathbb{R})$. Now, since P, Q,

R are collinear, it turns out, after rearrangement and elimination of variables, that there exists $\rho \in \mathbb{R}$ such that

$$P = \beta B + \gamma' C' + \rho(B + C'), \qquad Q = \gamma C + \alpha' A' + \rho(C + A')$$
$$R = \alpha A + \beta' B' + \rho(A + B').$$

Similarly,

$$P' = \beta' B' + \gamma C + \rho(B' + C), \qquad Q' = \gamma' C' + \alpha A + \rho(C' + A)$$
$$R' = \alpha' A' + \beta B + \rho(A' + B).$$

Now we produce six points on the lines AO, BO, CO, $A'O$, $B'O$, $C'O$ as follows:

$$A_1 = (\alpha + \rho) + \ell O, \qquad B_1 = (\beta + \rho)B + mO, \qquad C_1 = (\gamma + \rho)C + nO,$$
$$A_1' = (\alpha' + \rho) + \ell'O, \qquad B_1' = (\beta' + \rho)B + m'O, \qquad C_1 = (\gamma' + \rho)C + n'O,$$

where $\ell + m + n + \ell' + m' + n' = 0$, which ensures that

$$A_1 + B_1 + C_1 + A_1' + B_1' + C_1' = 0.$$

Then let

$$P_1 = P + (m + n')O, \qquad Q_1 = Q + (n + \ell')O, \qquad R_1 = R + (\ell + m')O,$$
$$P_1' = P' + (m' + n)O, \qquad Q_1' = Q' + (n' + \ell)O, \qquad R_1' = R' + (\ell' + m)O.$$

Therefore,

$$P_1 = B_1 + C_1', \qquad Q_1 = C_1 + A_1', \qquad R_1 = A_1 + B_1',$$
$$P_1' = B_1' + C_1, \qquad Q_1' = C_1' + A_1, \qquad R_1' = A_1' + B_1.$$

So we have similar properties to the initial configurations, except the grid lines are now two sets of four skew lines with any three of four in general position. By Segre's theorem (Theorem 19.12), $A_1 A_1'$ and $P_1 P_1'$ meet in a point L, $B_1 B_1'$ and $Q_1 Q_1'$ meet in a point M, $C_1 C_1'$ and $R_1 R_1'$ meet in a point N, and the points L, M, N are collinear in a line e_1. Now e_1 is the unique transversal to four associated lines of $\mathrm{PG}(4, \mathbb{R})$. It follows that the projection of e_1 to Σ (from O) is e or f. Without loss of generality, let us suppose that it is e. We can also repeat this argument for the opposite set of four associated lines in this figure, and so, by projection, this proves Schläfli's theorem. □

In 1905, Luigi Berzolari established a converse to Schläfli's theorem:

Theorem 19.14 (Berzolari, 1905)
Given a one-to-one correspondence between two simplices in $\mathrm{PG}(n, \mathbb{R})$ such that the secunda of intersection of corresponding faces are associated, there

is a nondegenerate quadric Q such that the simplices are self-polar with respect to Q.

Sketch of the proof We give some of the background for the proof given in Gerber (1975, Theorem 3.2 and its corollary). By Theorem 19.5, we may suppose that the first simplex is the fundamental one, whereas the second simplex is $(a_{11}, a_{12}, \ldots, a_{1,n+1})$, $(a_{21}, a_{22}, \ldots, a_{2,n+1})$, $\ldots (a_{n+1,1}, a_{n+1,2}, \ldots, a_{n+1,n+1})$. The quadric has *tangential equation*

$$\sum \sum a_{ij} x_i x_j = 0.$$

The tangential equation is the equation of the tangent spaces, that is, the dual quadric. If a quadric $\sum \sum c_{ij} x_i x_j$ is given then surround the matrix (c_{ij}) with a column $x_1 x_2 \ldots x_{n+1} x_0$ and a row $x_1 x_2 \ldots x_{n+1} x_0$: the determinant of the resulting $(n+2) \times (n+2)$ matrix is the tangential equation of the quadric. □

A theorem of Louis Kollros from 1946 is a variant of Theorem 19.11 and also a generalisation of Pascal's theorem (take $n = 2$ in the statement).

Theorem 19.15 (Kollros, 1947)
A non-degenerate quadric and a simplex are given in $\mathsf{PG}(n, \mathbb{R})$. The edges of the simplex meet the quadric in $n(n+1)$ points. These can be arranged in $n+1$ groups of n points so that each group contains points on n concurrent edges. The $n+1$ groups determine $n+1$ hyperplanes. One of these hyperplanes corresponds to each vertex of the simplex and to its opposite face. Each of these hyperplanes intersects the corresponding face of the given simplex in a secundum. Then these $n+1$ secunda are associated.

Proof First, we can assume that the given simplex is the canonical one with faces $x_i = 0$ for $i = 1, \ldots, n+1$. The quadric has equation

$$x_1^2 + \cdots + x_{n+1}^2 = \sum_{1 \leqslant i < k \leqslant n+1} (a_{ik} + a_{ik}^{-1}) x_i x_k.$$

Let $A_1, \ldots, A_{n+1}$ be the vertices of the simplex. Then the edge $A_1 A_2$ cuts the quadric in two points corresponding to the roots of the equation

$$x_1^2 - (a_{12} + a_{12}^{-1}) x_1 x_2 + x_2^2 = 0,$$

where the vertex A_1 is associated with the solution $\frac{x_1}{x_2} = a_{12}$, and A_2 is associated with the solution $\frac{x_2}{x_1} = a_{12}$. The hyperplanes containing n points situated on the n edges issuing from the vertex A_i have equations of the form

$$x_i = \sum_{j=1, j \neq i}^{n+1} a_{ij} x_j.$$

By cutting each of these hyperplanes by the face $x_i = 0$ opposite the vertex A_i, we will have $n + 1$ secunda, which, according to our theorem, must be associated. Indeed, a line is determined by $n - 1$ linear equations (the intersection of $n - 1$ hyperplanes), which we can write as

$$\sum_{s=1}^{n+1} b_{rs} x_s = 0, \quad r \in \{1, \ldots, n - 1\}, b_{rs} \neq b_{sr}.$$

These lines intersect the first secundum if the following determinant is zero:

$$D_1 = \begin{vmatrix} a_{12} & a_{13} & \cdots & a_{1,n+1} \\ b_{12} & b_{13} & \cdots & b_{1,n+1} \\ \vdots & \vdots & \ddots & \vdots \\ b_{n-1,2} & b_{n-1,3} & \cdots & b_{n-1,n+1} \end{vmatrix}.$$

Expand along the top row to obtain the following expansion of D_1:

$$D_1 = a_{12} B_{12} - a_{13} B_{13} + \cdots + (-1)^n a_{1,n+1} B_{1,n+1}.$$

So the B_{ik} are obtained as determinants of the above matrix with the i and k columns deleted of the $(n - 1) \times (n + 1)$ submatrix of the b_{rs}-coefficients. We can do this for the other vertex-face combinations to obtain formulae for the D_i. They will be of the form

$$D_i = a_{i2} B_{i2} - a_{i3} B_{i3} + \cdots + (-1)^n a_{i,n+1} B_{i,n+1}.$$

Now note that if we create the following alternating sum, each of the terms $a_{i,k} B_{i,k}$ appears there once positively and once negatively:

$$D_1 - D_2 + D_3 - \cdots + (-1)^n D_{n+1},$$

and so this expression is identically zero. So any line that intersects n of the $n + 1$ secunda also intersects the remaining one.   □

In 1955, Oene Bottema proved the converse of Louis Kollros' theorem (Theorem 19.15). Note that for $n = 2$, Bottema's theorem is the Braikenridge–Maclaurin theorem (Theorem 6.43).

Theorem 19.16 (Bottema, 1955)
Let two simplices $A_1 \ldots A_{n+1}$ and $B_1 \ldots B_{n+1}$ in $\mathrm{PG}(n, \mathbb{R})$, formed by hyperplanes $\alpha_1 \ldots \alpha_{n+1}$ and $\beta_1 \ldots \beta_{n+1}$, be so situated that the $n + 1$ secunda $\alpha_i \cap \beta_i$ are associated. Then the $n(n+1)$ points $A_i A_k \cap \beta_i$, $(i \neq k)$ lie on a non-degenerate quadric.

Sketch of the proof Bottema makes use of the Grassmann correspondence, and takes Plücker coordinates of the secunda. In particular, the hypothesis that they are associated means that these coordinates are *linearly related*.[2] So there exist $p_i \in \mathbb{R}$ $(i = 1, \ldots, n + 1)$ such that

$$\sum_{i=1}^{n+1} p_i \pi_{k\ell}^{(i)} = 0,$$

where $\pi_{k\ell}^{(i)}$ denotes these Plücker coordinates. The quadric in the conclusion of the result is

$$\sum p_i x_i^2 - \sum_{i>j} \left(p_i p_j \lambda_{ij} + \frac{1}{\lambda_{ij}} \right) x_i x_j = 0,$$

where the λ_{ij} are non-zero real numbers such that $\pi_{ij}^{(i)} = p_i \lambda_{ij}$. □

The following result of Baker was anticipated to some extent by Berzolari (1905) and is a generalisation of Hesse's theorem (Theorem 6.15) (which is the case $n = 2$ below).

Theorem 19.17 (Baker, 1936a)
For a given polarity ρ on $\mathrm{PG}(n, \mathbb{R})$, and a given simplex $P_0 P_1 \cdots P_n$, let Q_v be the pole of the hyperplane spanned by all the P_is except P_v. If $P_v \neq Q_v$ for every v, the $n + 1$ lines $P_v Q_v$ are 'associated' in the sense that every secundum that meets n of them meets the remaining one also.

Proof This proof appears in Beatty and Todd (1944) and again uses the equation of a *tangential quadric*. Let one simplex be the fundamental simplex, and let

$$\sum \sum a_{ij} x_i x_j = 0$$

be the *tangential equation* of the quadric associated to ρ, so that the other simplex is $Y_1, \ldots, Y_{n+1}$ where Y_1, being the pole of $x_1 = 0$, is $(a_{1,1}, a_{1,2}, \ldots, a_{1,n+1})$, and so forth. A point of the line $X_1 Y_1$ is of the form $(\lambda_1, a_{1,2}, \ldots, a_{1,n+1})$, where $\lambda_1 \in \mathbb{R}$. The condition for n such points, one on each of the lines $X_i Y_i$, to lie in a secundum is the same as the condition for the first $n - 1$ points and the whole line $X_n Y_n$ to lie in a hyperplane. That is,

---

[2] Plücker coordinates are not in a vector space; they are only well-defined as homogeneous coordinates. However, linear dependence of homogeneous coordinates means that the subspace spanned by the associated points has algebraic dimension less than the number of points, just as we observed in Theorem 1.5.

$$0 = \begin{vmatrix} \lambda_1 & a_{1,2} & \cdots & a_{1,n-1} & a_{1,n} & a_{1,n+1} \\ a_{2,1} & \lambda_2 & \cdots & a_{2,n-1} & a_{2,n} & a_{2,n+1} \\ \vdots & \vdots & \ddots & \vdots & \vdots & \vdots \\ a_{n-1,1} & a_{n-1,2} & \cdots & \lambda_{n-1} & a_{n-1,n} & a_{n-1,n+1} \\ a_{n,1} & a_{n,2} & \cdots & a_{n,n-1} & a_{n,n} & a_{n,n+1} \\ 0 & 0 & \cdots & 0 & 1 & 0 \end{vmatrix}$$

$$= \begin{vmatrix} \lambda_1 & a_{1,2} & \cdots & a_{1,n-1} & a_{1,n+1} \\ a_{2,1} & \lambda_2 & \cdots & a_{2,n-1} & a_{2,n+1} \\ \vdots & \vdots & \ddots & \vdots & \vdots \\ a_{n-1,1} & a_{n-1,2} & \cdots & \lambda_{n-1} & a_{n-1,n+1} \\ a_{n,1} & a_{n,2} & \cdots & a_{n,n-1} & a_{n,n+1} \end{vmatrix}.$$

This determinant is a relation between the parameters $\lambda_1, \ldots, \lambda_{n-1}$, but since $a_{i,j} = a_{j,i}$, the relation is symmetrical between the suffixes n and $n + 1$. Thus, every secundum which meets the lines $X_1 Y_1, \ldots, X_n Y_n$ meets $X_{n+1} Y_{n+1}$ too. $\square$

Exercises

19.1 Let Q be a quadratic form on $\mathbb{R}^d$. Show that Q is degenerate if and only if it has a non-trivial singular radical. (We use a similar definition as before for degeneracy, that is, Q factors over $\mathbb{C}$.)

19.2 Show that three lines of $PG(4, \mathbb{R})$ in general position have a unique transversal line.

20

The Klein Quadric and the Veronese Surface

> From Veronese's point of view, a 2-dimensional variety in 3-dimensional space was like a shirt crammed into a too-full suitcase. If you give it a little more room, in the form of extra dimensions, you can smooth out the creases and the pinch points.

Dana Mackenzie (2009, p. 78)

20.1 The Klein Correspondence

Let ℓ be a line in $\mathsf{PG}(3, \mathbb{R})$, with $\ell = XY$, where $X = (x_1, x_2, x_3, x_4)$, $Y = (y_1, y_2, y_3, y_4)$. Let

$$p_{ij} = p_{ij}(X, Y) = \begin{vmatrix} x_i & x_j \\ y_i & y_j \end{vmatrix} = x_i y_j - x_j y_i.$$

Then $(p_{12}, p_{13}, p_{14}, p_{23}, p_{42}, p_{34})$ represents a point of $\mathsf{PG}(5, \mathbb{R})$ and is independent of the choice of X and Y in the representation of the line ℓ. To see this, suppose that $\ell = X'Y'$. Then $X' = XA$, $Y' = YA$, for some $A \in \mathsf{GL}(2, \mathbb{R})$, and so $p_{ij}(X', Y') = \det(A)p_{ij}(X, Y)$, showing that the homogeneous coordinates agree:

$$(p_{12}(X, Y), p_{13}(X, Y), p_{14}(X, Y), p_{23}(X, Y), p_{42}(X, Y), p_{34}(X, Y))$$
$$= (p_{12}(X', Y'), p_{13}(X', Y'), p_{14}(X', Y'), p_{23}(X', Y'), p_{42}(X', Y'), p_{34}(X', Y')).$$

We say that $(p_{12}, p_{13}, p_{14}, p_{23}, p_{42}, p_{34}) = p(\ell)$ are the **Plücker coordinates** of ℓ, and we call the map $\ell \mapsto p(\ell)$ the **Plücker map**.

Let Q be the **Klein quadric**

$$\{(x_1, x_2, x_3, x_4, x_5, x_6) \colon x_1 x_6 + x_2 x_5 + x_3 x_4 = 0\}.$$

Lemma 20.1
$p(\ell) \in Q$, for all lines ℓ of $PG(3, \mathbb{R})$.

Proof This is just the assertion that $p_{12}p_{34} + p_{13}p_{42} + p_{14}p_{23} = 0$, which is straightforward to check. □

Lemma 20.2
The image of the Plücker map is the set of points of the Klein quadric Q.

Proof Let $(x_{12}, x_{13}, x_{14}, x_{23}, x_{42}, x_{34}) \in Q$. As a notational shorthand, reversing indices on x_{ij} will result in the negative value; $x_{ij} = -x_{ji}$. So, for example, $x_{21} := -x_{12}$. Now we have $x_{ij} \neq 0$, for some i, j, so let λ be x_{ij}. Put $x = \sum_{k \neq i} x_{ik}e_k$, $y = \sum_{k \neq j} x_{jk}e_k$ where e_k denotes the k-th canonical basis vector of $\mathbb{R}^4$. Then it is straightforward to compute that

$$p(\langle x, y \rangle) = \lambda(x_{12}, x_{13}, x_{14}, x_{23}, x_{42}, x_{34}),$$

and hence the image of the Plücker map is the whole set of points of Q. □

Let $f(x, y) = x_1y_6 + x_6y_1 + x_2y_5 + x_5y_2 + x_3y_4 + x_4y_3$, that is, the symmetric bilinear form associated to the Klein quadric.

Theorem 20.3 (Klein, 1868)
Two lines ℓ, m of $PG(3, \mathbb{R})$ meet if and only if $f(p(\ell), p(m)) = 0$.

Proof This follows from the observation that $f(p(\langle x, y \rangle), p(\langle z, w \rangle))$ is the determinant of the matrix with rows x, y, z, and w. □

Lemma 20.4
The Plücker map is one-to-one.

Proof By Theorem 20.3, if $p(\ell) = p(m)$, then $f(p(\ell), p(m)) = 0$, so ℓ meets m. Thus, $\ell = \langle x, y \rangle$ and $m = \langle x, z \rangle$, for some x, y, z. Thus, there exists c with $p_{ij}(x, y) = cp_{1j}(x, z)$, for all i, j. So $p_{ij}(x, y - cz) = 0$, for all i, j, which implies that $y = cz$ and so $\ell = m$. □

A subspace of $PG(5, \mathbb{R})$ is **singular** with respect to Q if all of its points lie in Q. Note that a singular subspace is absolute with respect to the polarity $\perp$ arising from Q. Now the image of a subspace S under the polarity is a subspace $S^\perp$ with complementary dimension. That is, $\dim S^\perp = 4 - \dim S$. So points and hyperplanes of $PG(5, \mathbb{R})$ are interchanged, lines and solids are interchanged, and planes are sent to planes. Therefore, a singular subspace has dimension at most 2, for a solid cannot be singular. The singular planes are often called

generators of the Klein quadric. One of the most interesting properties of Q is that its singular planes split into two equivalence classes. Define a reflexive binary relation $\sim$ on the set of singular planes via the following rule: two distinct singular planes π_1 and π_2 have $\pi_1 \sim \pi_2$ if and only if they intersect in a point. It turns out that this is an equivalence relation with precisely two classes. These two classes are often called the **Latin** and **Greek** planes. (It will not matter in which order we associate these two terms with the two classes.)

Theorem 20.5

(a) The Plücker image of a plane-pencil is a line of Q.

(b) The Plücker image of a bundle of lines is a (Latin) plane of Q.

(c) The Plücker image of the set of lines of a plane is a (Greek) plane of Q.

Proof First we prove (a). Let P be a point incident with a plane π of $\mathsf{PG}(3, \mathbb{R})$, and consider the lines incident with P and π. By Theorem 20.3, any two lines of this plane-pencil correspond to collinear points of Q. The lines of a plane-pencil can be written down as the image of a set of lines of the form PX where P is a fixed point, and $X = \lambda A + \mu B$ for two fixed points A and B. Moreover, the line AB is not incident with P. So let us write $P \colon (p_1, p_2, p_3, p_4)$, and $X \colon \lambda(a_1, a_2, a_3, a_4) + \mu(b_1, b_2, b_3, b_4)$. Then the Plücker coordinates of PX can be calculated as follows, and we can write them as linear combinations of other Plücker coordinates:

$$\begin{vmatrix} p_i & p_j \\ \lambda a_i + \mu b_i & \lambda a_j + \mu b_j \end{vmatrix} = \lambda \begin{vmatrix} p_i & p_j \\ a_i & a_j \end{vmatrix} + \mu \begin{vmatrix} p_i & p_j \\ b_i & b_j \end{vmatrix}.$$

In other words, a plane-pencil of lines consists of the lines whose Plücker coordinates are of the form $\lambda p_{ij} + \mu p'_{ij}$ where p_{ij} and p'_{ij} are the coordinates of two collinear points of $\mathsf{PG}(5, \mathbb{R})$. So we have proved (a).

For (b) and (c), notice that a bundle of lines on a point P (and the lines of a plane) is mapped to a set of pairwise collinear points of Q. Indeed, a bundle of lines is a maximal set of lines that are pairwise concurrent, and so its image is a maximal set of points of Q that are pairwise collinear. Such a set is precisely the points incident with a singular plane. Likewise, the set of lines of a plane is a maximal set of lines that are pairwise concurrent, and so the same is true.

Finally, the bundles of lines and the lines of planes map to different classes of planes of Q. To see this, note that a bundle of lines meets a plane of $\mathsf{PG}(3, \mathbb{R})$ either in no lies at all or in a plane-pencil of lines (depending on whether the plane lies on the vertex of the bundle or not). Therefore, the two image planes of the Klein quadric are inequivalent under $\sim$ because the planes would either intersect trivially or in a line (by (a)). $\qquad\square$

Table 20.1 *How lines linearly dependent on three linearly independent lines map under the Klein correspondence.*

PG(3, $\mathbb{R}$)	Klein quadric Q
Regulus	Plane meeting Q in a conic
Bundle of lines	Latin generator
Set of all lines of a plane	Greek generator
Union of two pencils having distinct points and planes but a common line	Plane meeting Q in two concurrent lines

One of the advantages of using the Klein correspondence to study the line geometry of PG(3, $\mathbb{R}$) is that the ambient projective space PG(5, $\mathbb{R}$) is larger in dimension, and so we can realise some of the interesting sets of lines of 3-space as naturally occurring sets of points of the Klein quadric. For example, a parabolic congruence corresponds to the points of the Klein quadric that lie in a solid, such that the solid meets the quadric in a quadric cone.

Worked Example 20.6
The image of a regulus of PG(3, $\mathbb{R}$) is the set of points of a plane of PG(5, $\mathbb{R}$) that meets the Klein quadric in a conic. □

Indeed, we had a classification in Theorem 16.9 of sets of lines of PG(3, $\mathbb{R}$) linearly dependent on three linearly independent lines. If we use the Klein correspondence, and Theorem 16.8, we are simply classifying the possible ways a plane of PG(5, $\mathbb{R}$) can meet the Klein quadric. So, essentially, we just need to think of all the ways we can place a quadratic form on PG(2, $\mathbb{R}$), and what the singular points look like (see Table 20.1):

Recall that a linear congruence of PG(3, $\mathbb{R}$) is a set of lines linearly dependent on four linearly independent lines, and a linear complex is a set of lines linearly dependent on five linearly independent lines. In the Klein correspondence, we can easily describe linear congruences and linear complexes as sets of points of the Klein quadric because linear dependence of lines corresponds to linear dependence of the underlying vectors of the corresponding points. So the image of a linear congruence is a set of points of the Klein quadric incident with a solid of PG(5, $\mathbb{R}$). Now, by the theory of bilinear and quadratic forms, a solid of PG(5, $\mathbb{R}$) can meet the Klein quadric in one of four different ways:

- the solid is non-degenerate and the intersection induces a hyperbolic quadric on the solid;
- the solid is non-degenerate and the intersection induces an elliptic quadric on the solid;
- the solid is degenerate and the intersection is a quadratic cone;

- the solid is degenerate and the intersection is the set of points on two singular planes that meet in a line.

This gives us Theorem 16.10, by taking the preimages of the points within the solids described above. To summarise, we have the following correspondences (see Table 20.2):

Table 20.2 *Congruences and complexes under the Klein correspondence.*

PG$(3, \mathbb{R})$	Klein quadric Q
Hyperbolic congruence	Solid meeting Q in a hyperbolic quadric
Elliptic congruence	Solid meeting Q in an elliptic quadric
Parabolic congruence	Solid meeting Q in a quadric cone
Degenerate congruence	Solid meeting Q in two generators meeting in a line
Special linear complex	Hyperplane meeting Q in a cone
General linear complex	Hyperplane meeting Q in a non-degenerate quadric

Worked Example 20.7 (Elliptic congruences revisited)
Theorem 16.12 classified elliptic congruences as regular spreads, and we will elaborate on the proof here. There are still some details for the reader to fill in, since a proper treatment would take us much further into the geometry of quadrics. In the Klein correspondence, we have a solid Σ meeting Q in an elliptic quadric. Now an elliptic quadric has no lines, or, in other words, no two points of it lie on a line of Q. So the preimage of the points of the elliptic quadric under the Klein correspondence is a set of skew lines. Moreover, these lines form a partition of the points of PG$(3, \mathbb{R})$: every generator of Q is incident with a unique point of the elliptic quadric. Next, if we take a point P of Q that does not lie in Σ, the points of the elliptic quadric in Σ that are collinear to P span a plane meeting Q in a conic. (To see why, the polar image of P under the polarity arising from Q is a hyperplane and meets Σ in a non-degenerate plane.) The preimage of this conic is a regulus (see Worked Example 20.6). □

Worked Example 20.8 (Parabolic congruences revisited)
Recall from Theorem 16.11 that if we take a projectivity ϕ between the points and planes incident with a line ℓ, then the plane-pencils defined by corresponding points and planes (under ϕ) form a parabolic congruence. We give a sketch of a proof here.

Let us apply the Klein correspondence to the statement above. We have a point L of the Klein quadric (the image of ℓ) and a collineation g preserving the Klein quadric that interchanges pairs of planes (Latins and Greeks) incident with L. Now the plane-pencils defined by corresponding points and planes are mapped to (some of the) lines of the Klein quadric incident with L. Moreover, these lines are self-polar, so lie in the hyperplane $L^{\perp}/L$. What we will exploit here is that there is a natural quotient map from the Klein quadric

Q to a hyperbolic quadric of $L^\perp/L$, thought of as $\mathsf{PG}(3, \mathbb{R})$. In vector form, this quotient map simply takes $x \in L^\perp$ and maps it to a coset $L + x$. A bilinear form $\tilde{B}$ on $L^\perp/L$ is inherited from the bilinear form B defining Q. Namely, $\tilde{B}(x, y) := B(L + x, L + y)$ (where $x, y \in L^\perp$), and this is a well-defined bilinear form. So Q *modulo* L is a hyperbolic quadric on which ϕ induces a projectivity from the lines of one regulus to the lines of its opposite regulus. Each such projectivity corresponds to a unique plane section. Indeed, a conic $\mathcal{K}$ on a hyperbolic quadric in 3-space defines a map from the lines on one regulus $\mathcal{R}_1$ to the lines of the other $\mathcal{R}_2$ via mapping a line m of $\mathcal{R}_1$ to the line of $\mathcal{R}_2$ on $m \cap \mathcal{K}$. This is a projectivity. Conversely, given a projectivity ψ between $\mathcal{R}_1$ and $\mathcal{R}_2$ let $m_1, m_2, m_3 \in \mathcal{R}_1$, and let π be the plane spanned by $m_i \cap m_i^\psi$, $i = 1, 2, 3$ and $\mathcal{K}$ be the conic in which π meets the hyperbolic quadric.

So, to summarise, the lines on L correspond to a conic-plane section of a hyperbolic quadric of $\mathsf{PG}(3, \mathbb{R})$, under the natural quotient map from the Klein quadric to $L^\perp/L$. So, in the preimage of this quotient map, the image of the original plane-pencils are lines incident with L, whose quotient modulo L is a conic. This means that we have a quadric cone of Q with vertex L, and so a parabolic congruence in the original setting. $\qquad\square$

Worked Example 20.9 (Linear complexes)
Recall from above that a special complex of $\mathsf{PG}(3, \mathbb{R})$ corresponds to a hyperplane meeting the Klein quadric Q in a cone. In particular, the points of this cone are precisely the points of Q collinear to the **vertex** V of this cone. (Indeed, we can obtain V by taking the defining hyperplane Π, and computing $\Pi \cap \Pi^\perp$, where $\perp$ is the polarity arising from Q.) So in the preimage, a special linear complex consists of a line (the preimage of V) and all lines concurrent to it. By Exercise 20.1, the isometry group of Q acts transitively on these degenerate hyperplanes, and so we have proved Theorem 16.13.

A general complex of $\mathsf{PG}(3, \mathbb{R})$ corresponds to a non-degenerate hyperplane section of the Klein quadric Q. Clearly, the points S of Q within such a hyperplane Π have the property that every generator of Q meets S in a line of Q. Therefore, a general complex is precisely a set S of lines of $\mathsf{PG}(3, \mathbb{R})$ such that, for every point P, the lines of S on P are a pencil. This yields the first part of Theorem 16.14. Moreover, all non-degenerate hyperplanes are equivalent under the isometry group of the Klein quadric (by Exercise 20.1).

For the second part of Theorem 16.14, consider the bilinear alternating form $B(x, y) := x_1 y_2 - x_2 y_1 + x_3 y_4 - x_4 y_3$. Consider a line ℓ that is the join of two points $x := (x_1, x_2, x_3, x_4)$ and $y := (y_1, y_2, y_3, y_4)$. For ℓ to be self-polar with respect to the null polarity arising from B, we require $(x_1 y_2 - x_2 y_1) + (x_3 y_4 - x_4 y_3) = 0$. The parentheses are deliberate, since they highlight that there are two line coordinate components for ℓ in this equation. From Equation 16.1,

$\ell_1 = x_1y_2 - x_2y_1$ and $\ell_6 = x_3y_4 - x_4y_3$. Therefore, a self-polar line ℓ for B must satisfy $\ell_1 + \ell_6 = 0$, or, in other words, the Plücker coordinates for ℓ lie on the hyperplane $X_1 + X_6 = 0$ of $PG(5, \mathbb{R})$. This hyperplane is nondegenerate. So a null polarity corresponds to non-degenerate hyperplanes under the Klein correspondence. □

We see in the above examples another simple consequence: (i) a linear complex is precisely the set of all lines of $PG(3, \mathbb{R})$ whose Plücker coordinates satisfy a homogeneous linear equation; (ii) a linear congruence is precisely the set of all lines of $PG(3, \mathbb{R})$ whose Plücker coordinates satisfy two linearly independent homogeneous linear equations. So we have finally proved Theorem 16.16.

* * *

The following is an interesting application of the Klein correspondence to a line-geometric characterisation of quadrics of $PG(4, \mathbb{R})$, due to Beniamino Segre. It can be thought of as a *four-dimensional* version of Pascal's theorem (Theorem 6.28).

Theorem 20.10 (Segre, 1945)
Five lines ℓ_1, ℓ_2, ℓ_3, ℓ_4, ℓ_5 of $PG(4, \mathbb{R})$ such that there is no transversal line to four of them and any three span $PG(4, \mathbb{R})$ lie on a quadric if and only if the five points $\ell_i \cap \langle \ell_{i-1}, \ell_{i+1} \rangle$ (subscripts taken modulo 5) lie on a hyperplane.

Proof For the ' $\implies$ ' direction, embed the quadric in the Klein quadric. After applying the Klein correspondence, the result is reduced to the following:

Let $A_1A_2A_3A_4A_5$ be five general points in $PG(3, \mathbb{R})$, such that any four span $PG(3, \mathbb{R})$ and a_1, a_2, a_3, a_4, a_5 are their (respective) polar planes with respect to a given (general or special) linear complex of lines. Then the five pairs of planes a_i and $A_{i-1}A_iA_{i+1}$ ($i = 1, 2, 3, 4, 5$, subscripts taken modulo 5) intersect in five lines having two common transversals. The five lines indicated in the hypothesis of the theorem are both in the given linear complex, and in the linear complex defined by the conditions of having $A_{i-1}A_iA_{i+1}$ as the polar plane of A_i, ($i = 1, 2, 3, 4, 5$). Hence, they are in the linear congruence formed by the common lines of the two complexes. This proves the ' $\implies$ ' direction.

Conversely, take three distinct points on each of ℓ_1, ℓ_2, ℓ_3, ℓ_4 and two distinct points on ℓ_5. The hypothesis on the lines (that there is no transversal line to four of them and any three span) implies that the conditions on a quadric containing those 14 points are linearly independent and therefore that there is a unique quadric Q containing those 14 points (and therefore the lines ℓ_1, ℓ_2, ℓ_3, ℓ_4). By the forward direction, Q contains ℓ_5 if and only if the five points $\ell_i \cap \langle \ell_{i-1}, \ell_{i+1} \rangle$ (subscripts modulo 5) lie on a hyperplane. $\qquad\qquad\qquad\qquad\qquad\qquad$ □

20.2 Rational Normal Curves

A curve of $PG(n, \mathbb{R})$ is *rational* if its points can be *parameterised* by polynomial functions in a variable from $\mathbb{R} \cup \{\infty\}$. For example, a conic is rational. The importance of *rational normal curves* in geometry stems from the fact that every rational curve is the projection of a rational normal curve. So, in some sense, rational normal curves are canonical rational curves.

A non-singular conic of $PG(2, \mathbb{R})$ is projectively equivalent to a curve obtained as the image of the map

$$v_2 : (x, y) \mapsto (x^2, xy, y^2)$$

from the points of $PG(1, \mathbb{R})$ to the points of $PG(2, \mathbb{R})$. Similarly, the twisted cubic of $PG(3, \mathbb{R})$ is obtained as the image of the map

$$v_3 : (x, y) \mapsto (x^3, x^2y, xy^2, y^3)$$

from the points of $PG(1, \mathbb{R})$ to the points of $PG(3, \mathbb{R})$.

More generally, we define v_d to be the map from $PG(1, \mathbb{R})$ to $PG(d, \mathbb{R})$ defined by

$$(x_0, x_1) \mapsto (x^d, x^{d-1}y, \ldots, xy^{d-1}, y^d),$$

and its image is the standard **rational normal curve** C of $PG(d, \mathbb{R})$. It could be alternatively defined as the common zero locus of the quadrics defined by the polynomials

$$f_{i,j} := X_i X_j - X_{i-1} X_{j+1}, \quad i, j \in \{1, \ldots, d-1\}.$$

Note that these polynomials are just 2×2 sub-determinants of the matrix

$$\begin{bmatrix} X_0 & X_1 & X_2 & \cdots & X_{d-2} & X_{d-1} \\ X_1 & X_2 & X_3 & \cdots & X_{d-1} & X_d \end{bmatrix}.$$

This is an example of a *determinantal variety*: those that can be expressed by a set of matrices with a given upper bound on their ranks. A curve projectively equivalent to the standard rational normal curve is a **rational normal curve**. Any $d + 1$ points of C, when written as homogeneous coordinates, are linearly independent since the Vandermonde determinant vanishes only if two of its rows coincide.[1] In fact, this is the *defining* property of a set of points of $PG(d, \mathbb{R})$ projectively equivalent to the rational normal curve.

Recall from Theorem 5.17 that five points of $PG(2, \mathbb{R})$, no three collinear, determine a conic. By Theorem 16.24, six points of $PG(3, \mathbb{R})$, no four coplanar, determine a twisted cubic. We have the following generalisation to rational normal curves.

Theorem 20.11

Through any $d + 3$ points of $PG(d, \mathbb{R})$, no $d + 1$ on a hyperplane, there is a unique rational normal curve containing these points.

Proof If $G(x, y)$ is a homogeneous polynomial of degree $d + 1$ with distinct zeros (μ_i, ν_i), $1 \leqslant i \leqslant d + 1$, in $PG(1, \mathbb{R})$ then the polynomials $H_i(x, y) = G(x, y)/(\mu_i x - \nu_i y)$, $1 \leqslant i \leqslant d + 1$, form a basis for the vector space of homogeneous polynomials of degree d in x and y (if there were a linear dependence $\sum a_i H_i(x, y) = 0$ then substituting (μ_i, ν_i) shows that $a_i = 0$). Thus, $\{(H_1(s, t), H_2(s, t), \ldots, H_{d+1}(s, t)) : s, t \in R, (s, t) \neq (0, 0)\}$ is a rational normal curve passing through all the coordinate points $(1, 0, \ldots, 0), \ldots, (0, \ldots, 0, 1)$ of $PG(d, \mathbb{R})$. In addition, if μ_i and ν_i are non-zero, then the points corresponding to $(s, t) = (1, 0)$ and $(s, t) = (0, 1)$ are $(\mu_1^{-1}, \ldots, \mu_{d+1}^{-1})$ and $(\nu_1^{-1}, \ldots, \nu_{d+1}^{-1})$, and any two points forming a frame with the coordinate points can be obtained uniquely in this way. Conversely, any rational normal curve through the coordinate points can be written in this way. $\qquad \square$

20.3 The Veronese Surface

The Veronese surface came about from Giuseppe Veronese's study of conics of the plane. Recall that a conic C in $PG(2, \mathbb{R})$ is the set of points that satisfy a homogeneous equation of degree 2:

$$C: aX^2 + bY^2 + cZ^2 + dXY + eYZ + fZX = 0.$$

There are various different classes of conics: ellipses, pairs of concurrent lines, double-lines. Seeking a better way to view the properties of conics and to make calculations simpler, Veronese observes that if we simply take the coefficients from the equation of a conic, we have a point (a, b, c, d, e, f) of $PG(5, \mathbb{R})$,

[1] Vandermonde determinant: $\begin{vmatrix} 1 & x_1 & x_1^2 & \cdots & x_1^{n-1} \\ 1 & x_2 & x_2^2 & \cdots & x_2^{n-1} \\ \vdots & \vdots & \vdots & \cdots & \vdots \\ 1 & x_n & x_n^2 & \cdots & x_n^{n-1} \end{vmatrix} = \prod_{1 \leqslant i < j \leqslant n}(x_j - x_i).$

and different points determine different conics. This correspondence is known as the *Veronese correspondence* between conics of $PG(2, \mathbb{R})$ and points of $PG(5, \mathbb{R})$.

> GIUSEPPE VERONESE (1854–1917) was a professor in algebraic geometry at Padua. He was a pioneer of both higher-dimensional geometry and non-Archimedean geometry, and is best remembered for the Veronesean map and surface in algebraic geometry. Veronese also served as a senator of the Kingdom of Italy.

There is another way to view the conic C, and it arises from the *quadratic Veronese map* v from the points of $PG(2, \mathbb{R})$ to the points of $PG(5, \mathbb{R})$:

$$v: (x, y, z) \mapsto (x^2, y^2, z^2, xy, yz, zx).$$

So C is the set of points (x, y, z) of $PG(2, \mathbb{R})$ such that

$$v(x, y, z)(a, b, c, d, e, f)^\top = 0.$$

That is, $v(x, y, z)$ lies in the hyperplane of $PG(5, \mathbb{R})$ with dual coordinates $[a, b, c, d, e, f]$. We will call the image of v in $PG(5, \mathbb{R})$ the *Veronese surface* $\mathcal{V}$ of $PG(5, \mathbb{R})$. As we have just seen, the hyperplanes of $PG(5, \mathbb{R})$ correspond to conics of $PG(2, \mathbb{R})$, but what we do not know yet is whether these conics are degenerate or not.

Theorem 20.12

The stabiliser of $\mathcal{V}$ in $PGL(6, \mathbb{R})$, in its action on hyperplanes, is permutationally isomorphic to the action of $PGL(3, \mathbb{R})$ on (non-empty) conics of $PG(2, \mathbb{R})$.

Proof By definition of the Veronese map, the action of $PGL(3, \mathbb{R})$ on conics of $PG(2, \mathbb{R})$ is *transferred* to an action on the hyperplanes of $PG(5, \mathbb{R})$. Exercise 20.3 completes the proof.   □

Corollary 20.13

The stabiliser of $\mathcal{V}$ has the following orbits on hyperplanes of $PG(5, \mathbb{R})$:

 (i) *tangent hyperplanes to $\mathcal{V}$,*
 (ii) *hyperplanes that meet $\mathcal{V}$ in the points of two conic-planes,*
 (iii) *hyperplanes that meet $\mathcal{V}$ in the points of one conic-plane,*
 (iv) *hyperplanes that meet $\mathcal{V}$ in the points of a rational normal curve admitting $PGL(2, \mathbb{R})$.*

Table 20.3 *Conics and hyperplanes.*

Conic of PG(2, $\mathbb{R}$)	Hyperplane of PG(5, $\mathbb{R}$)
point	tangent hyperplane
double-line	hyperplane meeting in a conic-plane
pair of lines	hyperplane, two conic-planes meeting in a tangent line to $\mathcal{V}$
non-degenerate conic	hyperplane meeting in a rational normal curve

Proof There are four orbits of PGL(3, $\mathbb{R}$) on non-empty conics of PG(2, $\mathbb{R}$), indicated by the first column of Table 20.3. Since we have a permutational isomorphism (Theorem 20.12) we can simply compute the image under v of a nice representative in each case. For instance, the double-line conic $xy = 0$ maps to the hyperplane $X_4 = 0$, and this hyperplane intersects $\mathcal{V}$ in a conic spanning a plane. □

Theorem 20.14
There are no lines contained in the Veronese surface.

Proof A line in PG(5, $\mathbb{R}$) is the intersection of four independent hyperplanes. So the preimage of these hyperplanes under v_2 is four conics. So we have a set of zeros (in PG(2, $\mathbb{R}$)) of four independent degree 2 polynomials, and it is not difficult to see that such a set is finite. Indeed, if two of the conics are non-degenerate, then they intersect in four (possibly imaginary) points. Likewise, if one of them was nondegenerate, and another was non-degenerate, then we would also have a finite intersection. If all of them are degenerate, then they each split into two linear factors, However, at least four linear factors in three variables are always dependent. □

Indeed, Theorem 20.14 can be extended: there are no curves of odd degree on the Veronese surface of PG(5, $\mathbb{R}$).

Let P_1, P_2, P_3, P_4, P_5 be five points of PG(2, $\mathbb{R}$). The set of conics passing through a point (x, y, z) of PG(2, $\mathbb{R}$) define a hyperplane in PG(5, $\mathbb{R}$), namely, $\langle v(x, y, z)\rangle^{\perp}$ where $\perp$ denotes the standard duality of PG(5, $\mathbb{R}$) arising from the standard inner product. So the conics passing through P_1, P_2, P_3, P_4, P_5 will correspond to the intersection of five hyperplanes of PG(5, $\mathbb{R}$) (see Table 20.4). The intersection of two hyperplanes is a solid of PG(5, $\mathbb{R}$), and, upon intersection with another hyperplane, the dimension might not change or we obtain a plane. After five intersections, the smallest object we can obtain in terms of

Table 20.4 *The Veronese map glossary.*

PG$(2, \mathbb{R})$	PG$(5, \mathbb{R})$
point	point
line	conic-plane
pair of lines	two conic-planes meeting in a point of $\mathcal{V}$
non-degenerate conic	rational quartic curve
pencil of lines	conic-planes on a point of $\mathcal{V}$

dimension is a point. So, in other words, the intersection of five hyperplanes is non-empty. The points in this intersection of hyperplanes correspond to conics of PG$(2, \mathbb{R})$ passing through P_1, P_2, P_3, P_4, P_5. So, via the Veronese correspondence, we have proved that there exists at least one conic containing our original five points. If the five hyperplanes are linearly independent, then their intersection is a point and the result is a unique conic through our five points.

Let's suppose that no four of our points are collinear. We have the following cases:

Three of our points are collinear: If three of our points, P_1, P_2, P_3 say, are collinear but no four are collinear, then the lines $P_1 P_2$ and $P_4 P_5$ are the loci of two degree 1 polynomials whose product yields a unique conic passing through all of our points.

No three collinear: To prove that five points, no three collinear, lie on a unique irreducible conic, it is sufficient to show that the image of those five points lies on no solid. First of all, note that the images of representative vectors for four points, no three collinear, under v are a linearly independent set. (Noting that a linear change of coordinates in $\mathbb{R}^3$ induces a linear change of coordinates in vector space $\mathbb{R}^6$ underlying the image space of v, it follows that we need only show that it holds for the fundamental quadrangle, and this is straightforward.) Therefore, there is a pencil of conics on four points, no three collinear, and this pencil contains an irreducible conic C. Second, consider five points, no three collinear. If the image of a representative vector fifth point is dependent upon the images of representative vectors for the other four, then every conic of the pencil on those four would pass through the fifth point, but there is no such point for the fundamental quadrangle; it is a contradiction. Hence, the images of representative vectors for five points under v are a linearly independent set, and so the points lie on a unique conic. That the conic is irreducible follows from the observation that reducible conics do not contain five points, no three collinear.

Exercises

20.1 Let G be the isometry group of the Klein quadric Q of PG(5, $\mathbb{R}$). Show that G has two orbits on hyperplanes of PG(5, $\mathbb{R}$). (Hint: A hyperplane is the polar image of a point under the polarity arising from Q. So it suffices to show that G has two orbits on the points of PG(5, $\mathbb{R}$): those lying in Q, and those not lying in Q.)

20.2 Show that the image of the tangent lines to a twisted cubic under the Klein correspondence is a normal rational quartic curve of the 4-space corresponding to the general linear complex defined by the twisted cubic.

20.3 Prove that the stabiliser of $\mathcal{V}$ in PGL(6, $\mathbb{R}$) is isomorphic to PGL(3, $\mathbb{R}$).

Appendix: Group Actions

Let G be a group and let Ω be a set. Then a *group action* is a function μ from $\Omega \times G$ to Ω, written $(\omega, g) \mapsto \omega^g$, such that the following hold:

(i) $(\omega^g)^h = \omega^{gh}$, for all $\omega \in \Omega$ and $g, h \in G$;

(ii) $\omega^1 = \omega$ for any $\omega \in \Omega$.

One benefit of this notation is the mnemonic property of obeying index laws. The **kernel** of a group G acting on a set Ω is the normal subgroup

$$G_{(\Omega)} := \{g \in G : \omega^g = \omega, \text{ for all } \omega \in \Omega\}.$$

If the kernel of the action of a group G on a set Ω consists only of the identity element, then we say that G acts *faithfully* on Ω.

Let G be a group acting on a set Ω, and let ω be an element of Ω. Then the **orbit of ω under** G is the subset of Ω defined by $\omega^G := \{\omega^g : g \in G\}$. So, for example, the orbit of 1 under the subgroup $\langle (1\,2\,3), (2\,3)(4\,5) \rangle$ of S_5, is $\{1, 2, 3\}$. We will consider two types of subgroups that arise from stabilising a set or completely fixing a set. Let G be a group acting on a set Ω, and let Σ be a subset of Ω. Then the *set-wise* stabiliser of Σ in G is the subgroup

$$G_\Sigma = \{g \in G : \sigma^g \in \Sigma, \text{ for all } \sigma \in \Sigma\}.$$

The *point-wise* stabiliser of Σ in G is the subgroup

$$G_{(\Sigma)} = \{g \in G : \sigma^g = \sigma, \text{ for all } \sigma \in \Sigma\}.$$

If $\Sigma = \{\sigma\}$, then we write G_σ for G_Σ.

Let G be a group acting on a set Ω. If G has just one orbit of Ω, namely the whole set itself, then we say that G is **transitive** on Ω. Alternatively, G is transitive on Ω if for any pair of points $\omega_1, \omega_2 \in \Omega$, there exists an element

$g \in G$ such that $\omega_1^g = \omega_2$. The map $G_\omega u \mapsto \omega^u$ is a bijection between right cosets and elements of the orbit ω^G. Thus:

Theorem 21.1 (The Orbit–Stabiliser Theorem)
Let G be a group acting on a set Ω and let $\omega \in \Omega$. Then there is a bijective correspondence between the elements of ω^G and the set of right cosets of G_ω in G.

References

Abhyankar, S. S. 1976. Historical ramblings in algebraic geometry and related algebra. *Am. Math. Mon.*, **83**, 409–48.

Abhyankar, S. S. 1990. *Algebraic Geometry for Scientists and Engineers* (Mathematical Surveys and Monographs, vol. 35). Providence, RI: American Mathematical Society.

Abram, W. A. 1876. Memorial of the late T. T. Wilkinson, F.R.A.S., of Burnley. *Trans. Hist. Soc. Lancs. Chesh.*, ser. 3, **4**, 77–94.

Adler, I. 1968. *Groups in the New Mathematics: An Elementary Introduction to Mathematical Groups Through Familiar Examples*. New York: John Day Company.

Adler, V. E. 2006. Some incidence theorems and integrable discrete equations. *Discrete Comput. Geom.*, **36**(3), 489–98.

Aigner, M., and Ziegler, G. M. 2014. *Proofs from the Book*, 5th ed. Including illustrations by K. H. Hofmann. Berlin: Springer-Verlag.

Akopyan, A. V., and Zaslavsky, A. A. 2007. *Geometry of Conics*. Mathematical World, 26. Providence, RI: American Mathematical Society. Trans. from the 2007 Russian original by A. Martsinkovsky.

Alberti, L. B., Kemp, M., and Grayson, C. 2005. *Della pittura (On painting)*. Reprint ed. London: Penguin Books.

Altshiller-Court, N., Jones, S. I., Frink, Jr., Orrin et al. 1932. Problems for solution: 3572–3578. *Am. Math. Mon.*, **39**(9), 548–50.

American Mathematical Monthly. 1957. A classification of monthly problems (1918–1950). *Am. Math. Mon.*, **64**(7), 65–75.

Andersen, K. 1990. Stevin's theory of perspective: The origin of a Dutch academic approach to perspective. *Tractrix*, **2**, 25–62.

Andersen, K. 1992. *Brook Taylor's Work on Linear Perspective*. Sources in the History of Mathematics and Physical Sciences, vol. 10. New York: Springer-Verlag. A study of Taylor's role in the history of perspective geometry, including facsimiles of Taylor's two books on perspective.

Andersen, K. 2007. *The Geometry of an Art*. Sources and Studies in the History of Mathematics and Physical Sciences. New York: Springer. The history of the mathematical theory of perspective from Alberti to Monge.

Andersen, K. 2013. Guidobaldo: The father of the mathematical theory of perspective. In A. Becchi, D. B. Meli, and E. Gamba (eds.), *Guidobaldo del Monte*

(1545–1607): Theory and Practice of the Mathematical Disciplines from Urbino to Europe, pp. 145–66. Berlin: epubl.

Apollonius de Perge. 2008. *Apollonius de Perge, Coniques. Tome 3: Livre V.* Scientia Graeco-Arabica, vol. 1/3. Berlin: Walter de Gruyter & Co. Greek and Arabic text established, translated, and annotated under the direction of Roshdi Rashed.

Apollonius de Perge. 2009. *Apollonius de Perge, Coniques. Tome 4: Livres VI, VII.* Scientia Graeco-Arabica, vol. 1/4. Berlin: Walter de Gruyter & Co. Greek and Arabic text established, translated, and annotated under the direction of Roshdi Rashed.

Apollonius de Perge. 2010. *Apollonius de Perge, Coniques. Tome 2.1: Livres II, III.* Scientia Graeco-Arabica, vol. 1/2. Berlin: Walter de Gruyter & Co. Greek and Arabic text established, translated, and annotated under the direction of Roshdi Rashed.

Argan, G. C., and Robb, N. A. 1946. The architecture of Brunelleschi and the origins of perspective theory in the fifteenth century. *J. Warburg Courtauld* (9) 96–121.

Arnold, L. 2003. Introduction: Of the mind and the eye. *Pacific Rim Report*, **27**, 2–16.

Artin, E. 1957. *Geometric Algebra*. New York: Interscience.

Artin, M. 1991. *Algebra*. Englewood Cliffs, NJ: Prentice-Hall.

Atiyah, Sir M. 2010. *VIII.6 Advice to a Young Mathematician*. Princeton, NJ: Princeton University Press. Pages 1000–1010.

Aubert, P. 1889. Sur une généralisation du théorème de Pascal donnant neuf points en ligne droite. *Nouv. ann. math.* (3) **VIII**, 529–35.

Auric, A. 1915. Généralisation du théorème de Kariya. *Nouv. ann. math.*, **15**(4), 222–25.

Bacon, R., and Burke, R. B. 1928. *Opus Majus, Volumes 1 and 2*. Philadelphia: University of Pennsylvania Press.

Baker, H. F. 1920. On a proof of the theorem of a double six of lines by projection from four dimensions. *Proc. Camb. Philos. Soc.*, **20**, 133–44.

Baker, H. F. 1936a. Polarities for the nodes of a Segre cubic primal in space of four dimensions. *Proc. Camb. Philos. Soc.*, **32**, 507–20.

Baker, H. F. 1936b. Verification of the Petersen–Morley theorem. *J. Lond. Math. Soc.*, **S1-11**(1), 24.

Baker, H. F. 1943. *An Introduction to Plane Geometry*. Cambridge: Cambridge University Press.

Baker, H. F. 1952. Note on the foundations of projective geometry. *Proc. Camb. Philos. Soc.*, **48**, 363–4.

Baralić, Đ. 2013. Around the Carnot theorem. http://arxiv.org/pdf/1308.6144v1.pdf.

Baralić, Đ. 2015. A short proof of the Bradley theorem. *Am. Math. Mon.*, **122**(4), 381–5.

Barbilian, G., and Vodă, V. Gh. (eds.). 1984. *Pagini Inedite*. Vol. II. Bucharest: Editura Albatros.

Beatty, S., and Todd, J. A. 1944. Problem 4079. *Am. Math. Mon.*, **51**(10), 599–600.

Bell, E. T. 1986. *Men of Mathematics*. Reprint of the 1937 ed. New York: Simon & Schuster.

Bellavitis, G. 1838. Saggio di geometria derivata. *Nuovi Saggi della Imperiale Regia Accademia di Scienze Lettere ed Arti in Padova*, **4**, 243–88.

Belting, H. 2007. *Renaissance Art and Arab Science*. Cambridge, MA: Belknap Press.

Beltrami, E. 1862. Intorno alle coniche dei nove punti e ad alcune quistioni che ne dipendono. *Memorie dell'Accademia delle Scienze dell'Istuto di Bologna*, **2**(2), 361–95.

Beltrami, E. 1863. Sulle coniche di nove punti. *Giornale di Matematiche*, **1**, 109–18.

Benedetti, G. B. 1585. De rationibus operationum perspectivae. In *Diversarum speculationum mathematicarum et physicarum liber*, pp. 119–40. Torino.

Berzolari, L. 1905. Sui sistemi di $n + 1$ rette dello spazio ad n dimensioni, situate in posizione di *Schläfli*. *Rend. Circ. Mat. Palermo*, **20**, 229–47.

Blaikie, J. 1905. Question 15584, Mathematical Questions, with Their Solutions. *Educational Times*.

Blondel, N.-F. 1673. *Résolution des quatres principaux problèmes d'architecture*. Imprimerie Royale Paris.

Blumenberg, H. 1985. *The Legitimacy of the Modern Age*. Cambridge, MA: MIT Press.

Blumenberg, H. 1989. *Höhlenausgánge*. Frankfurt: Suhrkamp.

Bobillier, É. 1827–8. Géométrie pure. Démonstration de divers théorèmes de géométrie. *Ann. math. pures appl.* **18**, 185–202.

Bobillier, É. 1829. Géométrie des courbes. Mémoire sur l'hyperbole équilatère. *Annales de Gergonne*, **19**, 340–59.

Bôcher, M. 1892–3. Some propositions concerning the geometric representation of imaginaries. *Ann. of Math.*, **7**(1/5), 70–2.

Bonnet, P. O. 1848. Mémoire sur la théorie générale des surfaces. *J. Éc. polytech.*, **19**, 1–146.

Bonola, Roberto. 1955. *Non-Euclidean Geometry: A Critical and Historical Study of Its Developments*. New York: Dover. Translation with additional appendices by H. S. Carslaw; supplement containing the G. B. Halsted translations of 'The science of absolute space' by J. Bolyai and 'The theory of parallels' by N. Lobachevski.

Boon, A. W. 1978–9. De rechte van Euler – een bewijs. *Vakblad Euclides*, **54**(2), 56–7.

Borwein, P. B. 1983. The Desmic conjecture. *J. Combin. Theory Ser. A*, **35**(1), 1–9.

Bosse, A. 1648. *Maniere Universelle De Mr Desargues pour pratiquer la perspective par petid-pied comme la gemetral ensembles les places et proportions les fortes et foibles touches, teintes ou couleurs*. Paris: Pierre Des-Hayes.

Bosse, A. 1672. *Regle universelle, pour décrire toutes sortes d'arcs rampans dans toutes les surjections que l'on puisse proposer, sans se servir des Axes, Des Foyers, ny du Cordeau*. Paris: A. Bosse.

Bottema, O. 1955. A generalization of Pascal's theorem. *Duke Math. J.*, **22**, 123–7.

Boutin, A. 1890a. Problèmes sur le triangle. J. de Math. spéc. (3) IV, 265–9.

Boutin, A. 1890b. Sur un groupe de quatre coniques remarquables du plan d'un triangle. *J. de Math. spéc.* (3) *IV*, 104–7, 124–7.

Bradley, C. 2011. Problems requiring proof. Article CJB/2011/182. http://people.bath.ac.uk/~masgcs/Article182.pdf.

Braikenridge, W. 1733. *Exercitatio Geometrica in Descriptione Linearum Curvarum*. London: Richard Hett and John Nourse.

Brannan, D. A., Esplen, M. F., and Gray, J. J. 2012. *Geometry*. Cambridge: Cambridge University Press.

Brianchon, C.-J. 1806. Memoirs sur les Surfaces courbes du second Degré. *Journal de l'École Polytechnique*, **VI**, 297–311.

Brianchon, C.-J. 1817. *Mémoire sur les lignes du second ordre*. Paris: Chex Bachelier, libraire.

Brianchon, C.-J., and Poncelet, J.-V. 1821. Recherches de la détermination d'une hyperbole équilaterè, au moyen de quatres conditions données. *Ann. math. pures appl.*, **XI**, 223–5.

Brigaglia, A. 1996. The influence of H. Grassmann on Italian projective N-dimensional geometry. In G. Schubring (ed.), *Hermann Günther Graßmann (1809–1877): Visionary Mathematician, Scientist and Neohumanist Scholar*, pp. 155–63. Dordrecht: Kluwer Academic Publishers.

Brocard, H. 1877. Question 234. *Nouv. corresp. math.*, **3**.

Bruno, G., and Gosnell, S. 2014. *On the Infinite, the Universe and the Worlds: Five Cosmological Dialogues (Giordano Bruno Collected Works)*, vol. 2. Huginn, Munnin & Co.

Buchanan, T. 1988. The twisted cubic and camera calibration. *Comput. Vis. Graph. Image Process.*, **42**(1), 130–2.

Butler, N. M. 1895. What knowledge is of most worth? *J. Educ.*, **42**(6 (1039)), 107–110.

Carathéodory, C. 1937. The most general transformations of plane regions which transform circles into circles. *Bull. Amer. Math. Soc.*, **43**(8), 573–9.

Carman, C. H. 2016. *Leon Battista Alberti and Nicholas Cusanus: Towards an Epistemology of Vision for Italian Renaissance Art and Culture*. London: Routledge.

Carnot, L. N. M. 1801. *De la corrélation des figures géométriques*. Paris: Duprat.

Carnot, L. N. M. 1803. *Géométrie de position*. Paris: Duprat.

Carnot, L. N. M. 1806. *Mémoire sur la relation qui existe entre les distances respectives de cinq points quelconques pris dans l'espace; suivi d'un essai sur la théorie des transversales*. Paris: Courcier.

Cartan, É. 1938. *Lecons sur la théorie des spineurs*. Paris: Hermann.

Casey, J. 1871. On cyclides and sphero-quartics. *Philos. T. R. Soc. Lond.*, **161**(0), 585–721.

Casey, J. 1885. *A Treatise on the Analytical Geometry of the Point, Line, Circle, and Conic Sections, Containing an Account of Its Most Recent Extensions, with Numerous Examples*. Dublin: Hodges, Figgis, & Co. See especially Nature XXXIII, pp. 172–3.

Casey, J. 1889. *A Treatise on Spherical Trigonometry*. Dublin. Hodges, Figgis, & Co. See especially XV + 165 S.

Casey, J. 1892. *A Sequel to the First Six Books of the Elements of Euclid Containing an Easy Introduction to Modern Geometry, with Numerous Examples*, 5th ed. London:. London. Longmans, Green and Co.

Castelnuovo, G. 1904. *Lezioni di geometria analitica e projettiva. Vol. I (Forme di I specie-Geometria analitica del piano-Curve di secondo ordine)*. Roma-Milano. VII u. 507 S. See also Lezioni di geometria analitica e proiettiva. *Monatsh. f. Mathematik und Physik*, **16**, A26–A27 (1905). https://doi.org/10.1007/BF01693810.

Cayley, A. 1859. A sixth memoir on quantics. *Phil. Trans.*, **149**, 61–90.

Cayley, A. 1861. A theorem in conics. *Q. J. Pure Appl. Math.*, **iv**, 131 – 133.

Cayley, A. 1870. A memoir on abstract geometry. *Philos. T. R. Soc. Lond.*, **160**(0), 51–63.

Cayley, A. 1872. On the non-Euclidian geometry. *Math. Ann.*, **5**(4), 630–4.

Cayley, A. May 1864. Question 1505, Mathematical Questions, with Their Solutions. *Educational Times*.

Cazamian, A. 1895. Sur le théorème de Carnot. *Nouv. ann. math. (3e series)*, **14**, 30–40.

Ceva, G. 1678. *De lineis rectis se invicem secantibus statica constructio*. Ex typographia L. Montiae.

Chasles, M. 1828a. Note sur une propriété générale des coniques, dont un cas particulier relatif à la parabole, a été démontré dans la Correspondence, tom. IV, p. 155. *Corresp. math. phys.*, **4**, 363–71.

Chasles, M. 1828b. Sur une propriété générale des coniques. *Corresp. math.*, v.

Chasles, M. 1828–9. Géométrie de situation. Démonstration de quelques propriétés du triangle, de l'angle trièdre et du tétraèdre, considérés par rapport aux lignes et surfaces du second ordre. *Ann. math. pures appl.*, **19**, 65–85.

Chasles, M. 1829. Propriétés générales des coniques. *Corresp. math. phys.*, **5**, 6–22.

Chasles, M. 1837. *Aperçu historique sur l'origine et le développement des méthodes en géométrie*. Brussels: pub. not identified.

Chasles, M. 1852. *Traité de Géométrie Supérieure*. Paris: Gauthier-Villars.

Chasles, M. 1865. *Traité des sections coniques*. Paris: Gauthier-Villars.

Chasles, M. 2010. *Cours de Geometrie Superieure (1847)*. Whitefish, MT: Kessinger Publishing.

Clawson, J. W. 1919. A theorem in the geometry of the triangle. *Am. Math. Mon.*, **26**(2), 59–62.

Clebsch, A. 1861. Über symbolische Darstellung algebraischer Formen. *J. Reine Angew. Math.*, **59**, 1–62.

Cohen, H. 1889. *Kant's Begründung der Ästhetik*. Berlin: F. Dümmler.

Cohl, T. 2014. A purely synthetic proof of Dao's theorem on six circumcenters associated with a cyclic hexagon. *Forum Geom.*, **14**, 261–4.

Collings, S. N. 1973. Reflections on a triangle. *Math. Gaz.*, **57**(402), 291–3.

Commandino, F. 1588. *Mathematicae collectiones*. Pisauri [Pesaro]: Apud Hieronymum Concordiam,.

Commandino, F. 1558. *In planisphaerium Ptolemaei commentarius*. Venetiis [Aldus].

Conti, P., and Traverso, C. 1995. A case of automatic theorem proving in Euclidean geometry: The Maclane 8_3 theorem. In G. Cohen, M. Giusti, and T. Mora (eds.), *Applied Algebra, Algebraic Algorithms and Error-Correcting Codes: 11th International Symposium, AAECC-11, Paris, France, July 1995, Proceedings*, pp. 183–93. Berlin: Springer-Verlag.

Coşniţă, C. 1941. Coordonnées barycentriques. *Ann. Roumaines Math.*, **4**, viii+176.

Court, N. A. 1943. Theorems, their converses and their extensions. *Natl. Math. Mag.*, **17**(5), 195–201.

Cousin, J. 1560. *Livre de perspective*. Paris: impr. de J. Le Royer.

Coxeter, H. S. M. 1949. *The Real Projective Plane*. New York: McGraw-Hill.

Coxeter, H. S. M. 1971. Inversive geometry. *Educ. Stud. Math.*, **3**(3/4), 310–21.

Coxeter, H. S. M. 1974. *Projective Geometry*, 2nd ed. Toronto: University of Toronto Press.

Coxeter, H. S. M. 1993. *The Real Projective Plane*, 3rd ed., with an appendix by George Beck. New York: Springer-Verlag.

Coxeter, H. S. M., and Greitzer, S. L. 1967. *Geometry Revisited*. New Mathematical Library, vol. 19. New York: Random House.

Crannell, A., Frantz, M., and Futamura, F. 2017. The image of a square. *Am. Math. Mon.*, **124**(2), 99–115.

Cremona, L. 1885. *Elements of Projective Geometry*, trans. C. Leudesdorf. Oxford: Clarendon Press.

Cusani, N. 1913 [1440]. *De Docta Ignorantia Libri Tres. Testo Latino con Note di Paolo Rotta*. Bari: Laterza.

da Vignola, G. B. c. 1530–73, published posthumously 1583. *La due regole della prospective practica*. Bologna: pub. not identified.

da Vinci, L., Richter, J. P., and Bell, R. C. 1970. *The Notebooks of Leonardo Da Vinci, Vol. II*. New York: Dover Publications.

Dandelin, G. 1822. *Mémoire sur quelques propriétés remarquables de la focale parabolique*. Nouveaux mémoires de l'Académie Royale des Sciences et Belles-Lettres de Bruxelles.

Dandelin, G. 1824/5. Géométrie pure. Recherches nouvelles sur les sections du cône et sur les hexagones inscrits et circonscrits à ces sections. *Ann. math. pures appl.*, **15**, 387–96.

Dandelin, G. 1826. *Mémoire sur l'hyperboloïde de révolution, et sur les hexagones de Pascal et de M. Brianchon*. Nouveaux mémoires de l'Académie Royale des Sciences et Belles-Lettres de Bruxelles.

Dandelin, G., and Gergonne. 1825/6. Géométrie pure. Usages de la projection stéréographique en géométrie. *Ann. math. pures appl. [Ann. Gergonne]*, **16**, 322–7.

Dao Thanh Oai. 2014. Dao's blog. http://oaithanhdao.blogspot.com.

Darboux, M. G. 1880. Sur le théorème fondamental de la géométrie projective. *Math. ann.*, **17**(1), 55–61.

Daston, L. J. 1986. The physicalist tradition in early nineteenth century French geometry. *Studies in History and Philosophy of Science Part A*, **17**(3), 269–95.

Daus, P. H. 1936. isotomic conjugates and their projective generalization. *Am. Math. Mon.*, **43**(3), 160–4.

de Fermat, P. 1679. *Varia opera mathematica*. Toulouse: Pech.

de La Hire, P. 1672. *Observations de Ph. de la Hire sur les points d'attouchement de trois Lignes droits qui touchent la Section d'un Cone sur quelques-uns des Diametres, et sur le centre de la mesme Section*. Paris: A. Bosse.

de La Hire, P. 1673. *Nouvelle Méthode en Géométrie pour les Sections des Superficies coniques et Cylindriques*. Paris: de La Hire and Thomas Moette.

de La Hire, P. 1685. *Sectiones Conicae in novum libros distributatae, in quibus quidquid hactenus observatione dignum cum a veteribus, tum a recentioribus Geometris traditum est, novis contractisque demonstrationibus explicatur; multis etiam & exquisitis propositionibus recens inventis illustratur*. Paris: Stephanum Michallet.

De Longchamps, G.-G. 1866. Étude de géométrie comparée, avec applications aux sections coniques et aux courbes d'ordre supérieur, particulièrement à une famille de courbes du sixième ordre et de la quatrème classe. *Nouv. ann. math. (2e série)*, **5**, 118–28.

De Morgan, A. 1867. Unsolved problem 2555, Mathematical Questions, with Their Solutions. *Educational Times*.

de Perga, Apollonius, and Heath, T. L. 2013. *Treatise on Conic Sections*. Cambridge Library Collection – Mathematics. Cambridge: Cambridge University Press.

de Villiers, Michael. 1996. A dual to Kosnita's theorem. *Math. Informatics Q.*, **6**(3), 169–71.

Deaux, R. 1957. *Introduction to the Geometry of Complex Numbers*. New York: Frederick Ungar. Trans. from the revised French edition by H. Eves.

della Francesca, P. [c. 1474] 2021. *De prospectiva pingendi: primo trattato matematico di prospettiva, primo manuale di disegno e di architettura*, ed. and trans. R. Sinisgalli. Cinisello Balsamo, Milan: Silvana.

Demir, H., Tezer, C., and Grivaux, J.-P. 1992. Problems and solutions: Solutions: E3422. *Am. Math. Mon.*, **99**(7), 679–81.

Desargues, G. 1951. *L'oeuvre mathématique de Desargues*, ed. R. Taton. Paris: Presses Universitaires de France.

Dieudonné, J. A. 1948. *Sur les groupes classiques*. Paris: Hermann.

Dijksterhuis, E. J. 1987. *Archimedes*. Princeton, NJ: Princeton University Press.

Diocles, Medicus, D. C., Toomer, G. J., and d'Ascalon, E. 1976. *Diocles on Burning Mirrors: The Arabic Translation of the Lost Greek Original*. Sources in the History of Mathematics and Physical Sciences. Berlin: Springer-Verlag.

Dobos, S. 2011. Cross ratio in use. *Math. Gaz.*, **95**(534), 444–53.

Dold-Samplonius, Y., Hermelink, H., Schramm, M. (trans.), and Archimedes. 1975. Uber einander beruhrende Kreise. *Archimidis Opera mathematica, Stuttgart*, **4**.

Droz-Farny, A. 1899. Question 14111. *Educational Times*, **71**, 89–90.

Durrande, J. B. 1824–5. Démonstration des propriétés des quadrilatères à la fois inscriptibles et circonscriptibles au cercle. *Ann. math. pures appl.*, **15**, 133–45.

Eddy, R. H., and Fritsch, R. 1994. The conics of Ludwig Kiepert: A comprehensive lesson in the geometry of the triangle. *Math. Mag.*, **67**(3), 188–205.

Edelstein, M., and Kelly, L. M. 1966. Bisecants of finite collections of sets in linear spaces. *Canad. J. Math.*, **18**, 375–80.

Edgerton, S. 2000. Die ideologische Wurzel der Zentralperspektive in der Renaissance: Warum Leon Battista Alberti Windows 1435 erfand. In N. Bolz, F. Kittler, and R. Zons (eds.), *Weltbürgertum und Globalizierung*, pp. 127–45. Munich: Wilhelm Fink Verlag.

Edgerton, S. Y. 2009. *The Mirror, the Window, and the Telescope: How Renaissance Linear Perspective Changed Our Vision of the Universe*. Ithaca, NY: Cornell University Press.

Eggar, M. H. 1998. Pinhole cameras, perspective, and projective geometry. *Am. Math. Mon.*, **105**(7), 618–30.

Ehrmann, J.-P., and van Lamoen, F. 2004. A projective generalization of the Droz-Farny line theorem. *Forum Geom.*, **4**, 225–7.

Einstein, A. 1905. Zur Elektrodynamik bewegter Körper. *Annalen der Physik*, **322**, 891–921.

Elkins, J. 2018. *The Poetics of Perspective*. Ithaca, NY: Cornell University Press.

Encyclopaedia Britannica. 1929. Camera obscura. In *Encyclopaedia Britannica: A New Survey of Universal Knowledge*, 14th ed., Vol. 4 Brain to Casting, pp. 658–60. London: Encyclopaedia Britannica Company.

Euler, L. 1766. Nouvelle méthode d'liminer les quantités inconnues des équations. *Mémoires de l'académie des sciences de Berlin*, **20**, 91–104.

Evans, L. S., and Rigby, J. F. 2002. Octagrammum mysticum and the golden cross-ratio. *Math. Gaz.*, **86**(505), 35–43.

Evelyn, C. J. A., Money-Coutts, G. B., and Tyrrell, J. A. 1974. *The Seven Circles Theorem and Other New Theorem*. London: Stacey International.

Eves, H. 1972. *A Survey of Geometry*, revised ed. Boston, MA: Allyn and Bacon. XVIII, 442.

Eves, H. 1983. Problem E2990. *Am. Math. Mon.*, **90**, 212.

F. G. M. 1912. *Exercices de géométrie comprenant l'exposé des méthodes géométriques et 2000 questions résolues*, 5ᵉ édition. Paris: de Gigord. XXIV + 1302 S. 8°.

Falco, C. M. 2007. Ibn al-Haytham and the origins of modern image analysis. *9th International Symposium on Signal Processing and Its Applications, Sharjah, United Arab Emirates, 2007*, pp. 1–2. DOI: http://doi.org/10.1109/ISSPA.2007.4555635.

Federici Vescovini, G. 1965. Contributo per la storia della fortuna di Alhazen in Italia: il volgarizzamento del ms. vat. 4595 e il 'Commentario terzo' del Ghiberti. *Rinascimento*, **2**(5), 17–50.

Federici Vescovini, G. 1980. Il problema delle fonti ottiche medievali del Commentario Terzo di Lorenzo Ghiberti. In R. Krautheimer (ed.), *Atti del Convegno colloque consacre a Lorenzo Ghiberti: Lorenzo Ghiberti nel suo tempo*, vol. II, pp. 349–87. Florence: Olschki.

Federici Vescovini, G. 1990. La fortune de l'Optique d'Ibn Haitham: le livre De aspectibus (Kitab al-Manazir) dans le Moyen Âge latin. *Archives internationales d'histoire des sciences*, **40**, 220–38.

Federici Vescovini, G. 1998. Alhazen Vulgarisé: Le 'De li aspecti' d'un manuscrit du Vatican (moitié du XIVe siècle) et le troisième Commentaire sur l'optique de Lorenzo Ghiberti. *Arabic Sciences and Philosophy: A Historical Journal*, **8**(1), 67–96.

Fermat, P. 1999. *Œuvres de Pierre Fermat. I*. Collection Sciences dans l'Histoire. [Science in History Collection]. Paris: Librairie Scientifique et Technique Albert Blanchard. La théorie des nombres [Number theory]. Trans. P. Tannery, with an introduction and commentary by R. Rashed, Ch. Houzel, and G. Christol.

Ferrers, N. M. 1857a. Note on Mr. Cayley's extension of Sir E. Pollock's theorem. *Q. J. Pure Appl. Math.*, **I**, 175–7.

Ferrers, N. M. 1857b. Note on reciprocal triangles and tetrahedra. *Q. J. Pure Appl. Math.*, **1**, 191–5.

Ferriot, L. A. S. 1838. *Application de la methode des projections a la recherche de certaines proprietes geometriques*. Paris: Bachelier.

Field, J. V. 1985. Giovanni Battista Benedetti on the mathematics of linear perspective. *J. Warburg Courtauld*, **48**, 71–99.

Field, J. V. 1997. *The Invention of Infinity*. Oxford: Oxford University Press.

Field, J. V., and Gray, J. J. 1987. *The Geometrical Work of Girard Desargues*. New York: Springer-Verlag.

Field, J. V., Lunardi, R., and Settle, T. B. 1989. The perspective scheme of Masaccio's *Trinity* fresco. *Nuncius Ann. Storia Sci.*, **4**(2), 31–118.

Fox, M. D., and Goggins, J. R. 2003. Morley's diagram generalised. *Math. Gaz.*, **87**(510), 453–67.

Frantz, M. 2011. A car crash solved – with a Swiss army knife. *Math. Mag.*, **84**(5), 327–38.

Frantz, M. 2012. A different angle on perspective. *College Math. J.*, **43**(5), 354–60.

Frégier. 1815–16. Géométrie analitique. Théorèmes nouveaux, sur les lignes et surfaces du second ordre. *Ann. math. pures appl. [Ann. Gergonne]*, **6**, 229–41.

Frégier. 1816/17. Géométrie analitique. Théorèmes nouveaux, sur les lignes et surfaces du second ordre. *Ann. math. pures appl. [Ann. Gergonne]*, **7**, 95–8.

Friedberg, A. 2006. *The Virtual Window: From Alberti to Microsoft*. Cambridge, MA: MIT Press.

Fritsch, R. 2015. Remarks on orthocenters, Pappus' theorem and butterfly theorems. *J. Geom.*, **107**(2), 305–16.

Gale, D. 1996. Mathematical entertainments. *Math. Intelligencer*, **18**(4), 29–32.

Gale, D. 1998. *Tracking the Automatic Ant and Other Mathematical Explorations: A Collection of Mathematical Entertainments Columns from the Mathematical Intelligencer*. New York: Springer.

Gallatly, W. 1910. *The Modern Geometry of the Triangle*. London: Hodgson. See especially 70 S. 8° (s. S. 574).

Gallucci, G. 1906. Studio della figura delle otto rette e sue applicazioni alla geometria del tetraedro ed alla teoria delle configurazioni. *Rendiconto dell'Accademia delle scienze fisiche e matematiche (Sezione della Societá reale di Napoli)*, **12**, 49–79.

Gauss, C. F. 1900. *Werke*. Göttingen: K. Gesellschaft Wissenschaft, Göttingtn.

Gerber, L. 1975. Associated and perspective simplexes. *Trans. Am. Math. Soc.*, **201**, 43–55.

Gergonne. 1825/6. Philosophie mathématique. Considérations philosophiques sur les élémens de la science de l'étendue. *Ann. math. pures appl. [Ann. Gergonne]*, **16**, 209–31.

Gergonne. 1826–7. Géométrie de situation. Recherches sur quelques lois générales qui régissent les lignes et surfaces algébriques de tous les ordres. *Ann. Gergonne*, **17**, 214–52.

Gergonne, J. D., and Querret, J.-J. 1824–5. Géométrie élémentaire. Démonstration de deux théorèmes de géométrie, desquels on peut déduire, comme cas particulier, le théorème de M. Hamett. *Ann. math. pures appl. [Ann. Gergonne]*, **15**, 84–9.

Gindikin, S. 2007. *Tales of Mathematicians and Physicists*, 2nd ed. Trans. from the Russian by A. Shuchat. New York: Springer.

Godfrey, C., and Siddons, A. W. 1908. *Modern Geometry*. Cambridge: Cambridge University Press. See especially 178 S. 8°..

Goldstein, B. R., and Smith, A. M. 1993. The medieval Hebrew and Italian versions of Ibn Mu'ādh's 'On twilight and the rising of clouds'. *Nuncius*, **8**(2), 611–13.

Goormaghtigh, R. 1930. Sur une généralisation du théorème de Noyer-Droz-Farny et Neuberg. *Mathesis*, **44**, 25.

Gorjian, I., Karamzadeh, O. A. S., and Namdari, M. 2015. Morley's theorem is no longer mysterious! *Math. Intelligencer*, **37**(4), 6–7.

Grabiner, J. V. 2009. Why did Lagrange 'prove' the parallel postulate? *Am. Math. Mon.*, **116**(1), 3–18.

Grace, J. H. 1902. On the zeros of a polynomial. *Proc. Camb. Philos. Soc.*, **11**, 352–6.

Grassmann, H. G. 1862. *Die Ausdehnungslehre. Vollständig und in strenger Form begründet*. Berlin: Enslin.

Grau, O. 2003. *Virtual Art: From Illusion to Immersion*. Leonardo (Series). Cambridge, MA: MIT Press.

Graustein, W. C. 1930. *Introduction to Higher Geometry*. New York: Macmillan. XV + 486 p. with ill.

Greenblatt, S. 2011. *The Swerve: How the Renaissance Began*. London: Random House.

Grinberg, D. n.d. The Lamoen theorem on the cross-triangle. *Unpublished notes – Geometry*. www.cip.ifi.lmu.de/~grinberg/geometry2.html.

Hamilton, H. 1758. *De Sectionibus Conicis, Tractatus Geometricus: In Quo, Ex Natura Ipsius Coni, Sectionum Affectioens [sic] Facillime Deducuntur Methodo Nova*. London: Impensis Gul. Johnston.

Hardy, G. H. 1920. Review of *The Theory of the Imaginary in Geometry*, by J. L. S. Hatton. *Math. Gaz.*, **10**(146), 77–9.

Harries, Karsten. 2001. *Infinity and Perspective*. Cambridge, MA: MIT Press.

Hatton, J. L. S. 1913. *The Principles of Projective Geometry Applied to the Straight Line and Conic*. Cambridge: Cambridge University Press. See especially X u. 366 S. 8° [Nature 22, 196].

Heading, J. 1958. *An Elementary Introduction to the Methods of Pure Projective Geometry*. London: Macmillan. See especially IX, 254 p.

Hermiz, R. 2015. English translation of the *Sphaerica* of Menelaus. Master's thesis, California State University San Marcos. https://bit.ly/41TLvAT.

Hesse, L. O. 1840a. De curvis et superficiebus secundi ordinis. *J. Reine Angew. Math.*, **20**, 285–308.

Hesse, L. O. 1840b. *De octo punctis intersectionis trium superficierum secundi ordinis (Dissertatio pro venia legendi)*. Dissertation, University of Königsberg.

Hesse, L. O. 1842. Ueber die Construction der Oberflächen zweiter Ordnung, von welchen beliebige neun Puncte gegeben sind. *J. Reine Angew. Math.*, **24**, 36–9.

Hilbert, D. 1891. *Projective geometrie*. Vorlesung, Sommessemester [Lecture, summer semester]. Göttingen: Lower Saxony State and University Library.

Hjelmslev, J. 1907. Neue Begründung der ebenen Geometrie. *Mathematische Annalen*, **64**, 449–74.

Hockney, D. 2001. *Secret Knowledge: Rediscovering the Lost Techniques of the Old Masters*. London: Thames & Hudson.

Hodge, W. V. D., and Pedoe, D. 1994. *Methods of Algebraic Geometry. Volume I. Book I: Algebraic Preliminaries. Book II: Projective Space*. Cambridge: Cambridge University Press.

Hogendijk, J. P. 1985. *Ibn al-Haytham's 'Completion of the Conics'*. Trans. from the Arabic. New York: Springer-Verlag.

Hon, G., and Goldstein, B. R. 2008. *From Summetria to Symmetry: The Making of a Revolutionary Scientific Concept*. Archimedes: New Studies in the History and Philosophy of Science and Technology, vol. 20. New York: Springer.

Hopkins, E. J. 1950. Some theorems on concurrence and collinearity. *The Mathematical Gazette*, **34**(308), 129–33.

Hopkins, J. 1978. *A Concise Introduction to the Philosophy of Nicholas of Cusa*. Minneapolis: University of Minnesota Press.

Hopkins, J. 1998. *De Beryllo (On [Intellectual] Eyeglasses)*, trans. from Nicolai de Cusa Opera Omnia, vol. XI, 1: *De Beryllo*, eds. H. G. Senger and K. Bormann (Hamburg: Meiner Verlag, 1988). In J. Hopkins, *Nicholas of Cusa: Metaphysical Speculations. Six Latin Texts Translated into English*. Minneapolis, MN: Arthur J. Banning Press. https://jasper-hopkins.info/DeBeryllo12-2000.pdf.

Horadam, A. F. 1970. *A Guide to Undergraduate Projective Geometry.* Oxford: Pergamon Press. See especially XIII, 349 p.

Huxley, G. L. 1959. *Anthemius of Tralles: A Study of Later Greek Geometry.* Greek, Roman and Byzantine Monographs 1. Cambridge, MA: Greek, Roman and Byzantine Studies.

Hyman, I. 1974. *Brunelleschi in Perspective.* Artists in Perspective Series. Englewood Cliffs, NJ: Prentice-Hall.

Ihde, D. 2008. Art precedes science: Or did the camera obscura invent modern science? In H. Schramm, L. Schwarte, and J. Lazardzig (eds.), *Instruments in Art and Science: On the Architectonics of Cultural Boundaries in the 17th Century,* vol. 2, pp. 383–93. Berlin: Walter de Gruyter.

Ivins Jr., W. M. 1938. On the rationalization of sight. *Metropolitan Museum of Art,* 8.

Jacobi, C. F. A. 1825. *De triangulorum rectilineorum proprietatibus quibusdam nondum satis cognitis.* Quibus memoriam anniversariam... Scholae Provincialis Portensis... pie celebrandam indicunt et ad recitationes et orationes discipulorum... audiendas invitant: Landesschule. Schwickert.

Jacobi, C. G. J. 1857. Über einen algebraischen Fundamentalsatz und seine Anwendungen. *J. Reine Angew. Math.,* **53,** 275–80.

Jeřábek, V. 1886. Question 504. *Mathesis,* **6.**

Johnson, P. 2000. *The Renaissance: A Short History.* Modern Library Chronicles. New York: Modern Library.

Jones, D. 1980. Quadrangles, butterflies, Pascal's hexagon, and projective fixed points. *Am. Math. Mon.,* **87**(3), 197–200.

Jordan, C. 1989. *Traité des substitutions et des équations algébriques.* Les Grands Classiques Gauthier-Villars. [Gauthier-Villars Great Classics]. Sceaux: Éditions Jacques Gabay. Reprint of the 1870 original.

Juel, C. 1934. *Vorlesungen über projektive Geometrie mit besonderer Berücksichtigung der v. Staudtschen Imaginärtheorie* (Die Grundlehren d. math. Wiss. in Einzeldarstell. mit besonderer Berücksichtigung d. Anwendungsgebiete. 42). Berlin: Julius Springer. XI, 287 S., 87 Fig.

Kantor, S. 1881. Ueber die Configurationen (3, 3) mit den Indices 8, 9 und ihren Zusammenhang mit den Curven dritter Ordnung. *Wien. Ber.,* **84,** 915–32.

Kaplansky, I. 2003. *Linear Algebra and Geometry,* revised ed. New York: Dover Publications. A second course.

Kariya, J. 1904. Un problème sur le triangle. *Enseign. Math.,* **6,** 130–2.

Kárteszi, F. 1976. *Introduction to Finite Geometries.* North-Holland Texts in Advanced Mathematics, vol. 2. Amsterdam: North-Holland Publishing. Trans. from the Hungarian by L. Vekerdi.

Kepler, J. 1604. *Ad Vitellionem paralipomena: quibus astronomiæ pars optica traditur; potissimum de artificiosa observatione et æstimatione diametrorum deliquiorumá; solis & lunæ.* apud Claudium Marnium & haeredes Joannis Aubrii.

Kiepert, L. 1869. Solution de question 864. *Nouv. ann. math.,* **8,** 40–2.

King, J. R., and Schattschneider, D. 1997. *Geometry Turned On: Dynamic Software in Learning, Teaching and Research.* Washington, DC: Mathematical Association of America.

Klein, C. F. 1868. *Über die Transformation der allgemeinen Gleichung des zweiten Grades zwischen Linien-Koordinaten auf eine kanonische Form.* PhD thesis, University of Bonn.

Klein, C. F. 1870. Zur Theorie der Liniencomplexe des ersten und zweiten Grades. *Math. Ann.*, **2**(2), 198–226.

Klein, C. F. 1871. Ueber die sogenannte Nicht-Euklidische Geometrie. *Math. Ann.*, **4**, 573–625.

Klein, C. F. 1873. Ueber die sogenannte Nicht-Euklidische Geometrie. *Math. Ann.*, **6**(2), 112–45.

Klein, C. F. 1883. Eine Uebertragung des Pascal'schen Satzes auf Raumgeometrie. *Math. Ann.*, **22**, 246–9.

Klein, C. F. 1893. A comparative review of recent researches in geometry. *Bull. Am. Math. Soc.*, **2**(10), 215–49.

Klein, C. F. 1968. *Gesammelte mathematische Abhnadlungen.* Vols. 1–3. Berlin: Springer-Verlag.

Kline, M. 1955. Projective geometry. *Sci. Am.*, **192**(1), 80–7.

Kline, M. 1972. *Mathematical Thought from Ancient to Modern Times.* New York: Oxford University Press.

Knobloch, E. 2000. Analogy and the growth of mathematical knowledge. In E. Grosholz and H. Breger (eds.), *The Growth of Mathematical Knowledge*, pp. 295–314. Dordrecht: Kluwer Academic.

Koenigsberger, L. 1904. *Carl Gustav Jacob Jacobi. Rede zu der von dem internationalen Mathematikerkongreß in Heidelberg veranstalteten Feier der hundertsten Wiederkehr seines Geburtstages, gehalten am 9. August 1904. Mit einem Bildnis C. G. J. Jacobis.* Leipzig: B. G. Teubner. 40 S. 4°; Deutsche Math.-Ver. 13, 405–53.

Kollros, L. 1947. Théorème de l'hyperespace analogue au théorème de Pascal. *Comment. Math. Helv.*, **19**, 316–19.

Kouřilová, T., and Röschel, O. 2013. A remark on Feuerbach hyperbolas. *J. Geom.*, **104**(2), 317–28.

Koyre, A. 1957. *From the Closed World to the Infinite Universe.* Baltimore, MD: Johns Johns Hopkins University Press.

Kung, S. 2005. A butterfly theorem for quadrilaterals. *Math. Mag.*, **78**(4), 314–16.

Kvasz, L. 2008. *Patterns of Change: Linguistic Innovations in the Development of Classical Mathematics.* Basel: Birkhäuser.

Labourie, F. 2008. What is. . . a cross ratio? *Notices Am. Math. Soc.*, **55**(10), 1234–5.

Laguerre, E. 1853. Sur la théorie des foyers. *Nouv. ann. math.*, **12**, 57–66.

Lalesco, T. 1937. *La géométrie du triangle: La géométrie d'Euler. La géométrie récente. Les théories générales. La métrique du triangle. 2ième édit. Avec une lettre de Emile Picard et une preface de Georges Tzitzeica.* Annales Roumaines de Mathématiques. Cahier 1. Paris: Libr. Vuibert.

Lambert, J. H. [1761] 2009. *Insigniores Orbitae Cometarum Proprietates (1761).* Whitefish, MT: Kessinger. Klett.

Lambert, J. H. 1946. *Iohannis Henrici Lamberti Opera Mathematica. Volumen Primum. Commentationes Arithmeticae, Algebraicae et Analyticae, Pars Prima.* Ed. A. Speiser. Zurich: Orell Füssli.

Lambert, J. H. 1948. *Iohannis Henrici Lamberti Opera Mathematica. Volumen Secundum. Commentationes Arithmeticae, Algebraicae et Analyticae, Pars Altera.* Ed. A. Speiser. Zurich: Orell Füssli.

Lamé, G. 1816–7. Géométrie analitique. Sur les intersections des lignes et des surfaces. Extrait d'un mémoire présenté à l'académie royale des sciences, en décembre 1816. *Ann. math. pures appl.*, **7**, 229–40.

Lamé, G. 1818. *Examen des différentes méthodes employées pour résoudre les problèmes de géométrie. Réimpression fac-similé*. Paris: A. Hermann. XII u. 124 S. 8°, 2 Taf.

Lee, H. L., Burton, L. J., Tan, K., Langr, J., and Ungar, P. 1949. Elementary problems and solutions: problems for solution E876–E880. *Am. Math. Mon.*, **56**(7), 473–4.

Legendre, A.-M. 1794. *Éléments de géométrie, avec des notes*. Paris: Didot.

Lehmer, D. N. 1911. On the combination of involutions. *Am. Math. Mon.*, **18**(3), 52–7.

Lemoine, É. 1889. Contribution à la géométrie du triangle. *Ass. Franç.*, **XVIII**, 197–222.

Lemoine, É. 1928. *Mathesis*, **42**, 435. This reference relates to an exam question once set by Lemoine.

l'Hôpital, G. F. A. Marquis de. 1707. *Traité analytiques des sections coniques*. Paris: Jean Boudot.

Lie, S. 1893. *Vorlesungen über continuirliche Gruppen mit geometrischen und anderen Anwendungen. Bearbeitet und herausgegeben von G. Scheffers*. Leipzig. B. G. Teubner. XII + 810 S. 8°.

Lo Bello, A. 2009. *The Commentary of a-Nayrizi on Books II–IV of Euclid's* Elements of Geometry: *With a Translation of That Portion of Book I Missing from MS Leiden Or. 399.1 but Present in the Newly Discovered Qom Manuscript Edited by Rüdiger Arnzen*. Studies in Platonism, Neoplatonism, and the Platonic Tradition, vol. 8. Leiden: Brill.

Lobachevski, N. I. 1946–51. *Complete Collected Works*, vols. I–IV (Russian), ed. V. F. Kagan. Moscow: GITTL.

Lorentz, H. A. 1904. Electromagnetic phenomena in a system moving with any velocity smaller than that of light. *Proc. Acad. Sci. Artist*, **6**, 809. https://doi.org/10.1016/B978-0-08-006995-1.50012-0.

Loria, G. 1908. Perspektive und darstellende Geometrie. in M. Cantor (ed.), *Vorlesungen über Geschichte der Mathematik*, vol. 4, pp. 577–637. Leipzig: BG Teubner.

Mackenzie, D. 2009. *What's Happening in the Mathematical Sciences*, vol. 7. Providence, RI: American Mathematical Society.

Mac Lane, S. 1936. Some interpretations of abstract linear dependence in terms of projective geometry. *Am. J. Math.*, **58**, 236–40.

Maclaurin, C. 1720. *Geometria Organica*. Manuscripta; History of Science, 16th to 19th Century. London: Impensis Gul. & Joh. Innys.

Maclaurin, C. 1735. V. A letter from Mr. Colin Mac Laurin, Math. Prof. Edinburg. FRS to Mr. John Machin, Astr. Prof. Gresh. & Secr. RS concerning the description of curve lines. Communicated to the Royal Society on December 21, 1732. *Philos. T. R. Soc. Lond.*, **39**(439), 143–65.

Maclaurin, C. 1748. *A Treatise of Algebra in Three Parts*. posthumous. London: Millar and Nourse.

Maleuvre, D. 2011. *The Horizon: A History of Our Infinite Longing*. Berkeley: University of California Press.

Manetti, A. 1970. *The Life of Brunelleschi*. University Park: Pennsylvania State University Press.

Mantel, M. W. 1889. Sur une projection imaginaire. *Mathesis*, **9**, 217–19.

Marchand, J. 1931. Sur une méthode projective dans certaines recherches de géométrie élémentaire. *Enseign. Math.*, **29**, 289–93.

Massey, L. 2016. *Picturing Space, Displacing Bodies: Anamorphosis in Early Modern Theories of Perspective*. University Park: Pennsylvania State University Press.

Mathieu, J.-J.-A. 1865. Étude de géométrie comparée, avec applications aux sections coniques. *Nouv. ann. math.*, 2e série, **4**, 529–37.

Maxwell, J. C. 1864. On the calculation of the equilibrium and stiffness of frames. *Lond., Edinb., Dubl. Philos. Mag. J. Sci.*, **27**(182), 294–9.

McGregor, R. T., Latham, M. L., and Cleland, W. E. 1922. Problems and Solutions: solutions: 2892. *Am. Math. Mon.*, **29**(8), 316–17.

McKenzie, J. L. 1881. Question 6871, Mathematical Questions, with Their Solutions. *Educational Times*, October.

Meskens, A. J. 2021. Chapter 10: Joannes della Faille and the beginning of projective geometry. In A. J. Meskens (with contributions from H. van Looy), *Between Tradition and Innovation: Gregorio a San Vicente and the Flemish Jesuit Mathematics School*, pp. 165–83. Leiden: Brill.

Migliari, R., and Salvatore, M. 2015. Il 'Teorema Fondamentale' del De Prospectiva Pingendi. In M. T. Bartoli and M. Lusoli (eds.), *Le teorie, le tecniche, i repertori figurativi nella prospettiva d'architettura tra il '400 e il '700: Dall'acquisizione alla lettura del dato*. Studi e saggi; 148. Florence: Firenze University Press.

Mills, S. 1984. Note on the Braikenridge–Maclaurin theorem. *Notes and Records Roy. Soc. Lond.*, **38**(2), 235–40.

Milne, J. J. 1911. *An Elementary Treatise on Cross-Ratio Geometry, with Historical Notes*. Cambridge: Cambridge University Press. XXII u. 288 S. 8°.

Minkowski, H. 1909. *Raum und Zeit*. Leipzig: BG Teubner.

Miquel, A. 1844. Mémoire de géométrie. *J. math. pures appl.*, **9**, 20–27.

Möbius, A. F. 1827. *Der barycentrisches Calcul*. Leipzig: Verlag von Johann Ambrosius Barth.

Möbius, A. F. 1828. Kann von zwei dreiseitigen Pyramiden eine jede in Bezug auf die andere um- und eingeschrieben zugleich heißen? *J. Reine Angew. Math.*, **3**, 273–8.

Möbius, A. F. 1848. Verallgemeinerung des Pascalschen Theorems, das in einen Kegelschnitt beschriebene Sechseck betreffend. *J. Reine Angew. Math. (Crelles Journal)*, **1848**(36), 216–20.

Möbius, A. F. 1855. *Theorie der Kreisverwandschaft in rein geometrischer Darstellung*. Leipzig: Hirzel.

Monte, G. del. 1600. *Perspectivae libri sex*. Pisauri, Girolamo Concordia.

Morley, F., and Morley, F. V. 1933. *Inversive Geometry*. New York: Ginn and Company.

Muratori, L. A. 1751. *Vita di Leon Battista Alberti*. In *Rerum Italicarum Scriptores*, vol. XXV, pp. 295–304. Milan.

Narducci, E. 1871. *Intorno ad una traduzione Italiana fatta nel secolo decimoquarto del trattato d'Ottica d'Alhazen matematico del secolo undecimo e ad altri lavori di questo scienziato nota*. Forni.

Needham, J. 1962. *Science and Civilisation in China. Vol. 4: Physics and Physical Technology. Part I: Physics*. With the collaboration of Wang Ling and the special co-operation of Kenneth Girdwood Robinson. New York: Cambridge University Press.

Neël, E. E. 1879. *Les applications de Blanchet: Théorèmes lieux géométriques et problèmes Librairie de L'Office de publicité.* Louvain: Peeters-Ruelens.

Nehring, O. 1942. Zyklishe Projektionen im Dreieck. *J. Reine Angew. Math.*, **184**, 129–37.

Neuberg, J. 1888. Sur les transformations quadratiques involutives. *Mathesis*, **8**, 177–83.

Neumann, P. M. 2011. *The mathematical writings of Évariste Galois.* Heritage of European Mathematics. Zurich: European Mathematical Society.

Newton, I. 1968. *The Mathematical Papers of Isaac Newton. Vol. II: 1667–1670* ed. D. T. Whiteside, with the assistance in publication of M. A. Hoskin. London: Cambridge University Press.

Newton, I. 1999. *The Principia: Mathematical Principles of Natural Philosophy.* Newly trans. from 3rd ed. (1726) by I. B. Cohen and A. Whitman, assisted by J. Budenz. Preceded by 'A guide to Newton's Principia' by I. B. Cohen. Berkeley: University of California Press.

Newton, I., and Motte, A. 1848. *The Mathematical Principles of Natural Philosophy… Translated… by Andrew Motte. To which are added Newton's System of the World; a Short Comment on, and Defence of the Principia, by W. Emerson. With the Laws of the Moon's Motion according to gravity, by John Machin… (The preface of Mr. Roger Cotes to the second edition.).* New York: Daniel Adee. Google eBook, https://archive.org/details/B-001-003-063/page/n19/mode/2up.

Ngoc, T. M. 2018. A purely synthetic proof of Dao's theorem on a conic and its applications. *Int. J. Comput. Discov. Math*, **3**, 145–52.

Nguyen, G. N. 2015. A proof of Dao's theorem. *Glob. J. Adv. Res. Class. Mod. Geom.*, **4**(2), 102–5.

Nicéron, J. F. 1638. *La perspective curieuse, ou magie artificielle des effets merveilleux.* Paris: F. Langlois.

Nixon, R. C. J. 1891. Question 10963, Mathematical Questions, with Their Solutions. *Educational Times*, **LV**.

Noyer, A. 1893. Sur les triangles autopolaires. *J. de Math. spéc, II.*, **4**, 39–44.

O'Hara, C. W., and Ward, D. R. 1937. *An Introduction to Projective Geometry.* IX + 298 p. Oxford: Clarendon Press.

Ostrom, T. G., and Sherk, F. A. 1964. Finite projective planes with affine subplanes. *Canad. Math. Bull.*, **7**(4), 549–59.

Panakis, I. c. 1965, undated. *2500 problems of geometric loci with their solutions.* vols. 1 and 2. Privately published.

Papelier, G. 1927. *Exercices de Géométrié Moderne.* Vol. II, Transversales. Paris: Vuibert.

Pappus of Alexandria. 1986. *Book 7 of the Collection.* Sources in the History of Mathematics and Physical Sciences, vol. 8. New York: Springer-Verlag. Part 1. Introduction, text, and translation, Part 2. Commentary, index, and figures, ed., trans., and commentary by A. Jones.

Pascal, B. 1639. *Essai pour les coniques.* Paris.

Paterson, A. M. 1970. *The Infinite Worlds of Giordano Bruno.* Monograph in the Bannerstone Division of American Lectures in Philosophy. Springfield, IL: Thomas.

Patrizi, F. 1591. *Nova de universis philosophia.* Mamarellus.

Pedoe, D. 1967. On (what should be) a well-known theorem in geometry. *Am. Math. Mon.*, **74**(7), 839–41.

Peirce, C. S., Hartshorne, C., and Weiss, P. 1931. *Collected Papers of Charles Sanders Peirce*, vols. 1–2. Cambridge, MA: Harvard University Press.

Pèlerin, J. 1504. *De artificiali perspectiva.* Toulouse: P. Jacques.

Petard, H. 1938. A contribution to the mathematical theory of big game hunting. *Am. Math. Mon.*, **45**(7), 446–7.

Petersen, J. 1898. Sur le théorème de Tait. *L'Intermédiaire des Mathématiciens*, **5**, 225–7.

Pickert, G. 1975. *Projektive Ebenen*, zweite auflage. Die Grundlehren der mathematischen Wissenschaften, Band 80. Berlin: Springer-Verlag.

Pickford, A. G. 1909. *Elementary Projective Geometry.* Cambridge: Cambridge University Press. XII u. 256 S. 8°.

Plato. 1941. *The Republic of Plato*, trans. C. F. McDonald. Oxford: Clarendon Press.

Poincaré, H. 1921. Rapport sur les travaux de M. Cartan. *Acta Math.*, **38**(1), 137–45.

Poincaré, H. 1905. Sur la dynamique de l'électron (On the Dynamics of the Electron). *Comptes Rendus de l'Académie des Sciences*, **140**, 1504–8.

Poncelet, J.-V. 1817–18. Théorie des pôlaires réciproques, et de réflexions sur l'élimination. *Ann. math. pures. appl.*, **8**, 201–32.

Poncelet, J.-V. 1817. Géométrie des courbes. Théorèmes nouveaux sur les lignes du second ordre. *Ann. math. pures. appl.*, **8**, pp. 1–14 (XI, pp. 10–11).

Poncelet, J.-V. 1822. *Traité des propriétés projectives des figures.* Paris: Gauthier-Villars.

Poncelet, J.-V. 1862–4. *Applications d'analyse et de géomeétrie.* Paris: Gauthier-Villars.

Poncelet, J.-V. 1995a. *Traité des propriétés projectives des figures. Tome I.* Les Grands Classiques Gauthier-Villars [Gauthier-Villars Great Classics]. Sceaux: Éditions Jacques Gabay. Reprint of the second (1865) edition.

Poncelet, J.-V. 1995b. *Traité des propriétés projectives des figures. Tome II.* Les Grands Classiques Gauthier-Villars [Gauthier-Villars Great Classics]. Sceaux: Éditions Jacques Gabay. Reprint of the second (1866) edition.

Prager, F. D. 1968. A manuscript of Taccola, quoting Brunelleschi, on problems of inventors and builders. *Proc. Am. Philos. Soc.*, **112**(3), 131–49.

Prasolov, V. V., and Tikhomirov, V. M. 2001. *Geometry.* Translations of Mathematical Monographs, vol. 200. Providence, RI: American Mathematical Society. Trans. from the 1997 Russian original by O. V. Sipacheva.

Primrose, E. J. F. 1953. Theorems on concurrence and collinearity. *Math. Gaz.*, **37**(319), 60.

Primrose, E. J. F. 1960. 2934. A certain type of theorem concerning two conics. *Math. Gaz.*, **44**(350), 286.

Pătraşcu, I. 2010 (March). *The dual theorem relative to the Simson's line.* viXra: 1003.0056.

Quadling, D. 2012. A curious misattribution: The early history of 'Simson's line'. *Math. Gaz.*, **96**(537), 420–7.

Rademacher, H., and Toeplitz, O. 1957. *The Enjoyment of Mathematics: Selections from Mathematics for the Amateur.* Princeton, NJ: Princeton University Press.

Ramus, P. 1569. *Scholarum mathematicarum libri unus et triginta*. Basel: Eusebius Episcopius.

Rehkämper, K. 2003. What you see is what you get: The problems of linear perspective. In H. Hecht, R. Schwartz, and M. Atherton (eds.), *Looking into Pictures: An Interdisciplinary Approach to Pictorial Space*, pp. 179–90. Cambridge, MA: MIT Press.

Retali, V. 1896. Lettre de M. V. Retali. *Periodico di Matematica*, 11, 70–1.

Reye, Th. 1882. Die Hexaëder- und die Octaëder-Configurationen ($12^6, 16^3$). *Acta Math.*, 1(1), 97–108.

Reye, Th. 1886. *Die Geometrie der Lage. Erste Abteilung. 3te Auflage [ed.]*. Leipzig: Baumgärtner. XIV u. 248 S.

Richter, J. P., and Da Vinci, L. 1970. *The Literary Works of Leonardo Da Vinci: Volume 1*, 3rd ed. New York: Phaidon.

Richter-Gebert, J. 2011. *Perspectives on Projective Geometry*. Heidelberg: Springer. A guided tour through real and complex geometry.

Riesinger, R. 2004. On Wallace loci from the projective point of view. *J. Geom. Graph.*, 8(2), 201–13.

Rigby, J. F. 1965. Affine subplanes of finite projective planes. *Canad. J. Math.*, 17, 977–1009.

Robinson, R. T. 1940. Theorems on perspectivity. *Math. Gaz.*, 24, 9–14.

Robson, A. 1923. 660. Morley's theorem (v. note 621). *Math. Gaz.*, 11(164), 310–11.

Robson, A. 1949. *An Introduction to Analytical Geometry*. Cambridge: Cambridge University Press.

Robson, A. 1953. 2376. Pascal's theorem: For the collector of projective proofs. *Math. Gaz.*, 37(322), 284.

Rosenbaum, R. A., and Rosenbaum, J. 1949. Some consequences of a well known theorem on conics. *Bull. Am. Math. Soc.*, 55, 933–5.

Rosenfeld, B. A. 2005. The analytic principle of continuity. *Am. Math. Mon.*, 112(8), 743–8.

Rowe, D. E. 1989. The early geometrical works of Sophus Lie and Felix Klein. In D. E. Rowe and J. McCleary (eds.), *The History of Modern Mathematics, Volume I: Ideas and Their Reception*, Proceedings of the Symposium on the History of Modern Mathematics, Vassar College, Poughkeepsie, New York, June 20–24, 1988, vol. 1, pp. 209–74. San Diego, CA: Academic Press.

Rupp, C. A. 1929. An extension of Pascal's theorem. *Trans. Am. Math. Soc.*, 31(3), 580.

Russell, J. W. 1893. *An Elementary Treatise on Modern Pure Geometry. With Numerous Examples*. Oxford Clarendon Press. XVI + 323 S.

Saccheri, G. 2011. *Euclide vendicato da ogni neo*. Mathematica. Centro di Ricerca Matematica Ennio De Giorgi (CRM), 1. Edizioni della Normale, Pisa. With a companion volume containing a facsimile reprint of the 1733 Latin original, intro., trans., and notes by V. De Risi.

Sachse, A. 1882. Beweis der vorigen Sätze. *Schlömilch Z.*, 27, 381–3.

Salmon, G. 1857. Geometrical notes. *Q. J. Pure Appl. Math.*, 1, 237–41.

Salmon, G. 1882. *Treatise of Analytic Geometry of Three Dimensions*, 2nd edn. Dublin: Hodges, Figgis, & Co.

Salmon, G. 1954. *A Treatise on Conic Sections: Containing an Account of Some of the Most Important Modern Algebraic and Geometric Methods*, 6th ed. New York: Chelsea Publishing.

Schläfli, L. 1858. An attempt to determine the twenty-seven lines upon a surface of the third order, and to divide such surfaces into species, in reference to the reality of the lines upon the surface. *Q. J. Pure Appl. Math.*, **2**, 55–65, 110–20.

Schläfli, L. 1866. Erweiterung des Satzes, dass zwei polare Dreiecke perspectivisch liegen, auf eine beliebige Zahl von Dimensionen. *J. Reine Angew. Math. (Crelles Journal)*, **65**(Jan), 189–97.

Schodde, C. 2013. Romans paint better perspective than Renaissance artists. *Found in Antiquity*, https://foundinantiquity.com/2013/08/09/pompeiian-fresco-painters-used-perspective-better-than-renaissance-artists/.

Schreck, K. R. 2016. *Monge's Legacy of Descriptive and Differential Geometry*. Boston, MA: Docent Press.

Schröcker, H.-P. 2017. Singular Frégier conics in non-Euclidean geometry. *J. Geom. Graph.*, **21**(2), 201–8.

Segre, B. 1945. A four-dimensional analogue of Pascal's theorem for conics. *Am. Math. Mon.*, **52**, 119–31.

Segre, C. 1887. Sulla varietà cubica con dieci punti doppii dello spazio a quattro dimensioni. *Torino Atti*, **22**, 791–801.

Segre, C. 1888a. Alcune considerazioni elementari sull' incidenza di rette e piani nello spazio a quattro dimensioni. *Rend. Circ. Mat. Palermo*, **2**, 45–52.

Segre, C. 1888b. Sulle varietà cubiche dello spazio a quattro dimensioni e su certi sistemi di rette e certe superficie dello spazio ordinario. *Torino mem.*, (2), XXXIX, 48 S.

Semple, J. G., and Kneebone, G. T. 1952. *Algebraic Projective Geometry*. Oxford: Clarendon Press.

Serret, P. 1862. De quelques analogies de la géométrie du plan à celle de l'espace. *J. math. pures appl.*, 377–406.

Servois. 1813/14. Géométrie pratique. Problème. Prolonger une droite accessible au-delà d'un obstacle qui borne la vue, en n'employant que l'équerre d'arpenteur, et sans faire aucun chaînage ? Solution. *Ann. math. pures appl. [Ann. Gergonne]*, **4**, 250–3.

Seydewitz, F. 1847. Construction und Classification der Flächen des zweiten Grades mittels projectivischer Gebilde. *Archiv der Mathematik*, **9**, 158–214.

Shephard, G. C. 1999. Isomorphism invariants for projective configurations. *Canad. J. Math.*, **51**(6), 1277–99.

Siebeck, F. H. 1865. Ueber eine neue analytische Behandlungsweise der Brennpunkte. *J. Reine Angew. Math.*, **64**, 175–82.

Silverman, J. H., and Tate, J. T. 2015. *Rational Points on Elliptic Curves*, 2nd ed. Cham: Springer International Publishing.

Simson, R. 1735. *Sectionum conicarum libri V.* Edinburgh: T. & W. Ruddimannos.

Smith, A. M. 2001. *Alhacen's Theory of Visual Perception: First Three Books of Alhacen's De Aspectibus*. Transactions of the American Philosophical Society, 91, pts 4–5. Philadelphia, PA: American Philosophical Society.

Smith, G. 2015. A projective Simson line. *Math. Gaz.*, **99**(545), 339–41.

Snapper, E. 1981. An affine generalization of the Euler line. *Am. Math. Mon.*, **88**(3), 196–8.

Son, T. H. 2014. A synthetic proof of Dao's generalization of Goormaghtigh's theorem. *Glob. J. Adv. Res. Class. Mod. Geom.*, **3**(2), 125–9.

Spencer, J. R. n.d. *Filarete's Treatise on Architecture: Being the Treatise by Antonio Di Piero Averlino, Known as Filarete. The Translation.* Yale Publications in the History of Art, vol. 1. New Haven, CT: Yale University Press.

Starr, A. T. 1961. *Scholarship Mathematics, Volume II: Geometry.* London: Pitman.

Steffens, B. 2007. *Ibn al-Haytham: First Scientist.* Greensboro, NC: Morgan Reynolds.

Steiner, J. 1828–9. Démonstration de quelques théorèmes de géometrie. *Annales de Gergonne*, **19**, 1–8.

Steiner, J. 1898. *Jacob Steiner's Vorlesungen über synthetische Geometrie. Zweiter Teil: Die Theorie der Kegelschnitte, gestützt auf projective Eigenschaften. Bearbeitet von H. Schröter; 3. Auflage, durchgesehen von R. Sturm.* Leipzig: B. G. Teubner. XVII + 537 S. 8°.

Steiner, J. 1827a. Geometrische Lehrsätze. *J. Reine Angew. Math.*, **2**, 190–3.

Steiner, J. 1827b. Vorgelegte Lehrsätze. *Crelle's Journal II*, 287–92.

Steiner, J. 1828a. Géométrie pure. Développement d'une série de théorèmes relatifs aux sections coniques. *Annales de Gergonne*, **19**, 37–64.

Steiner, J. 1828b. Questions proposées. Théorèmes sur l'hexagramum mysticum. *Annales de Gergonne*, **18**, 339–40.

Steiner, J. 1832. *Systematische Entwicklung der Abhangigkeit geometrischer Gestalten voneinander, mit Berücksichtigung der Arbeiten alter und neuer Geometer über Porismen, Projections-Methoden, Geometrie der Lage, Transversalen, Dualität und Reciprocität.* Berlin: Fincke.

Steiner, J. 1846. Teoremi relativi alle coniche inscritte e circoscritte. *J. Reine Angew. Math.*, **30**, 97–106.

Steiner, J. 1867. *Vorlesungen über synthetische Geometrie 2: Die Theorie der Kegelschnitte.* Leipzig: B. G. Teubner.

Stephanos, C. 1879. Sur les systèmes desmiques de trois tétraèdres. *Bull. sci. math. astron.*, **3**(2), 424–56.

Stephanos, C. 1882. Sur une configuration remarquable de cercles dans-l'espace. *C. R. Acad. Sci., Paris*, **93**, 578–80.

Stirling, J. 1717. *Lineae Tertii Ordinis Neutonianae, Sive, Illustratio Tractatus D. Neutoni De Enumeratione Linearum Tertii Ordinis.* Impensis Edvardi Whistler Bibliopolae Oxoniensis Oxford: Edward Whistler's edition of the Oxford Library.

Stolfi, J. 1991. *Oriented Projective Geometry: A Framework for Geometric Computations.* Boston, MA: Academic Press.

Strange, J. 1974. A generalization of Morley's theorem. *Am. Math. Mon.*, **81**, 61–3.

Struve, R., and Struve, H. 2016. An axiomatic analysis of the Droz–Farny line theorem. *Aequationes Math.*, **90**(6), 1201–18.

Sturm, C. 1826–7. Géométrie analytique. Mémoire sur les lignes du second ordre. *Ann. math. pures appl. [Ann. Gergonne]*, **17**, 173–98.

Sturm, P. 2011. A historical survey of geometric computer vision. In P. Real, D. Diaz-Pernil, H. Molina-Abril, A. Berciano, and W. Kropatsch (eds.), *Computer Analysis of Images and Patterns: 14th International Conference, CAIP 2011, Seville, Spain,*

August 29–31, 2011, Proceedings, Part I (Lecture Notes in Computer Science, vol. 6854), pp. 1–8. Heidelberg: Springer.

Sylvester, J. J. 1852. A demonstration of the theorem that every homogeneous quadratic polynomial is reducible by real orthogonal substitutions to the form of a sum of positive and negative squares. *Philos. Mag.*, **4**(23), 138–42.

Szilasi, Z. 2012. Two applications of the theorem of Carnot. *Ann. Math. Inform.*, **40**, 135–44.

Takasu, T. 1931. Sur la généralisation projective et affine de la droite simsonienne. *Sci. Rep. Tôhoku Univ., I. Ser.*, **20**, 213–51.

Tannery, P., and Henry, C. 1894. *De fermat oeuvres, Tome II. Correspondance.* Paris: Gauthier-Villars.

Taylor, B. 1719. *New Principles of Linear Perspective.* London: R. Knaplock.

Telesio, B. 1587. *De rerum natura iuxta propria principia, libri IX.* Biblioteca digital Dioscórides. Neapoli: apud Horatium Salvianum.

Terquem, O. 1842. Analyse d'ouvrages nouveaux. *Nouv. ann. math.*, **1**(1), 49–56.

Todd, J. A. 1936. Dual vectors and the Petersen-Morley theorem. *Math. Gaz.*, **20**(239), 184–5.

Todd, J. A., and Eves, H. 1942. E481. *Am. Math. Mon.*, **49**(4), 257–60.

Todhunter, I. 1855. *A Treatise on Plane Co-ordinate Geometry: With numerous examples.* Cambridge, MA: Macmillan.

Totten, J. 2006. Contributor profile: Toshio Seimiya. *Crux Mathematicorum*, **32**(3), 129.

Townsend, Reverend R. 1864. Question 1501, Questions, with Their Solutions. *Educational Times.*

Turner, J. S. 1925. An extension of the property of the circle known as 'Simson's line'. *Bull. Am. Math. Soc.*, **31**, 118.

Unguru, S. (ed.). 1977. *Witelonis Perspectivae Liber Primus [Book I of Witelo's Perspectiva]. An English Translation with Introduction and Commentary and Latin Edition of the Mathematical Book of Witelo's Perspectiva.* Warsaw: Polish Academy of Sciences Press.

Vajda, S. 1946. Generalised metrical theorems. *Math. Gaz.*, **30**(290), 122–5.

van Kempen, H. P. M. 2006. On some theorems of Poncelet and Carnot. *Forum Geom.*, **6**, 229–34.

van Lamoen, F. 1999. Bicentric triangles. *Nieuw Archief voor Wiskunde*, **17**(3), 363–72.

van Lamoen, F. 2000. Morley related triangles on the nine-point circle. *Am. Math. Mon.*, **107**(10), 941–5.

van Lamoen, F. 2001. Pl-perpendicularity. *Forum Geom.*, **1**, 151–60.

Vasari, G. 2008. *The Lives of the Artists*, trans. Julia Conaway Bondanella and Peter Bondanella. Oxford World's Classics. Oxford: Oxford University Press.

Veblen, O., and Young, J. W. 1918. *Projective Geometry.* vol. 2. New York: Ginn.

Veblen, O., and Young, J. W. 1965. *Projective Geometry*, vol. 1. New York: Ginn.

Veselovskii, I. N., and Rosenfeld, B. A. 1962. *Works of Archimedes [Russian].* Moscow: Gosudarstvennoe izdatel'stvo fiziko-matemat. lit.

Vesely, D. 2014. The role of perspective in the transformation of European culture. In R. Lupacchini and A. Angelini (eds.), *The Art of Science: From Perspective Drawing to Quantum Randomness*, pp. 49–70. Cham: Springer International Publishing.

Voelke, J.-D. 2005. *Renaissance de la géométrie non euclidienne entre 1860 et 1900.* Bern: Peter Lang.

Vogel, K. n.d. 'Aristaeus.' *Complete Dictionary of Scientific Biography*. www.encyclopedia.com/literature-and-arts/classical-literature-mythology-and-folklore/folklore-and-mythology/aristaeus.

Voigt, G. 1859. *Die Wiederbelebung des classischen Alterthums, oder das erste Jahrhundert des Humanismus*. Berlin: Georg Reimer.

Volkert, K. 2013. *Das Undenkbare denken*. Berlin: Springer-Verlag.

von Staudt, G. K. C. 1847. *Geometrie der Lage*. Nuremberg: Verlag von Bauer und Raspe (Julius Merz).

von Staudt, G. K. C. 1856. *Beiträge zur Geometrie der Lage, Erstes Heft*. Nuremberg: Verlag von Bauer und Raspe (Julius Merz).

von Staudt, K. G. C. 1831. *Über die Kurven II. Ordnung*. Königliche Studienanstalt zu Nürnberg: Programm. Campeschen Officin.

Walker, P. R. 2009. *The Feud That Sparked the Renaissance: How Brunelleschi and Ghiberti Changed the Art World*. New York: Harper Collins.

Wallace, W. ('Scoticus'). 1806. 1804 Question 86, article XXVI. In T. Leybourne (ed.), *New Series of the Mathematical Repository*, vol. 1. London: W. Glendinning.

Wallis, J. 1658. *Commercium epistolicum de quaestionibus mathematicis quibusdam mathematicus*. Oxford: Thomas Robinson.

Wallis, J. 1693. *De algebra tractatus, historicus [et] practicus*. Oxford: Sheldonian Theatre.

Wallis, J. 2003. *Correspondence of John Wallis (1616–1703). Vol. 1: 1641–1659*, ed. P. Beeley and C. J. Scriba with the assistance of U. Mayer and S. Probst. Oxford: Oxford University Press.

Waring, E. 1762. *Miscellanea analytica, de aequationibus algebraicis, et curvarum proprietatibus*. Cambridge: Gul. Thurlbourn and J. Woodyer.

Waring, E. 1772. *Proprietates algebraicarum curvarum* (written 1762). Cambridge: J. Woodyer.

Whewell, W. 2015. *History of the Inductive Sciences*, Scholar's Choice Edition. London: Creative Media Partners, LLC.

Whittaker, T. 1925. Nicholas of Cusa. *Mind, New Series*, **34**(136), 436–54.

Wilkinson, T. T. 1858. Notae Geometricae. *The Lady's and Gentleman's Diary*, 86–7.

Wilkinson, T. T. 1872. Question 2015, Mathematical Questions, with Their Solutions. *Educational Times*, **XVII**, p. 72 (solution p. 88).

Wittkower, R. 1953. Brunelleschi and 'proportion in perspective'. *J. Warburg Courtauld*, **16**(3/4), 275–91.

Wood, P. W. 1960. *The Twisted Cubic, with Some Account of the Metrical Properties of the Cubical Hyperbola*. Reprint of Cambridge Tracts in Mathematics and Mathematical Physics, No. 14. New York: Hafner Publishing.

Yamashita, C. 1954. An elementary and purely synthetic proof for the double six theorem of Schläfli. *Tôhoku Math. J. (2)*, **5**(3), 215–19.

Zajonc, A. 1995. *Catching the Light: The Entwined History of Light and Mind*. Oxford: Oxford University Press.

Zhao, Y. 2009. Cyclic quadrilaterals – The big picture. *Winter Camp 2009*. https://yufeizhao.com/olympiad/cyclic_quad.pdf

Index

Printed in the United States
by Baker & Taylor Publisher Services